# An Introduction to Ultra Wideband Communication Systems

## Prentice Hall Communications Engineering and Emerging Technologies Series

Theodore S. Rappaport, *Series Editor*

# An Introduction to Ultra Wideband Communication Systems

## Jeffrey H. Reed

PRENTICE HALL

PROFESSIONAL TECHNICAL REFERENCE

UPPER SADDLE RIVER, NJ • BOSTON • INDIANAPOLIS • SAN FRANCISCO
NEW YORK • TORONTO • MONTREAL • LONDON • MUNICH • PARIS • MADRID
CAPETOWN • SYDNEY • TOKYO • SINGAPORE • MEXICO CITY

The publisher offers excellent discounts on this book when ordered in quantity for bulk purchases or special sales, which may include electronic versions and/or custom covers and content particular to your business, training goals, marketing focus, and branding interests. For more information, please contact:

U. S. Corporate and Government Sales
(800) 382-3419
corpsales@pearsontechgroup.com

For sales outside the U. S., please contact:

International Sales
international@pearsoned.com

Visit us on the Web: www.phptr.com

Library of Congress Catalog Number: 2004117283

ISBN 0-13-148103-7
Text printed in the United States on recycled paper at R.R. Donnelley in Crawfordsville, Indiana
First printing, March 2005

# CONTENTS

# PREFACE

To a great extent, this book was inspired by the Defense Advanced Research Project Agency project called NETEX. The program manager, Stephen Griggs, recognized that far too little information about Ultra Wideband (UWB), particularly pulse-based UWB systems, appeared in the literature. Moreover, he believed that the UWB information available in the public domain is sometimes incorrect and misleading, particularly regarding interference issues with UWB. Thus, the program participants were encouraged to widely disseminate the results of the NETEX program to help clarify many of these outstanding issues. Much of the information presented in this book is a direct result of this program and other UWB research programs from Virginia Tech and the Army Research Lab.

UWB activity has picked up immensely since the Federal Communication Commission's 2002 decision to allow for the transmission of UWB and the subsequent standardization efforts with the Institute of Electrical and Electronics Engineers, Inc. Some see UWB as an enabling technology for new wireless applications that span from high-data-rate transmission of raw multimedia video to new location-aware, low-data-rate, and low-power communication of sensor data. Non-communication applications, such as through-the-wall imaging and ground-penetrating radar, also capture the imagination of researchers and entrepreneurs.

The controversy surrounding the standardization efforts illustrates the many debatable issues of UWB. From an academic perspective, many fundamental research issues remain unresolved, including the best modulation types for particular applications, efficient broadband antennas with a desirable form factor, propagation characteristics of UWB in various environments, the impact of UWB physical layer attributes on the networking layers, and the list continues. UWB will certainly be the subject of many theses and dissertations to come.

This book provides a broad technical view of UWB. Chapters 2 and 3 deal with simulation of propagation issues, modeling, and channel simulation, and much of this information is presented in the public domain for the first time. Chapter 4 focuses on antennas and outlines some surprising differences in antenna performance with respect to narrowband systems. Chapter 5 examines transmitter design issues and covers some of the basic modulation principles of UWB. Chapter 6 presents an overview of receiver design issues and explains how impulse UWB systems are particularly different from conventional carrier-based systems. Chapter 7 addresses the controversial interference issues of UWB; unlike traditional communications, UWB may intentionally transmit co-channel with other communication signals. Chapter 8 examines how to simulate UWB systems; such simulation requires more finesse because broadband signals encounter excessive simulation time if structured incorrectly. Chapter 9 describes how the physical layer capabilities of UWB impact the performance and design of upper layers. Finally, applications and case studies of existing UWB systems are addressed in Chapter 10.

We hope readers will find this book of interest and will check the web site (www. mprg.org/publications/Reed/UWBbook.shtml) for additional information about UWB and this book.

Jeffrey H. Reed
Virginia Tech, 2004

# ACKNOWLEDGMENTS

When writing a book, the authors usually have three groups of people upon whom they rely: those who support them financially, those who offer their advice, and those who contribute their skill.

We would like to acknowledge the Bradley Foundation for its financial support, through the Bradley Fellowship program, of Chris Anderson and Nathaniel August. We also offer our thanks to the National Science Foundation for its financial support, through an Integrative Graduate Education and Research Training grant (award DGE-9987586), for Michelle Gong and James Neel. Many of the authors also received funding via the Industrial Affiliates program of the Mobile and Portable Radio Research Group (MPRG) and through Virginia Tech VLSI for Telecommunications (VTVT).

We would especially like to thank our reviewers, whose comments and perspectives enriched this book: Robert Ulman of Army Research Labs, Krishna Balachandran of Lucent, Andy Molisch of Mitsubishi Electric Research Laboratories, Jiann-An Tsai of Samsung, Dennis Akos of Stanford University, and Laurence Milstein and Claudio da Silva of the University of California, San Diego. In addition, we offer our most sincere gratitude to the following reviewers, who graciously gave of their time to review multiple chapters (some the entire book): Jonathan Cheah of femto Devices, Bruce Fette and John Polson of General Dynamics, Robert Aiello and Jason Ellis of Staccato Communications, Steven Sarraf of the U.S. Army, and Mauro Biagi of the University of Roma "La Sapienza."

Our appreciation is extended to the students in the Spring 2003 advanced simulation course at Virginia Tech, who developed a simulation of a UWB communications system as a class project. The students involved in this effort were Ihsan Akbar, Ramesh Chembil, Jina Kim, Hyung-Jin Lee, Maruf Mohammad, James Neel, Muhammad Nizamuddin, and Sujay Sachindar. The results of their class project formed the first draft of the simulation chapter of this book. We also acknowledge students who helped with other research that led to the writing of this book: Ahmet Bayram, Vivek Bharadwaj, Brian Donlan, Jihad Ibrahim, Gaurav Joshi, Dave McKinstry, Ali H. Muqaibel, Joseph Noronha, Swaroop Venkatesh, and Liu Yang. We offer our thanks to Chris Sadler for helping with Matlab code used in some of the examples in the interference chapter.

Finally, special thanks goes Cyndy Graham of Virginia Tech's MPRG, who worked tirelessly to form something cohesive out of submissions from numerous authors (all with different working and writing styles), and to Lori Hughes, a freelance production assistant (lori@lorihughes.com), who deftly guided us through the production process. We also offer our utmost appreciation to Bernard Goodwin, Acquisitions Editor at Prentice Hall PTR, for his patience and his guidance.

# Chapter 1

# INTRODUCTION

Christopher R. Anderson,
Jeffrey H. Reed,
R. Michael Buehrer,
Dennis Sweeney,
and Stephen Griggs

## 1.1   Fundamentals

### 1.1.1   Overview of UWB

Ultra wideband (UWB) communication systems can be broadly classified as any communication system whose instantaneous bandwidth is many times greater than the minimum required to deliver particular information. This excess bandwidth is the defining characteristic of UWB. Understanding how this characteristic affects system performance and design is critical to making informed engineering design decisions regarding UWB implementation.

The very first wireless transmission, via the Marconi Spark Gap Emitter, was essentially a UWB signal created by the random conductance of a spark. The instantaneous bandwidth of spark gap transmissions vastly exceeded their information rate. Users of these systems quickly discovered some of the most important wireless system design requirements: providing a method to allow a specific user to recover a particular data stream, and allowing all the users to efficiently share the common spectral resource. The UWB technology of the time did not offer a practical answer to either requirement. These problems were solved during the evolution into carrier-based communications systems with regulatory bodies, such as the Federal Communications Commission (FCC) in the United States. The FCC is responsible for carving the spectrum into narrow slices, which are then licensed to various users.

1

This regulatory structure effectively outlawed UWB systems and relegated UWB to purely experimental work for a very long time.

Within the past 40 years, advances in analog and digital electronics and UWB signal theory have enabled system designers to propose some practical UWB communications systems. Over the past decade, many individuals and corporations began asking the FCC for permission to operate unlicensed UWB systems concurrent with existing narrowband signals. In 2002, the FCC decided to change the rules to allow UWB system operation in a broad range of frequencies.[1] In the proceedings of the FCC UWB rule-making process [14], one can find a vast array of claims relating to the expected utility and performance of UWB systems, some of them quite fantastic. Testing by the FCC, FAA, and DARPA has uniformly shown that UWB still conforms to Maxwell's Equations and the laws of physics.

UWB has several features that differentiate it from conventional narrowband systems:

1. Large instantaneous bandwidth enables fine time resolution for network time distribution, precision location capability, or use as a radar.

2. Short duration pulses are able to provide robust performance in dense multipath environments by exploiting more resolvable paths.

3. Low power spectral density allows coexistence with existing users and has a Low Probability of Intercept (LPI).

4. Data rate may be traded for power spectral density and multipath performance.

What makes UWB systems unique is their large instantaneous bandwidth and the potential for very simple implementations. Additionally, the wide bandwidth and potential for low-cost digital design enable a single system to operate in different modes as a communications device, radar, or locator. Taken together, these properties give UWB systems a clear technical advantage over other more conventional approaches in high multipath environments at low to medium data rates.

Currently, numerous companies and government agencies are investigating the potential of UWB to deliver on its promises. A wide range of UWB applications have been demonstrated [15, 16] but much more work needs to be done. Designers are still faced with the same two problems that Marconi faced more than 200 years ago: How does a particular user recover a particular data stream, and how do all the users efficiently share the common spectral resource? Additionally, now that wireless communications have progressed beyond the point where just making it work at all was sufficient, a designer must face a third and perhaps more important question: Can a UWB system be built with a sufficient performance or cost advantage over conventional approaches to justify the effort and investment?

---

[1]The FCC defines UWB as a signal with either a *fractional bandwidth* of 20% of the center frequency or 500 MHz (when the center frequency is above 6 GHz) [14]. The formula proposed by the FCC commission for calculating the fractional bandwidth is $2(f_H - f_L)/(f_H + f_L)$ where $f_H$ represents the upper frequency of the -10 dB emission limit and $f_L$ represents the lower frequency limit of the -10 dB emission limit.

## 1.1.2 A Brief History of UWB Signals
### Impulse UWB Signals

The modern era in UWB started in the early 1960s from work in time domain electromagnetics and was led by Harmuth at Catholic University of America, Ross and Robins at Sperry Rand Corporation, and van Etten at the United States Air Force (USAF) Rome Air Development Center [2,3]. Harmuth's work culminated in a series of books and articles between 1969 and 1990 [23–32]. Harmuth, Ross, and Robbins all referred to their systems as baseband radio. During the same period, engineers at Lawrence Livermore, Los Alamos National Laboratories (LLNL and LANL), and elsewhere performed some of the original research on pulse transmitters, receivers, and antennas.

A major breakthrough in UWB communications occurred as a result of the development of the sampling oscilloscope by both Tektronix and Hewlett-Packard in the 1960s. These sampling circuits not only provided a method to display and integrate UWB signals, but also provided simple circuits necessary for subnanosecond, baseband pulse generation [3, 17]. In the late 1960s, Cook and Bernfeld published a book [11] that summarized Sperry Rand Corporation's developments in pulse compression, matched filtering, and correlation techniques. The invention of a sensitive baseband pulse receiver by Robbins in 1972, as a replacement for the sampling oscilloscope, led to the first patented design of a UWB communications system by Ross at the Sperry Rand Corporation [45].

In parallel with the developments in the United States, extensive research into UWB was conducted in the former Soviet Union. In 1957 Astanin developed an X-band 0.5 ns duration transmitter for waveguide study at the A. Mozjaisky Military Air Force Academy, while Kobzarev et al. conducted indoor tests of UWB radars at the Radioelectronics Institute of the USSR Academy of Science [4]. As in the United States, development accelerated with the advent of sampling oscilloscopes.

By the early 1970s, the basic designs for UWB radar and communication systems evolved with advances in electronic component technology. The first ground-penetrating radar based on UWB was commercialized in 1974 by Morey at the Geophysical Survey Systems Corporation. In 1994, McEwan at LLNL developed the Micropower Impulse Radar (MIR), which provided a compact, inexpensive, low power UWB system for the first time [35].

Around 1989, the Department of Defense created the nomenclature ultra wideband to describe communication via the transmission and reception of impulses. The U.S. government has been and continues to be a major backer of UWB research. The FCC effort to authorize the use of UWB systems [14] spurred a great amount of interest and fear of UWB technology. In response to the uncertainty of how UWB systems and existing services could operate together, several UWB interference studies were sponsored by the U.S. government.

In 1993, Robert Scholtz at the University of Southern California wrote a landmark paper that presented a multiple access technique for UWB communication systems [38]. Scholtz's technique allocates each user a unique spreading code that determines specific instances in time when the user is allowed to transmit. With a viable multiple access scheme, UWB became capable of supporting not only radar and point-to-point communications but wireless networks as well.

With the advent of UWB as a viable candidate for wireless networks, a number of researchers in the late 1990s and early 2000s began detailed investigations into UWB propagation. These propagation studies, and the channel models developed from the measurement results, culminated in a number of notable publications by Cassioli, Win, Scholtz, Foerster, and Molisch [8, 12, 13, 18, 19, 36, 39, 40, 43, 44]. Additionally, the DARPA-funded Networking in Extreme Environments (NETEX) project began detailed investigations into indoor/outdoor UWB propagation modeling, characterization of the response of building materials to UWB impulses, and characterization of the antenna response to UWB signals.

Recently there has been a rapid expansion of the number of companies and government agencies involved with UWB, growing from a handful in the mid 1990s that included Multispectral Solutions, Time Domain, Aether Wire, Fantasma Networks, LLNL and a few others, to the plethora of players we have today. The FCC, NTIA (National Telecommunications and Information Adminstration), FAA, and DARPA, as well as the previously mentioned companies, spent many years investigating the effect of UWB emissions on existing narrowband systems. The results of those studies were used to inform the FCC on how UWB systems could be allowed to operate. In 2003, the first FCC certified commercial system was installed [37], and in April 2003 the first FCC-compliant commercial UWB chipsets were announced by Time Domain Corporation.

## 1.1.3  Types of UWB Signals

There are two common forms of UWB: one based on sending very short duration pulses to convey information and another approach using multiple simultaneous carriers. Each approach has its relative technical merits and demerits. Because Impulse UWB (I-UWB) is generally less understood than Multicarrier (MC-UWB), this book primarily focuses on impulse modulation approaches. The most common form of multicarrier modulation, Orthogonal Frequency Division Multiplexing (OFDM), has become the leading modulation for high data rate systems, and much information on this modulation type is available in recent technical literature.

Pure impulse radio, unlike classic communications, does not use a modulated sinusoidal carrier to convey information. Instead, the transmit signal is a series of baseband pulses. Because the pulses are extremely short (commonly in the nanosecond range or shorter), the transmit signal bandwidth is on the order of gigahertz. Note that the fractional bandwidth is greater than 20%, as shown in Figure 1.1. The unmodulated transmit signal as seen by the receiver, in the absence of channel effects, can be represented as

$$s(t) = \sum_{i=-\infty}^{\infty} A_i(t) p(t - iT_f) \qquad (1.1)$$

where $A_i(t)$ is the amplitude of the pulse equal to $\pm\sqrt{E_p}$, where $E_p$ is the energy per pulse, $p(t)$ is the received pulse shape with normalized energy, and $T_f$ is the frame

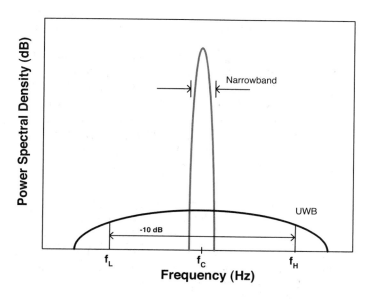

Figure 1.1: Comparison of the Fractional Bandwidth of a Narrowband and Ultra Wideband Communication System.

repetition time. (A UWB frame is defined as the time interval in which one pulse is transmitted.) We also define $T_p$ to be the duration of the pulse. Note that the pulse repetition rate $R_f = \frac{1}{T_f}$ is not necessarily equal to the inverse of the pulse width. In other words, the duty cycle of the transmitted signal is almost always less than 1. In this work we will refer to $s(t)$ as the transmit signal to avoid confusion with the received signal $r(t)$ that includes channel and antenna effects. Most practical systems will use some form of pulse-shaping to control the spectral content of each pulse to conform to regulatory limits.

## Multicarrier UWB Signals

Multicarrier communications were first used in the late 1950s and early 1960s for higher data rate HF military communications. Since that time, OFDM has emerged as a special case of multicarrier modulation using densely spaced subcarriers and overlapping spectra, and was patented in the United States in 1970 [9]. However, the technique did not become practical until several innovations occurred. First, the OFDM signal needs precisely overlapping but noninterfering carriers, and achieving this precision requires the use of a real-time Fourier transform [41], which became feasible with improvements in Very Large-Scale Integration (VLSI). Throughout the 1980s and 1990s, other practical issues in OFDM implementation were addressed, such as oscillator stability in the transmitter and receiver, linearity of the power amplifiers, and compensation of channel effects. Doppler spreading caused by rapid time variations of the channel can cause interference between the carriers and held back the development of OFDM until Cimini developed coded multicarrier modulation [10].

OFDM is now used in Asymmetric Digital Subscriber Line (ADSL) services, Digital Audio Broadcast (DAB), Digital Terrestrial Television Broadcast (DVB) in Europe, Integrated Services Digital Broadcasting (ISDB) in Japan, IEEE 802.11a/g, 802.16a, and Power Line Networking (HomePlug). Because OFDM is suitable for high data rate systems, it is also being considered for the fourth generation (4G) wireless services, IEEE 802.11n (high speed 802.11) and IEEE 802.20 (MAN) [34].

MC-UWB is very different from I-UWB. In multicarrier UWB, the complex baseband model transmitted signal has the form

$$s\left(t\right) = \sum_{i=1}^{N} d_i\left(t\right) e^{j2\pi i(T/T_s)} \tag{1.2}$$

where $N$ is the number of carriers, $T_s = NT_b$ is the symbol duration, and $d_i(t)$ is the symbol stream modulating the $i^{th}$ carrier. Figure 1.2 illustrates a comparison of the spectrum of I-UWB and MC-UWB transmissions.

## Relative Merits of Impulse Versus Multicarrier

The relative merits and demerits of I-UWB and MC-UWB are controversial issues and have been debated extensively in the standards bodies. One particularly important issue is minimizing interference transmitted by, and received by, the UWB system. MC-UWB is particularly well-suited for avoiding interference because its carrier frequencies can be precisely chosen to avoid narrowband interference to or from narrowband systems. Additionally, MC-UWB provides more flexibility and scalability, but requires an extra layer of control in the physical layer. For both forms of UWB, spread spectrum techniques can be applied to reduce the impact of interference on the UWB system.

I-UWB signals require fast switching times for the transmitter and receiver and highly precise synchronization. Transient properties become important in the design of the radio and antenna. The high instantaneous power during the brief interval of the pulse helps to overcome interference to UWB systems, but increases the possibility of interference from UWB to narrowband systems. The RF front-end of an I-UWB system may resemble a digital circuit, thus circumventing many of the problems associated with mixed-signal integrated circuits. Simple I-UWB systems can be very inexpensive to construct.

On the other hand, implementing a MC-UWB front-end can be challenging due to the continuous variations in power over a very wide bandwidth. This is particularly challenging for the power amplifier. In the case of OFDM, high-speed FFT processing is necessary, requiring significant processing power.

Another issue in the implementation of a UWB system is the general detection theory assumption that the system operates in an AWGN noise environment. Unfortunately, this is not always true for any real communication system and especially for UWB systems. There can be other signals that are within the UWB passband

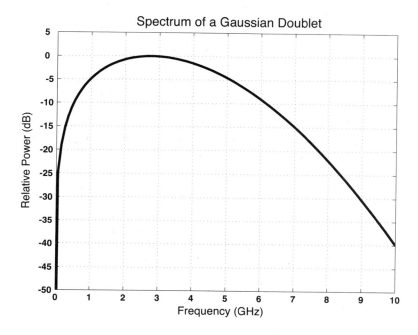

(a) Spectrum of a Gaussian Monocycle-Based Impulse UWB Signal

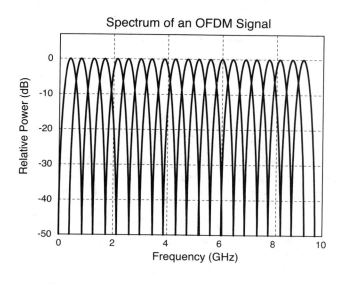

(b) Spectrum of an OFDM-based MC-UWB Signal

Figure 1.2: Comparison of Impulse and Multicarrier UWB Spectrums.

that do not have Gaussian noise statistics. These narrowband signals force a system to operate at higher transmit power or find a way to excise the in-band interference.

### 1.1.4 Regulatory, Legal, and Other Controversial Issues

On September 1, 1998, the FCC issued a Notice of Inquiry pertaining to the revision of Part 15 rules to allow the unlicensed use of UWB devices [14]. The FCC was motivated by the potential for a host of new applications for UWB technology: high-precision radar, through-wall imaging, medical imaging, remote sensors, and secure voice and data communications. Investigating the potential use of UWB devices presented a very different mode of operation for the FCC. Instead of dividing the spectrum into distinct bands that were then allocated to specific users/services, UWB devices would be allowed to operate overlaid with existing services. Essentially, the UWB device would be allowed to interfere with existing services, ideally at a low enough power level that existing services would not experience performance degradation. The operation of UWB devices in tandem with existing users is a significantly different approach to spectral efficiency than achieving the highest possible data rates in a channel with precisely defined bandwidths. In fact, many have questioned whether the operation of UWB devices is "efficient" in the strict sense of the word, or if it is instead an exercise in interference tolerance.

By May 2000, the FCC had received more than 1,000 documents from more than 150 different organizations in response to their Notice of Inquiry, to assist the FCC in developing an appropriate set of specifications. Specifically, the FCC was concerned about the potential interference from UWB transmissions on Global Positioning System (GPS) signals and commercial/military avionics signals. On February 14, 2002, the FCC issued a First Report and Order [14], which classified UWB operation into three separate categories:

1. Communication and Measurement Systems

2. Vehicular Radar Systems

3. Imaging Systems, including Ground Penetrating Radar, Through-Wall Imaging and Surveillance Systems, and Medical Imaging.

Each category was allocated a specific spectral mask, as shown in Figure 1.3. Table 1.1 summarizes the various UWB operational categories and their allocated bandwidths, along with restrictions on organizations that are allowed to operate in that particular mode.

The FCCs ruling, however, did not specifically address precision location for asset tracking or inventory control. These applications, known as location-aware communication systems, are a hybrid of radar and data communications that use UWB pulses to track the 2-D and 3-D position of an item to accuracies within a few tens of centimeters [15], as well as transmitting information about the item, such as its contents, to a centralized database system.

Note that the FCC has only specified a spectral mask and has not restricted users to any particular modulation scheme. As discussed previously, a number of

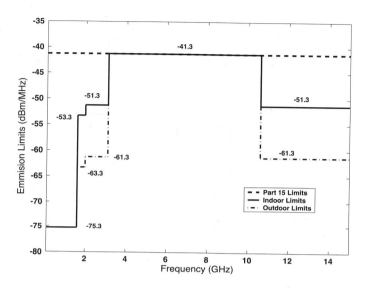

(a) Indoor UWB Communication Systems

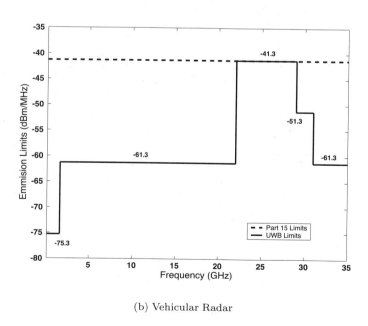

(b) Vehicular Radar

Figure 1.3: FCC Allocated Spectral Mask for Various UWB Applications (continued).

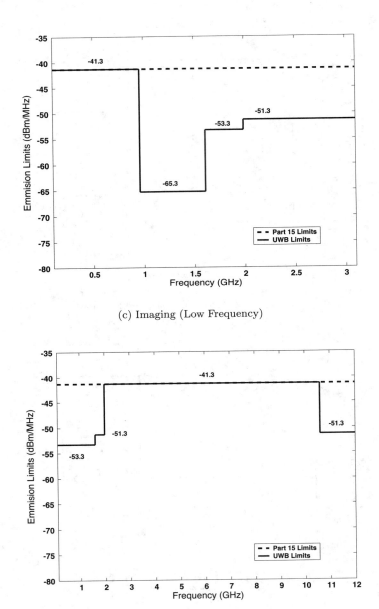

(c) Imaging (Low Frequency)

(d) Imaging (Mid Frequency)

Figure 1.3: (cont.) FCC Allocated Spectral Mask for Various UWB Applications

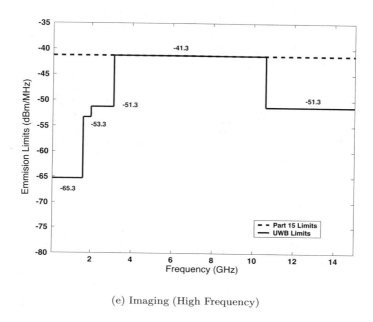

(e) Imaging (High Frequency)

Figure 1.3: (cont.) FCC Allocated Spectral Mask for Various UWB Applications.

Table 1.1: Summary of FCC Restrictions on UWB Operation (continued).

| Application | Frequency Band for Operation at Part 15 Limits | User Restrictions |
|---|---|---|
| Communications and Measurement Systems (sensors) | 3.1–10.6 GHz (different emission limits for indoor and outdoor systems) | None |
| Vehicular Radar for collision avoidance, airbag activation, and suspension system control | 24–29 GHz | None |
| Ground Penetrating Radar to see or detect buried objects | 3.1–10.6 GHz and below 960 MHz | Law enforcement, fire and rescue, research institutions, mining, construction |

Table 1.1: (cont.) Summary of FCC Restrictions on UWB Operation.

| Application | Frequency Band for Operation at Part 15 Limits | User Restrictions |
|---|---|---|
| Wall Imaging Systems to detect objects contained in walls | 3.1–10.6 GHz and below 960 MHz | Law enforcement, fire and rescue, mining, construction |
| Through-wall Imaging Systems to detect location or movement of objects located on the other side of a wall | 1.99–10.6 GHz and below 960 MHz | Law enforcement, fire and rescue |
| Medical Systems for imaging inside people and animals | 3.1–10.6 GHz | Medical personnel |
| Surveillance Systems for intrusion detection | 1.99–10.6 GHz | Law enforcement, fire and rescue, public utilities, and industry |

organizations are promoting multicarrier techniques, such as OFDM, as a potential alternative to I-UWB for high data rate communications.

Beyond the United States, other countries have been using a similar approach toward licensing UWB technology. In both Europe and Japan, initial studies have been completed, and regulations are expected to be issued in the near future that are expected to harmonize with the FCC mask.

## 1.2    What Makes UWB Unique?

### 1.2.1    Time Domain Design

UWB has a very unique set of design requirements, and attempting to apply the principles for traditional narrowband or even broadband communications to the design of I-UWB systems can be misleading. Analysis of I-UWB systems often means examining the impulse response of the system as opposed to the steady state response, particularly when examining the antenna response. Time domain effects can include frequency dependant pulse distortion imparted by RF components or the wireless channel, pulse dispersion produced by the antenna, or timing jitter generated by non-ideal oscillators. For traditional communication systems, these transient effects are only a small fraction of the symbol duration and may often be ignored. In I-UWB systems, these effects directly impact the performance of the overall communication system. For example, timing jitter will lead to imperfect correlation at the receiver or potential loss of data and system synchronization for modulation schemes where data is transmitted in the precise position of a pulse.

## 1.2.2   Impact of the Antenna

One of the challenges of the implementation of UWB systems is the development of a suitable antenna that would enhance the advantages promised by a pulsed communication system. I-UWB requires antennas that can cover multi-octave bandwidths in order to transmit pulses on the order of a nanosecond in duration with minimal distortion. Because data may be contained in the shape or precise timing of the pulse, a clean impulse response (that is, minimal pulse distortion) can be considered as a primary requirement for a good I-UWB antenna.

While it may be more intuitive for communication engineers to think of the performance of an antenna in terms of its frequency domain characteristics, the response of an antenna to a I-UWB pulse stream can best be described in terms of its temporal characteristics. An ideal UWB antenna needs to be relatively efficient across the entire frequency band with a Voltage Standing Wave Ratio (VSWR) of at most 2:1. To prevent distorting the pulse, an ideal UWB antenna should produce radiation fields with constant magnitude and a phase shift that varies linearly with frequency [5]. An antenna that meets these characteristics will radiate a signal which is only a time derivative of the input signal.

In reality, due to size and cost constraints, practical UWB antennas may not meet the previous requirements. It must also be noted that the antenna induced distortion can change with elevation and azimuth angle. Thus, we assume that such effects will ultimately be included in the assumed channel model. Chapter 3, "Channel Modeling," and Chapter 4, "Antennas," detail channel modeling and antenna effects, respectively.

## 1.2.3   Propagation and Channel Models

To perform systems-level engineering, UWB propagation characteristics must be considered. UWB differs from conventional communications in that the signal may be overlaid on top of interference. This interference must be considered in the link budget and, in fact, can often be the primary reason for performance limitations. Another issue is the introduction of large numbers of multipath signals that were not resolvable in narrowband communication systems. Measurements of typical UWB channels have revealed dense, multipath-rich environments, allowing for RAKE receivers that can harvest a tremendous amount of energy. Additionally, UWB propagation is highly dependent on the effect the antenna has on the shape and duration of the transmitted pulse.

UWB propagation measurements and modeling are the subjects of ongoing debate in the engineering community; as such, this book does not claim to resolve that debate. Rather, it discusses the basic concepts behind several UWB channel models and some of the differences between narrowband and UWB signal propagation.

## 1.2.4   Transmitter and Receiver Design

RF design for UWB systems is distinct from traditional narrowband or broadband systems in several ways. The extremely wide bandwidth of a UWB necessitates RF components that have flat frequency responses. Significant deviation, or ripple, in the frequency response of RF components as well as the nonlinearities present in all RF devices will introduce distortion to the UWB signal. UWB transmitted signals also have a very high peak-to-average power ratio (PAPR). As RF components are peak power limited, it becomes important to ensure that all RF devices have a power handling capacity at least as great as the peak power in the UWB signal.

Furthermore, the coexistence of UWB and existing services means that narrowband interfering signals will be detected by the receiver. These narrowband signals can either corrupt the pulse or saturate the RF front-end, decreasing the receiver's dynamic range and effectively limiting the range of the UWB system. Introducing notch filters at the receiver is a potential solution; pulse-shaping techniques, such as those described in [22], provide an alternative method for mitigating narrowband interferers without distorting the UWB waveform.

Most UWB receiver techniques require highly accurate synchronization with the transmitter as well as stable oscillators to maintain synchronization. With certain I-UWB modulation schemes, data may be conveyed by the precise position or timing of the pulse, and a loss of precise synchronization could result in a loss of data.

### 1.2.5   Difficulties in Using DSP Technology

Designing an I-UWB transmitter to broadcast short pulses is much simpler than designing a receiver to demodulate those pulses. For instance, assuming a pulse width of 250 picoseconds and 2 samples/pulse requires a sampling rate of 8 Gigasamples per second. Assuming 6 bits per sample, the receiver must process a data stream of 48 Gbps; at 8 bits per sample, the data stream increases to 64 Gbps. At the time of this writing, only the most technologically advanced FPGAs and ASICs are capable of handling such a huge amount of data.

Another problem is the limitations inherent in practical Analog to Digital Converters (ADCs). Most mass-produced commercial grade ADCs have analog input bandwidths[2] less than 1 GHz. Regardless of the sampling clock frequency, the ADC can only sample signals that fall within its input bandwidth. The highest performance commercially available ADCs can have input bandwidths, which extend into several GHz and have a maximum sampling clock frequency in the low GHz range. It is quite obvious, therefore, that in order to sample a UWB signal which lies in the 3.1–10.6 GHz range, the ADC must, at the very least, have an analog input bandwidth equal to or greater than the highest frequency component of the input signal (that is, an input bandwith of 10.6 GHz). The use of high-performance (and high-cost) FPGAs, DSPs, and ADCs are, however, an anathema to engineers who have heralded UWB as a low-cost, simple communication system.

### 1.2.6   Networking Issues

A primary driving application of UWB is a high rate Wireless Personal Area Network (WPAN) confined to a small coverage area (less than 10 m radius). The network should be a self-organized, dynamic, ad hoc network, which means the network

---

[2]Analog Input Bandwidth is defined as the frequency at which the sampled output of the ADC falls 3 dB below the full-scale input amplitude.

is formed without advanced planning and that users can join or leave at any time. Network security is also an important issue. Even though UWB signals may have a Low Probability of Intercept (LPI), it is still important to provide authentication, confidentiality, integrity, and availability. Variable modes of operation should allow for both long-range, low data rate communications and short-range, high-speed connections for multimedia or large data transfers.

UWB communications presents some unique challenges for a wireless network's Medium Access Control (MAC). As discussed in Chapter 9, "Networking," as the signal bandwidth becomes significantly greater than the data rate, a hybrid CDMA and Time Division Multiple Access (TDMA)-based MAC becomes a more optimal approach than a traditional TDMA MAC. This hybrid technique provides greater flexability and adaptability—an important advantage for UWB networks that may need to meet a variety of Quality of Service (QoS) requirements. Furthermore, the unique nature of I-UWB communications means that several additional features should be built into the MAC layer. Ranging information will assist in the formation of piconets by excluding users that fall outside a predetermined radius of operation. The need for strict synchronization between transmitter and receiver and the ability to generate accurate channel estimates must also be addressed by the MAC. Implementing a decentralized MAC provides the ability to incorporate UWB into consumer electronics and mobile phones that can operate over ad-hoc networks. Finally, different modes of operation, such as high data rate, long-range, or distributed sensor networks, each have somewhat different design constraints, suggesting that multiple approaches to the MAC design may be necessary to develop an optimal MAC layer for a particular application.

### 1.2.7  Future Directions

At the present time, the FCC is content to allow UWB devices to develop within the limitations of their First Note and Order [14]. As the technology matures, it is possible that the FCC may relax both the transmitted power level and bandwidth restrictions for UWB operation. Such modifications will most likely be a result of detailed investigations that demonstrate the minimal impact that higher power UWB devices will have on the QoS of existing users. In particular, major concerns still exist about the potential interference of UWB emissions to GPS and air traffic control signals.

A potential future application of UWB communications is low power, low data rate distributed sensor communications, similar to the 802.15.4/ZigBee standard. Because the duty cycle of I-UWB pulses is inherently very small, an I-UWB-based extension of the 802.15 standard would help to conserve valuable battery life [15]. Also, the extremely low power spectral density and short time duration of the pulse makes the transmitted signal difficult to detect and intercept, which is a definite advantage for ensuring a secure network.

Another potential application for I-UWB signals is the field of medicine. Microwave and radar monitoring of physiologic functions is an idea that has been around in concept since the 1970s [7, 20], but its development was hampered by the cumbersome and expensive technology of the time. With sufficiently short pulse duration (on the order of 100 picoseconds), an I-UWB radar would be capable of monitoring the movements of internal organs such as the heart or lungs without the need for direct skin contact or constraining the patient in space. Additionally, research is underway that analyzes the backscattered signals from a UWB pulse to detect cancer [6, 21]. Although I-UWB imaging may not provide the resolution of CT (Computed Tomography) or MRI (Magnetic Resonance Imaging) scans, it has the potential to cost-effectively provide critical information and determine, based on those results, whether further diagnostics are required.

## 1.3   The I-UWB System Model

### 1.3.1   Overview of the I-UWB System

This section presents the overall system model and notation convention that will be used throughout the book. The basic model for an unmodulated I-UWB pulse train was given in (1.1) and is repeated here

$$s(t) = \sum_{i=-\infty}^{\infty} A_i(t) p(t - iT_f) \qquad (1.3)$$

Note that we have assumed that the pulse is not distorted by the channel. Thus, $p(t)$ is the pulse that would be observed by the receiver in a distortionless and noiseless channel with infinite SNR. However, this is not necessarily equal to the pulse generated by the pulser circuitry, nor is it necessarily equal to the pulse launched by the transmitter. This is a unique feature of UWB systems and arises from the fact that the transmit and receive antennas can, and often do, distort the pulse shape.

Thus, in our system model we assume that the antenna-induced distortion is included in the received pulse $p(t)$. Recall also that the antenna-induced distortion can change with elevation and azimuth angles. Thus, we assume that such effects will ultimately be included in the assumed channel model.

### 1.3.2   Pulse Shapes

By far the most popular pulse shapes discussed in I-UWB communication literature are the Gaussian pulse and its derivatives, as they are easy to describe and work with. A Gaussian pulse is described analytically as

$$p(t) = \frac{1}{\sqrt{2\pi\sigma^2}} e^{(t-\mu)^2/(2\sigma^2)} \qquad (1.4)$$

where $\sigma$ is the standard deviation of the Gaussian pulse in seconds, and $\mu$ is the location in time for the midpoint of the Gaussian pulse in seconds. Note that the pulse width, $\tau_p$, is related to the standard deviation as $\tau_p = 2\pi\sigma$. An example is plotted in Figure 1.4 (a). The first derivative of a Gaussian pulse is also a commonly used analytical pulse shape, due to the fact that a UWB antenna may differentiate the generated pulse (assumed to be Gaussian) with respect to time,[3] leading to the following pulse shape

$$p(t) = \left(\frac{32k^6}{\pi}\right) t e^{-(kt)^2} \qquad (1.5)$$

where $k$ is a constant that determines the pulse width, and we have assumed $\mu = 0$. A third model uses the second derivative of a Gaussian pulse or

$$p(t) = \left(\frac{32k^2}{9\pi}\right)^{\frac{1}{4}} \left(1 - 2(kt)^2\right) e^{-(kt)^2}$$

These three pulse types are plotted in Figure 1.4. The time axis is arbitrary and depends on the values assumed above for $k$ and $\sigma$. We should also note that the current FCC rules make UWB transmission most practical in the 3.1–10.6 GHz band. As a result, the preceding pulse shapes may not be useful for commercial systems. Instead, the Gaussian modulated sinusoidal pulse is more practical. Specifically, the pulse shape

$$p(t) = \left(\frac{8k}{\pi}\right)^{\frac{1}{4}} \frac{1}{\sqrt{1 + e^{\frac{2\pi^2 f_c^2}{k}}}} e^{-(kt)^2} \cos(2\pi f_c t) \qquad (1.6)$$

where $f_c$ is the desired center frequency for the pulse. An example plot of this pulse is given in Figure 1.5.

### 1.3.3   Modulation Schemes

I-UWB systems allow for several modulation schemes, including Pulse Position Modulation (PPM) and Pulse Amplitude Modulation (PAM).[4] A detailed discussion of modulation schemes will be presented in Chapter 5, "Transmitter Design." We introduce them here in order to establish the notation and system model used throughout this book. The transmit signal in the case of amplitude modulation is represented by

$$s(t) = \sum_{i=-\infty}^{\infty} A_i(t) p(t - iT_f) \qquad (1.7)$$

where $A_i = d_i(t)$ now represents the amplitude of the $i^{th}$ pulse, which is dependent on the data $d_i(t)$ and the specific modulation scheme. A pulse position modulation scheme is represented by

$$s(t) = \sum_{i=-\infty}^{\infty} A p(t - iT_f - \delta d_i(t)) \qquad (1.8)$$

---

[3]Certain types of antennas, such as the Bicone antenna, can be configured to prevent differentiation of the generated pulse.

[4]Pulse Amplitude Modulation (PAM) is also referred to as Pulse Position Modulation (PPM), or BPSK in the literature. We have chosen to use the generic PAM, as it allows for a more general discussion of UWB modulation. PPM thus becomes a specal case of PAM (2-PAM).

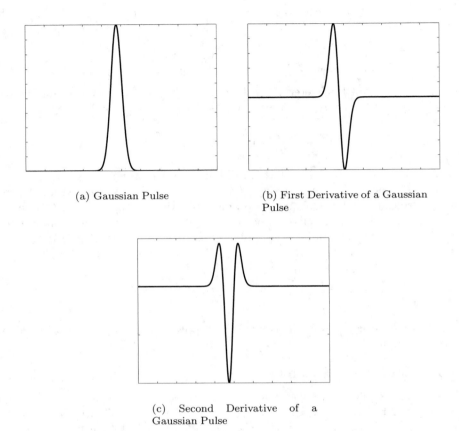

(a) Gaussian Pulse

(b) First Derivative of a Gaussian Pulse

(c) Second Derivative of a Gaussian Pulse

Figure 1.4: Example UWB Pulses.

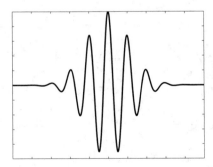

Figure 1.5: Example of a Sinusoidal Gaussian Pulse.

where $d_i(t)$ is the time modulation based on the information and $\delta$ is the base time increment.

As an example, let $d_i(t)$ be an antipodal binary bit stream consisting of +1's and −1's. The transmitted PAM signal will consist of a stream of positive and negative pulses (Figure 1.6a). The transmitted PPM signal will consist of pulses that are shifted either slightly before or slightly after their ideal positions in a regularly spaced pulse train (Figure 1.6b).

A key characteristic of UWB systems is their low power spectral density. The desire for low power spectral density (PSD) impacts the system model in two distinct ways. First, the pulse rate is often higher than the data rate. In other words, to obtain sufficient energy per symbol while maintaining sufficiently low PSD, multiple pulses will be associated with a single symbol. In this case the received signal is represented by

$$s\left(t\right) = \sum_{i=-\infty}^{\infty} A_{\left\lfloor \frac{i1}{N_s} \right\rfloor} p\left(t - iT_f\right) \tag{1.9}$$

where $N_s$ is the number of pulses per symbol, and the symbol rate is $R_s = \frac{R_p}{N_s}$. For PPM systems where multiple pulses per symbol are used, the received signal is represented as

$$s\left(t\right) = \sum_{i=-\infty}^{\infty} A p\left(t - iT_f - \delta d_{\left\lfloor \frac{i}{N_s} \right\rfloor}\right) \tag{1.10}$$

The received signal is then modeled as

$$r\left(t\right) = s\left(t\right) * h\left(t\right) + n\left(t\right) \tag{1.11}$$

where $h(t)$ represents the channel that possibly distorts the transmit signal and is assumed to have unit average energy and represents the convolution operation. That is, we scale out all gross attenuation effects and include them in the noise power. The noise is assumed to be Additive White Gaussian Noise (AWGN) with power $\sigma^2 = \frac{1}{SNR}$, where SNR is the average signal-to-noise ratio.[5] In the case where the channel is modeled using a Finite Impulse Response (FIR) filter approach (Chapter 3), the channel is modeled as

$$h\left(t\right) = \sum_{i=1}^{N_p} \beta_i \delta\left(t - \tau_i\right) \tag{1.12}$$

where $\beta_i$ is the amplitude and sign of the $i^{th}$ path, $\tau_i$ is the relative delay of the $i^{th}$ path, $\delta(t)$ is an impulse function, and $N_p$ is the number of paths. Note that in general we will model these parameters as random variables, as discussed in Chapter 3.

---

[5]Unfortunately, narrowband interferers located within the bandwidth of the I-UWB signal invalidate the AWGN assumption. For the sake of simplicity, however, we will adhere to the assumption of AWGN in order to provide a means to enable some selected comparison of I-UWB with traditional narrowband communcation systems.

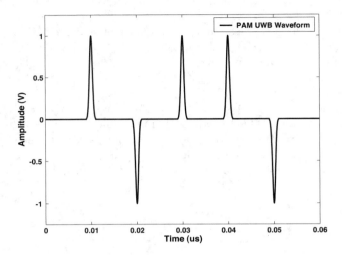

(a) Antipodal Pulse Amplitude Modulation: Positive Pulses Represent a +1 and Negative Pulses Represent a −1.

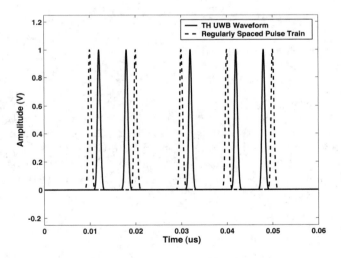

(b) Time Hop Modulation: Pulses Transmitted After Their Ideal Position Represent a +1 and Pulses Transmitted Prior to Their Ideal Position Represent a −1.

Figure 1.6: Example of Modulated UWB Signals Using the Data Sequence {1 −1 1 1 −1}.

## 1.3.4   Multiple Access Schemes

When a system has multiple users, we represent the transmit signal for user $k$ as $s_k(t)$. The total received signal is given by

$$r(t) = \sum_{i=1}^{K} s^{(k)}(t) * h^{(k)}(t) + n(t) \tag{1.13}$$

where $h^{(k)}(t)$ is the channel impulse response between the $k^{th}$ user and the receiver, $K$ is the total number of users considered, and $n(t)$ is AWGN.

In multiuser systems there are many forms of multiple access that will be discussed in Chapter 5. The signal model for TDMA-based (or random access methods) will not differ from that presented here. However, CDMA systems will require additional notation. When pseudorandom amplitude modulation is used to distinguish users, we represent the signal from the $k^{th}$ user as

$$s^{(k)}(t) = \sum_{i=-\infty}^{\infty} A_i^{(k)} p(t - iT_f) \tag{1.14}$$

where $A_i^{(k)} = c_i d_{\lfloor \frac{i}{N_s} \rfloor}^{(k)}$ and $c_i$ is the pseudorandom code value for the $i^{th}$ pulse. This value can repeat every symbol (short codes) so that $c_i = c_{i+N_s}$, or it can repeat at some much longer interval (long codes).

Pseudorandom codes can also be applied to PPM schemes, and are often referred to as time hopping or pseudorandom dithering. In this case the transmit signal of the $k^{th}$ user is

$$s^{(k)}(t) = \sum_{i=-\infty}^{\infty} A p\left(t - iT_f - c_i^{(k)} T_c - \delta d_{\lfloor i/N_s \rfloor}\right) \tag{1.15}$$

where the time hopping is accomplished with the sequence $c_i^{(k)}$, and $T_c$ is the fundamental hopping granularity.

## 1.3.5   Receiver Decision Statistic

The receiver estimates the most likely transmitted data symbol by using a decision statistic that is a function of the received signal

$$\widehat{d} = f(Z) \tag{1.16}$$

where $Z$ represents the output of the receiver, and the function $f(Z)$ depends on the modulation scheme and receiver structure, as discussed in Chapter 6, "Receiver Design Principles," and Chapter 7, "On the Coexistence of UWB and Narrowband Radio Systems." Additionally, in diversity systems with multiple receiver branches (such as multiple antenna structures or a RAKE receiver) the decision statistic $Z$ will be the sum of several statistics

$$Z = \sum_{i=1}^{L} Z_i \tag{1.17}$$

where $L$ is the number of diversity branches, either in time or space, and $z_i$ is the statistic calculated per diversity branch.

## 1.4   The MC-UWB System Model

### 1.4.1   Overview of the MC-UWB System

In the past several years, MC-UWB (also called frequency domain UWB) has received a significant amount of attention. The transmit MC-UWB signal $s(t)$ has the following complex baseband form

$$s(t) = A \sum_{r} \sum_{n=1}^{N} b_n^r p(t - rT_p) e^{(j2\pi n f_0(t-rT_p))} \qquad (1.18)$$

where $N$ is the number of subcarriers, $b_n^r$ is the symbol that is transmitted in the $r^{th}$ transmission interval over the $n^{th}$ subcarrier, and $A$ is a constant that controls the transmitted power spectral density and determines the energy per bit. The fundamental frequency is $f_0 = \frac{1}{T_p}$.

### 1.4.2   OFDM UWB

OFDM is a special case of multicarrier transmission that permits subcarriers to overlap in frequency without mutual interference and hence spectral efficiency is increased. Multiple users can be supported by allocating each user a group of subcarriers. OFDM-UWB is a novel system that has been proposed as a physical layer for high bit rate, short-range communication networks. Reliable communication systems achieve high throughput by transmitting multiple data streams in parallel on separate carrier frequencies. Unlike narrowband OFDM, the OFDM-UWB spectrum can have gaps between subcarriers. OFDM-UWB is one proposed physical layer standard for 802.15.3a Wireless Personal Area Networks.

OFDM-UWB uses a frequency coded pulse train as a shaping signal. The frequency coded pulse train is defined by

$$p(t) = \sum_{n=1}^{N} s(t - nT) e^{\left(-j2\pi c(n) \frac{1}{T_c}\right)} \qquad (1.19)$$

where $s(t)$ is an elementary pulse with unit energy and duration $T_s < T$, and $p(t)$ has duration $T_p = NT$. Each pulse is modulated with a frequency $f_n = \frac{c(n)}{T_c}$ where $c(n)$ is a permutation of the integers $\{1, 2, \ldots, N\}$. As shown in Chapter 6, the set $p_k(t) = p(t) e^{(j2\pi k f_0 t)}$ is orthogonal for $k = 1, 2, \ldots, N$.

## 1.5   Overview of the Book

This book is designed to be an interdisciplinary study of UWB communication systems. The development of channel models for UWB communication systems requires extensive data on propagation of UWB signals. Both experimental and simulation techniques can be used to examine the propagation of UWB signals in indoor and indoor/outdoor environments. In Chapter 2, "Channel Measurement and Simulation," time domain and frequency domain measurement methods for UWB channel sounding and their advantages and disadvantages are discussed. The electromagnetic simulation of UWB signal propagation in indoor environments is also addressed in Chapter 2. In Chapter 3, channel models and UWB link budgets are developed based on data collected from extensive UWB propagation measurements.

A critical component of UWB propagation, the antenna, is covered in Chapter 4. This chapter presents a detailed parametric study of time and frequency domain characteristics of both the antenna and scattering structures that must be considered in the performance estimates of UWB links. Additionally, mathematical modeling is presented that allows the antenna effects to be theoretically integrated into the UWB channel models.

As discussed previously in this chapter, transmitter and receiver design presents a unique challenge for UWB systems, particularly with the emphasis on low power, low cost devices. Chapter 5, "Transmitter Design," describes several widely used signal generation and modulation/signaling schemes unique to both I-UWB and MC-UWB. Chapter 6 provides a comprehensive review of a wide variety of UWB receiver architectures, with an emphasis on different mechanisms for optimally demodulating the received UWB signal.

UWB signals will encounter interference from many sources, primarily from relatively narrowband systems. In addition, UWB signals will also affect a large number of narrowband radios; of critical importance is the potential interference with GPS, E-911, and navigation bands. In Chapter 7, we assess, via analysis and simulations, the interference caused by UWB signals, as well as the impact of narrowband interference on a UWB receiver.

Simulation of UWB communication systems is unique in that it emphasizes the transient nature of UWB systems, and it is the focus of Chapter 8, "Simulation." This chapter covers several subjects important to efficient and accurate simulation of UWB communication systems, including challenges introduced by the simulation of UWB, architectural approaches to simulating UWB communication systems, and simulation models for UWB communication systems components.

Chapter 9, "Networking," addresses networking issues for networks of directly connected UWB nodes and larger ad hoc networks where network formation and multihop routing is required. In Chapter 9, we will examine several design issues related to UWB networks, including data link layer design, architectures of multihop ad hoc networks, and corresponding routing schemes, as well as related issues such as performance and Quality of Service (QoS) management.

Chapter 10, "Applications and Case Studies," explores the wide range of applications that can exploit the unique properties of I-UWB. Practical examples will be discussed from among some of the first commercial UWB products that feature several of the diverse set of UWB applications, including precision location, radar, imaging, distributed sensors, and high-speed communication. Appendix A of Chapter 10 provides a brief overview of the IEEE 802.15.3a preliminary standard that describes the MAC and PHY layers[6] of a wireless personal area network (WPAN).

---

[6]At the time of publication, the PHY layer has not been fully determined; two alternative candidates remain.

## Bibliography

[1] Available: http://www.aetherwire.com.

[2] T. W. Barrett, "History of Ultrawideband (UWB) Radar and Communications: Pioneers and Innovators," *Proc. Progress in Electromagnetics Symposium (PIERS) 2000*, Cambridge, MA, July 2000.

[3] T. W. Barrett, "History of Ultrawideband Communications and Radar: Part 1, UWB Communications," *Microwave Journal*, pp. 22-56, January 2001.

[4] T. W. Barrett, "History of Ultra Wideband Communications and Radar: Part 2, UWB Radar and Sensors," *Microwave Journal*, pp. 22-46, February 2001.

[5] A. Bayram, "A Study of Indoor Ultra-wideband Propagation Measurement and Characterization," Master's Thesis, Virginia Polytechnic Institute and State University, May 2004. Available: http://scholar.lib.vt.edu/theses/available/etd-05142004-171134.

[6] E. J. Bond, X. Li, S. C. Hagness, and B. D. Van Veen, "Microwave Imaging via Space-Time Beamforming for Early Detection of Breast Cancer," *IEEE Transactions on Antennas and Propagation*, vol. 51, no. 8, pp. 1690-1705, August 2003.

[7] C. G. Caro and J. A Bloice, "Contactless Apnoea Detector Based on Radar," *Lancet*, pp. 959-961, October 30, 1971, 2(7731).

[8] D. Cassioli, M. Z. Win, and A. F. Molisch, "The Ultra-Wide Bandwidth Indoor Channel: From Statistical Model to Simulations," *IEEE Journal on Selected Areas in Communications*, vol. 20, no. 6, pp. 1247-1257, August 2002.

[9] R.W. Chang, "Orthogonal Frequency Division Multiplexing," U.S. Patent 3,488,445, filed 1966, issued January 6, 1970.

[10] L.C. Cimini Jr., "Analysis and Simulation of a Digital Mobile Channel Using Orthogonal Frequency Division Multiplexing," *IEEE Transactions on Communications*, vol. 33, no. 7, pp. 665-675, July 1985.

[11] C. E. Cook and M. Bernfeld, *Radar Signals: An Introduction to Theory & Application*, New York: Academic Press, 1967.

[12] R. J.-M. Cramer, R. A. Scholtz, and M. Z. Win, "On the analysis of UWB Communication Channels," *IEEE MILCOM*, vol. 2, pp. 1191-1195, Atlantic City, NJ, November 1999.

[13] R. J.-M. Cramer, R. A. Scholtz, and M. Z. Win, "Evaluation of an Ultra-Wide-Band Propagation Channel," *IEEE Transactions on Antennas and Propagation*, vol. 50, no. 5, pp. 561-570, May 2002.

[14] "Revision of Part 15 of the Commission's Rules Regarding Ultra-Wideband Transmission Systems," First note and Order, Federal Communications Commission, ET-Docket 98-153, Adopted February 14, 2002, released April 22, 2002. Available: http://www.fcc.gov/Bureaus/Engineering_Technology/Orders/2002/fcc02048.pdf.

[15] R. J. Fontana, E. Richley, and J. Barney, "Commercialization of an Ultra Wideband Precision Asset Location System," *Proc. 2003 IEEE Conference on Ultra Wideband Systems and Technologies*, Reston, VA, May 2002.

[16] R. J. Fontana, A. Ameti, E. Richley, L. Beard, and D. Guy, "Recent Advances in Ultra Wideband Communications Systems," *Proc. 2003 IEEE Conference on Ultra Wideband Systems and Technologies*, Reston, VA, May 2002.

[17] R. J. Fontana, "A Brief History of UWB Communications." Available: http://www.multispectral.com/history.html.

[18] J. R. Foerster, "The Effects of Multipath Interference on the Performance of UWB Systems in an Indoor Wireless Channel," *IEEE 53rd Vehicular Technology Conference*, vol. 2, pp. 1176-1180, Rhodes, Greece, May 2001.

[19] A. Rajeswaran, V. S. Somayazulu, and J. R. Foerster, "Rake Performance for a Pulse Based UWB System in a Realistic UWB Indoor Channel," *IEEE International Conference on Communications*, vol. 4, pp. 2879-2883, May 2003.

[20] C. I. Franks, B. H. Brown, and D. M. Johnston, "Contactless Respiration Monitoring of Infants," *Medical Biological Engineering*, 14 (3), pp. 306-318, May 1976.

[21] S. C. Hagness, A. Taflove, and J. E. Bridges, "Three-Dimensional FDTD Analysis of a Pulsed Microwave Confocal System for Breast Cancer Detection: Design of an Antenna-Array Element," *IEEE Transactions on Antennas and Propagation*, vol. 47, No. 5, pp. 783-791, May 1999.

[22] M. Hamalainen, V. Hovinen, R. Tesi, J. H. J. Iinatti, and M. Latva-aho, "On the UWB System Coexistence With GSM900, UMTS/WCDMA, and GPS," *IEEE Journal on Selected Areas in Communications*, vol. 20, no. 9, pp. 1712-1721, December 2002.

[23] H. F. Harmuth, *Transmission of Information by Orthogonal Functions*, 1st ed., New York: Springer, 1969.

[24] H. F. Harmuth, *Transmission of Information by Orthogonal Functions*, 2nd ed., New York: Springer, 1972.

[25] H. F. Harmuth, "Signal Processing and Transmission by Means of Walsh Functions," U.S. Patent 3,678,204, dated July 18, 1972.

[26] H. F. Harmuth, "Sequency Filters Based on Walsh Functions for Signals with Two Space Variables," U.S. Patent 3,705,981, issued Dec. 12, 1972.

[27] H. F. Harmuth, *Sequency Theory*, New York: Academic Press, 1977.

[28] H. F. Harmuth, "Selective Reception of Periodic Electromagnetic Waves with General Time Variation," *IEEE Transactions on Electromagnetic Compatibility*, vol. EMC-19, no. 3, pp. 137-144, August 1977.

[29] H. F. Harmuth, "Frequency-Sharing and Spread-Spectrum Transmission with Large Relative Bandwidth," *IEEE Transactions on Electromagnetic Compatibility*, vol. EMC-20, no. 1, pp. 232-239, February 1978.

[30] H. F. Harmuth, *Nonsinusoidal Waves for Radar and Radio Communication*, New York: Academic Press, 1981.

[31] H. F. Harmuth, *Antennas and Waveguides for Nonsinusoidal Waves*, New York: Academic Press, 1984.

[32] H. F. Harmuth, *Radiation of Nonsinusoidal Electromagnetic Waves*, New York: Academic Press, 1990.

[33] E. C. Kisenwether, "Ultra Wideband (UWB) Impulse Signal Detection and Processing Issues," *IEEE Tactical Communications Conference*, vol. 1, pp. 87-93, Ft. Wayne, IN, April 1992.

[34] "Frequency Division Multiplexing," *IEEE Transactions on Communications*, vol. COM–33, no.7, pp. 665–675, July 1985. Available: http://www.multibandofdm.org/.

[35] T. E. McEwan, "Ultra-Wideband Radar Motion Sensor," U.S. Patent 5,361,070, issued November 1, 1994.

[36] A. F. Molisch, J. R. Foerster, and M. Pendergrass, "The Evolution of Wireless LANs and PANs-Channel Models for Ultrawideband Personal Area Networks," *IEEE Transactions on Wireless Communications*, vol. 10, no. 6, pp. 14-21, December 2003.

[37] R. Mulloy, "MSSI Successfully Installs FCC Certified Ultra Wideband (UWB) Asset Tracking System for Aircraft Engine Location and Identification," MSSI Press Release, Germantown, MD, September 2003.

[38] R. A. Scholtz, "Multiple Access with Time-Hopping Impulse Modulation," *IEEE MILCOM*, vol. 2, pp. 447-450, Boston, MA, October 1993.

[39] R. A. Scholtz, R. J.-M. Cramer, and M. Z. Win, "Evaluation of the Propagation Characteristics of Ultra-Wideband Communication Channels," *IEEE International Antennas and Propagation Society Symposium*, vol. 2, pp. 626-630, Atlanta, GA, June 1998.

[40] R. A. Scholtz, and J.-Y. Lee, "Problems in Modeling UWB Channels," *IEEE 36th Conference on Signals, Systems, and Computers*, vol. 1, pp. 706-711, November 2002.

[41]  S. B. Weinstein and P. M. Ebert, "Data Transmission by Frequency–Division Multiplexing Using the Discrete Fourier Transform," *IEEE Transactions on Communication Technology*, vol. COM–19, no.5, pp.628–634, October 1971.

[42]  M. Welborn, and B. Shvodian, "Ultra-Wideband Technology for Wirelsss Personal Area Networks-The IEEE 802.15.3/3a Standards," *IEEE Ultra Wideband Systems and Technology Conference*, Reston, VA, November 2003.

[43]  M. Z. Win and R. A. Scholtz, "On the Robustness of Ultra-Wide Bandwidth Signals in Dense Multipath Environments," *IEEE Communication Letters*, vol. 2, no. 2, pp. 51-53, February 1998.

[44]  M. Z. Win and R. A. Scholtz, "Characterization of Ultra-Wide Bandwidth Wireless Indoor Channels: A Communication Theoretic View," *IEEE Journal on Selected Areas in Communications*, vol. 20, no. 9, pp. 1613-1627, December 2002.

[45]  G. F. Ross, "Transmission and Reception System for Generating and Receiving Base-Band Duration Pulse Signals without Distortion for Short Base-Band Pulse Communication Systems," U.S. Patent 3,728,632, issued April 17, 1973.

# Chapter 2

# CHANNEL MEASUREMENT AND SIMULATION

Ahmad Safaai-Jazi,
Ahmed M. Attiya,
and Sedki Riad

## 2.1   Introduction

The development of channel models for UWB communication systems requires extensive data on UWB signal propagation. Both experimental and simulation techniques can be used to examine the propagation of UWB signals in indoor and indoor-outdoor environments. The advantage of experimental methods is that all system and channel parameters affecting the propagation of UWB signals are accounted for without preassumptions. However, these methods are usually expensive, time consuming, and limited by the characteristics of available equipment, such as sensitivity, bandwidth, dispersion, and the attenuation of the connecting cables. On the other hand, simulation techniques are free from the limitations of experimental approaches, are cost effective, and are less time consuming. Simulators can serve as a useful tool for predicting the behavior of UWB communication systems. The accuracy of simulation results depends on the amount of details included in the simulation model. However, more details lead to more complex computer programs, and require more computational time. Thus, a compromise between the required accuracy and the available computational resources should be made in designing simulators for UWB communication systems.

In narrowband wireless communication systems, the information signal modulates a very high frequency sinusoidal carrier; thus, along each propagation path the signal suffers very little distortion because the system elements such as antennas, reflecting walls, diffracting objects in the channel, and so on, have essentially

**29**

constant electromagnetic properties over the narrow bandwidth of the radiated signal. The only signal degradation is caused by multipath components. On the other hand, in UWB systems, the information signal may suffer significant distortion due to the transmitting/receiving antennas not meeting the necessary bandwidth requirements, and also due to the dispersive behavior of building materials in the propagation channel. Of course, multipath components are also present in UWB channels. But, unlike narrowband signals, UWB signals do not suffer fading due to the destructive interference of multipath components.

In this chapter, both measurement and simulation of UWB signal propagation are addressed. Section 2.2 surveys time domain and frequency domain measurement techniques. Important issues such as triggering, calibration, interference rejection, and noise are discussed. Typical indoor measurement results in both time domain and frequency domain are presented in Section 2.3. The role of antennas, as well as the impacts of building architecture and dispersive properties of building materials, are also discussed in this section. The electromagnetic simulation of UWB signal propagation in indoor environments is detailed in Section 2.4. Channel models, based on data collected from extensive UWB propagation measurements in two buildings on the Virginia Tech campus, are discussed in Chapter 3, "Channel Modeling."

## 2.2   Measurement Techniques

Measurements of UWB signal propagation can be carried out using a variety of methods that may be broadly divided into two categories: time domain and frequency domain techniques. Both measurement techniques are discussed, and their advantages, disadvantages, limitations, and related important issues, such as triggering, noise, and calibration are addressed.

### 2.2.1   Time Domain Measurement Techniques

Time domain techniques can be used as a direct way to characterize UWB communication channels. Ideally, the impulse response provides complete characterization of a device or a system over the entire frequency band of use. However, it is not possible to directly measure the true impulse response of a device or a system since that requires the availability of an ideal dirac-delta excitation signal. In practice, very short duration pulses are used for the time domain characterization of UWB channels. The shorter the duration of the pulses used, the wider the bandwidth for which the UWB propagation characteristics can be measured. The basic idea of time domain measurements is to excite one end of the UWB channel with a periodic train of very short duration pulses separated by a sufficiently long quiet period so that all multipath components are received at the other end during the quiet period (prior to the next occurrence of the pulse). On the receiving side, the signal is detected using a wideband detector (typically, a digital sampling oscilloscope). In the following sections, we first discuss, in detail, the time domain measurement method used for characterizing UWB channels. Then, we examine the main sources of errors in the measurements, and the precautions needed to minimize these errors.

## Time Domain Channel Measurement

The time domain measurement technique for UWB channel sounding, as shown in Figure 2.1, consists of a pulse generator, a digital sampling oscilloscope, a pair of transmitting and receiving antennas, and a triggering signal generator. The pulse generator represents the UWB signal source. This generator should be connected to the transmitting antenna through a low loss wideband cable to minimize signal degradation (attenuation and dispersion) prior to propagation through the antenna. When higher radiated powers are desired, a UWB power amplifier may be used at the feed point of the transmitting antenna. This amplifier should have a constant gain and a linear phase characteristic over the spectrum of the UWB signal in order to minimize signal distortion. The receiving antenna and the digital sampling oscilloscope constitute the receiver. To enhance the received signal power, a low noise amplifier may be used right at the output port of the receiving antenna. This amplifier should also have a constant gain and a linear phase characteristic over the spectrum of the UWB signal.

An important task in time domain measurements is the synchronization of both transmitting and receiving sides of the channel sounding system. To achieve this synchronization, a low jitter triggering signal should be maintained between the pulse generator and the digital sampling oscilloscope. A simple approach is to use a sample of the radiated pulse captured by means of a small antenna situated close to the transmitting antenna to trigger the sampling oscilloscope and thereby achieve synchronization. This method does not require an independent triggering generator, but the captured signal is typically weak and may not be strong enough to trigger the oscilloscope. Another disadvantage is that the captured signal is often cluttered by multiple reflections from the surrounding objects, and contains multiple triggering points, leading to false triggers. A more reliable triggering approach is to use a triggering generator with two outputs to synchronize the transmitter and the receiver. The triggering sequence is arranged such that a pretrigger signal is sent to the sampling oscilloscope, while a delayed trigger is sent to the transmitter. The time delay between the pretrigger and the delayed trigger signal should be adjusted to compensate for the time delay introduced by the triggering cables and propagation through the UWB channel being measured.

Due to their ultra wide bandwidths, UWB channel measurements are susceptible to interference and noise from various sources. Generally, two categories of noise, namely, narrowband noise and wideband noise, can be identified. Narrowband noise is usually due to electromagnetic interference from nearby narrowband systems. This type of noise usually takes the form of a sinusoidal waveform added to the received signal. To eliminate the narrowband noise, the received signal is first Fourier transformed to the frequency domain where it is bandpass filtered to remove the noise spectrum, and then the signal is converted back to time domain via the inverse Fourier Transform. The wideband noise, on the other hand, is typically in the form of random short pulses that are not repetitive. Averaging multiple acquisitions can significantly reduce this noise. Signal averaging is generally available as a built-in feature in sampling oscilloscopes.

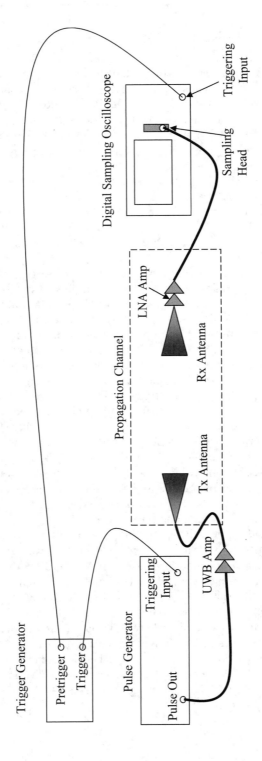

Figure 2.1: Schematic Diagram of the Time Domain Measurement Setup for UWB Channel Characterization.

Radiation from the electronic circuit of the pulse generator and leakage from the cable connecting the pulse generator to the transmitting antenna, where the signal level is high, may contribute significantly to the background noise level. This background noise level can be measured by replacing the transmitting antenna in Figure 2.1 with a matched termination while the pulse generator is on. This background noise level sets the minimum level below which received signals cannot be detected.

Another important issue is the calibration of the measurement setup. The purpose of calibration is to remove the influence of nonideal characteristics of the measurement equipment from the measured data. To measure the characteristics of a communication link consisting of the transmitting and receiving antennas and the propagation channel, the contribution of the measurement setup needs to be calibrated out. This requires measuring the response of the equipment while the link under measurement is replaced by a calibrated attenuator, as shown in Figure 2.2. This attenuator should have a fairly constant attenuation level, and a linear phase characteristic over the spectrum of the UWB signal. The attenuation of this attenuator is chosen to ensure the proper signal level at the sampling oscilloscope. The impulse response of the link can then be obtained by deconvolving the calibration measurement waveform from the corresponding measured waveform for the communication link.

## Sampling and Triggering Issues

Sampling and triggering are two critical aspects in direct time domain measurements. Figure 2.3 illustrates the principle of operation of a sampling oscilloscope. The real time signal to be measured is assumed to be periodic with a period $T$. The scope generates a train of $N$ sampling pulses to trigger its detection circuitry within a specified duration of the measured signal. The triggering signal required for generation of the sampling pulses should run at a slightly different period from that of the signal under measurement to ensure the acquisition of different sampled points on the waveform for the $N$ sampling pulse train. The period of sampling pulses is thus $T + (T/N)$. Each acquired sample corresponds to a point of the input waveform, so that the train of sampling pulses results in sequential sampled points of the measured waveform. The resulting sampled data represent the $y$-axis in the oscilloscope display. The corresponding $x$-axis is obtained via a ramp signal with a period of $(N + 1) T$, which is triggered at the beginning of the sampling train and is reset at its end. This brief description of the principle of operation of the sampling oscilloscope highlights the importance of triggering and the problems that may arise due to false triggering and jitter. False triggering signals may arise from the dispersion by triggering cables, the mismatch between the triggering cable and the triggering input, defects in the triggering cable, or the ringing effect of the trigger signal generator. The spectrum of the triggering signal is usually concentrated at the lower frequencies. Thus, the dispersion of triggering cables is not generally a major cause of false triggering for high-quality commercial coaxial cables. The shielding of the triggering cable should help prevent electromagnetic interference from, or to, the propagation channel. The connectors should also be selected and

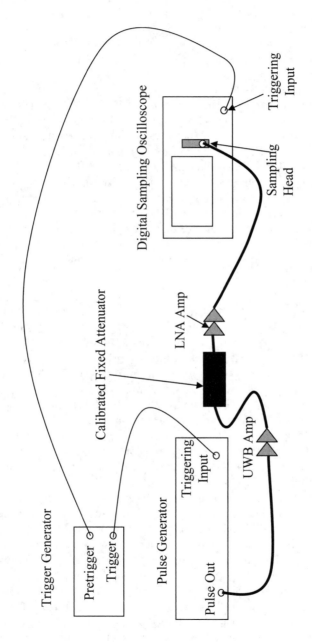

Figure 2.2: Calibration of Time Domain Measurement Setup.

34

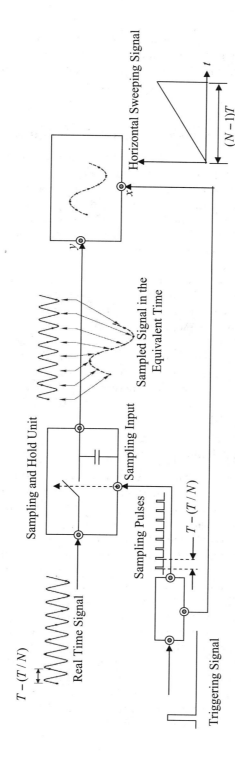

Figure 2.3: Principle of Operation of Sampling Oscilloscope: Sampled Points Connected by Interpolation to Produce a Continuous Waveform.

35

attached to the cable carefully to prevent multiple reflections between them and the triggering cable. The trigger generator should be designed with the least amount of vertical (amplitude) jitter and horizontal (time) jitter to ensure stability and consistency of the triggering and sampling process, and hence the overall precision of the time domain measurement.

## 2.2.2   Frequency Domain Measurement Techniques

Frequency domain characterization of UWB channels is based on measurements at different frequency points using a sweep harmonic generator. The chief advantage of the frequency domain technique over the time domain method is the availability of a much larger dynamic range, resulting in improved measurement precision. Each measurement is represented by a complex transfer function value described by its magnitude and phase terms. The inverse Fourier transform of the channel transfer function yields the impulse response of the channel, which is the information sought for UWB channel characterization.

Transfer function measurements are typically performed using a vector network analyzer (VNA), which is capable of measuring the complex ratio between the response of a device network under test to its excitation. A scalar network analyzer (SNA) provides an alternative for measuring the transfer function magnitude only. Phase measurements in UWB communication channels involving long propagation distances should be dealt with very carefully, as the available vector network analyzers were not originally designed for this purpose. Both the scalar and vector network analyzer approaches to channel characterization are surveyed in the following sections. The difficulties associated with direct phase measurements in UWB channel sounding are addressed as well.

### Scalar Frequency Domain Measurement

In the scalar frequency domain method, only the magnitude of the transfer function is measured. However, both magnitude and phase information are required to enable the conversion of the frequency domain data to the time domain impulse response. Hence, retrieving the phase information from the magnitude measurement, such as by using the Hilbert Transform, becomes an integral part of this approach. Figure 2.4 shows the setup for scalar frequency domain channel measurements using a scalar network analyzer. As shown in the figure, the synchronization between the source and the SNA is achieved by connecting the sweep output of the RF synthesizer to the sweep input of the SNA.

To calibrate the setup, a measurement without the UWB channel is performed. The calibration measurement is performed by connecting the line to the input of the transmitting antenna directly to the line at the output of the receiving antenna through a calibrated attenuator, as shown in Figure 2.5. The need for, and properties of, such an attenuator were discussed earlier in the time domain measurement section. The source calibration measurement is acquired and stored in the available SNA memory or that of an interfaced computer. Measurements on the UWB channel are then performed as shown in Figure 2.4, and the corresponding channel

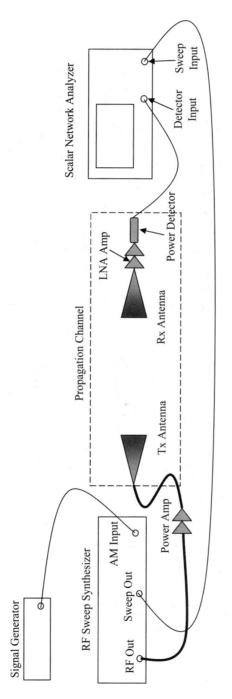

Figure 2.4: Setup for UWB Channel Characterization Using Scalar Frequency Domain Method.

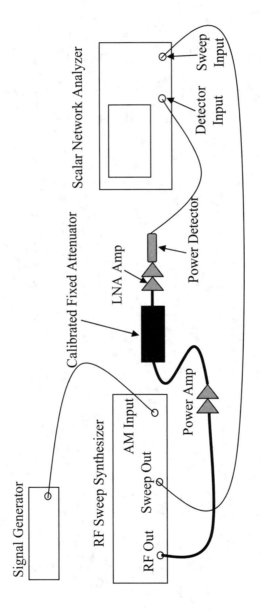

Figure 2.5: Calibration of Scalar Frequency Domain Measurement Setup.

transfer function is obtained by subtracting the prestored source calibration data in dB from the channel measurement data, also expressed in dB.

As discussed in the time domain measurement section, the background noise level sets the minimum level below which the received signal cannot be detected. Again, the background noise for this setup can be measured while replacing the transmitting antenna in Figure 2.4 with a matched load with the transmitter operating.

## Direct Measurement of Magnitude and Phase

In this approach, both magnitude and phase are measured directly. Figure 2.6 shows the schematic diagram of the measurement setup that can be used to obtain the complex frequency domain transfer function of the UWB channel. Here, the scalar network analyzer is replaced by a vector network analyzer in which the synchronization of all units is maintained internally, and no external synchronization is needed. The operation of the VNA is based on a superheterodyning mechanism rather than simple crystal detection or the thermocouple effect normally used in scalar measurements. Thus, the typical dynamic range of a VNA is much larger than that of a typical scalar network analyzer.

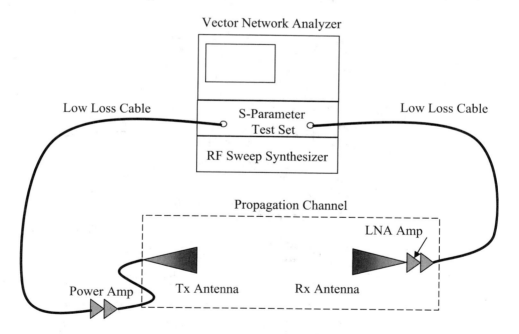

Figure 2.6: Measurement Setup for UWB Channel Characterization Using Vector Network Analyzer.

In conventional VNA language, measuring the channel transfer function is to perform a forward transmission complex scattering-matrix parameter $S_{21}$ measurement. It can be seen from Figure 2.6 that measurements of channels with long propagation distances require long cables to connect the VNA's $S$-parameter test set ports to the transmitting and receiving antennas at the two ends of the channel. Such long cables present a major limiting factor in this measurement configuration. As the frequency increases, cable losses increase "exponentially," causing serious reduction in the measurement's dynamic range. To overcome this drawback, one can use an electro-optic transmission system to replace the transmission end cable while maintaining a short cable connection at the receiving end. In this approach, the microwave signal modulates the optical transmission in an optical fiber cable, with the needed length determined by the channel length. The optical signal is demodulated at the end of the fiber cable and connected to the input of the power amplifier that feeds the transmitting antenna, as shown in Figure 2.7. The main advantage of using the optical fiber cable is its very low loss compared with traditional coaxial cables. Another advantage of an electro-optic transmission system is its immunity to electromagnetic interference. One obvious disadvantage is the high cost of wideband electro-optic transmission systems.

The calibration of the direct magnitude and phase measurement system can be carried out in exactly the same manner as in the scalar measurement system, as depicted in Figure 2.5. One difference is that typically VNAs have built-in processors

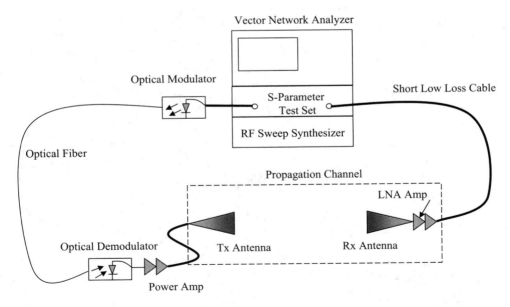

Figure 2.7: Setup for Frequency Domain UWB Channel Characterization Using VNA in Conjunction with Electro-Optic System.

for data reduction, and direct display and acquisition of final calibrated results. However, correction for the use of the calibration attenuator would be performed manually.

## Accuracy of Direct Phase Measurement with a VNA

In the previous section, we discussed the use of a VNA for direct measurements of magnitude and phase of the UWB channel. However, it should be noted that VNAs are generally designed for the measurement of complex $S$-parameters of small two-port networks rather than characterization of communication channels involving long propagation distances. Thus, it might be suspected that VNA-based measurements may not yield true characteristics of propagation channels, hence the need to understand and clarify the relationship between measured signals and true channel characteristics. In doing so, it is helpful to discuss briefly the principles of operation of a VNA, and to examine the effects of using long cables and long distances on UWB channel measurements.

A basic function of a VNA is to convert the measured signal to a lower fixed IF frequency, while preserving the magnitude and phase information during the conversion process. The phase can then be measured more easily at the IF frequency. Downconverters and a synchronized voltage tuned oscillator (VTO) are used to perform the frequency conversion. The input RF signal is divided in the $S$-parameter test set into two parts. For measuring $S_{21}$, the first part is used as a reference signal $a_1$, while the second part is used to feed port 1, as shown in Figure 2.8. The received signal $b_2$ is collected at port 2, thus $S_{21} = b_2/a_1$. For calibration, ports 1 and 2 are connected directly, which corresponds to a transmission coefficient of unity magnitude and a phase shift equal to zero. If the two-port network under test is connected to the VNA ports via long cables, the calibration procedure should be performed using the same cables. However, in this case, the time delay of the reference path is much less than that of the effective two ports. To avoid errors in the phase measurements, the same time delay should be realized in the path of the reference signal. This can be done by replacing the extension B in the reference path in Figure 2.8 with another cable that has the same time delay as the cable used in the measurement of the two-port network.

If the VNA is operating in the sweep mode, another source of error in direct phase measurements may result. This error would be due to the difference between the measured frequency and the frequency of the received signal due to the propagation delay time through the channel. In the sweep mode, the VNA frequency changes linearly with the sweep time. Due to the long channel path, the frequency of the signal at the end of the channel would be different from that at the beginning, resulting in errors in both the magnitude and the phase measurements. Typically, the magnitude function does not experience steep variations with frequency while the phase function typically does; hence, noticeable phase errors are observed. Sweep mode errors can be estimated and appropriate correction can be applied if channel delay (between the transmitting and receiving antennas) is accurately known. However, determining this delay accurately is virtually impossible.

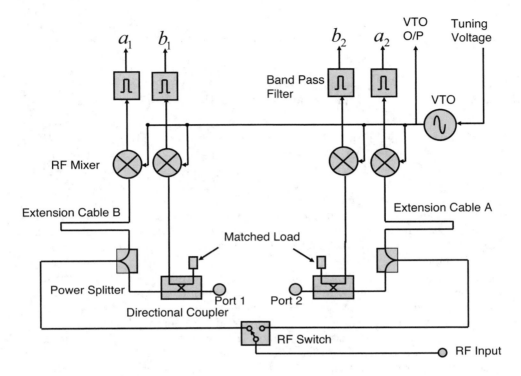

Figure 2.8: Block Diagram of S-Parameter Test Set.

To circumvent the sweep mode errors, the use of frequency stepping instead of frequency sweeping is highly recommended. In the frequency stepping case, the time of the frequency step should be larger than the time delay between the transmitting and receiving antennas. However, it should be taken into consideration that the measured signal is that recorded at the end of the time step rather than the average of measured signals received during the time step. If this condition is clearly satisfied, the measured signal would correspond to the actual magnitude and phase values.

## Phase Retrieval from Magnitude Measurements

With the difficulties associated with phase measurements, particularly when propagation distances between the transmitter and receiver are large, one appreciates the simplicity and ease with which magnitude measurements can be performed. However, the purpose of UWB channel sounding is to obtain the channel impulse response in the time domain, which requires full knowledge of the channel's complex transfer function in the frequency domain. This implies that knowledge of both the magnitude and phase of the transfer function is required. Here, we discuss how phase information can be derived from the magnitude data, which, as was discussed earlier, can be measured rather easily with a scalar network analyzer.

The phase retrieval method described in the following is based on using the Hilbert Transform, which requires that the impulse response of the UWB channel under test satisfies the causality and analyticity conditions. That is, the received signal cannot be detected any sooner than the corresponding free-space delay time, and for a finite energy pulse, the amplitude of the pulse decays to zero after a reasonable time period. Also, this method requires that the amplitude of each multipath component be smaller than the preceding one; that is, the amplitudes of multipath components decay successively. The latter requirement is satisfied for most line-of-sight scenarios and may not be satisfied in non-line-of-sight situations. For a causal analytic signal, the following Hilbert Transform relationship holds between the real and imaginary parts of its Fourier Transform

$$R(\omega) = X(\omega) + jY(\omega) = X(\omega) + jH(X(\omega)) \tag{2.1}$$

where $H(\;\;)$ denotes the Hilbert Transform. Thus, for causal analytic signals, the real part of the Fourier Transform is sufficient to reconstruct the missing imaginary part. However, in scalar frequency domain measurements, the measurable quantity is the magnitude of the complex Fourier Transform of the signal and not the real part. This can be remedied by using the natural logarithm operator to state the magnitude and phase of the transfer function as the real and imaginary parts of the transformed response

$$\widehat{R}(\omega) = \ln(R(\omega)) = \ln(|R(\omega)|) + j\Phi(\omega) \tag{2.2}$$

The function $\widehat{R}(\omega)$ represents the Fourier Transform of another signal that can satisfy causality and analyticity conditions if and only if all the poles and the zeros of $R(\omega)$ lie inside the unit circle in the complex plane [1–3]. In this case, (2.1) can be applied directly to (2.2), resulting in

$$\Phi(\omega) = H(\ln|R(\omega)|) \tag{2.3}$$

Thus, the complex spectrum of a causal analytic system can be expressed in terms of its magnitude. This gives

$$R(\omega) = |R(\omega)| \exp[jH(\ln|R(\omega)|)] \tag{2.4}$$

It is emphasized that the phase obtained from (2.3) represents the minimum phase of $R(\omega)$. If all the poles and the zeros of $R(\omega)$ do not lie inside the unit circle in the complex plane, a nonminimum phase contribution should be added to the minimum phase obtained by (2.3). This nonminimum phase contribution corresponds to the phase of an all-pass filter. Accordingly, (2.4) becomes

$$R(\omega) = |R(\omega)| \exp[jH(\ln|R(\omega)|)] G_{all}(\omega) \tag{2.5}$$

where $|G_{all}(\omega)| = 1$. It should be noted that the phase of this all-pass filter cannot be predicted by the Hilbert Transform because the logarithm of its magnitude is zero. The determination of the nonminimum phase contribution is the most complicated part of the phase retrieval process [3–5]. As far as the problem of UWB channel characterization is concerned, it has been found, by comparison with results obtained from time domain measurements, that the path loss exponent and time dispersion parameters for line-of-sight (LOS) scenarios when only the minimum phase contribution is considered do not differ too much from the actual results [6]. Thus, for LOS cases, (2.5) is used as the basis of analysis in scalar frequency domain measurements.

## 2.3    Measurement Results

In this section, measured results for UWB signal propagation in indoor environments are presented. These results are later used to develop UWB channel models. Different groups have carried out extensive propagation measurements for UWB channel modeling and characterization. Among them, researchers at the Communications Sciences Institute, University of Southern California, have been leading the UWB research in recent years. Also, researchers in the Department of Electrical and Computer Engineering at Virginia Tech have been actively involved in research on wireless communications during the past two decades, and more recently have carried out extensive studies of indoor UWB propagation and channel modeling. Here, the research results of these two groups, which for convenience are referred to as the USC group and the VT group, are presented.

### 2.3.1    Typical Results for Time Domain Measurements

The USC group has presented much of the earlier results for time domain channel propagation measurements [7,8]. However, the VT group has presented both time domain and frequency domain results for UWB channel propagation measurements [9–12]. Both groups have carried out measurements in modern laboratory and office buildings. The USC group used a quasi omni-directional diamond antenna, while the VT group used directive TEM horn antennas, as well as omni-directional biconical antennas.

To assess the effect of small-scale fading, the received signal at each receiving position was measured across a square array of points separated by a distance much smaller than the average distance between the transmitting and receiving antennas. Both groups used a square grid with a side length of 90 cm. The USC group considered 7 x 7 = 49 measurement points in the grid, while the VT group viewed 3 x 3 = 9 points in the same size grid as adequate. Figure 2.9 shows the blueprint for the measurement locations used by the USC group, while Figures 2.10 and 2.11 show the blueprints for the measurement locations used by the VT group. Figures 2.12a to 2.12c show the radiated pulses used in the measurements made by each group at a reference distance of nearly 1 meter between the transmitting and receiving antennas. The VT group used a pulse of smaller temporal width and thus larger bandwidth. Shorter pulses are more suitable for discriminating the multipath components. On the other hand, wider pulses generally contain more energy, thus enhancing the signal-to-noise ratio. The characteristics of the transmitting and receiving antennas also play an important role in the time domain measurements. In addition to the propagation channel, antennas may also contribute significantly

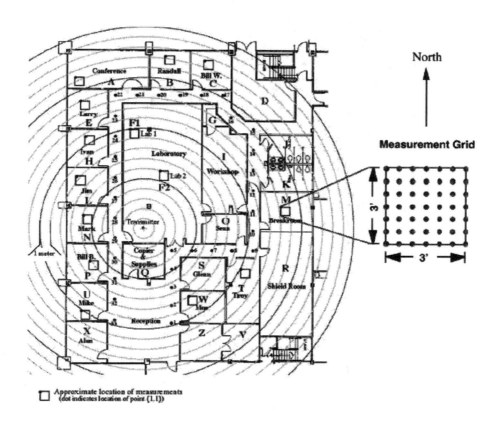

Figure 2.9: Diagram of Office Building Where Propagation Measurements of the USC Group Were Performed. The Concentric Circles Are Centered on the Transmitting Antenna and Are Spaced at 1 m Intervals.

SOURCE: M. Z. Win and R. A. Scholtz, "Characterization of Ultra-Wide Bandwidth Wireless Indoor Channel: From Statistical Models to Simulation," *IEEE Journal on Selected Areas in Communications*, [8]. © IEEE, 2002. Used by permission.

to the distortion of UWB signals. Therefore, in order to obtain a good estimate of the propagation channel characteristics we must minimize the dispersion effect of the antennas. It is well understood that antennas with wider bandwidths cause less dispersion, manifested by fewer ringing effects. Here, the bandwidth of an antenna is defined as the frequency range over which the gain, the input impedance, and the polarization of the antenna remain nearly constant, while the phase of the radiated far field varies linearly with frequency. In this regard, the TEM horn antenna with a radiated pulse as that shown in Figure 2.12b is the best example of the three

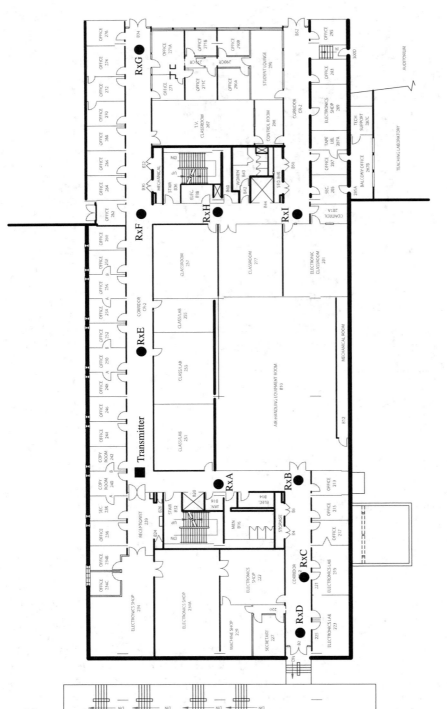

Figure 2.10: Blueprints for Several Locations in Whittemore Hall at Virginia Tech. Squares Represent Transmitter Locations, Circles Represent Receiver Locations. (Continued on next page.)

(a)

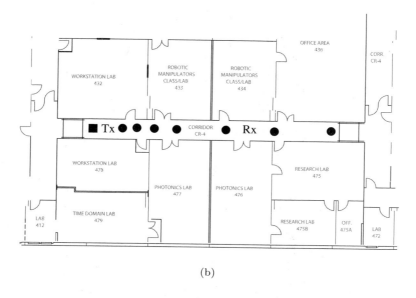

(b)

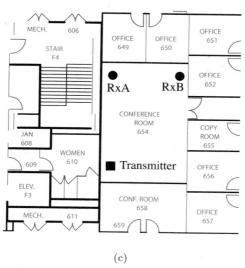

(c)

Figure 2.10: (cont.) Blueprints for Several Locations in Whittemore Hall at Virginia Tech. Squares Represent Transmitter Locations, Circles Represent Receiver Locations.

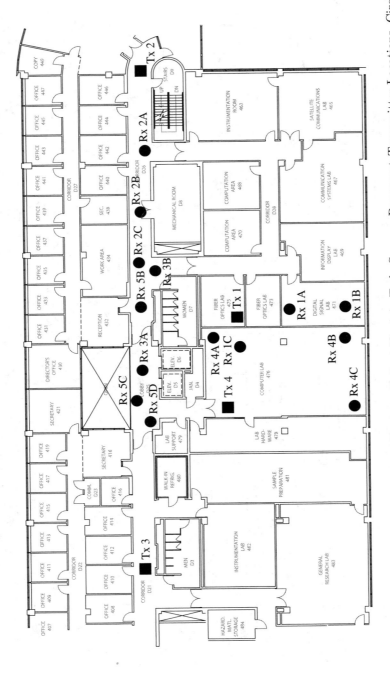

Figure 2.11: Blueprint for Fourth Floor of Durham Hall at Virginia Tech. Squares Represent Transmitter Locations, Circles Represent Receiver Locations.

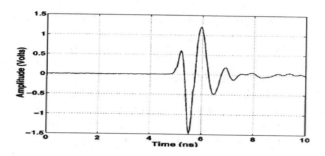

(a) Radiated Pulse of the USC Group

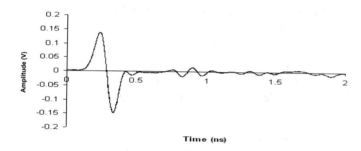

(b) Radiated Pulse of the VT Group Obtained with TEM Horn Antennas

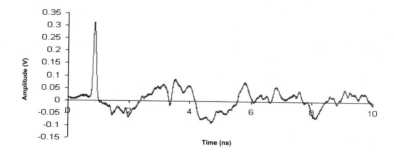

(c) Radiated Pulse of the VT Group Obtained with Biconical Antennas

Figure 2.12: Waveforms of Radiated Pulses Measured at Reference Point 1 m Away from the Transmitting Antenna.

pulses. The biconical antenna, which is viewed as a wideband antenna, has a non-linear phase behavior with frequency and thus causes signal distortion, as is noted in its radiated pulse shown in Figure 2.12c. The behavior of the diamond antenna, shown in Figure 2.12a, is an intermediate case between the TEM horn antenna and the biconical antenna. More details about the effects of the antenna structure on the UWB channel measurements are provided in Section 2.3.4.

Figure 2.13 shows the 1,000 ns-long measurement of the USC group, where two back-to-back multipath measurement cycles are captured by the receiver located in offices U, W, and M, as shown in Figure 2.9. The approximate distances between the transmitter and the location of the measurement grid located in offices U, W, and M are 10 m, 8.5 m, and 13.5 m, respectively. Figure 2.13 also shows that the response of the first probing pulse decays to negligible levels before the next pulse arrives. An excess delay on the order of 100 ns occurs in these situations. Substantial differences in the noise level at various locations throughout the building can also be observed from Figure 2.13. In particular, the profiles recorded in offices W and M

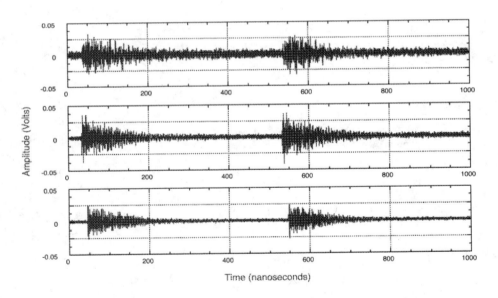

Figure 2.13: Two Back-to-Back Cycles of 1,000 ns-Long Averaged Multipath Measurement Captured by Receiver Located in Offices U (Upper Trace), W (Middle Trace), and M (Lower Trace) where the Measurement Grids are 10, 8.5, and 13.5 m Away from Transmitter, Respectively.

SOURCE: M. Z. Win and R. A. Scholtz, "Characterization of Ultra-Wide Bandwidth Wireless Indoor Channel: From Statistical Models to Simulation," *IEEE Journal on Selected Areas in Communications*, [8]. © IEEE, 2002. Used by permission.

have a substantially lower noise level compared with the profiles recorded in office U. A possible explanation may be given by examining the floor blueprint shown in Figure 2.9. It is noted that office U is situated at the edge of the building with a large glass window, and is therefore more vulnerable to external interference. On the other hand, offices W and M are situated roughly in the middle of the building. Furthermore, office W is situated in the vicinity of room R, and office M is adjacent to room R, which is shielded from electromagnetic radiation. Therefore, the external electromagnetic interference (EMI) is mostly attenuated by the shielded walls and multiple reflections by other regular walls, resulting in less noise in the central area of the building.

The VT group carried out time domain measurements using Gaussian-like pulses with a FWHM (full-width half-maximum) of 85 ps. Also, this group used two low noise wideband amplifiers, with a gain of 10 dB and a 3 dB bandwidth of 15 GHz, on the receiving side. Two different sets of measurements were performed using directional TEM horns and omni-directional biconical antennas. Both transmitter and receiver antennas were placed on plastic moving carts at an elevation of about 1.25 m above the floor. Styrofoam slabs were used to adjust the elevation without introducing reflections around the antenna. Nearly 400 profiles were collected on different floors of two different buildings.

Sample time domain measurement results by the VT group are shown in Figures 2.14 and 2.15. Figure 2.14a shows the raw data for a typical time domain measurement using two biconical antennas in the measurement setup. The presence of narrowband noise is clearly seen in the tail of the profile. Figure 2.14b shows the same profile after this narrowband noise has been filtered. The level of background noise is determined by the tail of the signal profile. The VT group set the threshold of the signal level at 6 dB above the background noise level. Accordingly, in the channel characterization analysis, the precursor and all signals below this threshold level are forced to zero. Figure 2.15 shows a zoomed profile in the first 20 ns time window to illustrate some details of multipath components. The antennas used in this measurement were TEM horns.

## 2.3.2   Typical Results for Frequency Domain Measurements

All frequency domain results presented here are based on amplitude measurements and the use of the Hilbert Transform for phase calculations. Figure 2.16 illustrates sample frequency domain responses by the VT group obtained with TEM horn antennas in the 2nd floor of Whittemore Hall at locations E, F, and G (see Figure 2.10a). The results for the same measurement locations, but obtained with omni-directional biconical antennas, are shown in Figure 2.17. The measurements in both cases were carried out over a frequency range of 100 MHz to nearly 12 GHz. It is noted that the frequency domain responses over this wide frequency range are characterized by multiple peaks and valleys formed due to constructive and destructive interference of multipath components. The results in Figures 2.16 and 2.17 are for a LOS scenario. Typical frequency domain responses for a NLOS scenario measured with TEM horn and biconical antennas are shown in Figure 2.18. It is noted that the signals for the NLOS scenario vary more rapidly with frequency than those of

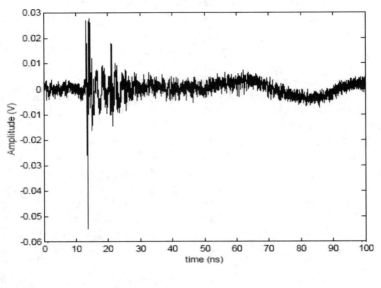

(a) Unfiltered Profile

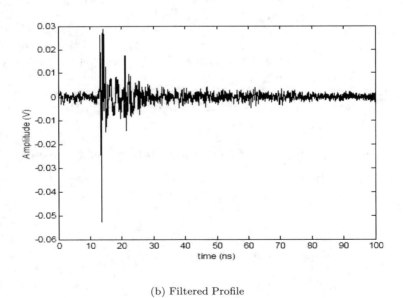

(b) Filtered Profile

Figure 2.14: Unfiltered and Filtered Time-Domain Profiles Measured with Biconical Antennas.

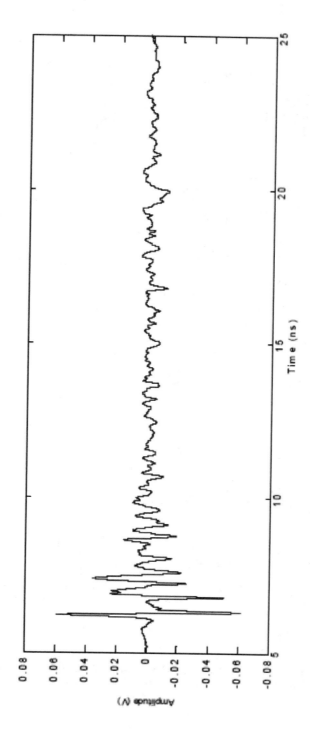

Figure 2.15: Time-Domain Profile Measured with TEM Horn Antennas in the Second Floor Hallway of Whittemore Hall at Virginia Tech.

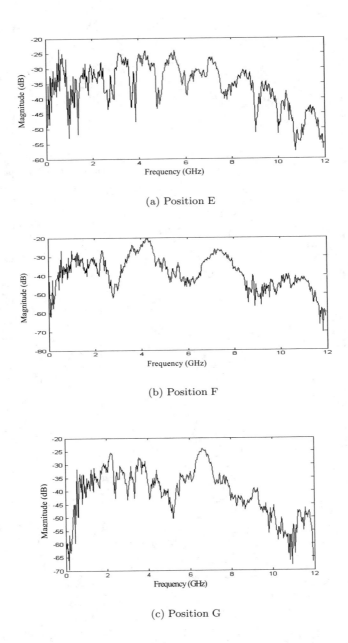

(a) Position E

(b) Position F

(c) Position G

Figure 2.16: Measured Frequency Domain Responses Obtained with TEM Horn Antennas in the Second Floor of Whittemore Hall at Different Locations.

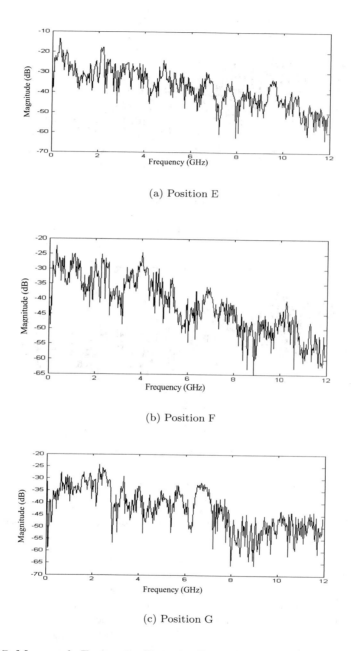

(a) Position E

(b) Position F

(c) Position G

Figure 2.17: Measured Frequency-Domain Responses Obtained with Biconical Antennas for the Same Locations as Figure 2.16.

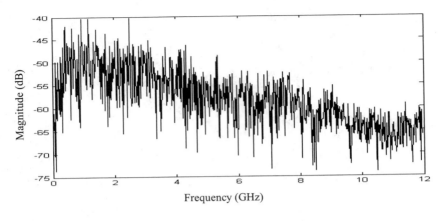

(a) With TEM Horn Antenna

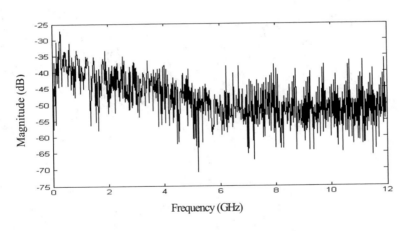

(b) With Biconical Antenna

Figure 2.18: Typical Frequency-Domain Responses for a NLOS Scenario.

the LOS. This is because in the NLOS case, more multipath components reach the receiver, creating the possibility of more constructive and destructive interference. Also, a comparison of responses measured with TEM horns and with biconical antennas for the NLOS scenario reveals that the average signal power level in the latter case is generally greater than that in the former case. This can be attributed to the fact that biconical antennas are omni-directional, resulting in more multipaths reaching the receiver.

### 2.3.3   The Role of Antennas

The antennas used in UWB links can have a significant impact on channel propagation measurements. Radiation properties of antennas, such as input impedance, gain, effective aperture area, polarization, and radiation pattern, and their behavior with frequency, can contribute to pulse dispersion and influence the waveform structure of the received signal, particularly with regard to the number of multipath components and their relative strengths. Only when all antenna properties satisfy certain requirements over the bandwidth of the UWB signal does the signal not suffer any degradation caused by the transmitting and receiving antennas. For example, the input impedance of a UWB antenna and its polarization and radiation pattern should remain unchanged over the bandwidth of the signal, whereas the gain may vary with frequency in such a manner that the radiated field becomes proportional to the time derivative of the signal feeding the antenna, as in a TEM horn. It is emphasized that a mere differentiation of the signal with respect to time is not generally regarded as distortion. Another important factor, which is not an issue in narrowband systems, is the phase of the radiated field. The phase characteristic must be linear in order to avoid further signal degradation. Thus, antennas intended for UWB applications are desired to be UWB in terms of the multitude of properties mentioned above.

Even if bandwidth requirements for antennas are ideally met, one should not assume that the propagation channel has a performance independent of the antennas. For example, considering two antennas, both satisfying the same requirements over the bandwidth of the signal but one directive and another omni-directional, the directive antenna generally produces fewer multipath components than the omni-directional one and therefore different waveforms for the respective received signals. This effect becomes more pronounced for NLOS scenarios. The choice of antennas, among other things, is dictated by the intended application. For point-to-point UWB applications, directive antennas are more suitable. On the other hand, for broadcast purposes and for mobile applications when antennas may not be at fixed locations, omni-directional antennas are a more appropriate choice. A more complete discussion of UWB antennas is presented in Chapter 4, "Antennas."

### 2.3.4   Impact of Building Architecture and Properties of Building Materials

The interior design of buildings, and the materials used in their construction, can have a significant impact on the propagation of UWB signals. Walls made of reinforced concrete or metallic studs severely attenuate the through-wall propagation of UWB signals, whereas some other walls, such as those made of wood, glass, and drywall, allow substantial transmission. An assessment of electrical properties of building materials in the UWB frequency band is very useful for simulation of UWB propagation in indoor environments and in designing indoor UWB systems. Knowing the complex dielectric constant of a wall, one is able to determine the insertion loss, as well as the distortion that a UWB signal may suffer upon propagating through it.

The VT group has conducted a comprehensive ultra wideband characterization of common building materials, including wooden doors, brick walls, concrete block walls, drywall, chip wood, glass, and fabric covered metal office partitions. The measurements were carried out in both the time and frequency domains using an insertion transfer function approach [13]. Figures 2.19 and 2.20 illustrate variations of the relative permittivity and insertion loss versus frequency for eight common building materials.

The architecture of an indoor environment also has a major impact on the propagation of UWB signals. For example, narrow hallways and corridors have the characteristics of lossy waveguides that guide the energy in the LOS direction, especially when directive antennas are used. This behavior is not observed in large rooms, where the received energy is mainly due to the direct path from the transmitting antenna to the receiving antenna. For NLOS propagation, the situation is completely different, and is more complicated. In this case, the signal travels along different paths, penetrates through different walls, and undergoes multiple reflections, refractions, and diffractions. All these phenomena greatly affect the shape of the received signal.

## 2.4    Electromagnetic Simulation of UWB Propagation in Indoor Environments

A communication propagation link starts from the feed point at the transmitting antenna and ends with the output terminals of the receiving antenna. Between the two antennas, electromagnetic waves propagate in the form of direct, refracted, reflected, and diffracted rays due to interactions with the surrounding media, which represent the channel. The characteristics of the channel are not independent from the properties of the transmitting and the receiving antennas, since received pulses depend on the direction of launched rays as well as the direction and the polarization of the received rays. Thus, the problem of simulating UWB propagation in indoor environments can be divided into two parts, one part dealing with transmitting and receiving antennas, and another part for the wave propagation in the channel.[1]

### 2.4.1    Simulation of Transmitting and Receiving Antennas

The simulation of a transmitting antenna should yield solutions for pulsed radiated electromagnetic fields in free space for all directions as a function of the input pulse and antenna characteristics. Assuming that a voltage signal $\nu_{TX}(t)$ is the input to a transmitting antenna located at $(x_s, y_s, z_s)$, then the radiated electric field at an observation point $(x_0, y_0, z_0)$ can be described as

$$\mathbf{e}_{rad}(x_0, y_0, z_0; t) = \nu_{TX}(t) * \mathbf{h}_{TX}(x_0, y_0, z_0; x_s, y_s, z_s; t) \qquad (2.6)$$

where $\mathbf{h}_{TX}(x_0, y_0, z_0; x_s, y_s, z_s; t)$ is the vector impulse response of the transmitting antenna at the observation point $(x_0, y_0, z_0)$, and "$*$" represents convolution in

---

[1]Several sentences in the first two paragraphs of Sections 2.4 and 2.4.1 as well as (2.7)–(2.12) are borrowed from [26] and reprinted here by permission of Wiley.

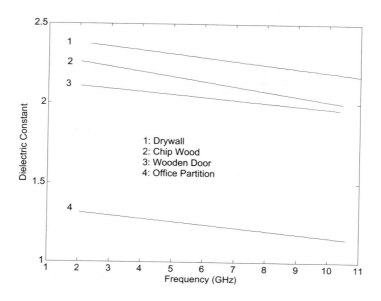

(a) Drywall, Chip Wood, Wooden Door, and Office Partition

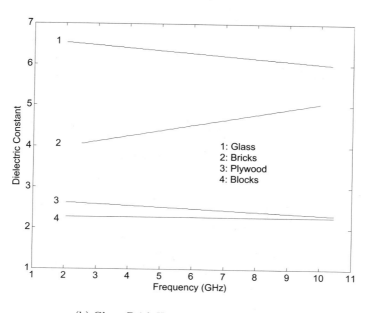

(b) Glass, Brick Wall, Plywood, Block Wall

Figure 2.19: Variations of Dielectric Constant versus Frequency.

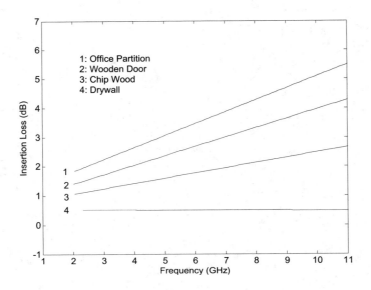

(a) Drywall, Chip Wood, Wooden Door, and Office Partition

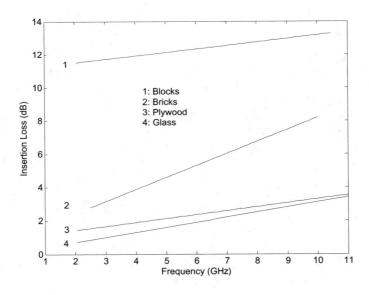

(b) Glass, Brick Wall, Plywood, Block Wall

Figure 2.20: Variations of Insertion Loss versus Frequency.

the time domain. In the far field region, the dependence of radiated fields on the distance between the antenna and the observation point, $r$, includes the spherical spread factor $1/4\pi r$, and the time delay $c/r$, where $c$ is the wave velocity in free space. Thus, in the far field region, (2.6) can be expressed as

$$\mathbf{e}_{rad}\left(x_0, y_0, z_0; \widehat{\mathbf{s}}_i, r; t\right) = \nu_{TX}\left(t - r/c\right) * \mathbf{f}_{TX}\left(\widehat{\mathbf{s}}_i; t\right) / \left(4\pi r\right) \tag{2.7}$$

where

$$r = \sqrt{\left(x_0 - x_s\right)^2 + \left(y_0 - y_s\right)^2 + \left(z_0 - z_s\right)^2} \tag{2.8}$$

$$\widehat{\mathbf{s}}_i = \left(\left(x_0 - x_s\right)\widehat{\alpha}_x + \left(y_0 - y_s\right)\widehat{\alpha}_y + \left(z_0 - z_s\right)\widehat{\alpha}_z\right) / r \tag{2.9}$$

and $\mathbf{f}_{TX}\left(\widehat{\mathbf{s}}_i; t\right)$ represents the impulse field pattern of the transmitting antenna. Also, in (2.9) $\widehat{\alpha}_x, \widehat{\alpha}_y$, and $\widehat{\alpha}_z$ are unit vectors in the Cartesian coordinate system.

The simulation of a receiving antenna should result in the received output pulse due to incident pulsed plane waves arriving at the antenna from any direction and with any polarization. The received voltage signal $\nu_{RX}\left(t\right)$ is

$$\nu_{RX}\left(t\right) = \mathbf{e}_{inc}\left(x_r, y_r, z_r; t\right) \oplus \mathbf{f}_{RX}\left(\widehat{\mathbf{s}}_r; t\right) \tag{2.10}$$

where $\mathbf{e}_{inc}\left(x_r, y_r, z_r; t\right)$ is a pulsed vector field incident on the receiving antenna located at $\left(x_r, y_r, z_r\right)$, $\mathbf{f}_{RX}\left(\widehat{\mathbf{s}}_r; t\right)$ is the impulse field pattern of the receiving antenna, and the symbol $\oplus$ indicates a combination of time domain convolution and scalar product operations.

## 2.4.2    Simulation of the UWB Channel

Nearly all channel simulations for narrowband signals are based on high-frequency techniques, such as ray tracing, geometrical theory of diffraction (GTD), or uniform theory of diffraction (UTD) [14–22]. This is due to the fact that most dimensions are much larger than the wavelength of the signal. For UWB signals, the problem is somewhat different in that the signal occupies a wide spectrum, and thus there is no narrowly defined wavelength with which the dimensions can be compared. Different approaches can be used to describe the spatial dimensions pertaining to UWB signals. For example, dimensions may be compared to the wavelength of the central frequency of the spectrum of the pulse, the wavelength of the peak energy frequency in the spectrum of the pulse, or the spatial width of the pulsed signal. However, for high data rate UWB communication systems, all of these reference lengths are generally much smaller than the dimensions in indoor environments. With this view in mind, UWB indoor channels can be simulated using high-frequency techniques without a significant loss of accuracy in simulation results.

Wang et al [23] introduced a hybrid technique based on combining ray tracing and the finite difference time domain (FDTD) method to simulate narrowband communication links in indoor environments. Schiavone et al [24] proposed a similar approach for simulating indoor UWB propagation. However, these methods require huge computational resources to simulate UWB channels in relatively simple indoor environments. Furthermore, the computational requirements increase cubically with frequency. Thus, these techniques are limited to simple environments, such as a small room, and to UWB signals with restricted upper frequency limits. Here, the simulation of UWB propagation in realistic indoor environments is addressed in a comprehensive manner.

As a first step toward UWB channel simulation, one can begin with the simplest case—when the propagation medium between transmitting and receiving antennas is free space. In this case, there are no reflected, refracted, and diffracted waves. The received signal is obtained from

$$\nu_{RX}\left(t\right) = \left[\nu_{TX}\left(t - r/c\right) * \mathbf{f}_{TX}\left(\widehat{s}_i; t\right) / \left(4\pi r\right)\right] \oplus \mathbf{f}_{RX}\left(\widehat{s}_r; t\right) \tag{2.11}$$

For the case of indoor communications, various components contribute to the received signal, as schematically shown in Figure 2.21. The transmitting antenna radiates its energy as a collection of ray tubes distributed in all directions according to its impulse radiation pattern and the input pulse signal. The received signal due to the LOS ray depends mainly on the radiation characteristics of the transmitting and receiving antennas, and is slightly affected by the surroundings. Other rays can reach the receiving antenna upon multiple reflections from the walls, the floor,

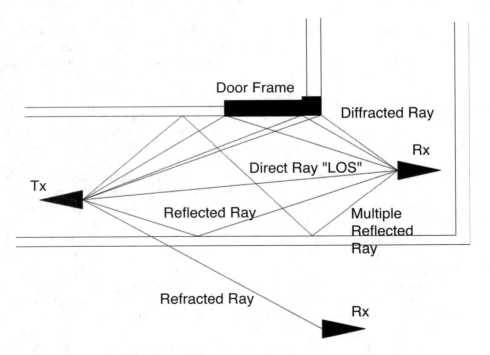

Figure 2.21: Schematic Diagram of Various Propagation Mechanisms for Indoor Communication Channel.

and the ceiling in the propagation environment. The received signal due to such multiple reflected rays depends on the number of reflections, the material of the reflecting surfaces, and the thickness and curvature of such surfaces. The thickness of each wall also causes internal multiple reflections inside the wall itself. The effect of these multiple internal reflections depends on the spatial length of the incident pulse and the thickness of the wall, and its complex dielectric constant. Diffraction at the edges of windows, door frames, and other scattering objects present in the medium gives rise to another group of rays, the diffracted rays. Each incident ray on an edge is diffracted as a cone of rays following Keller's diffraction cone [25]. The diffracted rays may directly reach the receiving antenna or undergo multiple reflections, and may even suffer multiple diffractions before reaching the receiving antenna.

The measured signal at the receiving antenna output can be formulated by combining the contributions from all types of rays,

$$
\nu_{RX}(t) = \sum_n \mathbf{e}_{inc\_n}(x_r, y_r, z_r; \widehat{\mathbf{s}}_{rn}; t) \oplus \mathbf{f}_{RX}(\widehat{\mathbf{s}}_{rn}; t)
$$

$$
= \sum_n \left[ \overline{\overline{\mathbf{Q}}}_n(x_r, y_r, z_r; x_{0n}, y_{0n}, z_{0n}; \widehat{\mathbf{s}}_{in}t) \cdot \right.
$$

$$
\left. (\nu_{TX}(t - r_n/c) * \mathbf{f}_{TX}(\widehat{\mathbf{s}}_{in}; t)/(4\pi r_n)) \right] \oplus \mathbf{f}_{RX}(\widehat{\mathbf{s}}_{rn}; t) \qquad (2.12)
$$

where the summation over $n$ accounts for all rays from the transmitting antenna. The first intersection point of the $n^{th}$ ray along the $\widehat{\mathbf{s}}_{in}$ direction is represented by $(x_{0n}, y_{0n}, z_{0n})$, $r_n$ is the path length for the $n^{th}$ ray from the transmitting antenna to its first intersection point, $\widehat{\mathbf{s}}_{rn}$ defines the direction of the $n^{th}$ ray upon reaching the receiving antenna, and $\overline{\overline{\mathbf{Q}}}_n(x_r, y_r, z_r; x_{0n}, y_{0n}, z_{0n}; \widehat{\mathbf{s}}_{in}t)$ is the channel dyadic impulse response at the receiving point due to the $n^{th}$ ray. This dyadic impulse response accounts for all reflections, diffractions, spreading factors, time delays, and so on. For a LOS ray, the dyadic impulse response is simply a unity matrix. Thus, the contribution of a LOS ray to the received signal is the same as that in (2.11). Equation (2.12) shows how difficult it is to separate channel characterization from the radiation properties of the transmitting and receiving antennas. Changing the antennas may cause significant changes in the received UWB signal for the same channel environment. Thus, the channel simulation should be done in conjunction with the simulation of the specific antennas used. Here, simulation results will be presented for the case where TEM horns are used as transmitting and receiving antennas. These antennas were used for indoor UWB propagation measurements by the VT group as described in Section 2.3.

## 2.4.3  Organization of the Electromagnetic Simulator

In the previous section, the individual parts of the simulator were discussed in some detail. These parts are now integrated to construct the UWB propagation simulator. The simulation begins with a pulse fed to the transmitting antenna and ends with the received signal at the terminals of the receiving antenna. The input data for the simulator include the antenna parameters and the channel parameters. Channel parameters are divided into geometrical parameters and electrical parameters. The geometrical parameters include the dimensions and the orientations of all the wall segments constituting the channel. The thickness of each segment is one of its geometrical parameters. Also the locations of transmitting and receiving antennas constitute part of the geometrical parameters of the problem. The electrical parameters of the channel are the real and the imaginary parts of the dielectric constant of each segment as functions of frequency over the spectrum of the incident pulse.

The first step of the simulation is to trace all the rays transmitted from the location of the transmitting antenna to the plane where the receiving antenna is located. The required information for the ray tracing procedure includes the location of the phase center of the transmitting antenna, the geometry of the channel, and the plane of the receiving antenna. The outcome of this step is the location of the reflection, refraction, and diffraction points in the channel; the parallel and perpendicular polarization vectors for the reflection and refraction rays; the soft and hard polarization vectors for the diffracted rays; the total path length from the transmitting antenna to a point on the receiving plane; and the location and the direction of arrival at the receiving plane.

The second step is to determine the rays captured by the receiving antenna. It is assumed that a ray reaching the receiving plane is launched from a point source located at a distance equal to the total path length between the transmitting antenna and the receiving point. The location of this assumed point source should satisfy the direction of arrival at the corresponding receiving point. A ray is transmitted from this assumed point source to the location of the phase center of the receiving antenna.

The third step of the simulation is to determine the electric field and its polarization at the first intersection of each ray upon leaving the transmitting antenna. In this step, the simulation of the transmitting antenna provides such a field. For a LOS ray from the transmitting antenna to the receiving antenna, this is the field that is directly received by the receiving antenna. However, for other rays, this field should be transformed to the frequency domain where reflection, refraction, and/or diffraction of the wave is treated. Then, the frequency domain received fields are converted back to the time domain.

The fourth step of the simulation involves calculation of the reflection, refraction, and diffraction dyadics at each frequency. The total dyadic for a ray is the multiplication of the dyadics of different segments along the ray path.

The last step of the simulation is to obtain the total field at the receiving antenna. The amplitude, polarization, and direction of arrival of the received field on each ray are calculated using the data for the corresponding transmitted ray and the total dyadic of the ray. The Fourier Transform of the total dyadic gives the channel dyadic impulse response at the receiving point due to this ray. Finally, by combining the received signals due to all captured rays, the simulation of the UWB communication link is completed.

## 2.4.4   Comparisons of Measurement and Simulation Results

Now that the electromagnetic simulator has been developed, this section will validate the results by comparing them to some experimental data. To facilitate this comparison, the input signal to the transmitting antenna is assumed to be the actual Gaussian-like pulse (generated by Pico Second Pulse Labs Pulse Generator, Model 4100) used by the VT group in their UWB indoor propagation measurements. Figure 2.22 shows the voltage waveform of the input pulse normalized to its peak value. The simulation results are presented for transmitting and receiving antennas being the TEM horns used in the measurements. The two horns are essentially identical with the following parameters: electrode length $l_0 = 36.4$ cm, half separating angle $A = 7.9°$, and half apex angle $B = 16.5°$. The horn antennas are packaged inside a dielectric foam structure of very low dielectric constant $\varepsilon_r \approx 1.07$. This packaging has a negligible effect on the amplitude of the radiated fields. The simulation of TEM horn antennas is discussed in detail in [9]. The simulation results are compared with the time domain measurements (by the VT group) in the 2nd floor of Whittemore Hall at locations R x E, R x F, and R x G (see Fig. 2.10a). As mentioned before, the accuracy of the simulation is greatly affected by the details of the input data. Here we present a sample of the required details to develop the simulation. The width of the hallway is 2.83 m. The dropped ceiling is made largely of a low dielectric constant material at a height of 2.54 m, above which metallic pipes and cables are used for utilities. Through several trials it was found that an effective height of nearly

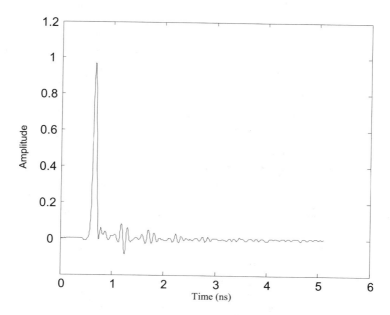

Figure 2.22: Input Pulse Signal to the Transmitting Antenna Normalized to Its Peak Value.

3 m for the ceiling yields a better agreement with the experimental results. The walls of the hallway are drywall supported by metallic studs. Metallic tiles coated with a ceramic layer cover the floor of the hallway. Thus, both the ceiling and the floor are assumed to be perfectly conducting surfaces. The overall thickness of a typical wall is nearly 0.1 m. The presented simulation is based on the assumption that the walls are completely composed of drywall. There are seventeen doors on the left side of the transmitter antenna and seven doors on the right. Almost all doors have the same dimensions and are made of the same materials. The height and the width of each door are 2.14 m and 0.9 m, respectively. The spacing between the centers of two successive doors is nearly 5.74 m. The distance between the transmitting antenna and the first door frame in the longitudinal direction is nearly 0.54 m. The distance between the starting point of the 10th frame and end point of the 9th frame in the longitudinal direction is nearly 1.7 m, and the distance between the starting point of the 11th frame and the end point of the 10th frame is nearly 3.4 m. The remaining doors on the right side are of the same periodic structure. The starting points of the right doorframes are at longitudinal distances of nearly 12.78 m, 17.95 m, 30.75 m, 33.02 m, 35.92 m, 41.13 m, and 44.03 m from the transmitting antenna. The door frames are perfectly conducting wedges of interior angle $\pi/2$.

Figures 2.23a to 2.23c compare the simulation results with the corresponding measured results at these three locations. As noted, there is generally good agreement between the measured and simulation results for all locations. The agreement for the first location, for which the distance between the transmitter and receiver is shortest, is best. The reasons may be due to internal multiple reflections in adjacent rooms and the presence of some tables and chairs in the hallway at larger distances. Ten multiple reflections were found to be adequate in the simulation. More multiple reflections do not increase the accuracy of simulation results.

## 2.5  Summary

The measurement and simulation of indoor UWB channels are examined in this chapter. Both time domain and frequency domain measurement techniques are discussed. The advantages and disadvantages of each technique are pointed out. Time domain measurements are more intuitive and clearly illustrate the transient behavior of the channel response. However, the signal-to-noise ratio for time domain measurements is lower than that of the frequency domain results because of the smaller dynamic range of the time domain measurement setup. The chief advantage of the frequency domain method is its much larger dynamic range. In this method, the measurand is the channel transfer function, which is a complex quantity, requiring both magnitude and phase information. The difficulties associated with phase measurements for UWB channels involving long propagation distances is examined. A scalar frequency domain technique is presented in which the phase information is extracted from the magnitude data by means of the Hilbert Transform. Other important issues, such as triggering, calibration, interference, noise, and jitter, are also addressed. Sampled measured results in both domains are presented.

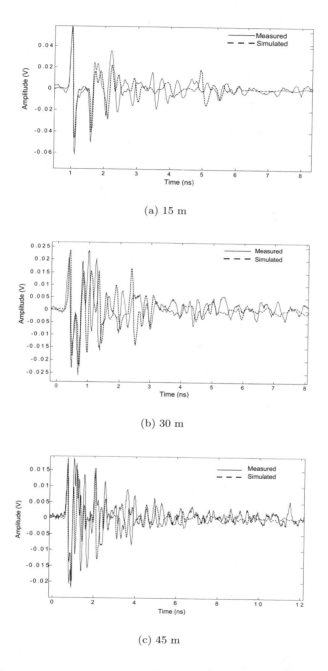

(a) 15 m

(b) 30 m

(c) 45 m

Figure 2.23: Comparison of Measured (Solid Line) and Simulated (Dashed Line) Received Signals in the Second Floor Hallway of Whittemore Hall at Distances of 15 m, 30 m, and 45 m.

Signal distortions due to the transmit and receive antennas, as well as the dispersive properties of building materials in the propagation channel, are pointed out.

The electromagnetic simulation of ultra wideband signal propagation in indoor environments is discussed. This simulation is based on ultra wideband ray tracing techniques and the uniform theory of diffraction. The simulation accounts for the radiation characteristics of the transmitting and receiving antennas and reflections, refraction, and diffraction of waves by the surrounding walls and objects in the channel. Frequency dependence of materials used in the structure of indoor channels, such as wood, drywall, bricks, and so on, can be accounted for in simulation of the channel. Sample simulation results for propagation of electromagnetic pulses with FWHM duration of less than 0.1 ns in a hallway are presented and compared with the corresponding measured results. It is noted that simulation and measured received signals are in good agreement.

# Bibliography

[1] B. P. Donaldson, M. Fattouche, and R. W. Donaldson, "Characterization of in-building UHF wireless radio communication channels using spectral energy measurements," *IEEE Transactions on Antennas and Propagation*, vol. 44, pp. 80-86, January 1996.

[2] J. Koh, Y. Cho, and T. K. Sarkar, "Reconstruction of non-minimum phase function from only amplitude data," *Microwave and Optical Technology Letters*, vol. 35, no. 3, pp. 212-216, November 2002.

[3] I. Páez, S. Loredo, L. Valle, and R. P. Torres, "Experimental estimation of wideband radio channel parameters with the use of a spectrum analyzer and the Hilbert transform," *Microwave and Optical Technology Letters*, vol. 34, no. 5, pp. 393-397, September 2002.

[4] T. K. Sarkar, "Generation of non-minimum phase from amplitude-only data," *IEEE Transactions on Microwave Theory Technology*, vol. 46, no. 8, pp. 1079-1084, August 1998.

[5] A. E. Yagle and A. E. Bell, "One- and two-dimensional minimum and non-minimum phase retrieval by solving linear systems of equations," *IEEE Transactions on Signal Processing*, vol. 47, no. 11, pp. 2978-2989, November 1999.

[6] A. Bayram, "A Study of Indoor Ultra-Wideband Propagation Measurement and Characterization," M.S. Thesis, Department of Electrical and Computer Engineering, Virginia Tech, May 2004.

[7] D. Cassioli, M. Z. Win, and A. F. Molisch, "The ultra-wide bandwidth indoor channel: from statistical model to simulations," IEEE Journal on Selected Areas in Communications, vol. 20, no. 6, pp. 1247-1257, August 2002.

[8] M. Z. Win and R. A. Scholtz, "Characterization of ultra-wide bandwidth wireless indoor channels: A communication-theoretic view," IEEE Journal on Selected Areas in Communications, vol. 20, no. 9, pp. 1613-1627, December 2002.

[9] R. M. Buehrer, A. Safaai-Jazi, W. A. Davis, and D. S. Sweeney, "Ultra-Wideband Propagation Measurements and Modeling," Final Report for DARPA-NETEX Program, Mobil and Portable Radio Research Group, Virginia Tech, January 31, 2004.

[10] A. H. Muquibel, A. Safaai-Jazi, A. M. Attiya, A. Bayram, and S. M. Riad, "Measurement and characterization of indoor ultra-wideband propagation," *Proc. 2003 IEEE Conference on Ultra Wideband Systems and Technologies*, pp. 295-299, November 16-19, 2003.

[11] A. H. Muquaible, "Characterization of Ultra Wideband Communication Channels," Ph.D. dissertation, The Bradley Department of Electrical and Computer Engineering Virginia Polytechnic Institute and State University, March 2003.

[12] R. M. Buehrer, W. A. Davis, A. Safaai-Jazi, and D. Sweeney, "Characterization of the ultra-wideband channel," *Proc. 2003 IEEE Conference on Ultra Wideband Systems and Technologies*, pp. 26-31, November 16-19, 2003.

[13] A. H. Muquibel and A. Safaai-Jazi, "A new formulation for characterization of materials based on measured insertion transfer function," *IEEE Transactions on Microwave Theory and Techniques*, vol. 51, no. 8, pp. 1946-1951, August 2003.

[14] C. Yang, B. Wu, and C. Ko, "A ray-tracing method for modeling indoor wave propagation and penetration," *IEEE Transactions on Antennas and Propagation*, vol. 46, no. 6, pp. 907-919, June 1999.

[15] J. H. Trang, W. R. Cheng, and B. J. Hsu, "Three-dimensional modeling of 900-MHz and 2.44-GHz radio propagation in corridors," *IEEE Transactions on Vehicular Technology*, vol. 46, no. 2, pp. 519-527, May 1997.

[16] U. Dersch and E. Zollinger, "Propagation mechanisms in microcell and indoor environments," *IEEE Transactions on Vehicular Technology*, vol. 43, no. 4, pp. 1058-1066, November 1994.

[17] G. E. Athanasiadou and A.R. Nix, "A novel 3-D indoor ray-tracing propagation model: The path generator and evaluation of narrow-band and wide-band prediction," *IEEE Transactions on Vehicular Technology*, vol. 49, no. 4, pp. 1152-1168, July 2000.

[18] V. Degli-Esposti, C. Lombardi, C. Passerini, and G. Riva, "Wideband measurement and ray tracing of the 1900-MHz indoor propagation channel: Comparison criteria and results," *IEEE Transactions on Antennas and Propagation*, vol. 49, no. 7, pp. 1101-1110, July 2001.

[19] T. Imai and T. Fujii, "Fast algorithm for indoor microcell area prediction system using ray-tracing method," *Electronics and Communication in Japan, Part 1*, vol. 85, no. 6, pp. 41-52, 2002.

[20] Z. Zhang, R. K. Sorensen, Z. Yun, M.F. Iskander, and J.F. Harvey, "A ray-tracing approach for indoor/outdoor propagation through window structure," IEEE Transactions on Antennas and Propagation, vol. 50, no. 5, pp. 742-748, May 2002.

[21] F. S. de Adana, O. G. Blanco, I. G. Diego, J. P. Arriage, and M. F. Catedra, "Propagation model based on ray tracing for design of personal communication system in indoor environments," *IEEE Transactions on Vehicular Technology*, vol. 49, no. 6, pp. 2105-2112, November 2000.

[22] S. Chen and S. Jeng, "An SBR/image approach for radio wave propagation in indoor environments with metallic furniture," IEEE Transactions on Antennas and Propagation, vol. 45, no. 1, pp. 98-106, January 1997.

[23] Y. Wang, S. Safavi-Naeini, and S. K. Chaudhuri, "A hybrid technique base on combining ray tracing and FDTD methods for site-specific modeling of indoor radio wave propagation," IEEE Transactions on Antennas and Propagation, vol. 48, no. 5, pp. 743-754, May 2000.

[24] G. A. Schiavone, P. Wahid, R. Palaniappan, J. Tracy, and T. Dere, "Analysis of ultra-wide band signal propagation in an indoor environment," *Microwave and Optical Technology Letters*, vol. 36, no. 1, pp. 13-15, January 2003.

[25] J. B. Keller, "Geometrical theory of diffraction," *Journal of the Optical Society of America*, vol. 52, pp. 116-130, 1962.

[26] A. M. Attiya and A. Safaai-Jazi, "Simulation of Ultra-Wideband Indoor Propagation," *Microwave and Optical Technology Letters*, vol. 42, no. 2, pp. 103-108, July 2004.

# Chapter 3

# CHANNEL MODELING

## R. Michael Buehrer

## 3.1 Introduction

The received signal in any communications system is an attenuated, delayed, and possibly distorted version of the signal that was transmitted plus noise and (possibly) interference. The relationship between the received signal and the transmitted signal is typically called the "channel." In order to evaluate and design wireless systems, we must create models of the channel. In the following sections, we will discuss how the channel is modeled for UWB systems. We should make a careful distinction about the type of modeling being examined. In general, there are two prevalent types of modeling of electromagnetic wave propagation. The first is what might be termed "site-specific modeling" or "deterministic modeling," and attempts to model the exact interaction of the EM wave in the specific environment of interest. This type of modeling is discussed in Chapter 2, "Channel Measurement and Simulation," for UWB signals and is often used to predict coverage patterns in wireless systems when detailed information concerning the environment is available. A second type of modeling attempts to model the relevant statistics of the received signal and may be called "statistical modeling." Statistical modeling, discussed in this chapter, is particularly useful in communication system development where the system must work in a wide variety of environments.

In our discussion of the UWB channel, we can conveniently divide the effects into three categories: (a) large-scale effects, (b) small-scale effects, and (c) undesired signals. The third category includes the impact of noise sources and interfering signals. We will defer the discussion of these until Chapter 6, "Receiver Design Principles." The phrase "large-scale" typically refers to the impact that the channel has on the transmit signal over large distances and generally includes only average attenuation effects due to distance and large objects that are in the propagation path. Conversely, "small-scale" typically refers to attenuation changes over small distances and the distortion to the waveform introduced by the channel. We will examine how the two effects are modeled separately.

## 3.2    What's Different About UWB?

Channel modeling for wireless systems is a well-investigated topic and is the subject of several texts [2, 3, 26]. Thus, an important first step in this chapter is to address the reasons why we need to revisit channel modeling for UWB. The main reason to revisit channel modeling is the extremely large bandwidth associated with UWB signals. Traditional channel models for path loss assume that diffraction coefficients, attenuation due to materials, and other propagation effects are constant over the band of interest. When the fractional bandwidth (see Chapter 1, "Introduction," for a definition of fractional bandwidth) is 0.01 or less, this is a safe assumption. Additionally, narrowband models often incorporate antenna effects, such as the effective aperture, into the path loss. Again, this is acceptable when the change in these antenna effects is negligible over the band. Neither of these assumptions is correct for a UWB system.

There are also small-scale assumptions that may no longer hold for UWB systems. Most importantly, narrowband, and even wideband, channel models assume that the received signal is the sum of delayed, phase shifted, and attenuated copies of the received signal. The interaction of these multiple signals results in fading and possible frequency distortion. However, it is assumed that the individual copies are not distorted. When UWB signals are being used, this may no longer be true. Individual components of the received signal may be distorted, thus introducing pulse-level frequency distortion in addition to the distortion seen in the total received signal. This means that we must revisit the traditional model.

In the following sections, we will address each of these assumptions and show how UWB affects traditional assumptions and how we need to either adjust our modeling approach or adjust our understanding of the modeling approach. Specifically, we will address the two main areas of channel modeling, large-scale modeling and small-scale modeling. Section 3.3 addresses large scale modeling, while Section 3.4 addresses small-scale modeling. The spatial behavior (important for multiple antenna systems) of UWB signals is addressed in Section 3.5. The impact of the discrete time channel model is examined in Section 3.6. Conclusions are made in Section 3.7. UWB channel modeling efforts are fairly recent and remain an active area of research [1]. Throughout this chapter, we will attempt to highlight recent work, although the goal is to provide a basic understanding rather than a comprehensive survey.

## 3.3    Large-Scale Channel Modeling

Large-scale channel modeling involves modeling the signal attenuation with distance. This large-scale attenuation is generally referred to as path loss. Path loss is a fundamental characteristic of electromagnetic wave propagation and is incorporated in the system design, viz link budgets, to predict expected received power. Traditionally, free-space path loss for narrowband signals is examined using the Friis transmission formula that provides a means for predicting the received power. For a constant frequency, the Friis transmission formula predicts that the received signal power will fall off with the square of the distance between the transmitter and receiver. Additionally, the formula predicts that the received signal power will decrease with the square of increasing frequency due to the assumption of constant gain antennas, which has little effect on narrowband systems. However, the large bandwidths of UWB signals, typically $> 500$ MHz, coupled with this definition of path loss, suggests that the channel introduces a path loss that varies across the band of the signal. This clearly would cause frequency dependent attenuation and thus distort the pulse shape. Thus, the Friis transmission formula needs to be examined more closely to justify its application to UWB.

## 3.3.1  Free-Space Path Loss Modeling: The Friis Transmission Formula

The basis for the Friis transmission formula is the flux density of a transmitting source. The flux density is given by

$$F = \frac{EIRP}{4\pi r^2} \tag{3.1}$$

where EIRP is the *Effective Isotropic Radiated Power*, which assumes that the power is radiated equally in all directions by the transmitter, and $r$ is the radius of a sphere, the surface of which the flux density is being calculated [4].

Equation (3.1) shows that the flux density assumes no frequency dependence, and that with a doubling of distance the flux decreases by a factor of four. This flux density can then be used to determine received power, $P_r$, by multiplying by $A_e$, the effective aperture of the receive antenna, resulting in

$$P_r = \frac{EIRP}{4\pi r^2} A_e \tag{3.2}$$

The Friis equation is typically stated in terms of the gains of the antennas, where the gain is related to the effective aperture of the antenna, $A_e$, by [4]

$$G = \frac{4\pi}{\lambda^2} A_e \tag{3.3}$$

Rearranging (3.3) to solve for $A_e$, and substituting the result into (3.2) gives

$$P_r = \frac{EIRP}{4\pi d^2} \left(\frac{\lambda^2}{4\pi}\right) G_r = EIRP \cdot G_r \left(\frac{\lambda}{4\pi d}\right)^2 \tag{3.4}$$

Further, EIRP can be expressed as $EIRP = P_t G_t$, where $P_t$ is the transmit power and $G_t$ is the transmit antenna gain. This results in the standard Friis transmission formula given as

$$P_r = \frac{P_t G_t G_t \lambda^2}{(4\pi d)^2} \tag{3.5}$$

The term $\left(\frac{4\pi d}{\lambda}\right)^2$ is typically defined as the path loss. The existence of $\lambda$ in the path loss equation is thus interpreted as frequency dependent path loss. However, this term is explicitly introduced as an antenna effect. To make this more obvious, it is instructive to consider another type of antenna, a constant aperture antenna. A constant aperture antenna has a flux density that is a function of the wavelength given by

$$F = P_t \left( \frac{4\pi A_{et}}{\lambda^2} \right) \left( \frac{1}{4\pi d^2} \right) = \frac{P_t A_{et}}{\lambda^2 d^2} \quad watts/m^2 \tag{3.6}$$

where $A_{et}$ is the effective aperture of the transmit antenna. This flux density can be used in the same manner as in (3.5) to give the expected received power

$$P_r = P_t A_{et} \left( \frac{1}{\lambda d} \right)^2 A_{er} \tag{3.7}$$

This result again shows frequency dependence, but here the received power increases with frequency. For systems with a constant gain antenna on one end of the link and a constant aperture antenna on the other end of the link, the received power is independent of frequency, as shown in (3.8) and (3.9).
Constant gain transmit/constant aperture receive:

$$P_r = P_t G_t \left( \frac{1}{4\pi d^2} \right) A_{er} \tag{3.8}$$

Constant aperture transmit/constant gain receive:

$$P_r = P_t \left( \frac{4\pi A_{et}}{\lambda^2} \right) \left( \frac{1}{4\pi d^2} \right) G_r = P_t A_{et} \left( \frac{1}{4\pi d^2} \right) G_r \tag{3.9}$$

The point of this development is that while the received power may be dependent on frequency, this is due to the antennas, not the path per se. The path loss, or more accurately the spreading loss, in free-space is not frequency dependent. This can be seen by taking line-of-sight (LOS) measurements using UWB signals with two different antennas [72]. A 200 ps Gaussian pulse was generated (Figure 3.1) and used in conjunction with wideband biconical antennas. The received pulse was measured at several distances in a LOS environment with any multipath being eliminated through time gating. The received pulses are plotted in Figure 3.2. The received waveforms are normalized by distance.[1] It is expected that if no frequency dependency exists in the path, all the pulses will retain the same pulse shape. Figure 3.2 shows that to within measurement error, the pulses do indeed maintain the same shape for distances between 1 m and 20 m. This experiment was repeated using the same generated pulse, but with TEM horn antennas at the transmitter and receiver. The results are plotted in Figure 3.3. Again, the pulse shape is consistent with distance. Note that the received pulse is not necessarily the same as the generated pulse. This is due to the fact that the antenna response is frequency

---

[1] Because power spreading loss is relative to the square of distance, the loss in voltage is expected to be relative to distance. Thus, normalizing the received signal values with respect to distance should result in similar voltage levels.

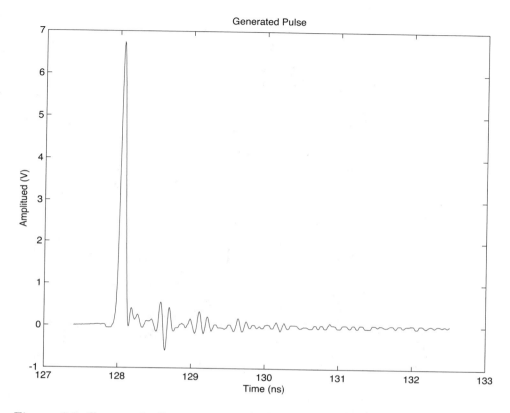

Figure 3.1: Generated Gaussian Pulse of ~200 ps Duration for Path Loss Experiment.

dependent. In fact, the two antennas have very different responses, as will be discussed in Chapter 4, "Antennas." However, the impact of the antennas is similar regardless of distance. Thus, it is not necessary to accommodate frequency dependence into the path loss for UWB signals, at least in LOS environments. Frequency dependence must be taken into account when considering the received power due to the antennas used, but not for the path loss, at least not for free-space or LOS propagation. This is a direct consequence of the fact that all antennas, regardless of their characteristics or frequency of operation, have their power flux density varying as $\frac{1}{d^2}$. This can be made more clear by defining the received power at some reference distance $d_o$ as $P_o$. The received power at any distance $d$ can then be calculated as

$$P_r(d) = P_o \left(\frac{d_o}{d}\right)^2 \tag{3.10}$$

where the frequency dependence due to antenna effects is captured entirely in the reference measurement, $P_o$.

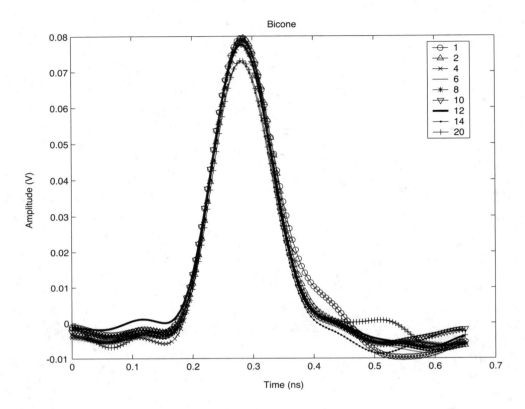

Figure 3.2: LOS-Received Pulses Normalized According to Their Respective Distances Using Bicone Antennas and Gaussian Pulse of Figure 3.1. (Legend Refers to Measurement Distance in Meters.)

SOURCE: R. M. Buehrer, A. Safaai-Jazi, W. A. Davis, and D. Sweeney, "Characterization of the UWB Channel," *Proc. IEEE Conference on Ultra Wideband Systems and Technologies* [5]. © IEEE, 2003. Used by permission.

### 3.3.2    Path Loss Modeling for Non-Free-Space Environments

The analysis given in the previous section provides the justification for applying the traditional path loss model to the analysis of UWB signals in free-space. The frequency dependence can be captured in a reference power, $P_o$, received at a close-in reference distance. In other words, because path loss is not frequency dependent, the traditional narrowband models apply. However, most environments of interest do not involve simple free-space propagation. Rather, the path between the transmitter and receiver will have other objects in the environment, perhaps completely blocking the LOS path. Even if the LOS path is clear of obstructions, objects surrounding the transmitter and receiver can greatly influence the average signal strength. It has

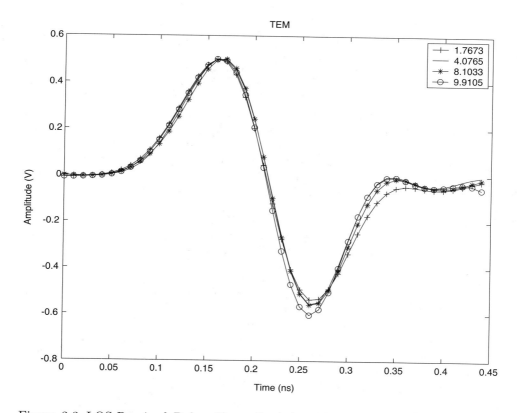

Figure 3.3: LOS-Received Pulses Normalized According to Their Respective Distances Using TEM Horn Antennas and Gaussian Pulse of Figure 3.1. (Legend Refers to Measurement Distance in Meters.)

SOURCE: R. M. Buehrer, A. Safaai-Jazi, W. A. Davis, and D. Sweeney, "Characterization of the UWB Channel," *Proc. IEEE Conference on Ultra Wideband Systems and Technologies* [5]. © IEEE, 2003. Used by permission.

been shown in many experiments and with theoretical models that the average path loss (both indoors and outdoors) for narrowband systems increases logarithmically with distance [23]

$$\overline{PL}(d) \propto \left(\frac{d}{d_o}\right)^n \tag{3.11}$$

where $d_o$ is again the reference distance, typically 1 m for indoor measurements, and $\overline{[\cdot]}$ represents an ensemble average. Specifically, the average received power can be modeled as

$$\overline{P_r}(d) = \overline{P_r}(d_o) \left(\frac{d_o}{d}\right)^n \tag{3.12}$$

where $P_r(d_o)$ is the received power at a reference distance. This is analogous to (3.10) with the difference being that the path loss exponent is not 2. The question is: How does UWB change this relationship? The main point at which frequency dependence enters the equation is through the antennas. By defining path loss relative to a reference power measurement, we have eliminated this factor. The antenna effects are subsumed into the reference measurement. The second place where frequency dependence can enter the equation is in the path loss exponent $n$. Due to the fact that diffraction, material penetration, and other effects are frequency dependent, the path loss exponent could change with frequency. One approach to this problem is to assume that the frequency and distance factors are independent [76].

$$PL(d, f) = PL(d) PL(f) \tag{3.13}$$

A second approach would be simply to make the path loss exponent $n$ a function of the frequency.

Currently, there is no clear consensus on the dependence of path loss on frequency (excluding antenna effects). It was found in [5] that frequency dependent behavior was not observable in the path loss. This can be seen in Figure 3.4 (taken from [5, 73]), which plots the path loss exponent and standard deviation versus frequency over the range of 1–10 GHz. The figure contains LOS and NLOS environments using bicone and TEM horn antennas. It can be seen that the path loss exponent does not vary with frequency. The standard deviation of the path loss exponent over the band was found to be less than 0.1. It should be noted, however, that the measurements were taken at relatively short distances (generally less than 10 m). It is very possible that larger distances may reveal frequency dependencies for NLOS channels due to the frequency dependency of many materials. We will discuss frequency dependence in more detail in Section 3.3.3. If frequency dependence is restricted to the antenna, the path loss can be defined as in the narrowband case using (3.11). Note that the reference measurement is important to define when determining the path loss model. The reference point can be defined to include multipath effects or can be defined to be the received power in free-space. Both are commonly used in conventional narrow/wideband path loss models. Defining the reference measurement to be free-space has the advantage that it is possible to directly calculate the reference point, thus eliminating the need for a reference measurement. This is somewhat more complicated in UWB systems than in narrowband systems. When defining the reference measurement to include environmental factors (such as multipath) the reference point becomes environment-specific. In such a case, statistical characterization of the reference may be useful.

Note that if free-space attenuation is to be used for the reference point in UWB, care must be taken in determining this value. It may be obtained via calculation, provided that sufficient information concerning the antennas is available. It may also be obtained via measurement, provided that all multipath effects are eliminated. For example, in [5] and [6] path loss was calculated with respect to free-space path

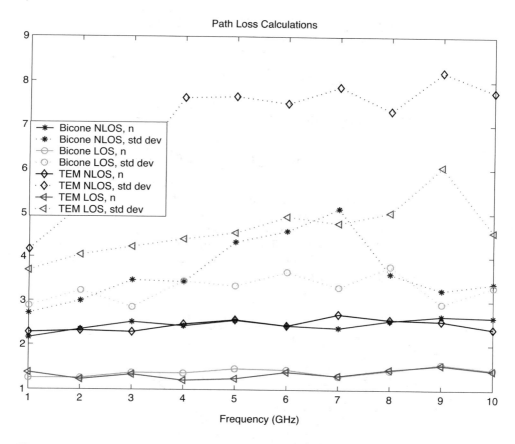

Figure 3.4: Path Loss Exponent and Standard Deviation Measured at Different Frequencies.

SOURCE: B. Donlan and R. M. Buehrer, "The UWB Indoor Channel: Large and Small-Scale Modeling," submitted to *IEEE Transactions on Wireless Communications* [73]. © IEEE, 2004. Used by permission.

loss at 1 m. The reference measurement was taken at $d_o = 1$ m, and time gating was used to obtain only the LOS component and thus eliminate multipath effects. *This makes the path loss calculations relative to free-space at 1 m.* Path loss can be calculated for any subsequent measurement taken at a distance $d$ with received power $P_r(d)$ using

$$\overline{PL(d)} = \frac{\overline{P_r(d_o)}}{\overline{P_r(d)}} \overline{PL(d_o)}$$

$$\overline{PL\,(d)}\,(dB) = \overline{P_r\,(d_o)}\,(dBm) - \overline{P_r\,(d)}\,(dBm)$$
$$+ \,\overline{PL\,(d_o)}\,(dB) \tag{3.14}$$

Combining (3.11) and (3.14) results in

$$\overline{PL\,(d)}\,(dB) = 10\log_{10}\left(\frac{\overline{P_r\,(d_o)}}{\overline{P_r\,(d)}}\right) + \overline{PL\,(d_o)}\,(dB)$$

$$= -10n\log_{10}\left(\frac{d_o}{d}\right) + \overline{PL\,(d_o)}\,(dB) \tag{3.15}$$

which allows $n$ to be determined using the *average* path loss, $\overline{PL}_{dB}$, and the distances, $d_o$ and $d$. This is typically done by taking several measurements at various distances and performing a linear regression to obtain a least squares fit to the data as in the narrowband case. An example is given in Figure 3.5 [6] for LOS, NLOS, bicone antennas, and TEM horn antennas. We will discuss commonly found values for the path loss exponent shortly.

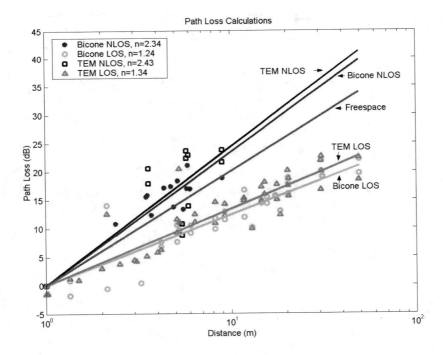

Figure 3.5: Example Path Loss Results: Directional and Omni-Directional Antennas for LOS and NLOS Environments.

Note that (3.15) represents the *average* path loss experienced at a distance $d$ relative to the distance $d_o$. The path loss observed at any given point will deviate from this average value due to variations in the environment. This variation has been shown to follow a log-normal distribution in many measurements [16, 26, 72]. Thus, the received signal power can be represented as

$$P_r(d)(dBm) = \overline{P_r(d_o)}(dBm) - 10n\log_{10}\left(\frac{d}{d_o}\right) + X_\sigma \qquad (3.16)$$

where $X_\sigma$ is a log-normal random variable (that is, it is a zero mean Gaussian in dB) with standard deviation $\sigma$.

As can be seen from (3.16), there are two main terms to determine from the environment: the path loss exponent and the standard deviation of the shadowing. There are two common approaches to obtaining these values. The first is to simply combine all measurements from similar environments (such as office NLOS) and determine the best fit to the data. This is the more common approach [5, 14, 25, 72]. A second approach is to determine the best fit for each specific building or room and create a statistical model for the path loss exponent and shadowing standard deviation. In this approach [17, 18], the two terms are modeled as Gaussian random variables. Thus, the path loss exponent is modeled as

$$n = \mu_n + \sigma_n X \qquad (3.17)$$

where $\mu_n$ is the mean of the path loss exponent, $\sigma_n$ is the standard deviation, and $X$ is a zero mean unit variance normal random variable. Similarly, the shadowing standard deviation is modeled as a Gaussian random variable. Note that in both cases the Gaussian distribution should be truncated to guarantee physically reasonable values.

### 3.3.3  Frequency Dependent Models

While it was previously shown that path loss is not frequency-dependent for free-space and, to a large extent, LOS scenarios, this is not necessarily true for NLOS scenarios. The NLOS environment provides the opportunity for substantial pulse interaction, possible frequency selectivity, and the introduction of frequency dependence into the channel. Note that if we consider the frequency domain transfer function of a single channel realization, we will clearly observe frequency-selective behavior. However, in large-scale modeling, we are concerned with the ensemble average of transfer functions in a particular environment at various distances. If the average transfer function shows frequency dependence, apart from antenna effects, we must accommodate this in the model. In other words, we can model the frequency-dependent path loss as [77]

$$PL(f) = E\left\{\int_{f-\frac{\Delta}{2}}^{f+\frac{\Delta}{2}} \left|H\left(\tilde{f}\right)\right|^2 d\tilde{f}\right\} \qquad (3.18)$$

where $H\left(\tilde{f}\right)$ is the measured channel transfer function, and $\Delta f$ is chosen to be small enough such that path loss is constant in the interval.

Again, there is no current consensus on this effect. While the measurements cited previously did not exhibit frequency dependence, other measurements have shown some frequency dependence in path loss. Specifically, the frequency-dependent decay of channel gain with increasing frequency is discussed in [51]. It was demonstrated from the measurements that the UWB channel shows, apart from a frequency selective fading pattern, a general decaying trend with increasing frequency. It is also shown that for the NLOS case, the decay is slightly steeper at higher frequencies than the LOS case. Specifically, the path loss at a given distance increases exponentially with frequency. The transfer function is given as

$$|H(f)| \propto f^{-k} \tag{3.19}$$

or equivalently the path loss varies as

$$\sqrt{PL(f)} \propto f^{-k} \tag{3.20}$$

where $0.8 \leq k \leq 1.4$ depending on the environment, and $f$ is given in GHz. This suggests that the path loss is a function of both frequency and distance. The most straightforward method is to model the two effects independently.

In [59] it was shown that the frequency-dependent path loss is well approximated by considering the path loss at an effective "center frequency" given by the geometric mean of the lower and upper band edge frequencies. Finally, [8] also observed some frequency-dependent behavior in path loss. Specifically, the authors modeled the average channel transfer function as being frequency-dependent according to

$$\log_{10}\left(\overline{\left|H(f)^2\right|}\right) \propto e^{-\delta f} \tag{3.21}$$

where $\delta$ is between 1 for LOS environments and 1.36 for NLOS environments with significant blockage of the path. Equivalently

$$\log_{10}(PL(f)) \propto e^{-\delta f} \tag{3.22}$$

Finally, it should be reiterated that if the antenna effects are included in "path loss," there will certainly be frequency-dependent effects, especially if constant gain antennas are used.

### 3.3.4 Partition Dependent Approaches

The information on the electromagnetic properties of building materials in the UWB frequency range provides valuable insights into assessing the capabilities and limitations of UWB technology. Reference [62] exhaustively examines propagation through 10 typical construction materials and their characterization for the frequency range of 0.5–15 GHz. Reference [60] discusses experimental results for the frequency dependent refractive index of dispersive materials between 5 and 85 GHz.

Material-dependent attenuation can be incorporated into a path loss model to potentially provide more accurate path loss modeling. For example, [61] discusses basic partition-based path loss models for the narrowband indoor LOS and NLOS cases at 5.85 GHz. The model assumes a path loss exponent of $n = 2$, but includes losses for each partition through which the signal passes. The result is a model that provides a smaller standard deviation. A similar approach was suggested in [63] for UWB. Specifically, the path loss is modeled as

$$\overline{PL\,(d)} = 20 \log \left( \frac{4\pi d}{\lambda} \right) + L_w\,(d - 4)\,u\,(d - 4) \tag{3.23}$$

where $L_w$ is a building loss due to partitions, $\lambda$ is wavelength at the center frequency, and $u\,(x)$ is a unit step function. A value of 0.7 dB/m is given for $L_w$ [63]. Unfortunately, the model is not extensively tested with measurements. While this approach may offer improved accuracy as compared to standard path loss exponent models, it has yet to gain much attention in the UWB community, possibly due to the fact that UWB is currently envisioned only for short-range communications. Also, it requires partition information for specific rooms, which makes the model more site-specific.

### 3.3.5   Large-Scale Modeling Studies

Many studies characterizing indoor path loss and shadowing experienced by UWB signals have now been done. The two environments most heavily studied are the indoor residential environment and the office environment. The models developed have primarily used the frequency-independent approach given in (3.11) and (3.16). Notable exceptions are the studies listed in Section 3.3.3. Table 3.1 summarizes the measurements reported in the literature. In general, it is found that LOS channels exhibit path loss exponents in the range 1–1.8, whereas NLOS channels exhibit path loss exponents in the range of 2–4.1. Specifically, in [16], [17], and [18] the traditional exponential path loss model was applied to UWB signals, and path loss results for a residential environment were reported for the LOS and NLOS scenarios. These measurements used a vector network analyzer (VNA) and examined the frequency range from 4.375–5.625 GHz. In [30] the residential setting was investigated, and large- and small-scale LOS and NLOS results were reported for 2–8 GHz. Other campaigns investigated the indoor office environment. The measurements in [51] used a VNA to investigate the frequency range from of 1–11 GHz, with path loss and amplitude statistics being calculated for LOS and NLOS environments. Large- and small-scale findings using a time domain (pulse-based) measurement system with a center frequency of 2 GHz and a bandwidth of 1.5 GHz were reported in [55]. Large- and small-scale results using a time domain measurement system were also reported in [9, 10]. The work in [5], [6], [7], [62], and [72] is based on both time domain and, to a lesser extent, frequency domain measurements taken as part of the DARPA NETEX project. There are many other existing measurement results including [8], [20], [21], [22], and [24] that present path loss exponents and shadowing variances.

### 3.3.6   Antenna Impact on Large-Scale Modeling

Ideally, we would like to establish channel models that are entirely independent of the antennas and the receiver structure. That is, the channel model should reflect

Table 3.1: Measured Path Loss Exponents ($n$) and Shadowing Standard Deviation ($\sigma$) in Published Measurement Studies. (Standard Deviation of $n$ and $\sigma$ Shown Where Available.)

| Researchers | $n$: Mean | $n$: Std. Dev. | $\sigma$(dB): Mean | $\sigma$(dB): Std. Dev |
|---|---|---|---|---|
| Virginia Tech (office) [5, 7, 62, 72, 73, 75] | 1.3–1.4 (LOS) 2.3–2.4 (NLOS) | | 2.5–3 (LOS) 2.6–5.6 (NLOS) | |
| AT&T (Res.) [16–18] | 1.7/3.5 (LOS/NLOS) | 0.3/0.97 | 1.6/2.7 | 0.5/0.98 |
| U.C.A.N. [8] | 1.4/3.2(soft)/4.1(hard) LOS/NLOS/NLOS | | 0.35 LOS/1.21(soft) / 1.87(hard) NLOS | |
| France Telecom [24] | 1.5/2.5 (LOS/NLOS) | | 4/4 (LOS/NLOS) | |
| CEA-LETI [21–23] | 1.6 (lab)1.7(flat) LOS 3.7 (office/lab/NLOS) 5.1 (flat/NLOS) | | | |
| Intel [14, 15, 30] (Resident.) | 1.7/4.1 (LOS/NLOS) | | 1.5/3.6 (LOS/NLOS) | |
| IKT, ETH Zurich [32] | 2.7–3.3 (on body) 4.1 (around the torso) | | | |
| Cassioli/Molisch/Win [9, 10] | 2.04 (d<11m) $-56 + 74\log(d)$ (d>11) | | 4.3 | |
| Oulu Univ. [20] | 1.04, 1.4, 1.8 LOS 3.2, 3.3, 3.9 NLOS | | | |
| Whyless [51] | 1.58/1.98 LOS/NLOS | | | |
| Time Domain [55] | 2.1 (LOS/NLOS) | | 3.6 | |

only the impact of the channel. However, this is extremely difficult to obtain because it requires the antenna effects to be completely removed from the received data. This can be accomplished to some degree by defining path loss as the power loss with respect to a reference measurement as opposed to transmit power. However, full independence requires knowledge of the antenna impact on the received signal at all azimuthal and elevation angles and knowledge of the three-dimensional angle-of-arrival of each signal path. This is, in general, difficult to obtain. The concept of the "double-directional" channel is presented in [28], which attempts to eliminate all antenna effects. However, it requires spatial grid measurements at both the transmitter and receiver. To date, this approach has not been applied to UWB measurements [28]. Thus, nearly all empirically-based channel models will contain a degree of antenna dependence. However, it is reasonable that antennas with broadly similar characteristics will exhibit similar path loss.

Perhaps the most important aspect of the antenna in this regard is its directionality. Because an omni-directional antenna collects more multipath energy than a directional antenna, it experiences lower "path loss" [5,74]. This is simply based on the fact that path loss is defined as the average power loss versus distance. However, we should understand that this is not a true channel characteristic as much as an antenna characteristic. This is a common theme in channel modeling for UWB. Care must always be taken when attributing observed behavior to the "channel." While we may reasonably model antenna effects as part of the channel, we must be cognizant of the antenna dependence of the models when applying them.

As an example, consider the data plotted in Figure 3.5 [72]. This plot shows measured path loss values (relative to a free-space reference at 1 m) in an indoor environment when using two different antennas. Specifically, the measurements represent four scenarios using two antennas (bicone and TEM horns) and two measurement environments (LOS and NLOS). The resulting path loss exponents and "shadowing" standard deviation values are shown in Table 3.2. We can observe two trends. As one would expect, the path loss exponent for NLOS environments is larger than for LOS environments. This is expected due to the fact that as distance increases in NLOS environments, more objects may block the path between the transmitter and receiver, thus leading to a decrease in the average received power, which decays faster than free-space (that is, $n > 2$).

Table 3.2: Example Large-Scale Path Loss Parameters for Different Antennas and Scenarios.

|  | Bicone | | TEM | |
|---|---|---|---|---|
|  | Total | | | |
|  | n | $\sigma$(dB) | n | $\sigma$(dB) |
| LOS | 1.3 | 2.6 | 1.3 | 2.8 |
| NLOS | 2.3 | 2.4 | 2.4 | 5.1 |

Secondly, we can observe that the path loss exponent experienced by TEM horn antennas is larger than by bicone antennas. Again, this is due to the fact that for the same transmit power the bicone antenna collects more multipath energy than the directional TEM horn antenna. It should be emphasized that these results [5, 72, 73] define path loss relative to a reference measurement, as opposed to transmit power. This eliminates the antenna gain from the path loss calculation. The result is interesting because it shows that in an indoor environment the additional gain due to directionality may be offset by larger "path loss," which is never the case in free-space. Note that similar results were presented in [29] for directional and omni-directional antennas.

### 3.3.7    Better than Free-Space Propagation

One item from Table 3.1 that may strike the reader as odd is that some LOS environments provide better path loss than free-space (that is, $n < 2$). However, this should be interpreted not as if the *spreading* loss associated with such channels is less than that observed in free-space. Rather, due to the additional power collected via reflections in indoor channels with an LOS path, the received signal power is greater than in free-space. The following analysis gives a specific example that demonstrates this effect [72, 73]. Note that this phenomenon is not unique to UWB, but has also been demonstrated for narrowband channels (such as [26, 27]).

Figure 3.6 shows an example UWB time domain measurement at a distance of 9 meters in an LOS scenario using TEM horn antennas. Note that in addition to the LOS path, there are reflections due to objects in the environment that are resolvable.

Figure 3.6 shows the received signal as well as the normalized cumulative received energy for two different time scales. The left figures show the entire received signal over approximately 70 ns, while the right figures show a close-up of 2 ns. The generated Gaussian pulse is shown in Figure 3.1, and TEM horn antennas are used. We can see that the LOS pulse of the received signal only accounts for about 25% of the total received energy of the signal. This by itself suggests that the path loss for this location would be better than free-space, but it is helpful to work out the exact path loss. The expected received power for free-space propagation, $P_r^{FS}$, at a distance, $d$, is given as

$$P_r^{FS} = P_0 - 20 \log_{10} \left( \frac{d}{d_0} \right) \tag{3.24}$$

where $P_0$ is a reference power taken at reference distance $d_0$. Similarly, the measured received power, $P_r^m$, at a distance, $d$, in an environment that may have a path loss exponent better than free-space, is given as

$$P_r^m = P_0 - 10 \left( 2 - \alpha \right) \log_{10} \left( \frac{d}{d_0} \right) \tag{3.25}$$

where $\alpha$ accounts for the deviation from free-space. The value of $\alpha$ and therefore the path loss exponent, $n$, can be determined by calculating the difference between the received powers given by (3.24) and (3.25), $\Delta P_r$, which is given as

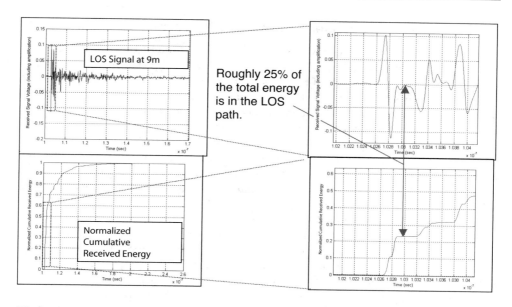

Figure 3.6: Received Signal (Top) and Cumulative Received Energy (Bottom) for LOS Example using TEM Horn Antennas. (Right plots are enlarged views of the left plots.) [72, 73]

SOURCE: B. Donlan and R. M. Buehrer, "The UWB Indoor Channel: Large and Small-Scale Modeling," submitted to *IEEE Transactions on Wireless Communications* [73]. © IEEE, 2004. Used by permission.

$$\Delta P_r = P_r^m - P_r^{FS} \tag{3.26}$$

Substituting values from Figure 3.6 for $P_r^m$, $P_r^{FS}$, and $\Delta P_r$, and combining (3.24), (3.25), and (3.26) gives

$$4.8 = 10\alpha \log_{10}\left(\frac{d}{d_0}\right),$$
$$\alpha = 0.5 \rightarrow n = 1.5 \tag{3.27}$$

Thus, the path loss exponent for this specific measurement, $n = 1.5$, is much better than free-space. Note that this represents only a single measurement. The true path loss exponent for an environment must be averaged over many such measurements. In environments that exhibit better than free-space propagation on average, several locations show similar behavior. However, we must reiterate that this does not mean that the signal actually propagates more efficiently; rather, the receiver simply collects more signal energy due to the waveguiding effect of the environment.

## 3.3.8   Receiver-Dependent Path Loss Models

One aspect that differentiates UWB systems from traditional narrowband and wideband systems is what is sometimes called energy capture. Specifically, in UWB systems, due to the high amount of time dispersion, capturing all the available energy is not always feasible. This will be discussed in more detail in the next section. The path loss that we have described to this point is sometimes called "total" path loss because it refers to the path loss observed by the total received signal [55, 72]. Depending on the receiver structure used, the total received signal energy may not be captured. As a result, it can be useful to define receiver-dependent path loss models. However, extreme care should be taken when defining such models, as they cannot be used with receiver structures that differ from those for which they are defined.

One type of receiver-dependent path loss model is what is sometimes called "peak path loss" [55, 72]. The phrase "peak path loss" is somewhat of a misnomer, but signifies that only the peak multipath amplitude of the received signal is used to calculate the received signal power. This value is then compared to the free-space path loss at 1 m. Thus, it represents the path loss of the strongest multipath component. As with the "total" path loss, local area averaging is used to determine the path loss. This metric is helpful in conjunction with total path loss in order to obtain a clearer picture of the experienced path loss. This is particularly useful for UWB because many receivers may not be able to capture the large amount of diffuse signal energy. A receiver that captures only the most significant multipath component observes a much larger average path loss, which is represented by what is called "peak path loss."

Table 3.2 gives the calculated total path loss parameters $n$ and $\sigma$ fit to the same measurement data discussed in Figure 3.5 [72]. As a comparison, Table 3.3 presents path loss parameters using the same measurement data for the path loss of the strongest multipath component. *Again, note that the path loss is relative to free space at a distance of 1 m.* The LOS path loss based on this definition will always be 2 because it only considers the strongest path (that is, the LOS path). Note that the path loss is significantly greater than the total path loss given in Table 3.2. This is due to the fact that we specifically eliminate the multipath collecting effect and focus only on the strongest path between the transmitter and receiver. Again, this path loss model would be applicable only to receiver structures that capture only a

Table 3.3: Large-Scale Path Loss Parameters for Path Loss of the Dominant Multipath Component for Different Antennas and Scenarios.

|        | Bicone    |              | TEM   |              |
|--------|-----------|--------------|-------|--------------|
|        | Total     |              |       |              |
|        | n         | $\sigma$(dB) | n     | $\sigma$(dB) |
| LOS    | 2         | 0.71         | 2     | -            |
| NLOS   | 2.7–4.3   | 2.97–3.98    | 3.35  | 6.3          |

single path. The link budget for such a receiver could use the "peak path loss" model to determine the received signal power. An alternative for such receiver structures would be to use the total path loss and simply incorporate an energy capture factor in the link budget to account for the receiver architecture. This is discussed in Section 3.3.10. A model that combines the total received energy (power), the power in the strongest component, and the rms delay spread is given in [35]. Specifically, the authors argue that the received power of the strongest component $P_s(d)$ at distance $d$ can be modeled as

$$P_s(d) = P(d) \left[ 1 - e^{\frac{-t_o}{\tau_{rms}}} \right] \tag{3.28}$$

where $P(d)$ is the total received power at distance $d$, $t_o$ is the time resolution of the receiver, and $\tau_{rms}$ is the rms delay spread. Further, the rms delay spread increases linearly with distance, leading to the approximate relationship

$$P_s(d) \approx P(d) \frac{d_t}{d} \tag{3.29}$$

where $d_t$ is a breakpoint distance of approximately 1 m. Further measurements in [35] demonstrate that

$$P_s(d) \propto \left( \frac{d_o}{d} \right)^3 P_s(d_o) \tag{3.30}$$

and

$$P(d) \propto \left( \frac{d_o}{d} \right)^2 P(d_o) \tag{3.31}$$

which validates the relationship in (3.29).

### 3.3.9   Shadowing

Referring to (3.16), the term $X_\sigma$, which accounts for the variation in the received signal power about its mean value, is typically termed "shadowing." This is usually modeled as a log-normal random variable with standard deviation $\sigma$. Typical values are given in Tables 3.1 and 3.2. Figure 3.7[2] shows some example CDFs for the deviation of the measured received power from the calculated average for an indoor channel with bicone antennas in LOS and NLOS environments [72]. This deviation in general has been shown to fit a log-normal distribution fairly well [16–18, 26]. It should be noted that while this variation has traditionally been termed "shadowing," because it typically stems from variations in the large-scale structure of the environment, we must be careful when applying this term to indoor data sets. Specifically, in LOS environments, the variation in received signal power is not necessarily due to the shadowing of the receiver by large objects as much as variation in the mutipath profile. LOS channels with a significant amount of clutter around the

---

[2]Figure 3.7 is adapted from Figure 3.4 in [73].

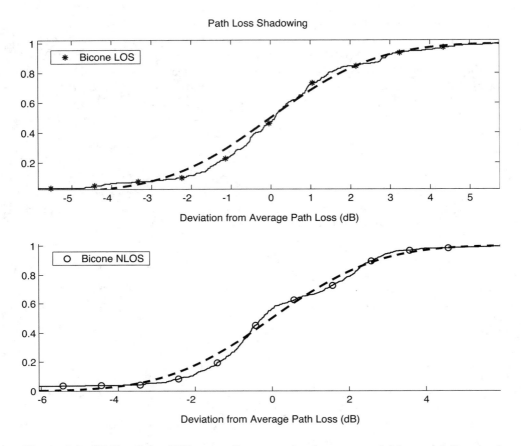

Figure 3.7: CDFs of the Difference Between the Average and Measured Received Power and Fit to a Log-Normal Distribution for Two Examples (Represents Shadowing).

transmitter or receiver may exhibit higher received signal power than simpler LOS channels. For example, hallways tend to result in strong multipath echoes that can be captured by the receiver, resulting in higher average signal power. Note that the standard deviation of the shadowing itself varies from environment to environment. Measurements have found this value to be between 1.5 dB and 4 dB (see Table 3.1). Like the path loss exponent, some researchers suggest modeling the shadowing parameter as a random variable. For example, [18] proposes modeling the standard deviation, $\sigma$, as a truncated Gaussian random variable. Thus, the path loss at distance $d$ is modeled as

$$PL\left(d\right) = \overline{PL}_o + 10 \cdot \mu_n \log_{10}\left(\frac{d}{d_o}\right) + 10\sigma_n X_1 \log_{10}\left(\frac{d}{d_o}\right)$$
$$+ X_2\mu_\sigma + X_2 X_3 \sigma_\sigma \tag{3.32}$$

where $\overline{PL}_o$ is the average measured path loss at the reference distance $d_o$, $\mu_n$ is the mean value at the path loss exponent, $\sigma_n$ is the standard deviation of the path loss exponent, $\mu_\sigma$ is the mean value of the shadowing standard deviation, and $\sigma_\sigma$ is the standard deviation of the random shadowing standard deviation. $X_1$, $X_2$, and $X_3$ are zero mean, unit norm Gaussian random variables. This model has also been proposed for the IEEE 802.15.4a study group [77].

### 3.3.10  Link Budget Calculations

In this last section dealing with large-scale channel modeling, we examine how it is used in the development of link budgets for UWB systems. Due to the extremely wide bandwidths involved, the UWB link budget can differ significantly from narrowband link budget relations. Thus, we will discuss these potential differences. However, before we discuss UWB links, let us first briefly review traditional link budget design assuming free-space propagation. We will relax the free-space restriction shortly. For the current discussion, the free-space model allows us to highlight the differences between traditional link budget design and UWB link budget design.

The goal of a link budget is to determine the range of a system (that is, maximum distance between the transmitter and receiver) for a required reliability and power limitation. For a digital communication system, the reliability is typically specified as an $E_b/N_o$ (that is, ratio of energy per bit to noise power spectral density) requirement for some distance or percentage of the coverage area. Thus, a standard link budget can be written as shown in Table 3.4. The received signal power in dBm is equal to the transmit power in dBm plus the gains of the transmit and receive antennas (in dBi or dB relative to an isotropic antenna) minus the path loss (again in dB). As noted previously, what is traditionally termed the path loss $\left(\left[\frac{4\pi d}{\lambda}\right]^2\right)$ is actually a combination of the spreading loss due to wave propagation and the conversion of the gain of the receive antenna to its effective aperture area. The received energy per bit is determined by integrating the received power over the duration of the received pulse, which can be approximated as the average power multiplied by the symbol (bit[3]) duration. The noise power spectral density is calculated from Boltzman's constant and the system noise temperature, and the received $E_b/N_o$ is calculated from the preceding quantities. However, this is the *average* received $E_b/N_o$ that ignores fading and is appropriate for free-space propagation. For non-free-space propagation, a fading margin is typically required. A fading margin must be determined to guarantee that the $E_b/N_o$ is achieved over some percentage of the coverage area. In this free-space example, the margin is not included, but in general the margin will need to be included. As will be shown in Section 3.5, due

---

[3]If the system uses binary modulation, the symbol and bit energy are the same. If the system uses $M$-ary modulation, we must adjust for the number of bits per symbol.

Table 3.4: Conventional Free Space Link Budget.

| Term | Value (example) | Comment |
|---|---|---|
| $P_t$ | -20dBm | Transmit power in dB relative to a mW |
| $G_t$ | 6dBi | Gain relative to isotropic radiator at operating frequency |
| $L_p$ | 100dB | $(4\pi r/\lambda)^2$ – Free space path loss given constant gain antennas at a distance $d$ (assume $fc = 2.5\text{GHz}$ and $r = 1\text{km}$) |
| $G_r$ | 6dBi | Gain relative to isotropic radiator at operating frequency |
| $P_r$ | $-108$dBm | $(P_t + G_t - L_p + G_r)$ |
| $k$ | $-228.6$dBW/Hz/K | Boltzman's constant |
| $T_s$ | 24.6dBK | System Noise Temperature (290K) |
| $N_o$ | $-204$dBW/Hz | Noise PSD $= k * T_s$ |
| $R_b$ | 50dB | Bit Rate (100kbps) |
| $E_b$ | $-148$dBmJ | $P_r - R_b$ |
| $E_b/N_o$ | 26dB | $E_b - 30 - N_o$ |
| $M_F$ | 0dB | Fading margin |
| $E_b N_o$ | 26dB | $E_b/N_o$ achieved |

to the reduced fading of UWB signals, this margin will be substantially reduced for UWB systems.

## UWB Link Budget

It is natural to apply the preceding analysis to UWB systems. However, there are several problems with this approach (specifically for I-UWB systems):

1. In narrowband systems, the average power and the peak power are typically the same, and even when they are not, average power is usually the metric of interest. However, due to the potential for low duty cycle pulses, this is not necessarily the case for UWB communications. Average power may or may not be useful depending on the pulse duty cycle.

2. Due to the wide bandwidth of the pulse, the transmit and receive antennas must be treated differently than narrowband link budgets. First, the pulse may be distorted in the transmit/receive process, which complicates any gain calculation. Second, the gain of the antenna will likely change over the frequency band of the pulse.

3. The traditional path loss equation is a combination of spreading loss and the effective aperture of the receive antenna. The latter can also change over the bandwidth of the pulse.

4. Due to pulse distortion from the channel, it is not always possible to use a matched-filter receiver. Thus, the filter used will not typically capture all the energy in the received pulse. This should also be considered in the link budget.

One proposed method to accommodate these differences is to make three basic changes to the traditional link budget approach [78]. First, the transmit and receive antenna gains are removed from the budget. Second, the impact of the effective aperture of the antenna is removed from the path loss. Instead, the antenna gains and effective apertures are replaced with a single term labeled the "antenna-pulse coupling gain" or $G_{AP}$. This term captures the gain in either energy or peak power in the received pulse with respect to the generated pulse. Whether energy or power is examined depends on the modulation scheme used, and we will discuss this shortly. In some cases the two measures may be equal, but not in general. It should be clear that this gain is very different from the traditional antenna gain. It is different because it defines an antenna pair as opposed to a single antenna, and because it is *pulse specific*. Just as the traditional "gain" of an antenna is frequency specific, $G_{AP}$ is pulse specific. Additionally, it captures the receive antenna aperture which is normally lumped into path loss. As a result, the "gain" actually has a negative dB value in general and has units of square meters (m$^2$).

The second modification to the traditional link budget is that path loss now accounts only for spreading loss, not effective aperture at the receive antenna. This is consistent with the path loss modeling discussed earlier. Thus, we can define path loss as

$$L_p = 4\pi r^2 \tag{3.33}$$

where $r$ is the distance between the transmitter and receiver. This simply represents the power distributed over the area of a sphere of radius $r$; thus, it has units of square meters (m$^2$).

A third difference in the I-UWB link budget is that it should be in terms of either pulse energy or peak pulse power. In traditional link budgets, we deal with *average* power. However, average power may not be useful in UWB because low duty cycle pulses are possible. Also, we are interested in the impact of the antenna on the pulse, not just the impact on average power. Because both the pulse shape as well as peak power could be affected, we allow for $G_{AP}$ to be defined either in terms of peak pulse power or pulse energy. The former would be more applicable to threshold detector receivers, while the latter would be more applicable to correlator receivers.

The resulting link budget is given in Table 3.5. The term $G_{AP}$ can either be determined directly from a time domain line-of-sight measurement or derived from an antenna $s_{21}$ measurement. Specifically, for a pulse energy-based link budget, the gain should be determined as

Table 3.5: Proposed Free Space UWB Link Budget—Correlator Detector.

| Term | Value (example) | Comment |
|------|-----------------|---------|
| $E_t$ | −101dBmJ | Transmit energy in one pulse in dB |
| $G_{AP}$ | −30dBi | Energy or power gain of the link (excluding path) from tx and rx antennas at 1m [note that this is *pulse dependent*] |
| $L_p$ | 31dB | $4\pi d$ - Free space spreading loss at a distance $d$ |
| $E_r$ | −161dBmJ | $(E_t + G_{AP} - L_p)$ Received energy per pulse |
| $k$ | −228.6dBW/Hz/K | Boltzman's constant |
| $T_s$ | 26dBK | System Noise Temperature |
| $N_o$ | −203dBW/Hz | Noise PSD = $k * T_s$ |
| $N_s$ | 0dB | Pulses per bit (1 pulse/bit) |
| $E_b$ | −161dBmJ | $E_r + N_s$ |
| $E_b/N_o$ | 12dB | $E_b - 30 - N_o$ |
| $M_F$ | 0dB | Fading margin (much lower for UWB) |
| $\rho$ | −7dB | Energy capture loss |
| $E_b/N_o$ | 5dB | $E_b/N_o$ achieved |

$$G_{AP} = 4\pi r^2 \frac{\varepsilon_p^{rx}}{\varepsilon_p^{tx}}$$

$$= 4\pi r^2 \frac{\int_{-\infty}^{\infty} b_r^2(t)\, dt}{\int_{-\infty}^{\infty} a_t^2(t)\, dt} \tag{3.34}$$

where $\varepsilon_p$ represents the pulse energy, $a_t(t)$ is the generated pulse, and $b_r(t)$ is the received pulse measured at a distance $r$ m in free-space or calculated as (see Chapter 4)

$$b_r(t) = \frac{k}{4\pi r} \frac{d}{dt} \left\{ a_t(t) * h_{tx}\left(t - \frac{r}{c}\right) * h_{rx}(t) \right\} \tag{3.35}$$

where the combined response of $h_{tx}(t)$ and $h_{rx}(t)$ can be determined for the antenna pair of interest from an $s_{21}$ measurement and k= $\frac{\mu}{\sqrt{50}}$. Additionally, note that we have neglected the vector nature of the general problem. We have assumed that (3.35) is calculated with both antennas aligned for maximum response. This is analogous to narrowband antenna gain, which is typically defined as the maximum gain versus azimuth and elevation angles. Additionally, traditional link budget designs assume proper polarization alignment or include a polarization loss factor.

Thus, $G_{AP}$ represents the energy gain of the pulse due to the antennas. Note that this term is *not waveform independent*. **That is, it is not merely a factor of the antenna, unlike the traditional antenna gain.** It is *waveform specific*. As such, care must be taken in its use.

The received pulse energy in dBmJ (that is, dB relative to one mJ) is then calculated as

$$E_r(dBmJ) = E_t(dBmJ) + G_{AP}(dBm^2) - L_p(dBm^2) \qquad (3.36)$$

The noise power spectral density $N_o$ is calculated as before, and $E_p/N_o$ dB is simply $E_r$ (dB) $-$ $N_o$ (dB). Of course, this is the energy per pulse, not necessarily per bit. In many systems multiple pulses may be transmitted per bit. Thus, this value must be multiplied by the number of pulses per bit $N_s$ (added in dB) to obtain the energy per bit. The result then represents the available $E_b/N_o$ or the SNR of a matched filter output.

Due to pulse mismatch, which is much more likely in UWB due to both the channel and limitations in replicating the received pulse, we must add a factor $\rho$ (negative in dB or less than unity in linear), which represents the fraction of the energy captured by the receiver. This factor can either be due to per path distortion or overall temporal dispersion of the channel. For example, consider a RAKE receiver that uses only two to three correlators. Based on the discussion in Section 3.4.8, we know that in many environments, such a receiver only captures $\sim$20% of the total energy. Thus, a factor of $\rho = 0.2$ ($-7$dB) should be included in this case, provided the path loss value implies total energy capture. Various receiver structures will have different values of $\rho$. The final result is the $E_b/N_o$ available to the detector.

### Receiver Structure Dependence

This development has been specifically in terms of energy because we are typically interested in the received energy per bit. Traditional link budgets can obtain $E_b$ from the average received power as $E_b = P_r T_b$. However, this is not necessarily true for UWB; thus, we have set up the budget in terms of the energy per pulse. If the receiver is based on threshold detection or is otherwise concerned only with the peak power of the pulse, the link budget should be worked in terms of peak power. In that case, we should define the pulse antenna coupling gain as

$$\begin{aligned} G_{AP} &= 4\pi r^2 \left. \frac{P_p^{rx}}{P_p^{tx}} \right|_{rm} \\ &= 4\pi r^2 \frac{\max\left\{b_r^2\left(t\right)\right\}}{\max\left\{a_t^2\left(t\right)\right\}} \qquad (3.37) \end{aligned}$$

where again $b_r\left(t\right)$ is the received signal at a distance of $r$ meters, $P_p^{rx}$ and $P_p^{tx}$ are the peak power of the received pulse measured at $r$ meters and generated pulse respectively. Note that the notation $[]|_{rm}$ signifies that the gain is measured at $r$ meters. This should coincide with the reference distance used for the path loss calculations.

As a final note, we should mention that the preceding analysis is more applicable to pulse-based I-UWB rather than to MC-UWB. With MC-UWB, we can resort to traditional narrowband approaches for each carrier.

## Link Budgets for Systems Other than Free-Space

The analysis in the previous section considered only free-space propagation. For non-free-space environments, as discussed in Section 3.3.2, a traditional approach is to model the average received power using a log-distance power decay law. That is, we can represent the average received pulse energy at a distance $r$ as

$$\overline{\varepsilon_r(r)} = \varepsilon_o \left(\frac{r_o}{r}\right)^n \tag{3.38}$$

where $\varepsilon_0$ is the pulse energy received at a reference distance assuming free-space. This is the traditional log-distance path loss model [26] that was discussed in Section 3.3.2. Note that the average is a spatial average. If we define our reference distance as 1 m (typical for indoors), we would redefine total path loss as

$$L_p = 10 \log(4\pi) + 10n \log(r) \tag{3.39}$$

where $n$ is the path loss exponent defined as the path loss relative to a 1 m free-space measurement. This reference distance should be the same used to calculate $G_{AP}$. Finally, we must include a fade margin in non-free-space environments as discussed previously.

## Validation

A validation of this approach was performed as part of a UWB measurement and modeling project through DARPA's NETEX program [72,78]. Specifically, measurements were taken at various distances to compare the results with predictions from the link budget approach. Specifically, the received pulse energy and the peak power of the received pulse were examined and compared to what was predicted. The antennas used at both the transmitter and receiver were wideband bicone antennas. $G_{AP}$ was determined from two separate experiments. First, the $s_{21}$ parameter of the link between the two antennas was measured. This measurement was used to determine the combined antenna frequency response

$$H_{tx}(j\omega) \cdot H_{rx}(j\omega) = \frac{4\pi r}{(j\omega)} \cdot S_{21}(j\omega) \cdot e^{+j\omega\frac{r}{c}} \tag{3.40}$$

Using this combined frequency response, $B_r(\omega)$ was determined from $A_t(\omega)$, the generated pulse plotted in Figure 3.8. Using the time domain versions of the generated and predicted received pulse, $G_{AP}$ was determined from (3.34). Next, $G_{AP}$ was determined from time domain measurements at 1 m, where the LOS pulse was time gated from the received signal. The resulting values of $G_{AP}$ from the two approaches are listed in Table 3.6 [78]. Note that two distinct time domain measurements were used to create two estimates using the time domain approach and

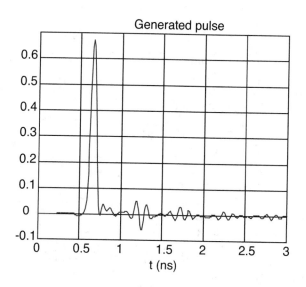

(a) Time Domain

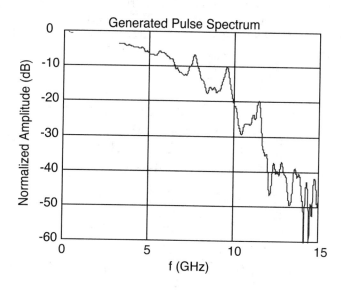

(b) Frequency Domain

Figure 3.8: Generated Pulse.

Table 3.6: $G_{AP}$ Values for Bicone Antennas and Pulse Given in Figure 3.8 and Figure 3.9.

| Technique | Value |
|---|---|
| Using Frequency Domain Measurement of $S_{21}(j\omega)$ and using Equation (3.34) and (3.35) | $-27.2$dB |
| Direct Measurements and Using Equation (3.34) | $-27.4$dB (a) $-26.0$dB (b) |

are listed as values (a) and (b) in the table. Note that the value is a large negative value (in dBm$^2$). This is to be expected because it captures the effective receive aperture. All the values are within 1.4 dB.

Using the measured values of $G_{AP} = -27.4$ dB and $G_{AP} = -26.0$ dB, the predicted received pulse energy was compared to the measured pulse energy at several distances and is given in ble 3.7 [78]. The measured and predicted values match very closely, with the measured and predicted values coming within 1 dB of each other in all cases. As previously mentioned, the energy gain in the pulse is not necessarily the best metric for every modulation scheme or receiver structure. Thus, we are also interested in the peak power gain in the pulse as defined in (3.37). This metric was measured for a pair of bicone and TEM horn antennas using the pulse given in Figure 3.8. The received pulses are given in Figures 3.9 and 3.10 and the results are presented in Table 3.8 [78]. We can see that the values are similar for

Table 3.7: Predicted and Measured Received Pulse Energy Using $G_{AP}$.

| Measurement Set A | | |
|---|---|---|
| $d$ (m) | $E_{rx}$ Meas. (dBmJ) | $E_{rx}$ Link Budget Est. |
| 1.1 | $-93.9$ | $\sim$ |
| 2.0 | $-99.5$ | $-99.9$ |
| 4.3 | $-106.0$ | $-106.6$ |
| 6.0 | $-108.4$ | $-109.5$ |
| 14.0 | $-115.8$ | $-116.8$ |

| Measurement Set B | | |
|---|---|---|
| $d$ (m) | $E$ (dBmJ) | Link Budget Est. |
| 1 | $-92.6$ | $\sim$ |
| 2 | $-98.8$ | $-98.6$ |
| 4 | $-105.1$ | $-104.6$ |
| 6 | $-108.4$ | $-108.1$ |
| 14 | $-116.5$ | $-115.5$ |

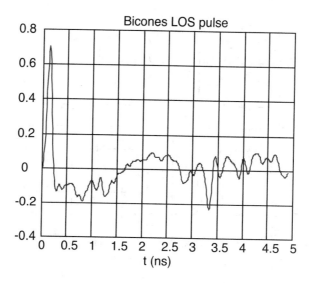

(a) Time Domain

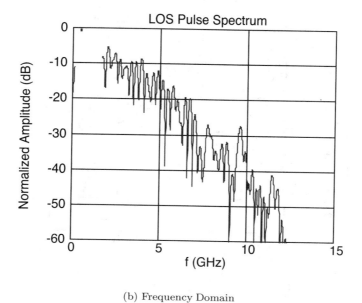

(b) Frequency Domain

Figure 3.9: Received Pulse Using Bicone Antennas.

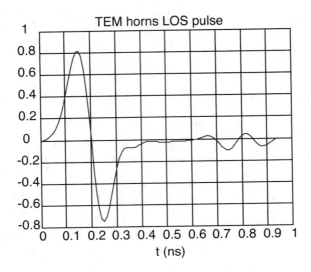

(a) Time Domain

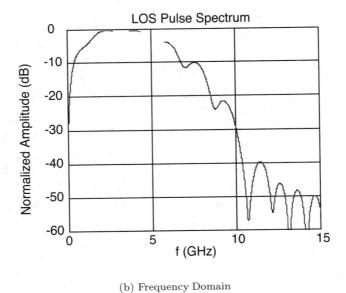

(b) Frequency Domain

Figure 3.10: Received Pulse Using TEM Horn Antennas.

Table 3.8: Measured $G_{AP}$ Energy and Power Values for Bicone and TEM Horn Antennas.

| Antenna | $G_{AP}$ for Energy (avg) | $G_{AP}$ for Peak Power (avg) |
|---|---|---|
| Bicone Antennas | $-26.7$dB | $-28.1$dB |
| TEM Horn Antnenas | $-24.7$dB | $-27.3$dB |

the bicone antennas (that is, peak gain is similar to energy gain). However, for the TEM horn antenna, the energy gain is approximately 3 dB higher than the peak power gain. Additionally, the energy gain of the TEM horn is 3 dB higher than the bicone. This is reflective of the fact that the TEM horn is a directive antenna. Thus, the energy gain is greater than the bicone. However, because the pulse shape is substantially changed (compare Figure 3.10 with Figure 3.8), the peak power gain is less (that is, a single peak has been transformed into two peaks).

## 3.4   Small-Scale Channel Modeling

In the previous section we concentrated on what is termed statistical modeling of the large-scale characteristics of the received signal. This is primarily concerned with the loss in received signal power versus distance (and possibly frequency) between the transmitter and receiver. It was noted that even when modeling relatively simple signal parameters, UWB presents some interesting challenges, and care must be taken when applying a model to a UWB system.

We now turn to a more complex part of channel modeling, the statistical modeling of the small-scale characteristics. The small-scale characteristics include the small-scale fading in a local environment as well as the distortion of the transmitted waveform due to multipath. UWB signals are often proposed for short-range applications (d<10m). As a result, when discussing fading (that is, the loss in received signal power) there is some ambiguity in the definition of large-scale and small-scale. In this text when we discuss signal fading we will typically restrict the term "small-scale" to refer to fading within a 1 m$^2$ area. Large-scale fading will refer to fading over distances much greater than 1 m.

Again, in this section we are concerned with statistical modeling of the received signal characteristics. While the statistical properties of the large-scale characteristics are necessary for proper link budget design, the statistical properties of the small-scale characteristics are needed for proper receiver design. Specifically, understanding small-scale fading and signal correlation over a small area aids in the evaluation of the receiver, diversity mechanisms, and potential multiple antenna applications. Additionally, understanding the waveform distortion suffered by a UWB signal is vital in designing the modulation scheme and the signal demodulator, as well as for evaluating receiver structures.

The received signal from a UWB transmission can be represented as shown in Chapter 4. Specifically, in (4.44) the received signal is given as

$$b_r(t) = \frac{\mu}{\sqrt{R_r R_t}} \frac{\partial}{\partial t} \int_{\Omega_t} \int_{\Omega_r}$$
$$\left[ h_{r_R}(\theta_r, \phi_r; t) \dot{\otimes} \overrightarrow{C} \dot{\otimes} h_{t_R}(\theta_t, \phi_t; t) \otimes a_t(t) \right]$$
$$d\Omega_r d\Omega_t \tag{3.41}$$

where $\overrightarrow{C}$ is the multipath channel impulse response (CIR) being studied, $a_t(t)$ is the generated pulse, $h_{t_R}$ and $h_{r_R}$ are the realized effective lengths of the transmit and receive antennas respectively, and $\otimes$ is the combination of time domain convolution and dot product. Please refer to Chapter 4 for a full development of this equation and an explanation of all variables. Unfortunately, this is an overly complicated expression and would be exceptionally difficult to evaluate and use for modeling. Therefore, one approach to channel modeling is to subsume the angular dependencies of the antenna responses into the channel impulse response and to consider only the antenna response at the boresite. Therefore, the resulting channel impulse response will be antenna dependent but will still offer a general characterization of the channel as would be seen by similar antennas (similar in this case being most significantly the spatial pattern of the antenna). For the receiving antenna, some distortion is expected as a function of angle-of-arrival (AOA). The significance of neglecting pulse distortion such as that introduced by the receiving antenna is discussed in Section 3.6. Applying these assumptions, the received signal can be approximated by

$$b_r(t) = \frac{1}{4\pi r} \frac{\mu}{50} \frac{\partial}{\partial t} \left[ a_t \left( t - \frac{r}{c} \right) \otimes h_t(t) \otimes h_c(t) \otimes h_r(t) \right] \tag{3.42}$$

where the vector nature of the responses is now incorporated into the scalar time domain impulse response of the channel, and $h_t \otimes h_r(t)$ is the combined response of the antennas at the boresite. It is now desirable to extract the channel impulse response, $h_c(t)$, from the measured response. Knowledge of the channel impulse response, $h_c(t)$, allows a general analysis of UWB channels independent of the generated pulse, $a_t(t)$, and *some* of the specific antenna properties. The convolution of both antenna responses with the generated pulse would be the observed signal if the channel were not present (or equivalently, if the channel impulse response were equal to an impulse). If the channel is free-space, the impulse response is simply an impulse at a delay equal to the propagation time and the received pulse is

$$b_r^{fs}(t) = \frac{1}{4\pi r} \frac{\mu}{50} \frac{d}{dt} \left\{ a_t(t) \otimes h_t(t) \otimes \delta \left( t - \frac{r}{c} \right) \otimes h_r(t) \right\}$$

A direct unobstructed LOS path with no reflectors/diffractors in the environment is very nearly equivalent to free-space. Therefore, if a received pulse from only the LOS path can be isolated from pulses arriving along other paths through time gating, this pulse can be used to determine the channel impulse response. These combined effects can then be removed from the received signal by deconvolving the LOS pulse from the received signal to estimate the channel impulse response where

$$b_r(t) = b_r^{fs}(t) \otimes h_c(t) \qquad\qquad (3.43)$$

It must be noted that the angular dependencies of the antennas, and thus the vector nature of the problem, are subsumed into the channel response. Therefore, the channel impulse response that is calculated is not strictly only due to the channel. The measured channel impulse response for directional antennas, such as TEM horns, would not be influenced significantly by reflectors/diffractors behind the antennas. However, the channel impulse response for omni-directional, in one plane only, antennas would be affected more by reflectors/diffractors in the environment than by directional antennas. Additionally, by only considering the antenna response at the boresite we are assuming that the pulse distortion due to the antenna is the same from all transmit/receive directions. While this is clearly a big assumption, it is justified by the fact that the strongest paths will be those transmitted and received via the boresite. Broadly speaking, all omni-directional antennas will behave in a similar fashion. The same could be said of directional antennas, with the same approximate gain. Thus, the characterizations of the "channel" for antennas such as TEM horns and bicones should be performed separately but may be representative of many antenna structures.

Channel modeling thus depends on how the channel parameters are obtained from the measurements, that is, on deconvolution. Deconvolution is the process of separating two signals that have been combined by convolution. Several deconvolution techniques have been developed [52], often for specific types of signals or for use with a specific application. Deconvolution can be performed in the time domain or the frequency domain. Different techniques will emphasize different aspects of the deconvolved signal and can offer different advantages depending on the further analysis desired.

In the frequency domain, the most straightforward deconvolution method is known as inverse filtering [53]. Another technique that has been widely used is the Van-Cittert deconvolution technique. This iterative method can be performed in either the time or frequency domain. Reference [54] proposes using this technique in the frequency domain and also proposes criteria to optimize the necessary number of iterations. Regardless of which frequency domain technique is used, such a process results in a bandlimited impulse response. The impact of this will be discussed shortly.

A third deconvolution technique often considered is the CLEAN algorithm, which is a time domain technique. The CLEAN algorithm was first introduced in [33] and has been used by many researchers in UWB channel measurement analysis [41, 55, 72]. The algorithm is important in channel modeling because it assumes that the impulse response being found is not bandlimited but is rather a sum of scaled and time-delayed impulses. This is consistent with classic small-scale channel models and thus makes the CLEAN algorithm very useful. We will discuss the algorithm in more detail in Section 3.6.1.

Examples of impulse responses generated by each deconvolution method for a specific measurement set are shown in Figure 3.11 for comparison [41]. The inverse filtering impulse response and the Van-Cittert impulse response are nearly identical, and the CLEAN impulse response shows the same primary features of the channel. Estimates of the received signal can be generated by convolving the impulse response with the LOS pulse used in the deconvolution. The measured signal is plotted with the estimates of this signal using the impulse responses from each of the three deconvolution techniques in Figure 3.12 [41]. Visually, the measured signal, the inverse filter reproduced signal, and the Van-Cittert reproduced signal

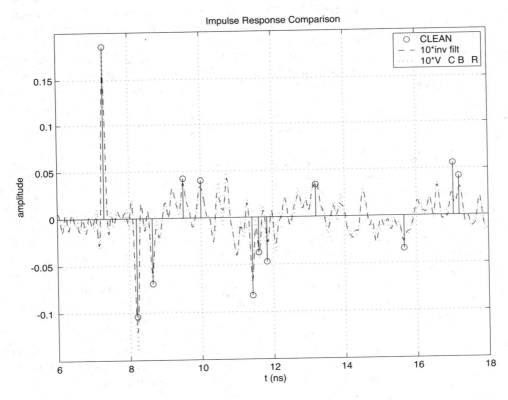

Figure 3.11: Comparison of Example Impulse Responses Generated by the CLEAN Algorithm, Inverse Filtering, and the Van-Cittert Techniques with Bennia-Riad Criteria. (Note: The Latter Two Responses have been Scaled by 10 to Allow for Easier Visual Comparison.)

SOURCE: D. McKinstry, "Ultra-Wideband Small-Scale Channel Modeling and its Application to Receiver Design" [41]. © D. McKinstry, 2002. Used by permission.

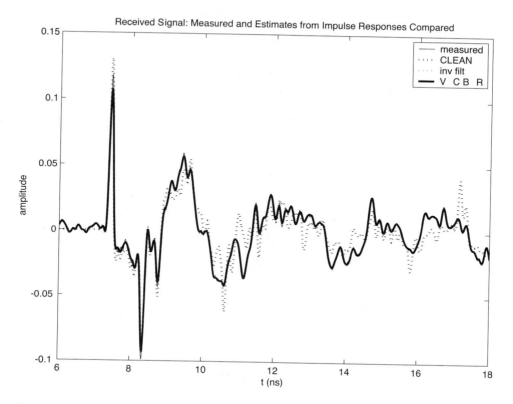

Figure 3.12: Comparison of Measured Signal with Estimates of the Received Signal Based on the Impulse Responses Generated by the CLEAN Algorithm, Inverse Filtering, and the Van-Cittert Techniques with Bennia-Riad Criteria.

SOURCE: D. McKinstry, "Ultra-Wideband Small-Scale Channel Modeling and its Application to Receiver Design" [41]. © D. McKinstry, 2002. Used by permission.

are nearly identical. Noticeable differences are seen between the measured signal and the CLEAN reproduced signal, but the signals still agree reasonably well.

## 3.4.1   Statistical Modeling of the Channel Impulse Response

As discussed previously, the small-scale effects of the wireless channel are commonly described by a simple scalar impulse response as shown in (3.43). More specifically, the small-scale channel is typically modeled as a time-varying linear filter where the received signal is given by [26]

$$r\left(t\right) = s\left(t\right) \otimes h\left(t, \tau\right) + n\left(t\right) \tag{3.44}$$

where $s(t)$ is the transmitted signal that is assumed to be the received LOS pulse in the case of UWB, $h(t, \tau)$ is the time-varying channel impulse response, and $n(t)$ is additive white Gaussian noise. The impulse response can change as a function of time (or as a function of spatial variation) due to the motion of the transmitter or receiver and/or changes in the channel itself. The channel is typically modeled using a tap-delay line approach [26, 38]. This approach was first proposed in [38], and well-established for indoor channels in [44, 45]. Thus, the channel model can be given as

$$h(t, \tau) = \sum_{k=0}^{N(t)-1} \beta_k(t)\, p_k(t)\, \delta\left(\tau - \tau_k(t)\right) \tag{3.45}$$

where $\beta_k(t)$, $\tau_k(t)$, and $p_k(t)$ are the time-varying amplitude, delay, and polarity of the $k^{th}$ path respectively. $N(t)$ is the number of multipath components, which is also in general a function of time. Note that for narrowband (or wideband) channels, the polarity, $p_k$, is typically subsumed into a phase term, $e^{j\theta(t)}$ due to the use of the complex baseband. In UWB modeling researchers often ignore phase since it is assumed that baseband pulses are being sent. If the pulses are upconverted or downconverted, a complex baseband model may be appropriate, and the phase term could be reintroduced. If the channel is assumed to be static over the interval of interest, the time-invariant model of the channel can be used

$$h(\tau) = \sum_{k=0}^{N-1} \beta_k p_k \delta\left(\tau - \tau_k\right) \tag{3.46}$$

The main goal of small-scale channel models is to statistically characterize the amplitudes, delays, and polarities of the multipath components of the channel. We will discuss the standard models shortly.

Besides specific statistical characterization of the multipath parameters, the channel can also be characterized by coarse statistics, such as mean excess delay, RMS delay spread, and maximum excess delay, that describe the time dispersive properties of the channel. These are useful as single number descriptions of the channel to estimate the performance and potential for intersymbol interference (ISI). These values tend to increase with greater transmitter/receiver separation [77]. The mean excess delay is defined as [26]

$$\tau_m = \frac{\sum_k \beta_k^2 \tau_k}{\sum_k \beta_k^2} \tag{3.47}$$

The RMS delay spread of the channel is defined as

$$\tau_{RMS} = \sqrt{\frac{\sum_k \beta_k^2 \tau_k^2}{\sum_k \beta_k^2} - \tau_m^2} \tag{3.48}$$

The number of paths is the number of significant multipath components that form the channel and is also used as a shorthand way of characterizing the time dispersion of the channel. The key to any model that assumes the discrete channel presented previously is the joint statistical characterization of the path amplitudes and delays. Various wireless channel models for doing this are discussed next. The models were originally developed to characterize indoor wideband channels. Researchers have found that the general approaches are also useful for UWB channels.

## 3.4.2   Saleh-Valenzuela Model

The most common statistical model for the discrete indoor channel impulse response is the Saleh-Valenzuela model [39]. It should be noted that although the Saleh-Valenzuela model was developed for NLOS channels, it has also been applied to LOS channels where it is perhaps less valid, unless LOS components are specifically added [14]. The Saleh-Valenzuela model is a tapped-delay line model given by

$$h\left(t\right) = \sum_{l=0}^{L}\sum_{k=0}^{K}\beta_{k,l}\delta\left(t - T_l - \tau_{k,l}\right) \tag{3.49}$$

where $L$ is the number of paths per cluster and $K$ is the number of clusters. The basic assumption behind this model is that the multipath components arrive in clusters. The cluster arrivals are described by a Poisson process, and thus, the cluster interarrival times $T_l$ are described by exponential random variables

$$p\left(T_l|\,T_{l-1}\right) = \Lambda\exp\left[-\Lambda\left(T_l - T_{l-1}\right)\right], \quad l > 0 \tag{3.50}$$

where $\Lambda$ is the mean cluster arrival rate. Within a cluster, the path arrivals are also described by a Poisson process, so that the distribution of the interarrival times is described as

$$p\left(\tau_{k,l}|\,\tau_{(k-1),l}\right) = \lambda\exp\left[-\lambda\left(\tau_{k,l} - \tau_{(k-1),l}\right)\right], \quad k > 0$$

where $\lambda$ is the mean ray arrival rate. The average power of both the clusters and the rays within the clusters is assumed to decay exponentially such that the average power of a multipath component at a given delay, $T_l + \tau_{k,l}$, is given by

$$\overline{\beta_{k,l}^2} = \overline{\beta_{0,0}^2}e^{-T_l/\Gamma}e^{-\tau_{k,l}/\gamma} \tag{3.51}$$

where $\overline{\beta_{0,0}^2}$ is the expected value of the power of the first arriving multipath component, $\Gamma$ is the decay exponent of the clusters, and $\gamma$ is the decay exponent of the rays within a cluster. In addition to the average power decay, a common practice with small-scale channel models is to normalize them such that the total *average* power is unity.

   The amplitude of each path is assumed to be a random variable about an exponential mean. Several distributions have been proposed for the amplitudes, although the original model assumed a Rayleigh distribution. The polarity is assumed to be a binary random variable with equal probability. When a random phase is used, the phase is assumed to be uniformly distributed. The Rayleigh distributed amplitudes and random phase model come from the assumption that several paths arrive at delays that are not resolvable to the measurement system used. This is a more valid assumption for wideband channels but is questionable for UWB channels. As a result, many researchers have found that log-normal or Nakagami distributions provide a better fit [71]. This will be discussed in more detail in Section 3.4.9.

The use of log-normal random variables is especially convenient when they are also used for the cluster amplitudes. This suggests two independent log-normal variables to represent the amplitude variations of the clusters and rays. However, these random variables can be combined as a single log-normal random variable. The polarity of the path is represented as an equi-probable binary random variable, $p_{k,l}$, taking on the values $\pm 1$. For log-normal amplitudes, the path amplitudes are given by

$$\beta_{k,l} = p_{k,l} 10^{(\mu_{k,l} + X_{\sigma,k,l})/20} \tag{3.52}$$

where

$$\mu_{k,l} = \frac{20 \ln\left(|\overline{\beta_{0,0}}|\right) - 10 T_l/\Gamma - 10\tau_k/\gamma}{In(10)} - \frac{\sigma^2 \ln(10)}{20} \tag{3.53}$$

and

$$X_{\sigma,k,l} = N\left(0, \sigma^2\right) \quad (\sigma \text{ is in db}) \tag{3.54}$$

To summarize, this model is described by five parameters:

1. $\Lambda$: The mean cluster arrival rate.

2. $\lambda$: The mean ray arrival rate.

3. $\Gamma$: The cluster exponential decay factor.

4. $\gamma$: The ray exponential decay factor.

5. $\sigma$: The standard deviation of the log-normal distributed path powers.

An example CIR for the Saleh-Valenzuela model is given in Figure 3.13.

### 3.4.3    $\Delta$-K Model

The $\Delta$-K model has also been used to model indoor wideband channels, and like the Saleh-Valenzuela model is based on the assumption that multipath components arrive in clusters. The probability that a path arrives at any given delay is higher by a factor of $K$ if a path has arrived within the past $\Delta$ seconds. By increasing the arrival rate when a path has recently arrived, the paths tend to arrive in clusters. The arrival times thus follow a modified, two-state Poisson process, and the inter-arrival times follow an exponential distribution where the arrival rate is based on the state. When in state S-1, the mean arrival rate is given by $\lambda$. The transition to state S-2 is triggered when a path occurs. In S-2 the mean arrival rate is given by $K\lambda$. If after $\Delta$ seconds a path has not arrived, transition back to S-1 occurs.

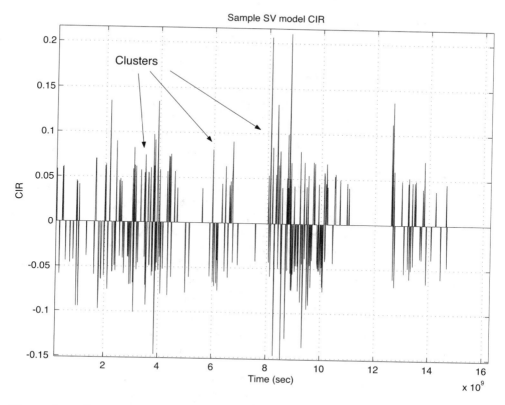

Figure 3.13: Sample Saleh-Valenzuela Model CIR.

Exponential energy decay is assumed here to describe the expected value of the energy in a path at a given delay. The polarity is assumed to be $\pm 1$ with equal likelihood, and amplitude fading is assumed to be log-normal such that the amplitude of a path is given by (3.52)–(3.54).

For computer simulation, this model is often implemented using a discrete form. In the discrete version, the time axis is divided into bins, and the probability of a path arriving in a given bin is based on whether a path arrived in the previous bin (the probability being higher by a factor of $K$ if a path was present).

### 3.4.4    Single Poisson Model

This is a simplified version of both of the previously introduced models and assumes that only one cluster is present in the impulse response (or equivalently, no clustering of paths). The arrivals of paths are treated as a Poisson process with an arrival rate of $\lambda$. Also, the decay of the paths is assumed to be exponential, with a decay time constant of $\gamma$. The amplitudes of the paths are modeled as a log-normal random variable with parameter $\sigma$.

### 3.4.5    Modified Poisson Model

A modified version of the single Poisson arrival time model with dominant early multipath components was developed to more accurately represent LOS channels [41]. The model is based on the characteristics observed in the measurement data. The first multipath components have significantly more energy than the later arriving components. The number of dominant components, $M$, is randomly chosen to be two, three, or four with equal likelihood. This agrees with the trends seen in some measurements. The interarrival times of the dominant components are exponentially distributed and given by

$$p\left(\tau_k \mid \tau_{k-1}\right) = \lambda_1 \exp\left[-\lambda_1\left(\tau_k - \tau_{k-1}\right)\right], \quad 0 < k < M \tag{3.55}$$

where $\lambda_1$ is the mean arrival rate of the dominant components. The amplitude of each dominant component is given by a log-normal random variable with unit mean energy and fading parameter $\sigma_1$.

The first component arrives at zero delay ($\tau_0 = 0$) and has a positive polarity ($p_0 = +1$) because this is assumed to be the LOS component, which will not be inverted. All other paths have an equi-probable positive or negative polarity.

The interarrival times of the weaker components also follow an exponential interarrival, but with a different mean arrival rate $\lambda_2$.

$$p\left(\tau_k \mid \tau_{k-1}\right) = \lambda_2 \exp\left[-\lambda_2\left(\tau_k - \tau_{k-1}\right)\right], \quad k \geq M \tag{3.56}$$

The mean energies of the weaker components follow a traditional exponential decay. The first component from this group has mean energy $W$ dB less than the mean of the dominant components, and the respective means of the later components are taken relative to this first weak path. The weak path amplitudes are also log-normal variables with a different fading parameter, $\sigma_2$, such that

$$\mu_k = -W - \frac{10\left(\tau_k - \tau_M\right)/\gamma}{In\left(10\right)} - \frac{\sigma_2^2 In\left(10\right)}{20}, \quad k \geq M \tag{3.57}$$

### 3.4.6    Split-Poisson Model

The Saleh-Valenzuela model is based on the generation of multiple exponentially decaying clusters. However, much measurement data indicates that very few clusters may exist, perhaps due to the limited range used for UWB systems and measurements. An example taken from [72] is shown in Figure 3.14. The data seems

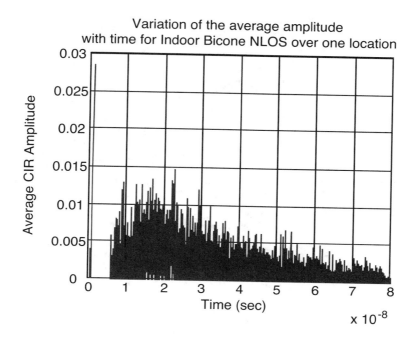

Figure 3.14: Average CIR Amplitude Over One Location From Example Measurement—Two Clusters Can Be Clearly Seen.

SOURCE: S. Venkatesh, J. Ibrahim, and R. M. Buehrer, "A New 2-Cluster Model for Indoor UWB Channel Measurements," *IEEE International Symposium on Antennas and Propagation* [75]. © IEEE, 2004. Used by permission.

to indicate that the average CIR consists basically of two clusters, the first short cluster containing several strong paths that decay quickly and the longer second cluster containing slowly decaying paths. Based on the previous observation, the Split-Poisson model was proposed in [72, 75, 76]. This model assumes two clusters of Poisson arrivals, one delayed by $\tau_1$ relative to the other. The first cluster is generated using a set of parameters $\lambda_1, \sigma_1, \gamma_1$ while the second cluster is generated using a separate set of parameters $\lambda_2, \sigma_2, \gamma_1$. The overall CIR is created by adding a delayed version of the second cluster to the first cluster. Also, in order to maintain continuity in the decay of energy in the overall CIR the first cluster is weighted higher than the second cluster by a factor $\alpha$. This is shown in Figure 3.15. Each of these parameters is estimated from the data as described in [75, 76].

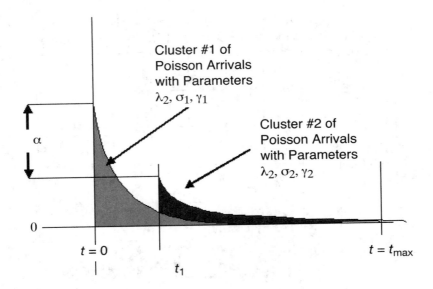

Figure 3.15: Illustration of the Two Cluster Model.

SOURCE: S. Venkatesh, J. Ibrahim, and R. M. Buehrer, "A New 2-Cluster Model for Indoor UWB Channel Measurements," *IEEE International Symposium on Antennas and Propagation* [75]. © IEEE, 2004. Used by permission.

### 3.4.7   Effect of Model Parameters

We now illustrate how different model parameters affect the RMS delay spread and mean excess delay of the generated channel impulse response. For simplicity we consider the single Poisson model and begin with the parameters shown in Table 3.9. To examine the parameter impact we consider the measured cumulative histogram of the RMS delay spread, mean excess delay, and number of multipath components when the parameters $\lambda$, $\gamma$, and $\sigma$, are (respectively) doubled, keeping the rest of the parameters fixed. Note that only paths within 20 dB of the strongest path are

Table 3.9: "Original" Poisson Parameters Used for Examining the Affect of Parameter Variation.

| Parameter | Value |
|---|---|
| $1/\lambda$ | 1.9 nsec |
| $\gamma$ | 5 nsec |
| $\sigma$ | 4 |

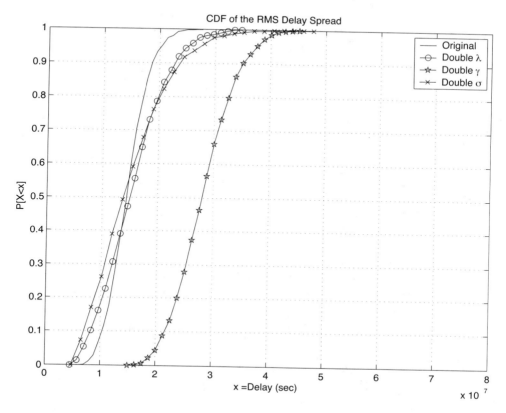

Figure 3.16: Effect of Poisson Model Parameters on the RMS Delay of the Generated Model CIR.

considered. Figures 3.16, 3.17, and 3.18 show the effect of doubling these parameters individually. We observe the following trends:

1. Increasing $\lambda$ (Poisson Arrival Rate) increases the mean excess delay and the number of paths. However, there is an increase in the variance of the RMS delay spread but no increase in the mean value. This is because the RMS delay spread depends strongly on the distribution of energy in the paths rather than the number of paths. If the energy distribution of the paths in the CIR remains the same, the mean square delay spread defined by

$$\tau_{MS}^2 = \frac{\sum_k \beta_k^2 \tau_k^2}{\sum_k \beta_k^2} \tag{3.58}$$

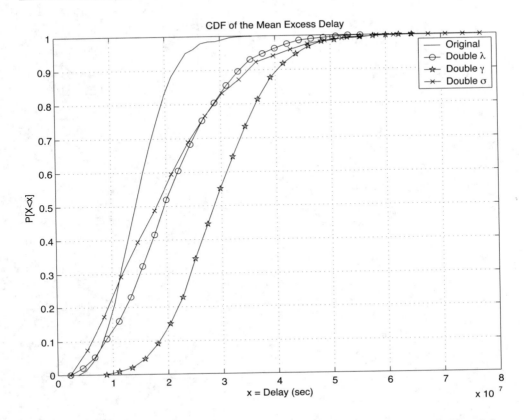

Figure 3.17: Effect of Poisson Model Parameters on the Mean Excess Delay of the Generated CIR.

effectively increases in proportion to the square of the mean excess delay

$$\tau_m^2 = \left( \frac{\sum_k \beta_k^2 \tau_k}{\sum_k \beta_k^2} \right)^2 \tag{3.59}$$

Recalling that the RMS delay spread is defined as

$$\tau_{RMS} = \sqrt{\tau_{MS}^2 - \tau_m^2} \tag{3.60}$$

we see that there isn't a significant change in the average RMS delay spread. The large variance in the RMS delay spread and mean excess delay is due to a larger number of paths, which is in turn due to the larger arrival rate.

2. Increasing $\gamma$, the time constant of exponential energy decay, increases the RMS delay spread, mean excess delay, and the number of paths because it takes longer for the paths to die out. Increasing $\gamma$ not only increases the

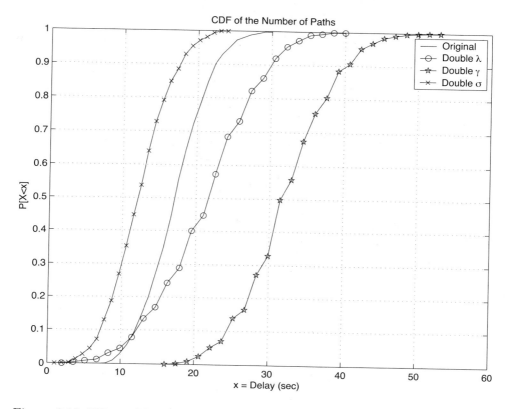

Figure 3.18: Effect of Poisson Model Parameters on the Number of Paths in the Generated CIR.

number of paths but also changes the energy distribution of the CIR. Hence, it also increases the variance of both the mean excess delay and the RMS delay spread.

3. Increasing $\sigma$ (proportional to the standard deviation of small-scale variation) decreases the slope of the CDF curves of the RMS delay spread and the mean excess delay. It also decreases the number of paths. This last item is due to the fact that a larger variance in the amplitude leads to a larger number of paths falling below the 20 dB threshold set for this example.

We can also examine the impact that the parameters of the Saleh-Valenzuela model have on the same statistics. For the Saleh-Valenzuela model, the parameters that may be varied are $\{\lambda, \Lambda, \gamma, \Gamma, \sigma\}$. Figures 3.19, 3.20, and 3.21 show the effect of doubling the parameters $\{\gamma, \Gamma, \sigma\}$ individually from the "original" parameters given in Table 3.10. We observe the following trends:

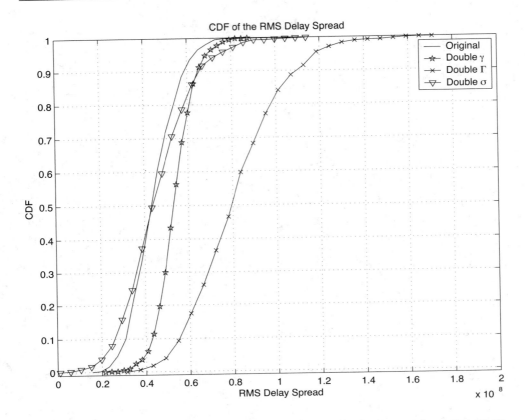

Figure 3.19: Effect of Doubling Saleh-Valenzuela Parameters $\{\gamma, \Gamma, \sigma\}$ on CIR RMS Delay Spread.

1. Increasing $\Gamma$ (cluster decay time constant) increases the RMS delay spread, mean excess delay, and the number of paths. It is also decreases the slopes of the CDFs of the RMS delay spread and mean excess delay. This can be interpreted as a larger region of support for the PDF, that is, a larger range of values.

2. Increasing $\gamma$ (time constant of exponential energy decay of rays) increases the RMS delay spread, mean excess delay, and the number of paths because it takes longer for the paths to die out.

3. Increasing $\sigma$ (proportional to the standard deviation of small-scale variation) decreases the slope of the CDF curves of the RMS delay spread and the mean excess delay. It also decreases the number of paths. This last item is due to the fact that a larger variance in the amplitude leads to a larger number of paths falling below the 20 dB threshold set for this example.

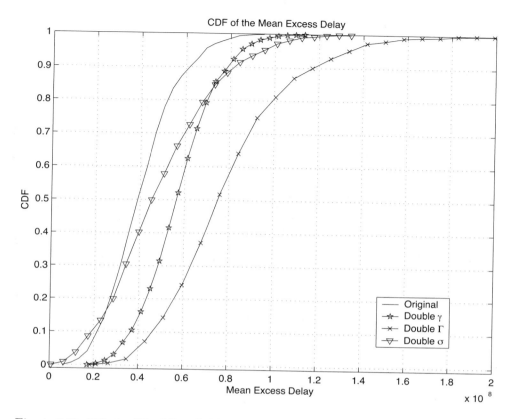

Figure 3.20: Effect of Doubling Saleh-Valenzuela Parameters $\{\gamma, \Gamma, \sigma\}$ on CIR Mean Excess Delay.

Figures 3.22, 3.23, and 3.24 show the effect of doubling the parameters $\{\lambda, \Lambda\}$ individually from the original parameters shown in Table 3.10. We observe the following trends:

1. Increasing $\Lambda$ (cluster arrival rate) does not significantly affect the RMS delay spread but increases the mean excess delay and the number of paths.

2. Increasing $\lambda$ (arrival rate of rays) does not significantly affect either the RMS delay spread or the mean excess delay but increases the number of paths. As mentioned previously, changing the arrival rate does not affect the energy distribution of paths that affects the RMS delay and the mean excess delay.

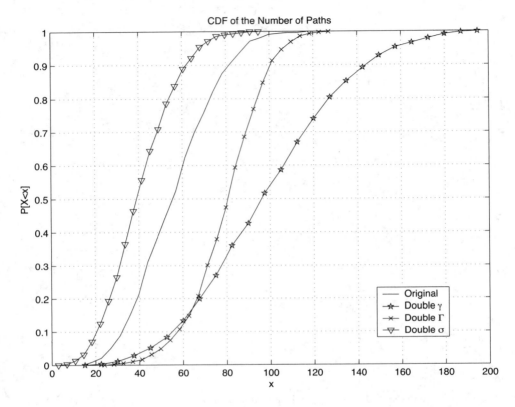

Figure 3.21: Effect of Doubling Saleh-Valenzuela Parameters $\{\gamma, \Gamma, \sigma\}$ on CIR Number of Paths.

Table 3.10: "Original" Saleh-Valenzuela Model Parameters for Examining the Affect of Parameter Variation. (Indoor Bicone NLOS Channel.)

| Parameter | Value |
|-----------|-------|
| $1/\Lambda$ | 5 nsec |
| $1/\lambda$ | 1 nsec |
| $\Gamma$ | 5 nsec |
| $\gamma$ | 2 nsec |
| $\sigma$ | 4 |

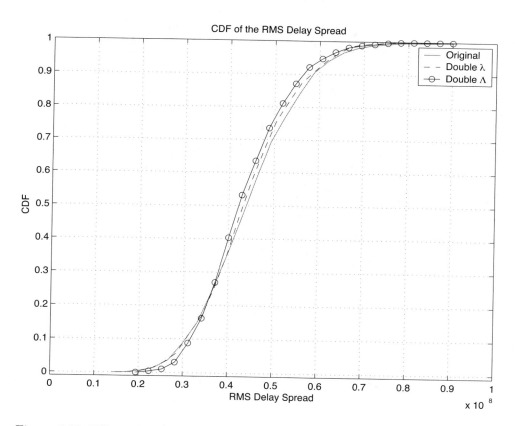

Figure 3.22: Effect of Doubling Saleh-Valenzuela Parameters $\{\lambda, \Lambda\}$ on CIR EMS Delay Spread.

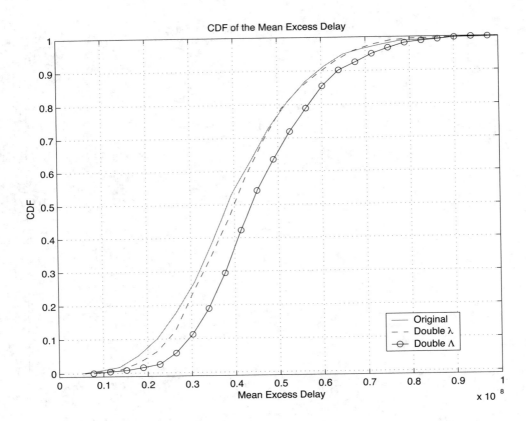

Figure 3.23: Effect of Doubling Saleh-Valenzuela Parameters $\{\lambda, \Lambda\}$ on CIR Mean Excess Delay.

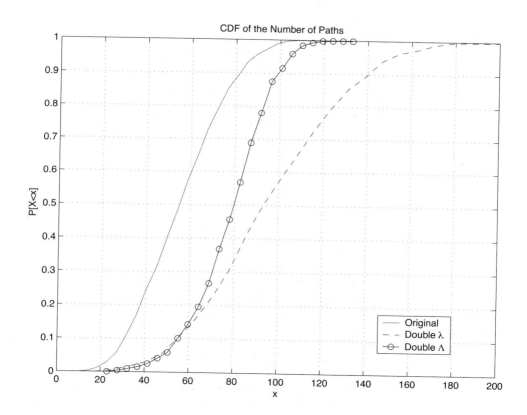

Figure 3.24: Effect of Doubling Saleh-Valenzuela Parameters $\{\gamma, \Gamma, \sigma\}$ on CIR Number of Paths.

### 3.4.8   Stochastic Tapped-Delay Line Model

A similar model that has been proposed in the literature is termed a stochastic tapped-delay line (STDL) model [10]. Specifically, the approach is derived from measurements in an indoor environment and assumes that the average energy follows a single exponential decay. The power delay profile is written as

$$\overline{h^2(\tau)} = \overline{\beta_1^2}\delta(\tau - \tau_1) + \sum_{k=2}^{N} r\overline{\beta_1^2}\exp\left(-(\tau_k - \tau_1)/\varepsilon\right)\delta(\tau - \tau_k) \tag{3.61}$$

The variable $r$ is the power ratio of the second path to the first and is random following a log-normal distribution. The decay constant $\varepsilon$ is also modeled as a log-normal random variable. The previous expression provides the average power gains for each delay. The specific gain value is selected from a Gamma distribution with a mean energy that follows the exponential distribution given previously and an $m$ value that is selected from a truncated Gaussian distribution. The model was used to compare total signal energy with measurements in [10], and good agreement was found.

### 3.4.9   Amplitude Statistics

It is important to estimate the distribution of the path amplitudes while fitting the empirical data to a specific channel model. For example, the original Saleh-Valenzuela model assumed that the path amplitudes follow a Rayleigh distribution. This is due to the fact that a large number of multipath components combined in each resolvable path, resulting in a complex Gaussian random variable. However for the case of UWB baseband pulses, this cannot necessarily be assumed. The following path amplitudes can be defined from two different perspectives: global[4] amplitude distributions at different excess delays can be examined and local[5] amplitude distributions at different excess delays can be examined. Note that the amplitude distribution describes the variation of the received amplitude from the average delay profile. As mentioned earlier, the average power delay profile is commonly modeled as an exponential decay. Four probability distributions have been discussed in the literature to represent these local variations: (1) the log-normal distribution, (2) the Weibull distribution, (3) the Nakagami distribution, and (4) the Rayleigh distribution.

Log-normal variables are convenient because large-scale fading is also modeled as log-normal. The log-normal cumulative distribution function (CDF) can be written as

$$F(x \mid \mu, \sigma) = \frac{1}{\sigma\sqrt{2\pi}} \int_0^x \frac{e^{\frac{-(In(t)-\mu)^2}{2\sigma^2}}}{t} dt \tag{3.62}$$

---

[4]By "global" we refer to the distribution over an entire set of measurements from different locations.

[5]By "local" we refer to the distribution over a set of local area measurements.

This is equivalent to the normal distribution when the amplitudes are expressed in a log scale $20 \log_{10}(A)$. The received signal amplitudes are expressed in dB and the mean and the standard deviation is calculated using dB values.

In [7], a Weibull distribution was found to represent the data best. The Weibull CDF is given as

$$F(x \mid a, b) = \int_0^x abt^{b-1} e^{-at^b} \, dt = 1 - e^{-axb} I_{(0,\infty)}(x) \tag{3.63}$$

The Weibull parameters $a$ and $b$ are known as the *normalization factor* and the *shape factor*, respectively. The Weibull distribution simplifies to the Rayleigh distribution when $b = 1$. The amplitudes in this case are linear and not expressed in dB. The Weibull distribution is also sometimes used to fit the variation of the power of the channel impulse response about a decaying exponential mean [8].

The Rayleigh CDF is common for narrowband systems and is also used to model late arriving path amplitudes in UWB models. This represents the observed phenomenon that later arriving paths are more numerous, causing more fading. The Rayleigh CDF is given as

$$F(x|b) = \int_0^x \frac{t}{b^2} e^{\left(\frac{-t^2}{2b^2}\right)} \, dt$$

$$= 1 - e^{-x^2/2b^2} \tag{3.64}$$

where $b \geq 0$. The Rayleigh distribution is popular in discussing pure scattering models in the absence of a strong LOS component. It has one parameter ($b$) which is obtained using a maximum likelihood estimate from the sample data.

Finally, a very versatile distribution that is commonly used to model the amplitude in wireless channels is the Nakagami probability distribution. The Nakagami probability distribution function has the following form

$$f(x) = \frac{2m^m x^{2m-1}}{\Gamma(m) \Omega^m} \exp\left(-\frac{mx^2}{\Omega}\right) \quad x \geq 0 \tag{3.65}$$

where $m \geq \frac{1}{2}$ and $\Omega \geq 0$ are shape and scale parameters given by $m = \frac{\left(E[X^2]^2\right)}{var[X^2]}$ and $\Omega = E[X^2]$. Note that for $m > 1$, the Nakagami distribution can be converted to and from a Rician distribution using [77]

$$K = \frac{\sqrt{m^2 - m}}{m - \sqrt{m^2 - m}} \tag{3.66}$$

and

$$m = \frac{(K+1)^2}{(2K+1)} \tag{3.67}$$

where $K$ is the Rice factor of the Rician distribution.

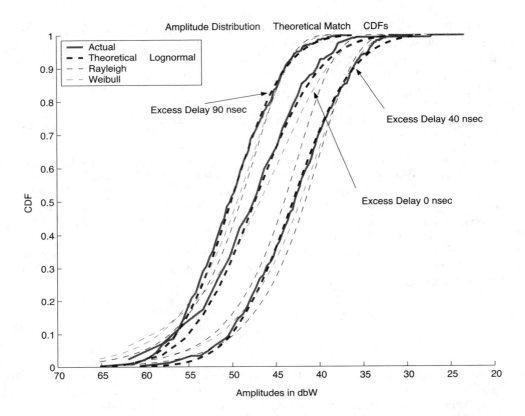

Figure 3.25: Global Amplitude Statistics Matched to Three Different Distributions at Different Excess Delays for the Gaussian Pulse. (It is Seen that the Log-Normal is the Best Fit in Most Cases.)

As an example, consider Figure 3.25, which plots data taken from [72]. Global amplitude statistics are plotted for various delays for an indoor office environment. The measured cumulative histograms are compared with three distributions. In this example, the data is found to match a log-normal distribution.

As mentioned, a unique feature of UWB channels is the large degree of time dispersion. This leads to an important metric for UWB receiver design: energy capture. The large amount of time dispersion makes high energy capture challenging. Thus, we want our channel model to reflect this behavior to allow for proper receiver design. Figure 3.26 illustrates example energy capture statistics for a RAKE receiver with a varying number of correlators corresponding to measurements of several channels [73]. It can be seen that the type of channel and the type of antenna has a large impact on energy capture. Channels with greater time dispersion require significantly more correlator resources (that is, RAKE fingers) in order to achieve the same energy capture. This must be reflected in the channel model that

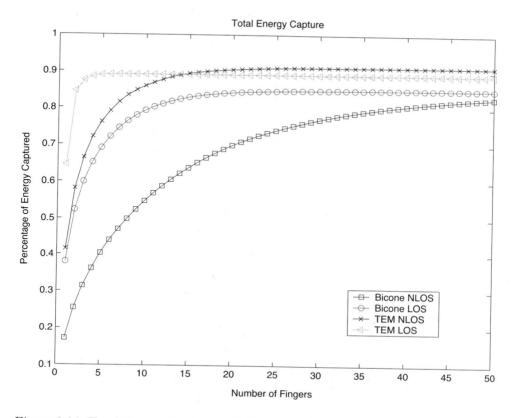

Figure 3.26: Total Energy Capture with Increasing Number of Rake Fingers for Example Data Sets.

SOURCE: B. Donlan and R. M. Buehrer, "The UWB Indoor Channel: Large and Small-Scale Modeling," submitted to *IEEE Transactions on Wireless Communications* [73]. © IEEE, 2004. Used by permission.

is used. As an example, consider a comparison of four different channel models for an NLOS channel using the bicone antennas shown in Figure 3.27. The figure plots energy capture versus the number of fingers for the Saleh-Valenzuela, Single Poisson, Split-Possion, and Δ-K models. Each of the models is parameterized such that they have the same average rms delay spread and mean excess delay. However, they have different energy capture statistics. For this particular data set, the Split-Poisson model matches the measurement data better than the other models. Such statistics should be considered when choosing a model, especially if a RAKE receiver architecture is being used.

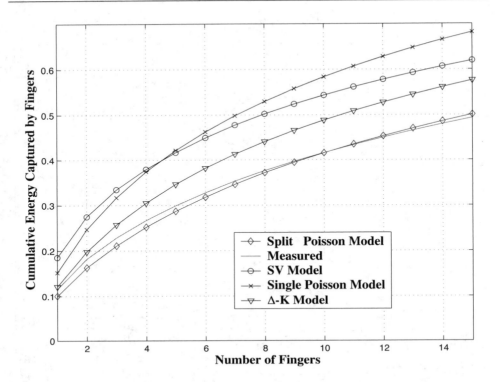

Figure 3.27: Average Cumulative Rake Finger Energy Capture for Measured Indoor Bicone NLOS Data and Various Models for Different Numbers of RAKE Fingers.

SOURCE: S. Venkatesh, J. Ibrahim, and R. M. Buehrer, "A New 2-Cluster Model for Indoor UWB Channel Measurements," submitted to *IEEE Transactions on Communications* [76]. © IEEE, 2004. Used by permission.

## 3.4.10   Summary of Measurement Campaigns and Modeling Efforts

A growing number of measurement campaigns and channel modeling efforts has been carried out to characterize the UWB channel. This work includes time domain and direct pulse measurements as well as frequency domain measurements using vector network analyzers. The results from the only known published measurement campaign taken in an outdoor environment are given in [48] by researchers originally at the University of Southern California and the Time Domain Corporation. The measurements were taken in a dense forest environment using a digital sampling oscilloscope as the receiver. The transmitter contains a pulse generator that creates UWB pulses with an approximate bandwidth of 1.3 GHz every 500 nanoseconds. The sampling rate of the receiver was 20.48 GHz (48.828 ps time resolution). A pulse repetition time of 500 ns was deemed sufficiently long to ensure that the multipath response of the previous pulse had decayed. Values for the mean excess delay, RMS delay spread, path loss exponents, and forestation losses are given. Outdoor results are also presented in [72].

Indoor measurements in a modern office building were taken by the same researchers using basically the same measurement system [49, 50]. Data was collected in 14 different rooms and hallways at 49 locations on a 7 x 7 square grid measuring 3 feet by 3 feet. The receive antenna was located 120 cm from the floor and 150 cm from the ceiling. Some initial analysis of the basic parameters of the channel was presented along with the measurement results. The authors and others use these results to further analyze and propose more specific channel models in several subsequent publications. Several papers [11–13] have been published offering analysis of this measurement campaign. Researchers from AT&T Labs used this data to create a model that treats channel parameters as random variables [9, 10]. The underlying model is similar to those described previously, but the variation in the model parameters from environment to environment is captured through random variables.

A TH-PPM-based channel sounder developed by Time Domain Corporation was introduced in [37]. The instrument was intended to be used to develop channel models that describe large-scale attenuation (that is, path loss), the number of time resolvable multipath components, their arrival times and amplitude statistics, and the variability of propagation paths as a function of distance. The implementation described involves transmitting a 500 ps pulse (1.5 GHz bandwidth centered at 2 GHz) at a 10 MHz repetition frequency. Reference [37] uses measurements taken with this system to calculate specific channel impulse responses, path loss values, and RMS delay spread values. From their measurements, it was found that path loss and RMS delay spread both increase with distance, and the two channel metrics are highly linearly correlated (RMS delay spread was found to be more linearly correlated to path loss than distance).

Intel researchers performed measurements from 2–8 GHz in residential environments. Reference [25] reports some of the results of these measurements. Some of the measurement work (performed in a townhouse) is also described in [30]. Both a time domain measurement system (digital sampling oscilloscope) and a frequency domain measurement system using a Vector Network Analyzer (VNA) were used.

Researchers at IMST in Germany [51] measured office scenarios including corridor, in-office, and between offices using a VNA. The frequency sweep of the VNA was from 1 to 11 GHz. Biconical horn antennas at 1.5 m above the floor were positioned on a 150 x 30 grid (with 1 cm separation between grid points) at each measurement location. Researchers at CEA-LETI in France performed frequency domain measurements from 2-6 GHz in an office scenario for ranges up to 10 meters [3, 22]. Omni-directional, conical monopole antennas were used, and the receive antenna was placed at locations on a 10 x 10 grid (with 10 cm separation between grid points) for each measurement location. Other researchers at AT&T Labs [17] report the results of measurements taken in 23 different homes using a VNA yielding over 300,000 collected power delay profiles from 712 locations in 23 homes at distances from 1 to 15 meters. Identical conical monopole antennas were used over the frequency range 4.375–5.625 GHz. Researchers at the University of Oulu in Finland [20] used a VNA to make channel measurements in a university building. Measurements were made over 2–8 GHz using conical antennas at three heights and three horizontal positions at each location. The different measurement campaigns are summarized in Table 3.11.[6]

A large set of indoor and outdoor measurements was taken by Virginia Tech as part of the DARPA NETEX program [5–7, 61, 72]. The measurements taken were both in the time domain and the frequency domain, and two types of antennas were used (bicones and TEM horns). The measurements characterized large-scale fading (discussed in Section 3.5) as well as small-scale behavior. A more detailed summary of this campaign is presented in Table 3.12.[7]

## 3.5   Spatial Behavior and Modeling of UWB Signals

### 3.5.1   Introduction

The previous sections described the large-scale and small-scale temporal variations of the UWB channel. Specifically, the discussion centered on the statistics of the received signal and models that result in similar statistics. Another important aspect of wireless channels is the spatial variation. It is important to know how the received signal will vary in local area both in time and space. These characteristics are important for understanding applications that attempt to exploit multiple antennas as well as for understanding the channel behavior in a mobile scenario. Models that attempt to capture spatial behavior must model the spatial correlation, spatial fading statistics, and angle-of-arrival statistics in addition to the standard temporal characteristics.

One of the attractive features of UWB is its immunity to multipath fading. Due to the narrow temporal pulses, the interaction between multipath components is severely limited as compared to narrowband signals. This substantially reduces multipath fading. The spatial fading characteristics (that is, the fading variation in a local area) of UWB signals were examined in [11], [12], [13], [25], [72], [79] and [80]. A common procedure to measure fading in a local area is to take measurements at a number of points in a local area (typically a 1 m$^2$ grid). Using such measurements three related aspects of the UWB channel are examined. First, the robustness of the UWB signal to local fading is studied by examining the cumulative distribution function of the received signal power (or energy) in the measurement area. Second, the spatial correlation is studied by examining the correlation between received signals at two spatially separated locations (assuming that the channel is stationary). Finally, the angle-of-arrival (AOA) statistics are also studied. These aspects are usually examined in combination with the temporal aspects of the channel. The AOA statistics are particularly useful in creating two-dimensional channel models as we will discuss.

---

[6]Table 3.11 is adapted from a table in [6].
[7]Table 3.12 is adapted from a table in [6].

Table 3.11: A Summary of Small-Scale Channel Measurement Campaigns.

| Researchers | Frequency Range | Environment | Notes |
|---|---|---|---|
| USC/TDC [48] | $BW = 1.3$ GHz | outdoor forest | • mean excess delay, RMS delay spread, path loss, and forestation loss are given |
| USC/TDC [36, 42] | $BW = 1.3$ GHz | indoor (TDC office) | • 7x7 local area grid in 14 rooms<br>• many publications based on these measurements |
| TDC [37, 48, 55] | 1.25–2.75 GHz | indoor (TDC office and homes) | • 429 profiles collected<br>• 6 to 70 m range in office<br>• statistics on office data given |
| Intel [25, 30] | 2–8 GHz | indoor (residential) | • 2 receive antennas used<br>• measured along baselines to test AOA |
| IMST [51] | 1–11 GHz | indoor (office) | • corridor, in office, inter-office scenarios<br>• 150x30 (icm) spacing in local area grid |
| CEA-LETI [21–23] | 2–6 GHz | indoor (office) | • up to 10 m range<br>• 10x10 (10cm) spacing local area grid |
| AT&T Labs [58, 82, 83] | 4.375–5.625 GHz | indoor (residential) | • 300,000 profiles at 712 locations in 23 homes from 1 to 15 m |
| University of Oulu [20] | 2–8 GHz | indoor (university) | • 3 heights and 3 horizontal positions at each location |
| Virginia Tech [5–7, 62, 72, 73, 75] | 0.1–12 GHz | indoor (office/university) | • 800 time domain and 400 frequency domain measurements<br>• bicone and TEM horn measurements<br>• spatial characteristics also examined |

131

Table 3.12: Small-Scale Channel Characteristics From [72] Using 15 and 20 dB Thresholds.

| | Bicone | | | | TEM | | | |
|---|---|---|---|---|---|---|---|---|
| | 15 | | 20 | | 15 | | 20 | |
| | NLOS | LOS | NLOS | LOS | NLOS | LOS | NLOS | LOS |
| Mean Excess Delay | 1.60E−08 | 5.19E−09 | 2.01E−08 | 1.05E−08 | 2.36E−09 | 5.52E−10 | 5.59E−09 | 1.22E−09 |
| Max Excess Delay | 6.57E−08 | 2.84E−08 | 7.86E−08 | 5.68E−08 | 1.61E−08 | 2.65E−09 | 4.31E−08 | 1.24E−08 |
| RMS Delay Spread | 1.37E−08 | 5.41E−09 | 1.62E−08 | 8.50E−09 | 3.27E−09 | 7.53E−10 | 7.09E−09 | 1.70E−09 |
| Number of Paths | 72.8415 | 24.2753 | 153.9571 | 64.5884 | 28.7333 | 6.4188 | 99.1556 | 15.7607 |
| Inverted Paths | 49.00% | 47.61% | 49.30% | 48.68% | 50.71% | 39.54% | 49.81% | 43.93% |
| Inverted Energy | 44.23% | 45.02% | 45.36% | 45.63% | 34.26% | 24.19% | 37.67% | 25.97% |

## 3.5.2   Spatial Fading

Spatial fading (sometimes called local fading) refers to the variation in received power over a local area. To define spatial fading we first must define the metric to be examined. For spatial fading in UWB we will examine the entire received signal energy. Note that this is equivalent to comparing received power for a fixed observation interval. The total received energy $\varepsilon_{i,j}^{l}$ at a position $(i, j)$ of location $l$ is calculated as

$$\varepsilon_{i,j}^{l} = \int_{0}^{T} \left| r_{i,j}^{l}(t) \right|^{2} dt \qquad (3.68)$$

where $r_{i,j}^{l}(t)$ is the received signal at the grid location, $(i, j)$ at the measurement location $l$, and $T$ is the observation duration. A local fade at a location $l$ and position $(i, j)$ can be defined as [50]

$$F_{i,j}^{l}[dB] = 10 \log_{10} \left( \varepsilon_{i,j^{l}} \right) - 10 \log_{10} \left( \varepsilon_{ref} \right) \qquad (3.69)$$

where $\varepsilon_{ref}$ is defined as the energy in the LOS path (that is, excluding multipath) measured by the receiver at a reference distance of 1 m from the transmitter. An example measurement local area (that is, location) is shown in Figure 3.28. The example uses 49 points equi-spaced over an area of approximately 1 m$^2$ (15 cm separation between consecutive grid points).

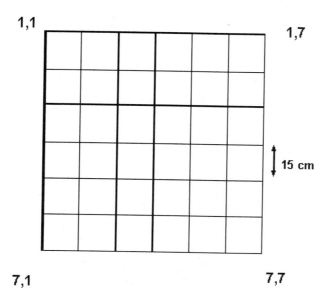

7 x 7 Grid with 49 Measurement Points

Figure 3.28: Typical Spatial Fading Measurements Grid Consisting of 29 Points.

First- and second-order statistics of the total received energy can be calculated for each location as follows. The sample mean is calculated as

$$\widehat{\mu}_l = \frac{1}{N} \sum_{i,j} F_{i,j}^l \tag{3.70}$$

while the sample standard deviation is calculated as

$$\widehat{\sigma}_l = \sqrt{\frac{1}{N-1} \sum_{i,j} \left(F_{i,j}^l - \widehat{\mu}_l\right)^2}$$

Example cumulative distribution functions of received signal energies are plotted for six locations in Figure 3.29, taken over an approximately 1 m$^2$ measurement grid consisting of 49 points [72]. From the plots, we can clearly see the immunity of UWB signals toward local fading. This lack of variation was first shown in [50] and

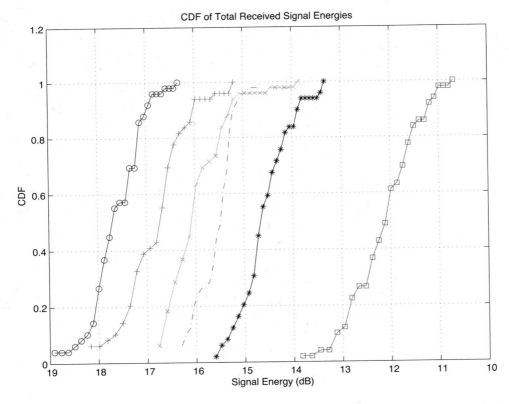

Figure 3.29: Example Estimated Cumulative Distribution for the Total Received Signal Energy at Different Distances—Very Little Fading is Exhibited at Each Location.

is in dramatic contrast to the spatial fading observed in narrowband signals [26]. Thus, we note that signal fading is significantly reduced as compared to traditional narrowband systems. This means that diversity mechanisms are not as necessary in UWB systems, and fading margins can be substantially reduced.

### 3.5.3    Spatial Fading of Signal Components

As mentioned previously, a classic receiver structure for channels with resolvable multipath is the RAKE receiver [43]. Because a RAKE receiver has a limited number of correlators and thus only captures a fraction of the energy, the fading seen by a RAKE receiver may be very different from the fading observed over the entire signal. This is an important consideration. While the total received signal energy may exhibit little variation, this may be irrelevant if a receiver cannot capture the entire received signal energy. Thus, it is important to understand the variation in portions of the received signal. Specifically, one should characterize the fading of the dominant signal components. This would correspond to the fading observed by a RAKE receiver with a limited number of correlators. The immunity to fading seen in the total received signal energy may be mitigated when only a portion of the received signal is captured.

In particular, examining the statistics of a single time delay bin provides insight into the performance of a single finger RAKE receiver collecting energy at a specific delay as it moves through the environment. As an example analysis, consider the time delay bin of the received signal with the highest average energy (over the spatial grid) and the statistics for that bin calculated as discussed in [11], [12], [42], and [72]. An example measured CDF [72] for a Gaussian pulse of approximate duration 200 ps is shown in Figure 3.30. It is clearly observed that the variance of the received signal energy for a single dominant component (that is, a 1 finger RAKE receiver) is much higher than the variance of the total energy. An example set of measurements for a larger number of dominant signal components is presented in Table 3.13 and Figure 3.31 (taken from [72]). We can see that while the variance of the total received signal can be extremely small (~0.5 dB), the variance of a single dominant component is more pronounced (~3 dB). Adding additional components quickly reduces the variation. Again, the spatial model must account for this variation, especially when RAKE receiver structures are to be examined.

### 3.5.4    Spatial Correlation

Spatial correlation can be defined in various ways, but a standard definition is the amount of correlation between received waveforms separated by some distance, $d$. Spatial correlation is an important parameter of the channel when using multiple antennas at the transmitter or receiver. The use of multiple antenna arrays is a classic technique in communications for combating mulitpath fading, obtaining SNR improvements, reducing the impact of interference, and more recently, for increasing the achievable channel capacity. Antenna technologies such as diversity techniques and multiple input multiple output (MIMO) applications require very low spatial correlation, whereas beamforming applications require high spatial

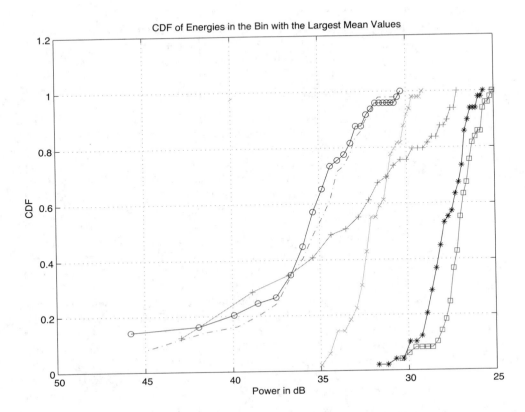

Figure 3.30: Example Cumulative Distribution Functions of Received Signal Energy for a Single-Finger RAKE Receiver Measured Over a Local Area at Various Distances [72].

Table 3.13: Standard Deviation for Energies Collected by RAKE Receiver with Different Fingers for 200 ps Gaussian Pulse.

| Fingers | Standard Deviation (dB) |
|---------|-------------------------|
| 1 | 3.67 |
| 2 | 2.81 |
| 5 | 2.18 |
| 10 | 1.67 |
| 15 | 1.43 |
| Entire Signal | 0.58 |

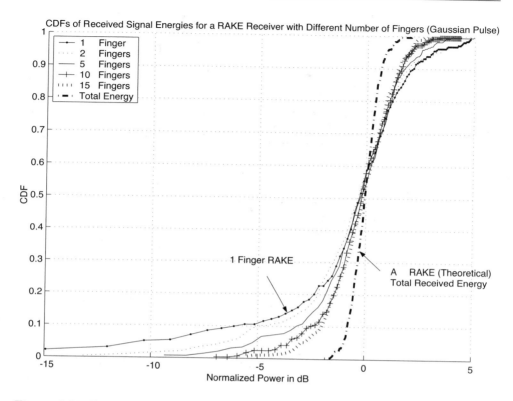

Figure 3.31: Example Cumulative Histograms for RAKE Receivers with Multiple Fingers for the Gaussian Pulse. (Measurements for Each Pulse are Normalized to Unit Average Energy.)

correlation. Several researchers have investigated the spatial correlation of UWB channels [11, 12, 25, 42, 72]. On the surface, one would expect that low spatial fading would correspond to high spatial correlation. However, this is not the case. UWB channels generally exhibit low spatial correlation.

A number of impulse response profiles collected in the same local area may be similar because the main features of the channel are essentially similar within the local area. However, due to the fine resolution of UWB pulses, the relative delays between paths observed by different antennas can be quite different. This can be true even for small spatial movements. If the multipath components arrive from a single direction, we would expect that the received signal at one antenna would simply be a time shifted version of the signal seen at a second antenna. However, if the received signal is coming from several directions, the signal seen at a neighboring antenna will be very different. This can be understood by examining Figure 3.32, which demonstrates an example receive geometry for three antennas and a single multipath component. Figure 3.33 presents an example of a measured

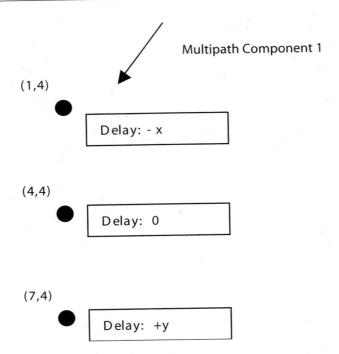

Figure 3.32: Illustration of Delays at Three Different Points on the Grid Referenced to the Center of the Grid.

received signal at the three locations shown in Figure 3.32. It can be seen from the shift in the profiles that (1,4) is closest to the transmitter and (7,4) is farthest away. As the signal propagates across the three sensors, the main signal components are simply shifted in time, as they come from a single direction. However, the weaker components clearly come from various directions and thus do not show a simple time shift.

Spatial fading is dependent on the entire received signal, which is the sum of many resolvable multipath components. Because the interaction between the components is limited due to the short duration of the pulses used, the fading is limited. However, the relative positions of the paths can change substantially over a short distance, leading to low spatial correlation despite the low spatial variation.

It was noted in [72] and [79] that the spatial correlation in UWB signals is very low when considering the entire received signal. However, it was found that the spatial correlation is significantly higher in the early part of the received signal than in the entire received signal. This implies that the early part of the signal is dominated by components arriving from the same direction, whereas the components in the later part of the received signal come from several directions. This leads to low correlation in the later-arriving components of the received signal. Because this

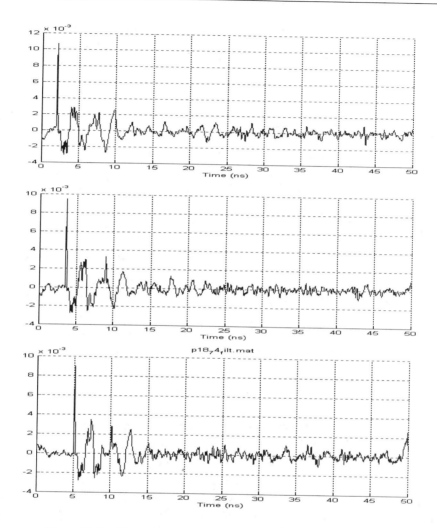

Figure 3.33: Sample LOS Signal at (1,4), (4,4), and (7,4).

correlated portion was relatively short (on the order of 20 ns), the overall signal correlation was heavily influenced by the second part of the received signal.

Figure 3.34 plots an example set of correlation measurements taken over a 1 m² grid with a 200 ps pulse [72]. The example demonstrates that there is a marked difference in spatial correlation between the first arriving components and the later arriving components. From the plots we can see that the initial portion of the received pulse can be highly spatially correlated (relatively). This means that the direct path and the first few multipath components exhibit high correlation. This, however, does not extend to the entire profile. The longer the time duration of the

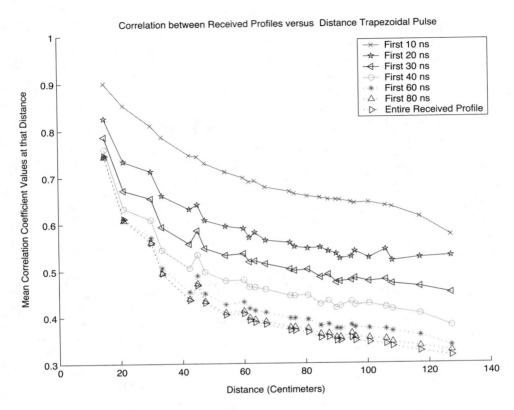

Figure 3.34: Correlation Coefficients Versus Distance for Different Lengths of the Profile (Gaussian Pulse).

received profile used for correlation, the smaller the coefficient values. Thus, we can conclude that while UWB signals as a whole exhibit low spatial correlation, portions of the signal can exhibit strong spatial correlation. This has some impact on the use of diversity arrays. If a RAKE receiver is used that collects only a portion of the signal energy, it is possible that diversity antennas will be ineffective. Again, any channel model used to simulate UWB signals should take this into account.

### 3.5.5    A Two-Dimensional Channel Model for UWB Indoor Propagation

Most proposed statistical models for the UWB indoor multipath channel include characteristics related only to the time-of-arrival (TOA). Notable exceptions are [11], [12], [42] [46], [47], and [72], which present statistics for the Angle-of-Arrival (AOA) in addition to the TOA for a collection of UWB indoor channel measurements. In order to use statistical models in simulating or analyzing the performance of systems employing spatial diversity combining, MIMO, or other multiantenna techniques, information about AOA statistics is required in addition to TOA information. Ideally, it would be desirable to characterize the full space time nature of the channel.

In this section, we discuss possible spatial models in terms of TOAs and AOAs of multipath components. Two-dimensional models have been previously proposed for wideband channels, but few models have been presented for UWB channels [11, 72]. The standard approach is to use the temporal models discussed previously and add AOA information. In [9], a model was developed that modeled cluster AOA and TOA independently. Each cluster from the Saleh-Valenzuela model is assumed to come from a separate direction and be uniformly distributed over $[0, 2\pi]$. In [72] and [76] it is proposed that the Split-Poisson model be extended by adding AOA information to the two clusters. To illustrate the basic ideas behind spatial modeling, we will discuss the models of [72] and [76] further.

From the measured indoor bicone NLOS data it was observed in [72] that the following trends emerged in the distribution of the AOA of multipath components:

1. The AOAs corresponding to the (stronger) earliest arriving paths (first 20 ns) appear to arrive principally from the same direction, as shown in Figures 3.35 and 3.36. Therefore, the AOAs of the first cluster were modeled using a Laplacian distribution with a mean value equal to the LOS direction.

2. The AOAs corresponding to the subsequent diffuse multipath do not appear to be concentrated in any specific direction. Thus, the second cluster AOAs were modeled using a uniform distribution on $[0, 2\pi]$.

The mean and variance of the Laplacian AOA distribution for the first cluster were determined empirically using measurement data. The two-dimensional CLEAN algorithm [11, 72] was used to determine the TOA and AOA of each mulitpath component.

These observations motivated an extension of the Split-Poisson channel model into a two-dimensional (space time) model that also takes the AOA information into account. In order to account for the spatial dimension in this model, each path in the first cluster is assigned an AOA taken from a Laplacian distribution (the mean corresponds to the LOS direction, and the standard deviation was found to be approximately five degrees). Each path in the second cluster is assigned an AOA taken from a uniform distribution over $[0°, 360°]$ as shown in Figure 3.37.

The model was compared with measurements in [72] and [76] in order to examine the performance of the two-dimensional model. The spatial correlation of both the measured and model received signals over distance were examined. The correlation was calculated separately for the early and late arriving paths. The pulse used to generate the received signals was a 500 ps Gaussian pulse. Figure 3.38 shows the results of this comparison. A good match between the model and the measured data was observed. The correlation is stronger in the first cluster because the variance of the AOAs is small. The second cluster, however, shows little spatial correlation since the paths come from $360°$. Note that a similar model was proposed for 802.15.4a [77]. Specifically, the SV model was proposed for time-of-arrival modeling. Individual

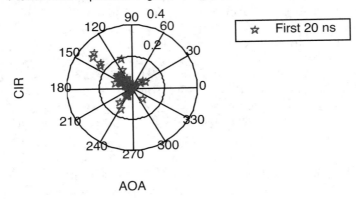

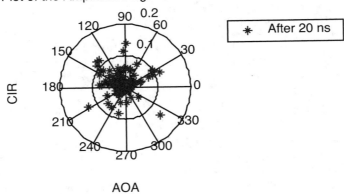

Figure 3.35: Example AOA Distribution for the First 20 ns and After 20 ns (Polar Plot).

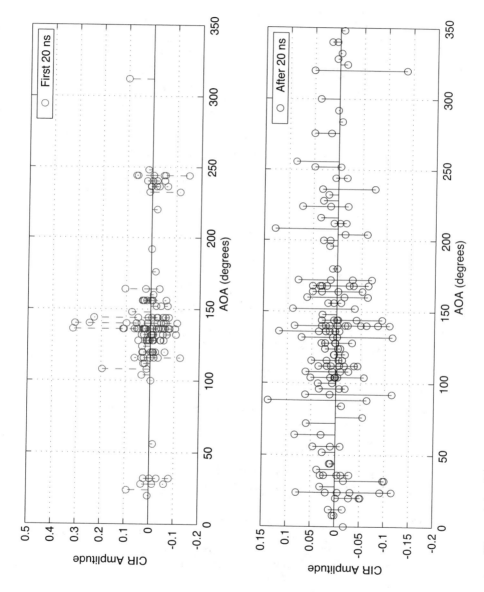

Figure 3.36: Example AOA Distribution for the First 20 ns and After 20 ns (Linear Plot).

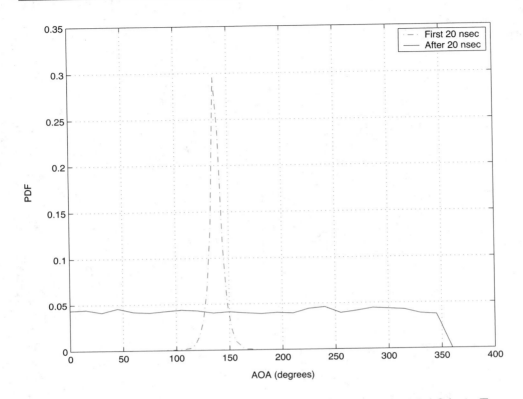

Figure 3.37: Example Laplacian and Uniform Distributions to Model AOAs in Two Cluster (Split-Poisson) Model.

clusters are modeled as uniformly distributed, while the angular power spectrum is modeled using a Laplacian distribution.

## 3.6   Impact of Frequency Distortion on Discrete Channel Modeling

One major assumption with the traditional wideband channel model is that the received signal is simply the sum of scaled and time shifted versions of the LOS pulse (see Equation (3.46)). In general, this is not the case for UWB signals. Instead, the received signal will be the sum of distorted and time shifted pulses. Pulse distortion occurs due to the pulse passing through materials that are frequency dependent, from diffraction [62] or from the antenna itself (that is, the pulse emitted by the antenna is not necessarily consistent over all transmit/receive angles).

In this section, we examine the impact of using a discrete-tap channel model in the presence of per path distortion or other channel effects which violate the

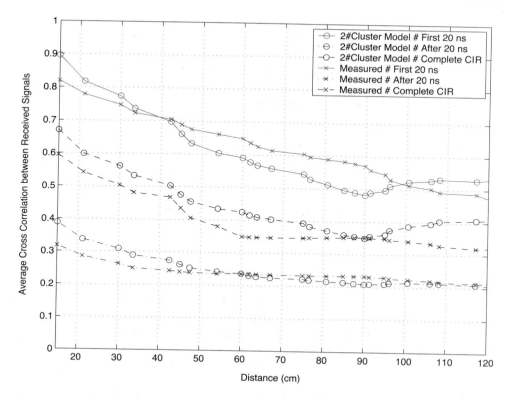

Figure 3.38: Example Results of the Spatial UWB Channel Model Compared to Measurement Data. The Average Correlation Coefficient between Adjacent Signals in the Profile Is Plotted Versus the Distance between the Points.

SOURCE: S. Venkatesh, J. Ibrahim, and R. M. Buehrer, "A New 2-Cluster Model for Indoor UWB Channel Measurements," submitted to *IEEE Transactions on Communications* [76]. © IEEE, 2004. Used by permission.

assumption that the received signal is simply a scaled and time shifted version of the LOS signal [74].

## 3.6.1   The CLEAN Algorithm

In obtaining a UWB channel model, often the first step is to extract the channel impulse response (CIR) from individual measurements. The CLEAN algorithm, first introduced in [33] and well-established in the radio astronomy and microwave communities, is often applied to UWB measurements [34–36]. The CLEAN algorithm is often preferred because of its ability to produce a discrete CIR in time.

In other words, it assumes that the channel is a series of impulses, consistent with the tapped-delay line channel model [8].

To extract the CIR from measurements of the received waveform $y(t)$, we first need to take an LOS measurement to obtain the template for the CLEAN algorithm. The algorithm searches the received waveform iteratively with the template to find the maximum correlation [5]. The steps involved are:

1. Calculate the autocorrelation of the template $r_{ss}(t)$ and the cross-correlation of the template with the received waveform $r_{sy}(t)$.

2. Find the largest correlation peak in $r_{sy}(t)$, record the normalized amplitudes $\alpha_k$ and the relative time delay $\tau_k$ of the correlation peak.

3. Subtract $r_{ss}(t)$ scaled by $\alpha_k$ from $r_{sy}(t)$ at the time delay $\tau_k$.

4. If a stopping criterion (a minimum threshold on the peak correlation) is not met, go to step 2. Otherwise, stop.

### 3.6.2   Impact of Frequency Dependent Distortion

Due to the extremely wide bandwidth of UWB pulses, pulse distortion can occur when a pulse passes through materials that are frequency dependent. Several materials were characterized in [60] and [62]. The impact on the output of the CLEAN algorithm is examined here. As an example, consider a pulse passing through plywood. The nondistorted 2 GHz-wide pulse is plotted in Figure 3.39. The received pulse after passing through a brick wall and the reconstructed pulse using the CLEAN algorithm are plotted in Figure 3.40 (left and right, respectively). As can be observed, the pulse is significantly distorted, and CLEAN represents the distortion very well. However, as shown in Figure 3.41, which plots the impulse response determined by CLEAN, it must use multiple channel taps to do this despite the fact that only one path exists. This clearly biases the number of "paths" reported. Similar plots for wallboard, cloth partition, a wooden door, concrete blocks, styrofoam, and plywood are given in [66]. In many cases, the frequency distortion is significant, resulting in many additional "paths" as interpreted by CLEAN. Thus, the discrete channel model can be used to represent path-specific distortion. However, the model must include "phantom paths" in order to accommodate the distortion. Care must be taken when using such a model when pulse distortion is expected because we must interpret the model output correctly.

### 3.6.3   Impact of Reflections

To examine the impact of reflections, measurements were taken in an anechoic chamber to characterize the distortion caused by reflections from common shapes [72]. The "free-space" signal was used as the template, and the distorted pulse as the received signal, to run the CLEAN algorithm. In general, the CLEAN algorithm generates multiple taps to represent the distorted pulse. It behaves like an FIR filter to approximate the frequency distortion of the pulse. As a result, the CLEAN algorithm is accurate at regenerating the distorted pulse, although it will again

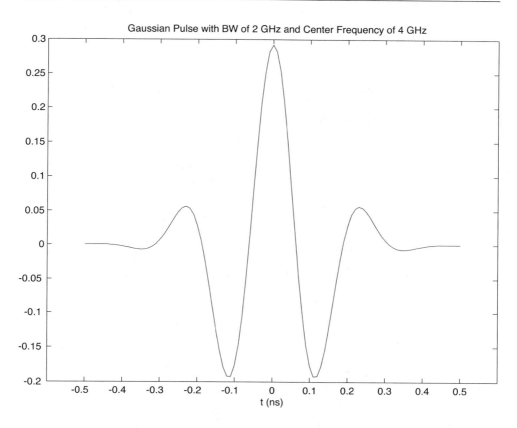

Figure 3.39: Simulated LOS Pulse.

create several "phantom paths." It is also noted that the angle of reflection also has a big impact on pulse distortion. The pulse reflected by the same material but at different angles can be quite different.

As an example, consider a pulse reflected by a hemisphere [72]. The transmit and receive antennas were both pointed at the hemisphere to ensure that the hemisphere was in the boresight of both antennas. Additionally, nonreflected (that is, direct) paths were removed by taking a measurement without the hemisphere present. The received pulses without the reflection and with the reflection are plotted in Figure 3.42. The reflection causes two effects. First, it inverts the pulse. Second, it also rings at the resonant frequency of the sphere. As a result, the received pulse is followed by a slow oscillation. While the "true" CIR is a single path, CLEAN will attempt to model the ringing by introducing several "phantom paths" as shown in Figure 3.43. Additionally, there is a limit to how well CLEAN can represent the ringing. After a number of iterations, the CLEAN algorithm cannot improve the accuracy using shifted and scaled versions of the template pulse. The use of the discrete model

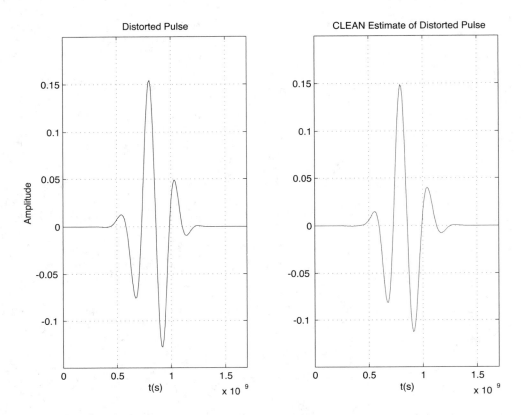

Figure 3.40: Distorted Pulse and CLEAN Estimated Pulse after Passing Through a Brick Wall.

SOURCE: B. Donlan and R. M. Buehrer, "The UWB Indoor Channel: Large and Small-Scale Modeling," submitted to *IEEE Transactions on Wireless Communications* [73]. © IEEE, 2004. Used by permission.

introduces additional multipath components in other cases as well. For example, the pulse shape transmitted or received at various angles will be different from that observed on the boresite. If the boresite pulse shape is assumed for channel modeling as described earlier, it will also cause "phantom paths" to represent pulses arriving at angles other than the boresite. However, these paths will generally have less energy due to the reduced antenna gain in those directions [41, 72].

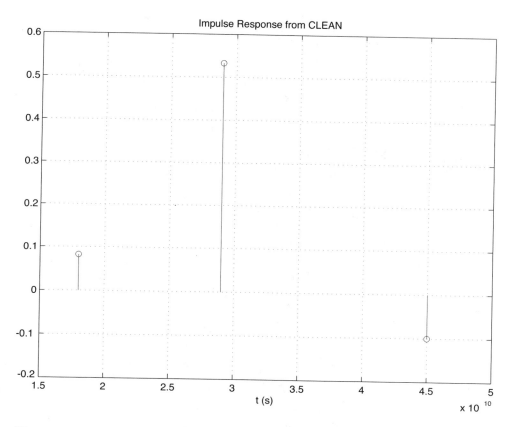

Figure 3.41: CLEAN Generated Impulse Response for Single Pulse Passing Through a Brick Wall.

SOURCE: B. Donlan and R. M. Buehrer, "The UWB Indoor Channel: Large and Small-Scale Modeling," submitted to *IEEE Transactions on Wireless Communications* [73]. © IEEE, 2004. Used by permission.

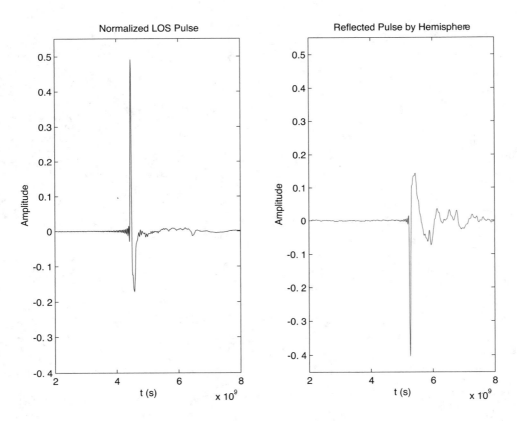

Figure 3.42: Signal Reflected by a Hemisphere.

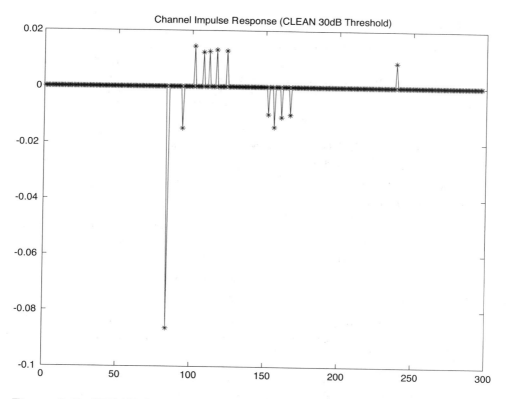

Figure 3.43: CLEAN-Generated Impulse Response for Signal Reflected by a Hemisphere.

## 3.7    Summary

In this chapter, we have examined statistical channel modeling for UWB signals, including large-scale and small-scale models, as well as spatial models. Additionally, we briefly examined the impact of the discrete channel model on UWB modeling, particularly in the presence of per path frequency distortion. We have shown that UWB allows for the application of traditional channel modeling approaches as long as small modifications are made to the large-scale path loss model (separation of antenna and path effects) and provided that care is taken in the interpretation of small-scale models (discrete paths may not be physical multipath components, but may simply represent pulse distortion).

# Bibliography

[1] A. F. Molisch, J. R. Foerster, and M. Pendergrass, "Channel Models for Ultra-wideband Personal Area Networks," *IEEE Wireless Communications Magazine*, vol. 10, no. 6, pp. 14-21, December 2003.

[2] H. L. Bertoni, *Radio Propagation for Modern Wireless Systems*, Upper Saddle River, NJ: Prentice Hall, 2000.

[3] J. D. Parsons, *The Mobile Radio Propagation Channel*, 2nd ed., New York, NY: John Wiley and Sons, 2000.

[4] W. Stutzman and G. Thiele, *Antenna Theory and Design*, New York, NY: John Wiley and Sons, 1981.

[5] R. M. Buehrer, A. Safaai-Jazi, W. A. Davis, and D. Sweeney, "Characterization of the UWB Channel," *Proc. IEEE Conference on Ultra Wideband Systems and Technologies*, pp. 26-31, Reston, VA, November 2003.

[6] B. Donlan and R. M. Buehrer, "The Indoor UWB Channel," *Proc. IEEE Vehicular Technology Conference—Spring*, Milan, Italy, May 2004.

[7] D. M. McKinstry and R. M. Buehrer, "UWB small-scale Channel Modeling and System Performance," *Proc. IEEE Vehicular Technology Conference—Fall*, pp. 6-10, Orlando, FL, September 2003.

[8] A. Alvarez, G. Valera, M. Lobeira, R. Torres, and J. L. Garcia "New Channel Impulse Response Model for UWB Indoor System Simulations," *Proc. Spring 2003 Vehicular Technology Conference*, pp. 1-5, Seoul, Korea, April 2003.

[9] D. Cassioli, M. Z. Win, and A. F. Molisch, "A Statistical Model for the UWB Indoor Channel," *Proc. IEEE Vehicular Technology Conference—Spring*, vol. 2, pp. 1159-1163, Rhodes, Greece.

[10] D. Cassioli, M. Z. Win, and A. F. Molisch, "The Ultra-Wide Bandwidth Indoor Channel: From Statistical Study to Simulations," *IEEE Journal on Selected Areas in Communications*, vol. 20, no. 6, pp. 1247-1257, August 2002.

[11] R. J.-M. Cramer, R. A. Scholtz, and M. Z. Win, "Spatio-Temporal Diversity in Ultra-Wideband Radio," *Proc. IEEE Wireless Communications and Networking Conference*, vol. 2, pp. 888-892, New Orleans, LA, 1999.

[12] R. J.-M. Cramer, R. A. Scholtz, and M. Z. Win, "Evaluation of an Ultra-Wide-Band Propagation Channel," *IEEE Transactions on Antennas and Propagation*, vol. 50, issue 5, pp. 561-570, May 2002.

[13] R. J.-M. Cramer, R. A. Scholtz, and M. Z. Win, "Evaluation of an Indoor Ultra-Wideband Propagation Channel" IEEE document no. P802.15-02/286-SG3a and P802.15-02/325-SG3a), Available at http://grouper.ieee.org/groups/802/15/pub/2002/Jul02/.

[14] J. Foerster, "Channel Modeling Sub-Committee Final Report," IEEE document no. 802-15-02/490r1-SG3a), Available at http://grouper.ieee.org/groups/802/15/pub/2002/Nov02/.

[15] J. Foerster and Q. Li, "UWB Channel Modeling Contribution from Intel," (doc: IEEE P802.15-02/279-SG3a), Available at http://grouper.ieee.org/groups/802/15/pub/2002/Jul02/.

[16] S. S. Ghassemzadeh, L. J. Greenstein, A. Kavcic, T. Sveinsson, and V. Tarokh, "UWB Indoor Path Loss Model for Residential and Commercial Buildings," *Proc. IEEE Vehicular Technology Conference—Fall*, vol. 5, pp. 3115-3119, 2003.

[17] S. S. Ghassemzadeh, R. Jana, C. W. Rice, W. Turin, and V. Tarokh, "A Statistical Path Loss Model for In-Home UWB Channels," *Proc. IEEE Conference on Ultra Wideband Systems and Technologies*, pp. 59-64, Baltimore, MD, May 2002.

[18] S. S. Ghassemzadeh and V. Tarokh, "UWB Path Loss Characterization In Residential Environments," *Proc. IEEE Radio Frequency Integrated Circuits Symposium*, pp. 501-504, June 2003.

[19] H. Hashemi, "The Indoor Radio Propagation Channel," *Proc. IEEE*, vol. 81, no. 7, pp. 943-968, July 1993.

[20] V. Hovinen et al., "A Proposal for a Selection of Indoor UWB Path Loss Model," IEEE document no. 02280r1P802.15, Available at http://grouper.ieee.org/groups/802/15/pub/2002/Jul02.

[21] J. Keignart and N. Daniele, "Subnanosecond UWB Channel Sounding in Frequency and Temporal Domain," *IEEE Conference on Ultra Wideband Systems and Technologies*, pp. 25-30, Baltimore, MD, May 2002.

[22] J. Keignart, J. B. Pierrot, N. Daniele, and P. Rouzet, "UWB Channel Modeling Contribution from CEA-LETI and ST Microelectronics," IEEE document no. P802.15-02/444-SG3a, Available at http://grouper.ieee.org/groups/802/15/pub/2002/Nov02/.

[23] J. Keignart and N. Daniele, "Channel Sounding and Modeling for Indoor UWB Communications," *Proc. International Workshop on Ultra Wideband Systems*, 2003.

[24] P. Pagani, P. Pajusco, and S. Voinot, "A Study of the Ultra-Wideband Indoor Channel: Propagation Experiment and Measurement Results," in COST273., TD(030)060, January 2003.

[25] C. Prettie, D. Cheung, L. Rusch, and M. Ho, "Spatial Correlation of UWB Signals in a Home Environment," *Proc. IEEE Conference on Ultra Wideband Systems and Technology*, pp. 65-69, Baltimore, MD, 2002.

[26] T.S. Rappaport, *Wireless Communications: Principles and Practice*, 2nd ed., Upper Saddle River, NJ: Prentice Hall, 2002.

[27] P. Bartolome and G. Vellejo, "Site Measurements for Installation of an Indoor Radio Communication System," *Proc. IEEE Vehicular Technology Conference*, pp. 57-60, Secaucus, NJ, May 1993.

[28] M. Steinbauer, A. F. Molisch, and E. Borek, "The Double-Directional Channel," *IEEE Antennas and Propagation Magazine*, vol. 43, no. 4, pp. 51-63, August 2001.

[29] J. A. Dabin, et al., "The Effects of Antenna Directivity on Path Loss and Multipath Propagation in UWB Indoor Channels," *Proc. IEEE Conference on Ultra Wideband Systems and Technologies*, pp. 305-309, Reston, VA, November 2003.

[30] L. Rusch, C. Prettie, D. Cheung, Q. Li, and M. Ho, "Characterization of UWB Propagation from 2 to 8 GHz in a Residential Environment," *IEEE Journal on Selected Areas in Communications*, submitted for publication.

[31] S. Y. Seidel and T. S. Rappaport, "914 MHz Path Loss Prediction Models for Indoor Wireless Communications in Multi-floored Buildings," *IEEE Transactions on Antennas and Propagation*, vol. 40, no. 2, pp. 207-217, February 1992.

[32] T. Zasowski, F. Althaus, M. Stager, A. Wittneben, and G. Troster, "UWB for Noninvasive Wireless Body Area Networks: Channel Measurements and Results," *Proc. IEEE Conference on Ultra Wideband Systems and Technologies*, pp. 285-289, Reston, VA, November 2003.

[33] J. A. Hogbom, "Aperture Synthesis with a Non-Regular Distribution of Interferometer Baselines," *Astronomy and Astrophysics Supplement Ser.*, vol. 15, 1974.

[34] R. J.-M. Cramer, "An Evaluation of Ultrawideband Propagation Channels," Ph.D. dissertation, Dept. of Electrical and Computer Engineering, University of Southern California, December 2000.

[35] K. Siwiak, H. Bertoni, and S. M. Yano, "Relation Between Multipath and Wave Propagation Attenuation," *Electronics Letters*, vol. 39, no. 1, pp. 142-143, January 2003.

[36] R. A. Scholtz, M. Z. Win, and J. M. Cramer, "Evaluation of the Characteristics of the Ultra-Wideband Propagation Channel," *Proc. Antennas and Propagation Society International Symposium*, vol. 2., pp. 626-630, 1998.

[37] P. Withington, R. Reinhardt, and R. Stanley, "Preliminary Results of an Ultrawideband (Impulse) Scanning Receiver," *Proc. IEEE Military Communications Conference*, vol. 2, pp. 1186-1190, 1999.

[38] G. L. Turin, F. D. Clapp, T. L. Johnston, S. B. Fine, and D. Lavry, "A Statistical Model of Urban Multipath Propagation," *IEEE Transactions on Vehicular Technology*, vol. VT-21, pp. 1-9, February 1972.

[39] A. A. Saleh and R. A. Valenzuela, "A Statistical Model for Indoor Multipath Propagation," *IEEE Journal on Selected Areas in Communications*, vol. SAC-5, no. 2, pp. 128-137, February 1987.

[40] H. Suzuki, "A Statistical Model for Urban Radio Propagation," *IEEE Transactions on Communications*, vol. COM-25, no. 7, pp. 673-680, July 1977.

[41] D. McKinstry, "Ultra-Wideband small-scale Channel Modeling and its Application to Receiver Design," Master's Thesis, Dept. of Electrical and Computer Engineering, Virginia Tech, 2002.

[42] M. Z. Win and R. A Scholtz, "Characterization of Ultra-Wide Bandwidth Wireless Indoor Channels: A Communication—Theoretic View," *IEEE Journal on Selected Areas in Communications*, vol. 20, no. 9, December 2002.

[43] R. Price and P. Green, "A Communication Technique for Multipath Channels," *Proc. IRE*, vol. 46, pp. 555-570, March 1958.

[44] H. Hashemi, "Impulse Response Modeling of Indoor Radio Propagation Channels," *IEEE Journal on Selected Areas in Communications*, vol. 11, no. 7, pp. 967-978, September 1993.

[45] H. Hashemi, "The Indoor Radio Propagation Channel," *Proc. of the IEEE*, vol. 81, no. 7, pp. 943-968, July 1993.

[46] R. J.-M. Cramer, M. Z. Win, and R. A. Scholtz, "Impulse Radio Multipath Characteristics and Diversity Reception," *Proc. IEEE International Conference on Communications*, pp. 1650-1654, Atlanta, GA, June 1998.

[47] R. J.-M. Cramer, M. Z. Win, and R. A. Scholtz, "Evaluation of the Multipath Characteristics of the Impulse Radio Channel," *Proc. IEEE International Symposium On Personal, Indoor and Mobile Radio Communications*, vol. 2, pp. 864–868. 1998.

[48] M. Z. Win, F. Ramirez-Mireles, R. A. Scholtz, and M. A. Barnes, "Ultra-wide Bandwidth (UWB) Signal Propagation for Outdoor Wireless Communications," *IEEE Vehicular Technology Conference*, vol. 1, pp. 251-255, 1997.

[49] R. A. Scholtz and M. Z. Win, "Impulse Radio," *Personal Indoor Mobile Radio Conference*, Invited Tutorial, September 1997.

[50] M. Z. Win, R. A. Scholtz, and M. A. Barnes, "Ultra-wide Bandwidth Signal Propagation for Indoor Wireless Communications," *IEEE International Conference on Communications*, vol. 1, pp. 56-60, 1997.

[51] J. Kunisch and J. Pamp, "Measurement Results and Modeling Aspects for the UWB Radio Channel," *IEEE Conference on Ultra Wideband Systems and Technologies*, pp. 19-23, 2002.

[52] S. M. Riad, "The Deconvolution Problem, an Overview," *Proc. of the IEEE*, vol. 74, no. 1, pp. 82-85, January 1986.

[53] R. G. Vaughan and N. L. Scott, "Super-Resolution of Pulsed Multipath Channels for Delay Spread Characterization," *IEEE Transactions on Communications*, vol. 47, no. 3, pp. 343-347, March 1999.

[54] A. Bennia and S. M. Riad, "Filtering Capabilities and Convergence of the Van-Cittert Deconvolution Technique," *IEEE Transactions on Instrumentation and Measurement*, vol. 41, no. 2, pp. 246-250, April 1992.

[55] S. M. Yano, "Investigating the Ultra-wideband Indoor Wireless Channel," *IEEE Vehicular Technology Conference*, vol. 3, pp. 1200-1204, 2002.

[56] M. Pendergrass and W. Beeler, "Emperically Based Statistical Ultra-Wideband (UWB) Channel Model," IEEE document no. IEEE 802-15-02/240SG3a, Available at http://grouper.ieee.org/groups/802/15/pub/2002/Jul02/.

[57] P. Odling, O. Borjesson, T. Magesacher, and T. Nordstrom, "An Approach to Analog Mitigation of RFI," *IEEE Journal on Selected Areas in Communications*, vol. 20, issue 5, pp. 974-986, June 2002.

[58] W. Turin, R. Jana, S. S. Ghassemzadeh, C. W. Rice, and V. Tarokh, "Autoregressive Modeling of an Indoor UWB Channel," *IEEE Conference on Ultra Wideband Systems and Technologies*, pp. 71-74, Baltimore, MD, May 2002.

[59] V. S. Somayazulu, J. R. Foerster, and S. Roy "Design Challenges for Very High Data Rate UWB Systems," *Signals, Systems and Computers, 2002, Conference Record of the Thirty-Sixth Asilomar Conference*, vol. 1, no. 3-6, pp. 717-721, November 2002.

[60] D. Kralj and L. Carin, "Ultra-Wideband Characterization of Lossy Materials: Short-pulse Microwave Measurements," *Microwave Symposium Digest, 1993. IEEE MTT-S International*, vol. 314-18, pp. 1239-1242, June 1993.

[61] G. D. Durgin, T. S. Rappaport, and H. Xu, "Partition-Based Path Loss Analysis for In-Home and Residential Areas at 5.85 GHz," *Global Telecommunications Conference*, vol. 2, no. 8-12, pp. 904-909, November 1998.

[62] A. Muqaibel, A. Safaai-Jazi, A. Bayram, and S. M. Riad, "Ultra Wideband Material Characterization for Indoor Propagation," *Antennas and Propagation Society International Symposium*, vol. 4, pp. 623-626, June 2003.

[63] K. Siwiak and A. Petroff, "A Path Link Model for Ultra Wide Band Pulse Transmissions," *Proc. IEEE Vehicular Technology Conference—Spring*, pp. 1173-1175, May 2001.

[64] P. L. Rice, A. G. Longley, K. A. Norton, and A. P. Barsis, "Transmission Loss Predictions for Tropospheric Communication Circuits," *NBS Tech Note 101*, Two volumes: issued May 7, 1965; revised May 1, 1966; revised January 1967.

[65] T. Okumura, E. Ohmori, and K. Fukuda, "Field Strength and its Variability in VHF and UHF Land Mobile Service," *Review of the Electrical Communication Laboratory*, vol. 16, no. 9-10, pp. 825-873, September-October 1968.

[66] M. Hata, "Empirical Formula for Propagation Loss in Land Mobile Radio Services," *IEEE Transactions on Vehicular Technology*, vol. VT-29, no. 3, pp. 317-325, August 1980.

[67] D. Akerberg, "Properties of a TDMA Picocellular Office Communication System," *Proc. IEEE Global Communications Conference*, pp. 1343-1349, December 1988.

[68] K. Haneda and J. Takada, *UWB Indoor Propagation Channel Measurement Based on Deterministic Approach,* Department of International Development Engineering, Tokyo Institute of Technology. Available: http://www.mobile.ss.titech.ac.jp/~haneda/ps/03uwbst_p.pdf.

[69] R. C. Qiu, "A Study of the Ultra-Wideband Wireless Propagation Channel and Optimum UWB Receiver Design," *IEEE Journal on Selected Areas in Communications*, vol. 20, no. 9, December 2002.

[70] A. Alvarez, G. Valera, M. Lobeira, R. Torres and J.L. Garcia, "New Channel Impulse Response Model for UWB Indoor System Simulations," *IEEE Vehicular Technology Conference—Spring*, vol. 1, pp. 22-25, April 2003.

[71] J. Kunisch and J. Pamp, "An Ultra-Wideband Space-Variant Multipath Indoor Radio Channel Model." Available: http://www.whyless.org/files/public/WP5_uwbst2003_jk.pdf.

[72] R. M. Buehrer, W. A. Davis, A. Safaai-Jazi, and D. Sweeney, "Ultra-Wideband Propagation Measurements and Modeling," DARPA NETEX Program Final Report, January 31, 2004. Available: http://www.mprg.org/people/buehrer/ultra/darpa_netex.shtml.

[73] B. Donlan and R. M. Buehrer, "The UWB Indoor Channel: Large and Small-Scale Modeling," submitted to *IEEE Transactions on Wireless Communications*, July 2004.

[74] J. A. Dabin, N. Ni, A. M. Haimovich, E. Niver, and H. Grebel, "The Effects of Antenna Directivity on Path Loss and Multipath Propagation in UWB Indoor Wireless Channels," *Proc. UWBST 2003*, 2003.

[75] S. Venkatesh, J. Ibrahim, and R. M. Buehrer, "A New 2-Cluster Model for Indoor UWB Channel Measurements," *IEEE International Symposium on Antennas and Propagation*, vol. 1, pp. 946-949, June 2004.

[76] S. Venkatesh, J. Ibrahim, and R. M. Buehrer, "A New 2-Cluster Model for Indoor UWB Channel Measurements," submitted to *IEEE Transactions on Communications*, June 2004.

[77] A. Molisch, "Status of Channel Modeling," IEEE document no. P802.15-04-0346-00-004a/r0, Available at ftp://ftp.802wirelessworld.com/15/04.

[78] R. M. Buehrer, W. A. Davis, and S. Licul, "Link Budget Design for UWB Systems," submitted to *IEEE Transactions on Communications*, June 2004.

[79] V. Bharadwaj and R. M. Buehrer, "Spatial Fading Characteristics of UWB Signals in Indoor Environments," submitted to *IEEE Transactions on Communications*, September 2004.

[80] V. Bharadwaj and R. M. Buehrer, "Spatial Diversity for SIR Improvement in UWB Systems," to appear *IEEE Communications Letters*, January 2005.

[81] L. Yang and R. M. Buehrer, "On the Impact of Discrete Channel Modeling on UWB Systems," submitted to *IEEE Transactions on Wireless Communications*, September 2004.

[82] S. S. Ghassemzadeh, L. J. Greenstein, T. Sveinsson, and V. Tarokh, "An Impulse Response Model for Residential Wireless Channels," *Global Telecommunications Conference 2003*, vol. 3, pp. 1211-1215, Dec. 1-5, 2003.

[83] S. S. Ghassemzadeh, L. J. Greenstein, T. Sveinsson, and V. Tarokh, "A Multipath Intensity Profile Model for Residential Environments," *Wireless Communications and Networking 2003*, vol. 1, pp. 150-155, March 16-20, 2003.

# Chapter 4

# ANTENNAS

William A. Davis
and Stanislav Licul

Ultra wideband (UWB) systems are limited by the response capabilities of the antennas used in the systems. This chapter emphasizes the modeling of both the antenna and scattering structures that must be considered in the performance estimates of UWB technology and links.

Classically, communication links have been modeled by the Friis transmission formula [1]

$$P_{rcv} = P_{xmt} \frac{G_t G_r \lambda^2}{(4\pi r)^2} \tag{4.1}$$

The basic concepts of antenna gain, $G$, and effective antenna aperture, $G\lambda^2/4\pi$, are difficult to define in a wideband sense, and new concepts of transmission are needed for understanding the transient behavior of UWB systems. The transient performance may be related to equivalent frequency performance metrics, and insight may be obtained by viewing both time and frequency domain representations.

Because the concepts of power and gain are frequency dependent, it becomes useful to look at the signal performance rather than power transmission, particularly for correlation reception applications. From antenna theory, we can define the effective length of an antenna, shown in Figure 4.1a, in terms of the current distribution on the antenna due to an input current $I_t$ at a given frequency as [2]

$$\mathbf{h}(\theta, \phi; \omega) = -\frac{\widehat{\mathbf{r}} \times \widehat{\mathbf{r}} \times}{I_t} \int_V \mathbf{J}(\mathbf{r}') \, e^{j\beta \widehat{\mathbf{r}} \cdot \mathbf{r}'} dv' \tag{4.2}$$

where the frequency is included in both the current distribution $\mathbf{J}$ and the wave number $\beta = 2\pi/\lambda$, while the direction of the radiation is included in the properties of the unit vector $\widehat{\mathbf{r}}$. The polarization properties are included in the vector nature of the current $\mathbf{J}$. The typical use of a conjugation of $\mathbf{h}$ in (4.2) has not been used to simplify the connection with the transient performance. Normally, effective length is

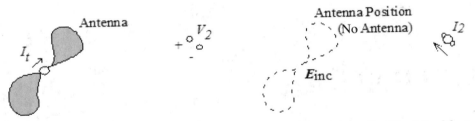

(a) The Transmit Problem. Receive        (b) The Incident Wave Problem. Incident
    Voltage at Test Antenna                   Field Due to Current into Test Antenna.

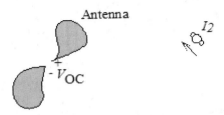

(c) The Open-Circuit Receive Problem. Open-Circuit
    Voltage Due to Current into Test Antenna.

Figure 4.1: The Three Problems Used in the Reciprocity Development of a Communications Link.

defined for linearly polarized antennas, but the form of (4.2) provides a general form for arbitrary antenna polarization. In the following sections, appropriate definitions are developed for this effective length and are related to commonly used antenna parameters. These definitions are followed by the measured performance of a variety of antennas, as well as a simple characterization for the antennas. Many aspects such as gain, pulse width, pulse amplitude, and a minimal model for the antenna properties are presented.

This chapter presents the basic properties of antennas from both frequency and time domain viewpoints. The effective length is developed and related to the need in transient link computations. Section 4.2 presents methods for representing antennas and scatterers in a minimal fashion. The basic concept is a pole-residue representation based on the theory of the singularity expansion method [18]. Several examples are included to demonstrate the modeling for a minimal representation of the antenna and scattering structures.

## 4.1  Basic Properties of Antennas

This section identifies the classic terms used to characterize antennas. We begin with a discussion of antenna effective length and the link equation based on reciprocity.

Some extra coverage is provided for the use of reciprocity due to misuse of the term in communications.

## 4.1.1  Reciprocity and Antenna Effective Length

Reciprocity is often incorrectly used by communications engineers to claim that the receive and transmit properties of an antenna are identical. The fundamental error in these arguments relates to neglecting a test-probe antenna used to measure the radiated electric field. To obtain the desired relationships, a bit more care is required. Three problems are needed to establish the reciprocity relationships required in the development of gain. First, we must establish a transmit problem (Figure 4.1a) to interact with the receive problem of interest (Figure 4.1c). We must establish an additional receive problem defining the incident field (Figure 4.1b).

The $I_2$ current of Figures 4.1(b) and (c) represents an infinitesimal dipole located at the same position as the measured radiated voltage, $V_2$, in Figure 4.1(a). We may simply write the reciprocity theorem for any two problems as [2]

$$\int_V \mathbf{J}_1 \cdot \mathbf{E}_2 \, dv = \int_V \mathbf{J}_2 \cdot \mathbf{E}_1 \, dv \tag{4.3}$$

where $\mathbf{J}$ and $\mathbf{E}$ represent current density and electric field intensity, respectively. Applying reciprocity to problems a and b, and also to problems a and c, we have

$$\int_{Antenna} \mathbf{J}_a \cdot \mathbf{E}_{inc} dv' = \int_{Test} \mathbf{J}_b \cdot \mathbf{E}_a dv = -I_2 V_2$$

$$\int_{Gap} I_t \mathbf{E}_c \cdot d\ell' = -I_t V_{oc} = \int_{Test} \mathbf{J}_b \cdot \mathbf{E}_a dv = -I_2 V_2 \tag{4.4}$$

Equating $(I_2 V_2)$ in both equations we have

$$V_{oc} = -\frac{1}{I_t} \int_{Antenna} \mathbf{J}_a (\mathbf{r}') \cdot \mathbf{E}_{inc} dv' \tag{4.5}$$

The details now involve expanding this form for the open-circuit voltage and relating all of the terms to obtain the expressions for the link.

We can write the incident field as a plane wave emanating from the $\hat{\mathbf{r}}$ direction as

$$\mathbf{E}_{inc} (\mathbf{r}') = \mathbf{E}_i (\theta, \phi) \, e^{i\beta \hat{\mathbf{r}} \cdot \mathbf{r}'} = -\hat{\mathbf{r}} \times \left( \hat{\mathbf{r}} \times \mathbf{E}_i (\theta, \phi) \, e^{i\beta \hat{\mathbf{r}} \cdot \mathbf{r}'} \right) \tag{4.6}$$

where $\mathbf{E}_i$ contains all of the angular information and is perpendicular to the direction $\hat{\mathbf{r}}$. The quantity $\beta = 2\pi/\lambda$ represents the propagation constant of space. This form allows the open-circuit voltage to be written in the simpler far-field form of

$$V_{oc} = \left[ \frac{1}{I_t} \int_{Antenna} \mathbf{J}_a (\mathbf{r}') \, e^{j\beta \hat{\mathbf{r}} \cdot \mathbf{r}'} \, dv' \right] \cdot \hat{\mathbf{r}} \times (\hat{\mathbf{r}} \times \mathbf{E}_i (\theta, \phi)) \tag{4.7}$$

or

$$V_{oc} = -\mathbf{h} (\theta, \phi) \cdot \mathbf{E}_i (\theta, \phi) \tag{4.8}$$

where **h** is the effective length of the antenna given by

$$\mathbf{h}\left(\theta,\phi\right)=-\widehat{\mathbf{r}}\times\left(\widehat{\mathbf{r}}\times\frac{1}{I_t}\int_{Antenna}\mathbf{J}_a\left(\mathbf{r}'\right)e^{j\beta\widehat{\mathbf{r}}\cdot\mathbf{r}'}\,dv'\right) \tag{4.9}$$

The usual conjugation has not been used for **h** to be consistent with the transient use of effective length. The integral in the effective-length definition can be further reduced to the radiation terms of the antenna by noting that the radiated far field, $\mathbf{E}_{rad}\left(\mathbf{r}\right)$, is given by

$$\mathbf{E}_{rad}\left(\mathbf{r}\right)=j\omega\mu\frac{e^{-j\beta r}}{4\pi r}\widehat{\mathbf{r}}\times\left(\widehat{\mathbf{r}}\times\int_{Antenna}\mathbf{J}_a\left(\mathbf{r}'\right)e^{j\beta\widehat{\mathbf{r}}\cdot\mathbf{r}'}dv'\right) \tag{4.10}$$

Substituting (4.10) into (4.9), we obtain

$$\mathbf{h}\left(\theta,\phi\right)=-\frac{\mathbf{E}_{rad}\left(\mathbf{r}\right)}{j\omega\mu\left(e^{-j\beta r}/4\pi r\right)I_t} \tag{4.11}$$

to give the radiated electric field of an antenna as

$$\mathbf{E}_{rad}\left(\mathbf{r}\right)=-j\omega\mu\frac{e^{-j\beta r}}{4\pi r}I_t\,\mathbf{h}\left(\theta,\phi\right) \tag{4.12}$$

In the time domain, the form of (4.9) for the effective length with an input current of $\delta\left(t\right)$ becomes

$$\mathbf{h}\left(\theta,\phi;t\right)=-\widehat{\mathbf{r}}\times\left(\widehat{\mathbf{r}}\times\int_{Antenna}\mathbf{J}_\delta\left(\mathbf{r}';t+\widehat{\mathbf{r}}\cdot\mathbf{r}'/c\right)dv'\right) \tag{4.13}$$

The corresponding open-circuit voltage and radiated field are then

$$v_{oc}\left(t\right)=-\mathbf{h}\left(\theta,\phi;t\right)\ast\mathbf{E}_{inc}\left(\theta,\phi;t\right) \tag{4.14}$$

$$\mathbf{E}_{rad}\left(\theta,\phi;t\right)=-\mu\frac{1}{4\pi r}\frac{\partial}{\partial t}\left[i_t\left(t\right)\ast\mathbf{h}\left(\theta,\phi;t-r/c\right)\right] \tag{4.15}$$

where the compound operator, $\ast$, denotes time convolution and a vector dot product and $c$ is the speed of light. The reader should observe that the time derivative only appears in the radiation form and not in the received open-circuit voltage.

## 4.1.2   Directivity, Gain, and Related Definitions

In the frequency domain, it is common to use antenna gain or area to represent the properties of the antenna rather than effective length. In this section, we relate the effective length to these factors.

The frequency-domain received power of a link may be defined in terms of the incident power density, $S_{aver}$, and effective area, $A_r$, by

$$Power\ Received=P_r\left(\theta,\phi\right)=A_r\left(\theta,\phi\right)S_{aver}\left(\theta,\phi\right) \tag{4.16}$$

where the incident power density is given by $|\mathbf{E}_i|^2/\eta$. To find the desired form of the effective area, we first consider the circuit representation of the antenna in Figure 4.2 given by the open-circuit voltage, $V_{oc}$, the radiation resistance, $R_{rad}$, the loss resistance, $R_{loss}$, and the receiver impedance, $Z_r$, along with the antenna reactance, $X_a$. For convenience, we represent the total antenna resistance as

$$R_a = R_{rad} + R_{loss} \tag{4.17}$$

With this circuit definition, we may write the received power in terms of the RMS open circuit voltage as

$$P_r(\theta, \phi) = \frac{R_r |V_{oc}|^2}{|Z_r + Z_a|^2} \tag{4.18}$$

where $Z_a = R_a + jX_a$. Because both the antenna loss and the impedance mismatch between the antenna and the receiver are included in the received power, $P_r$, we break out both the efficiency, $e$, and the mismatch loss factor, $q$. The mismatch loss factor, $q$, is defined as the ratio of the power delivered to the receiver compared to the power available from the antenna. This can be written in terms of the power-wave reflection coefficient [3] or directly in terms of the circuit quantities to obtain the mismatch factor

$$q = 1 - \left|\frac{Z_r - Z_a^*}{Z_r + Z_a}\right|^2 = \frac{4 R_r R_a}{|Z_r + Z_a|^2} \tag{4.19}$$

Efficiency is defined as

$$e = \frac{R_{rad}}{R_{rad} + R_{loss}} = \frac{R_{rad}}{R_a} \tag{4.20}$$

to obtain

$$P_r(\theta, \phi) = e\, q\, \frac{|V_{oc}|^2}{4 R_{rad}} \tag{4.21}$$

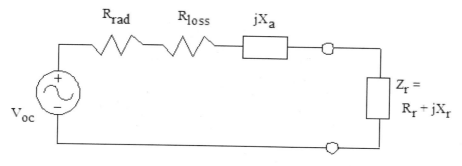

Figure 4.2: The Circuit Equivalent of the Antenna for Power Evaluation.

The effective area is obtained by inserting the definition of $V_{oc}$ as

$$P_r(\theta, \phi) = e\,q\,\frac{|\mathbf{h} \cdot \mathbf{E}_{inc}|^2}{4\,R_{rad}} \tag{4.22}$$

The polarization factor, $p(\theta, \phi)$, of the antenna relates the alignment of effective length and the incident field as

$$p(\theta, \phi) = \frac{|\mathbf{h} \cdot \mathbf{E}_{inc}|^2}{|\mathbf{h}|^2\,|\mathbf{E}_{inc}|^2} \tag{4.23}$$

showing the importance of defining the dot product for polarization effects. The resultant received power becomes

$$P_r(\theta, \phi) = e\,q\,p(\theta, \phi)\,\frac{|\mathbf{h}|^2\,|\mathbf{E}_{inc}|^2}{4\,R_{rad}} \tag{4.24}$$

or

$$P_r(\theta, \phi) = e\,q\,p(\theta, \phi)\,A_{em}(\theta, \phi)\,S_{inc}(\theta, \phi) \tag{4.25}$$

The corresponding maximum effective area, sometimes also called the collecting aperture or effective radiating aperture, is

$$A_{em}(\theta, \phi) = \frac{\eta}{4\,R_{rad}}\,|\mathbf{h}(\theta, \phi)|^2 \tag{4.26}$$

where $\eta$ is the characteristic impedance of space, typically given by $120\pi\ \Omega$. The realized effective area associates the efficiency and the mismatch with the area as

$$A_R = e\,q\,\frac{\eta}{4\,R_{rad}}\,|\mathbf{h}(\theta, \phi)|^2 \tag{4.27}$$

The effective area is often related to the gain in a simple formula using the wavelength of the signal. To develop this relationship, we substitute for the effective length to obtain the maximum effective area in terms of the radiated field as

$$A_{em}(\theta, \phi) = \frac{\eta}{4\,R_{rad}}\left|\frac{\mathbf{E}_{rad}(\mathbf{r})}{j\omega\mu\,\Psi(r)\,I_t}\right|^2,\ \Psi(r) = \frac{e^{-j\beta r}}{4\pi r} \tag{4.28}$$

Expanding and rearranging, we have

$$A_{em}(\theta, \phi) = \frac{1}{4\beta^2\frac{1}{4\pi}}\frac{\frac{|\mathbf{E}_{rad}(\theta,\phi)|^2}{\eta}}{\frac{1}{4\pi r^2}\,|I_t|^2\,R_{rad}} = \frac{\lambda^2}{4\pi}\frac{\frac{|\mathbf{E}_{rad}(\theta,\phi)|^2}{\eta}}{\frac{1}{4\pi r^2}\,|I_t|^2\,R_{rad}} \tag{4.29}$$

where $\omega\mu = \beta\eta$ and $\beta = 2\pi/\lambda$. The directive gain of an antenna is defined as

$$D(\theta, \phi) = \frac{\text{Average Power Density in } (\theta, \phi)}{\text{Average Radiated Power/Area}} = \frac{\frac{|\mathbf{E}_{rad}(\theta,\phi)|^2}{\eta}}{\frac{1}{4\pi r^2}\,|I_t|^2\,R_{rad}} \tag{4.30}$$

to give

$$A_{em}(\theta, \phi) = \frac{\lambda^2}{4\pi} D(\theta, \phi) \tag{4.31}$$

The received power can thus be written as

$$P_r(\theta, \phi) = e\, qp(\theta, \phi) \frac{\lambda^2}{4\pi} D(\theta, \phi) S_{inc}(\theta, \phi) \tag{4.32}$$

with $S_{inc}$ representing the incident power intensity (Poynting vector). The corresponding realized effective area of the antenna is

$$A_R(\theta, \phi) = e\, q \frac{\lambda^2}{4\pi} D(\theta, \phi) = \frac{\lambda^2}{4\pi} G_R(\theta, \phi) \tag{4.33}$$

for the direction $(\theta, \phi)$. The polarization factor, $p(\theta, \phi)$, has been left for separate inclusion because it represents an interaction between the antennas and is not a property of a single antenna. The realized gain, $G_R$, includes both the efficiency and mismatch loss of the antenna. If the polarization match to a given incident field is included, the gain is denoted as the partial realized gain, $g_R$ [4].

Communications engineers commonly desire to specify certain terms separately, particularly polarization and mismatch, and also to restrict the evaluation to the main beam of the antenna. Thus, we summarize the maximum effective area in the main-beam direction without consideration of the polarization or mismatch as

$$A_{em} = \frac{\lambda^2}{4\pi} D \tag{4.34}$$

with $D$ defining the directivity of the antenna. It is common practice to denote the angular maximums of the areas and gains without the angle designations as shown in (4.34).

### 4.1.3 A Link Model Using $S$-Parameters

If we wish to put the received-signal form in an $S$-parameter context, we need to evaluate the resultant voltage transferred to a reference load impedance, $Z_o$, or equivalently to the receiver, $Z_r$. In this context, we have the power wave at the receiver (normalized load voltage) in the frequency domain as

$$b_r = \frac{V_r}{\sqrt{R_r}} = \frac{1}{\sqrt{R_r}} \frac{R_r}{Z_r + Z_a} V_{oc} = \frac{1 - s_a}{2\sqrt{R_r}} V_{oc} \tag{4.35}$$

or in the time domain as

$$b_r(t) = \frac{1}{2\sqrt{R_r}} [1 - s_a(t)] * v_{oc}(t) \tag{4.36}$$

The quantity $s_a$ represents the reflection coefficient of the antenna and is given as $(Z_a - Z_r^*) / (Z_a + Z_r)$. The power wave concept was introduced in 1965 [3] to aid in the analysis of microwave structures using circuit concepts. The form is a normalized voltage wave; the square of this normalized voltage wave represents the traveling power, that is,

$$P_r = |b_r|^2 \tag{4.37}$$

In terms of the effective length, the received signal is given by

$$b_r(t) = -\frac{1}{2\sqrt{R_r}} [1 - s_a(t)] * \mathbf{h}(\theta, \phi; t) \stackrel{.}{*} \mathbf{E}_i(\theta, \phi; t)$$

$$= -\frac{1}{\sqrt{R_r}} \mathbf{h}_R(\theta, \phi; t) \stackrel{.}{*} \mathbf{E}_i(\theta, \phi; t) \tag{4.38}$$

where we define the realized effective length, $\mathbf{h}_R$, as

$$\mathbf{h}_R(\theta, \phi; t) = \frac{1 - s_a(t)}{2} * \mathbf{h}(\theta, \phi; t) \tag{4.39}$$

Farr [5] defines the normalized impulse response of the antenna as $\mathbf{h}_R$ normalized by $\sqrt{\eta/R_r}$ to give a ratio of $b_r$ to the normalized electric field.

The link between two line-of-sight (LOS) antennas is easily established with the definition of effective length. The incident field can be written in terms of a transmit effective length as

$$\mathbf{E}_i(\theta, \phi; \omega) = -j\omega\mu\,\Psi(r)\,I_t\,\mathbf{h}_t(\theta, \phi) \tag{4.40}$$

or in the time domain as

$$\mathbf{E}_i(\theta, \phi; t) = -\mu\frac{1}{4\pi r}\frac{\partial}{\partial t}I_t(t) * \mathbf{h}_t(\theta, \phi; t - r/c) \tag{4.41}$$

For an $S$-parameter form, we must convert the current to a power wave as well as provide for the mismatch to the transmitter impedance. Again, we find a form of the effective length given by

$$\mathbf{E}_i(\theta, \phi; t) = -\mu\frac{1}{4\pi r}\frac{\partial}{\partial t}\frac{a_t(t)}{\sqrt{R_t}} * [1 - s_t(t)] * \mathbf{h}_t(\theta, \phi; t - r/c) \tag{4.42}$$

or in terms of the realized effective length

$$\mathbf{E}_i(\theta, \phi; t) = -\mu\frac{1}{2\pi r}\frac{\partial}{\partial t}\frac{a_t(t)}{\sqrt{R_t}} * \mathbf{h}_{t_R}(\theta, \phi; t - r/c) \tag{4.43}$$

The quantity $a_t(t)$ represents the square root of the instantaneous available power from the source. Combining terms, we have the received power wave as

$$b_r(t) = \frac{\mu}{\sqrt{R_r R_t}}\frac{\partial}{\partial t}\left[\mathbf{h}_{r_R}(\theta_r, \phi_r; t) \stackrel{.}{*} \frac{1}{2\pi r}\mathbf{h}_{t_R}(\theta_t, \phi_t; t - r/c) * a_t(t)\right] \tag{4.44}$$

where the angles for the receive and transmit directions are simply the negative of each other. The LOS propagation loss is given by the $1/4\pi r$ term and the time delay of $r/c$. The frequency domain form of (4.44) is given by

$$
b_r\left(\omega\right) = \frac{j\omega\mu}{\sqrt{R_r R_t}}\left[\mathbf{h}_{r_R}\left(\theta_r,\phi_r;\omega\right)\cdot\frac{e^{-j\beta r}}{2\pi r}\,\mathbf{h}_{t_R}\left(\theta_t,\phi_t;\omega\right)a_t\left(\omega\right)\right] \tag{4.45}
$$

For a multipath propagation problem, the full context of the received signal is given by

$$
b_r\left(t\right) = \frac{2\mu}{\sqrt{R_r R_t}}\frac{\partial}{\partial t}\int_{\Omega_t}\int_{\Omega_r}\left[\mathbf{h}_{r_R}\left(\theta_r,\phi_r;t\right)\overset{*}{*}\overrightarrow{\mathbf{C}}\left(\theta_r,\phi_r;\theta_t,\phi_t;t\right)\right.
$$
$$
\left.\overset{*}{*}\mathbf{h}_{t_R}\left(\theta_t,\phi_t;t\right)*a_t\left(t\right)\right]d\Omega_r d\Omega_t \tag{4.46}
$$

where $\overrightarrow{\mathbf{C}}$ is a dyadic representing the multipath channel. Because the multipath components travel over multiple directions, the integrals over the angular extent of each antenna are required to completely capture the signal contributions. The separation of these multipath components using this model is beyond the subject of this chapter and will not be discussed further. In addition, in a practical system $\overrightarrow{\mathbf{C}}$ requires evaluation on a statistical basis. It is common to neglect the angles in processing the data by simply evaluating the channel term using directional or omni-directional antennas and neglecting the vector and angular nature of the problem. The antenna effects are then removed by an estimate based on LOS measurements. In this manner, it is reasonable to approximate (4.46) as

$$
b_r\left(t\right) \approx \frac{2\mu}{\sqrt{R_r R_t}}\frac{\partial}{\partial t}\left[h_{r_R}\left(t\right)*C\left(t\right)*h_{t_R}\left(t\right)*a_t\left(t\right)\right] \tag{4.47}
$$

and characterize the channel by $C\left(t\right)$, including both the channel properties and the polarization effects.

In (4.27) we found that the realized effective area is given by

$$
A_R = eq\,\frac{\eta}{4\,R_{rad}}\left|\mathbf{h}\left(\theta,\phi\right)\right|^2 \tag{4.48}
$$

The efficiency and mismatch can be evaluated in the frequency domain as

$$
\sqrt{qe} = \left|\frac{2\sqrt{R_r R_a}}{Z_a + Z_r}\sqrt{\frac{R_{rad}}{R_a}}\right| = \left|\frac{\left(1 - s_a\right)}{2}\sqrt{\frac{R_{rad}}{R_r}}\right| \tag{4.49}
$$

Substituting (4.36) for the $q$ and $e$ terms, we have

$$
A_R = \left(\frac{\left(1 - s_a\right)}{2}\sqrt{\frac{R_{rad}}{R_r}}\right)^2\frac{\eta}{R_{rad}}\left|\mathbf{h}\left(\theta,\phi\right)\right|^2 = \frac{\eta}{R_r}\left|\mathbf{h}_R\left(\theta,\phi\right)\right|^2 \tag{4.50}
$$

which suggests that both the mismatch and efficiency are included within the definition of the realized effective length.

Before concluding this section, it is worth noting again that the transmission of the link contains a time derivative. It is common in some communications definitions of link budgets to describe the distance in terms of wavelength and to incorporate the effect of frequency dependence as part of the path loss, though in reality it is not actually a path loss property. In the previous section we related the effective length to the realized gain and effective area commonly used in the frequency-domain description of antenna systems. The result of this relationship is the Friis transmission formula [1], given in the frequency domain by

$$P_r = P_t \frac{G_{t_R} G_{r_R} \lambda^2}{(4\pi r)^2} p(\theta, \phi) \qquad (4.51)$$

with the realized antenna gains used for completeness. The quantity $1/(4\pi r^2)$ is commonly called the free-space loss, and $\lambda^2/4\pi$ represents the conversion from gain to effective antenna aperture.

### 4.1.4  Link Budget Concepts

Communications systems often require benchmarks for comparison. The form of (4.44) or (4.46) for the LOS and full responses provide the foundation of the needed terms. For a peak detection system, only the maximum signal is required for the link budget estimate. For such a situation, some researchers provide peak signal level antenna patterns. However, many newer and proposed systems use correlation detection processes, or possibly multichannel, narrowband transmission. In these cases, the power spectrum of the input is typically known in terms of the square of the input power wave, $|a_t|^2$. With these receivers, the total power across the operating band of interest represents the energy of the system. Thus, for a given input spectral content, a potential useful measure of performance is a link gain defined as

$$G_L = 4\pi r^2 \frac{\int_t |b_r|^2 \, dt}{\int_t |a_t|^2 \, dt} = 4\pi r^2 \frac{\text{Energy Received}}{\text{Energy Input}} \qquad (4.52)$$

The $4\pi r^2$ provides the surface area of a sphere surrounding the antenna system and represents the spherical spreading loss. With such a definition we can represent the signal as

$$\mathcal{E}_r = \frac{\mathcal{E}_t G_L}{L_p} \qquad (4.53)$$

where the $\mathcal{E}$ terms represent the energy and $L_p = 4\pi r^2$. The quantity $G_L$ contains all the properties of the antennas, the time derivative of the transient link, and multipath effects, if present in the system. The basic restriction on the use of this form is a limitation on the spectral content of the signal. Within this context, the form is applicable to both impulse (I-UWB) and multicarrier (MC-UWB) systems.

It is common to express communication link quantities in decibels. Thus, (4.53) can be expressed as

$$\mathcal{E}_r (\text{dB}) = \mathcal{E}_t (\text{dB}) + G_L (\text{dB-m}^2) - 10 \log (4\pi r^2) \qquad (4.54)$$

or as

$$\mathcal{E}_r \, (\text{dB}) = \mathcal{E}_t \, (\text{dB}) + G_L \, (\text{dB-m}^2) - 10.99 - n20 \log (r) \qquad (4.55)$$

where the $n$ provides a path loss exponent defined in terms of the LOS component, with $r$ in meters. The separation of the path loss into a frequency independent term, including $n$, assumes that the properties of the channel environment are frequency independent. This assumption can be a major limitation of such a form when a wideband spectral response of the channel is considered, neglecting frequency dependent fading.

### 4.1.5  Fundamental Limits of Antennas

In the design of antennas, a typical design requirement is to maintain a small dimension. Fundamental limits are typically derived in terms of a lower bound of the energy transfer relative to the center frequency of an antenna. Chu originally developed his result in 1948 [6] in terms of the unloaded quality factor, $Q$, of the antenna given in terms of stored energy divided by the average energy dissipated per cycle as

$$Q = \frac{1 + 2 (\beta a)^2}{(\beta a)^3 \left[ 1 + (\beta a)^2 \right]} \qquad (4.56)$$

Several authors have attempted to improve on this result, typically adding errors that lead to incorrect conclusions. Thus, we retain (4.56) as a guide to the performance of the antenna. For a lower unloaded $Q$, one must consider either of two options: (1) to add loss to improve the match versus frequency or (2) reduce the match requirement. Typically, $Q$ is associated with a 3 dB impedance bandwidth. However, a much higher impedance mismatch may be acceptable, particularly for receive systems.

For antennas that need to be used at very low frequencies, it may be useful not to consider a match, but rather to take advantage of the flat frequency response of the open circuit voltage response or the input current feed of an antenna. In such cases, the mismatch term drops out, and for low frequencies the only limitation becomes the time-domain derivative of the response. To take advantage of these flat responses at low frequencies the antenna must be used in an active form, with the antenna feeding a high impedance receiver or being fed from a low impedance source.

## 4.2  Antenna Measurements and Modeling in the Time Domain

### 4.2.1  Basic Responses

Classic antenna measurements include pattern and impedance versus frequency. The pattern is often measured for both co-polarized and cross-polarized components of the antenna as defined in terms of the expected antenna performance. As an alternative to impedance, voltage-standing-wave ratio (VSWR) or return loss are often used as a measure of how closely an antenna is matched to a reference impedance. Each of these terms requires further consideration for transient analysis. However, we first review the basic concepts in the frequency domain to provide a common foundation.

The pattern is expressed in terms of the electric or magnetic field received at a fixed distance in the far field as the angular position is changed about the antenna. It is also common to normalize the pattern to the maximum value. Historically, patterns were measured in amplitude only, though modern ranges provide useful phase information.

The classic dipole antenna consists of a half wavelength of wire, fed at the center. The pattern, $F$, of such an antenna is a figure eight, denoted by

$$F(\theta, \phi) = \frac{\cos\left(\frac{\pi}{2}\cos\theta\right)}{\sin\theta} \tag{4.57}$$

In a more general form, and in the context of our previous development of effective length, the normalized pattern is given as

$$F(\theta, \phi) = \frac{|h(\theta, \phi)|}{\max_{\theta,\phi} |h(\theta, \phi)|} = \frac{|E(\theta, \phi)|}{\max_{\theta,\phi} |E(\theta, \phi)|} \tag{4.58}$$

for the electric field magnitude. The pattern can also be defined for each polarization or in terms of the power. Three-dimensional patterns are sometimes desired, but they are very time consuming to measure and difficult to display. To overcome these limitations, it is common to plot polar plots in planar or conical cuts of the antenna. Planar cuts are often selected to correspond to major axes of the antenna and are defined in terms of planes in which the electric or magnetic field intensities are parallel, called the E plane and H plane, respectively. In this sense, both co-polarized and cross-polarized fields can be plotted relative to the expected polarization. A conical cut provides a plot versus azimuth for a specified elevation angle. A typical plot for the dipole is displayed in Figure 4.3.

The input impedance of the antenna provides the necessary information for the design of a matching section for maximum power transfer or for the evaluation of the quality of the match to a transmission system. The impedance is often plotted on a Smith chart or in terms of either reflection coefficient magnitude or voltage-standing-wave ratio (VSWR) to denote the quality of the match to a reference impedance as in Figure 4.4.

### 4.2.2   UWB Performance

For UWB applications, we need to evaluate the antenna in terms of the modulation format of interest. Both I-UWB systems and MC-UWB systems must be considered. For pulse systems, instantaneous wideband performance is required, with a linear phase response. The linear phase response is necessary to avoid dispersion that creates pulse distortion at a nanosecond delay. For MC-UWB systems, each

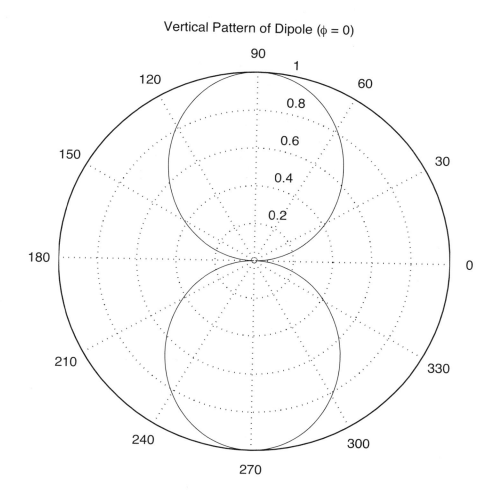

Figure 4.3: The Pattern of a Dipole Antenna in the E Plane (elevation). The Pattern is Independent of $\phi$ and the Azimuthal Cut is Omni-Directional.

channel is relatively narrowband compared to UWB, and dispersion relative to the pulse shape is no longer critical, requiring only gain and related parameters as in classic analysis. Synchronization is still important for MC-UWB systems and is simply obtained by incorporating pilot-carrier techniques for each channel. In the latter situation, only the frequency amplitude response of the basic channels is necessary for the system. For the pulse system, the pulse amplitude and shape of the transmission waveform and the reflection properties of the antenna are the basic representations for the antenna response. These terms include the information

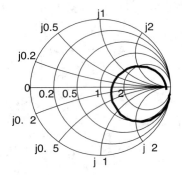

(a) The Smith Chart View

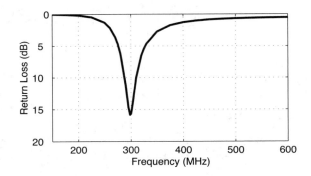

(b) Return Loss

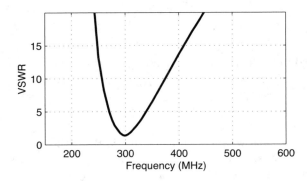

(c) Voltage Standing Wave Ratio (VSWR)

Figure 4.4: The Impedance of a Dipole Antenna for a Variety of Forms.

necessary for the MC-UWB system as well as the transient system. Insight is also provided into the operation of the antenna and channel that is often missed with the frequency-domain representation.

The properties of the resonant monopole, cavity-backed Archimedean spiral (CBAS), and TEM horn antennas are investigated in terms of their impulse and frequency responses and their usefulness in UWB technology based on duration, type, and amplitude of the radiated transient waveform. Additional data on the log-periodic toothed trapezoid (LPTTA), Vivaldi, and biconical antennas are also summarized. The frequency response needed for MC-UWB systems is also provided.

To perform the measurements for radiation performance, two identical antennas were measured in a LOS link using an HP8510 network analyzer. A 20 GHz spectrum was used, corresponding to a 50% pulse width of 50 ps. The impulse response, transmission coefficient, and reflection coefficient were all measured using the measurement setup illustrated in Figure 4.5.

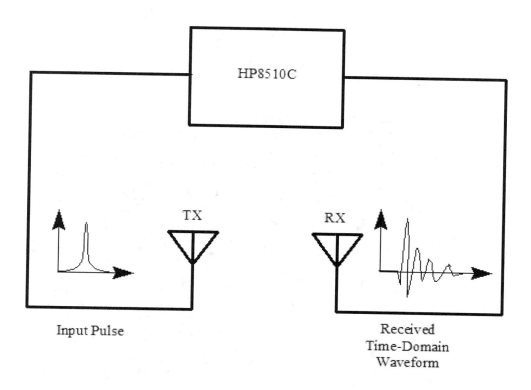

Figure 4.5: Measurement Setup for the Frequency and Impulse Response.

## Resonant Monopole—A Narrowband Example

The mechanisms of dipole radiation, excited by a short pulse, have been investigated by many authors, and some of the results have been reported in [8–13]. Only the note by Bantin [10] discusses the possibility of using a pair of 2 meter long dipoles as a link for UWB communications. For UWB antenna and link characterization, the time domain approach provides more insight than a classical frequency domain analysis [8]. The monopole antenna structure performance is similar to a dipole due to the image of the monopole in a ground plane depicted in Figure 4.6.

The typical frequency response shown in Figure 4.7 for a 5 GHz monopole pair highlights the transmission performance. The frequency response shows the most efficient transmission at the resonant frequency around 5.0 GHz, with a signal amplitude decay of 1 dB/GHz as the frequency increases from 4 GHz to 13 GHz. The Friis transmission (4.51) suggests that this decay represents a constant gain antenna above the basic resonance, with the decay due to the inverse frequency relation between gain and effective aperture. However, care must be taken with such a claim to ensure that the input reflection is also low over these frequencies. If the reflection is not low we may actually find the antenna has a general gain increase as the frequency is increased, with a mismatch that counters this increase to give us an effective constant gain on the boresight. The overall dipole response on the

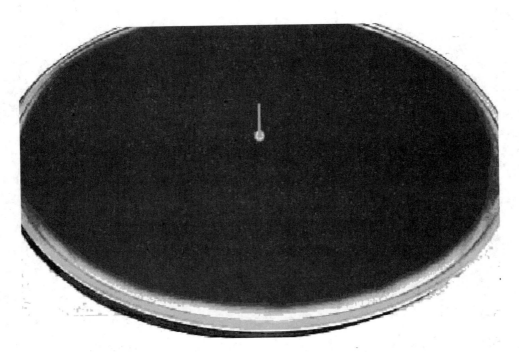

Figure 4.6: Monopole on a Finite Ground Plane.

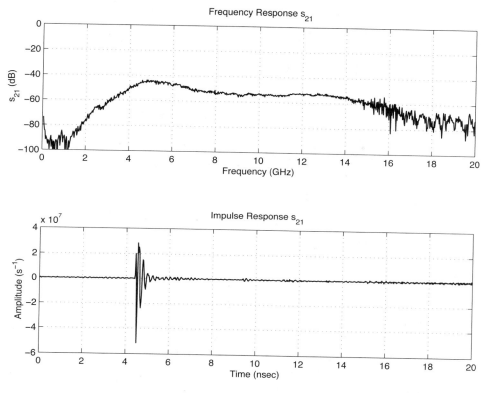

Figure 4.7: The Frequency and Transient Response Waveforms of the Monopole Link.

SOURCE: S. Licul, "Unified Frequency and Time-Domain Antenna Modeling and Characterization," Ph.D. dissertation [7]. © S. Licul, 2004. Used by permission.

boresight does appear to have an equivalent constant gain effect, but also has a substantial mismatch loss (see Figure 4.8) as the frequency increases.

The corresponding impulse response is shown in Figure 4.7. The measured waveform is not a clear doublet, but has a damped ringing that would be expected about the monopole resonance. The damped ringing has a duration of 1.00 ns. The peak amplitude is lower than the amplitude of classic UWB antennas due to the mismatch. Late time effects of diffraction from the back edge of the finite ground plane for our monopole example occur at 9.5 ns, but are small. Each monopole was tuned to a resonant frequency of 5 GHz.

Four loss mechanisms may exist in this link: pattern, space loss, gain to aperture loss, and reflection at the antenna terminals. All mechanisms contribute to the overall equivalent efficiency of the system. Even with these limitations, the

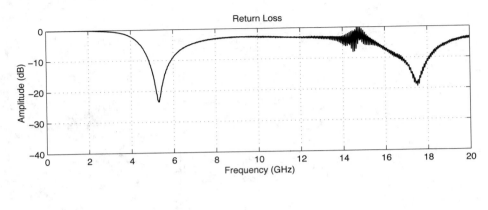

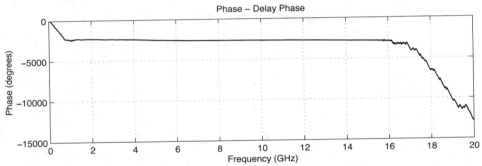

Figure 4.8: The Monopole Return Loss and Transmission Phase.

SOURCE: S. Licul, "Unified Frequency and Time-Domain Antenna Modeling and Characterization," Ph.D. dissertation [7]. Used by permission.

experimental results indicate that two resonant monopoles may be used in a UWB communication link with pulse modulation. The main drawbacks relate to efficiency and a slight damped ringing. The efficiency can be improved by using a fat or sleeve dipole/monopole arrangement.

## Cavity-Backed Archimedean Spiral (CBAS)—A Frequency Independent Antenna

The Archimedean spiral shown in Figure 4.9 is a classic frequency-independent antenna [1]. The phase of the antenna varies nonlinearly with frequency and can be explained by physically tracing the path of a transmitted pulse along the spiral. As the pulse travels from the center of the spiral towards the edge, radiation occurs. The higher-frequency signal components are radiated near the center, followed by the lower-frequency components towards the outside, causing frequency dispersion.

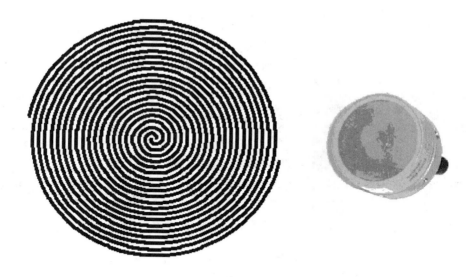

Figure 4.9: Geometry of an Archimedean Spiral Antenna.

SOURCE: S. Licul, "Unified Frequency and Time-Domain Antenna Modeling and Characterization," Ph.D. dissertation [7]. © S. Licul, 2004. Used by permission.

This antenna has a wide operating frequency range of 2.0 to 18.0 GHz, but frequency dispersion makes it unacceptable for UWB pulse applications. For MC-UWB applications, it provides circular polarization and would meet the basic need as long as a pilot carrier is transmitted with each carrier to ensure synchronization. Indeed, the MC-UWB application is consistent with the frequency-independent nature of the antenna that is generally used for narrowband applications that can be tuned within the band of the antenna.

The presence of the frequency dispersion of the cavity-backed Archimedean spiral is illustrated in Figure 4.10. The antenna demonstrates a ringing with a frequency shift. The frequency shift is due to the phase delays experienced by the different frequency components of the radiated field, giving rise to a chirp duration of 13.5 ns. Figure 4.11 shows the frequency match and transmission phase response. This type of antenna has the desired magnitude frequency response over the entire frequency band. However, the phase response has an undesired nonlinear behavior above the 2 GHz lower frequency that produces severe dispersion with a frequency chirp.

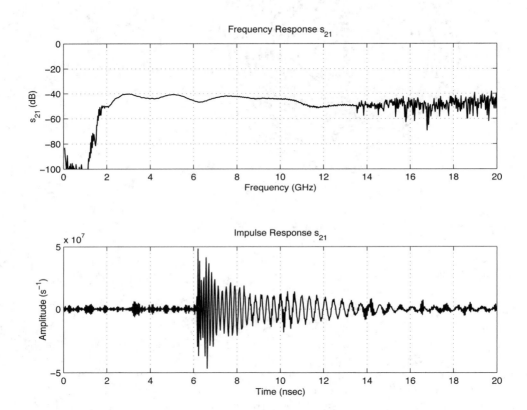

Figure 4.10: The Frequency and Transient Response Waveforms of the Cavity-Backed Archimedean Spiral Link.

SOURCE: S. Licul, "Unified Frequency and Time-Domain Antenna Modeling and Characterization," Ph.D. dissertation [7]. © S. Licul, 2004. Used by permission.

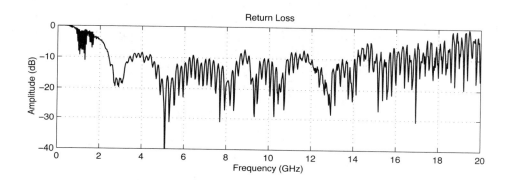

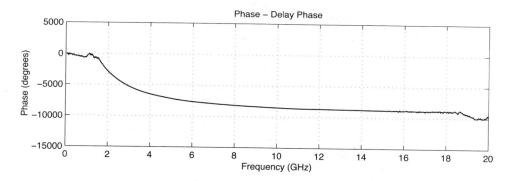

Figure 4.11: The Cavity-Backed Archimedean Spiral Return Loss and Transmission Phase.

SOURCE: S. Licul, "Unified Frequency and Time-Domain Antenna Modeling and Characterization," Ph.D. dissertation [7]. © S. Licul, 2004. Used by permission.

## TEM Horn—A UWB Antenna

The TEM (transverse electric and magnetic) horn is a traveling-wave antenna. TEM horns have been traditionally used for low-dispersion launching and receiving of UWB pulses [14]. Research conducted since the late 1970s has led to a further reduction of dispersion caused by the TEM horn abrupt edges. Figure 4.12 shows the basic TEM horn geometry.

In order to suppress the reflection from the TEM horn edges, the antenna is commonly loaded with chip resistors or a conductive film. Some authors have shown experiments with tapered resistive coatings that absorb the wave at the tip of the horn [15]. Figure 4.13 shows a TEM horn mounted over a ground plane, having a operating frequency range of approximately 1 to 18 GHz.

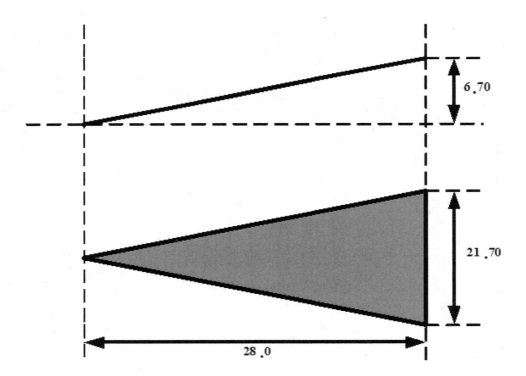

Figure 4.12: Geometry of the TEM Horn Antenna.

SOURCE: S. Licul, "Unified Frequency and Time-Domain Antenna Modeling and Characterization," Ph.D. dissertation [7]. © S. Licul, 2004. Used by permission.

Figure 4.13: The TEM Horn Antenna.

SOURCE: S. Licul, "Unified Frequency and Time-Domain Antenna Modeling and Characterization," Ph.D. dissertation [7]. © S. Licul, 2004. Used by permission.

The TEM horn impulse response shown in Figure 4.14 is a highly damped sinusoid with a duration of $\frac{1}{2}$ ns. The TEM horn reflection frequency response is shown in Figure 4.15. The reflection amplitude is low over the entire operating range.

Finally, the TEM horn can be used as a source for radiation through a tapered aperture. A tapered aperture reduces reflections of the aperture edges, thus enabling the incident wave to have a smoother transition. This configuration also achieves lower side lobes.

## Summary of Potential UWB Antenna Structures

The antennas presented represent classic narrowband, frequency-independent (wideband), and UWB performance. The antennas presented are compared to several other antennas in Table 4.1.

The basic structure, transient impulse response, and frequency response are shown for each antenna. The Archimedean spiral is an American Electronic Laboratories model AST-1492AA with a range of 2.0–18.0 GHz; the ridged TEM horn

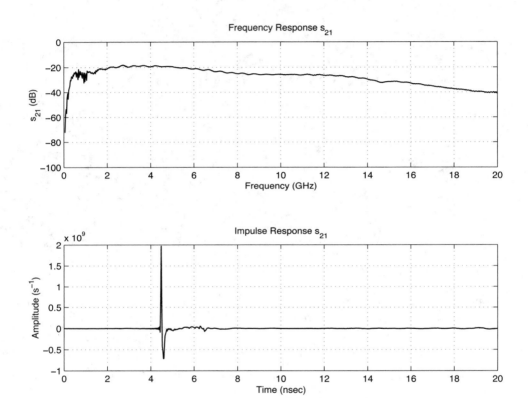

Figure 4.14: The Frequency and Transient Response Waveforms of the TEM Horn Antenna Link.

SOURCE: S. Licul, "Unified Frequency and Time-Domain Antenna Modeling and Characterization," Ph.D. dissertation, [7]. © S. Licul, 2004. Used by permission.

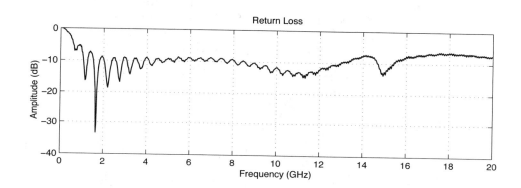

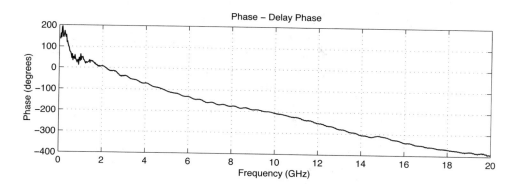

Figure 4.15: The TEM Horn Return Loss and Phase.

SOURCE: S. Licul, "Unified Frequency and Time-Domain Antenna Modeling and Characterization," Ph.D. dissertation [7]. © S. Licul, 2004. Used by permission.

Table 4.1: Plots for Ridged TEM, LPD, and Monopole.

SOURCE: S. Licul, "Unified Frequency and Time-Domain Antenna Modeling and Characterization," Ph.D. dissertation [7]. © S. Licul, 2004. Used by permission.

| Antenna | Geometry | Normalized Impulse Response | Frequency Response (dB) |
|---|---|---|---|
| Monopole (5GHz) | | | |
| TEM Horn | | | |
| Archimedean Spiral | | | |
| Ridged TEM | | | |
| Biconic | | | |
| Vivaldi | | | |
| LPD (log-periodic) | | | |
| Monopole (1 to 1.1 GHz) | | | |

Table 4.2: Properties of Different I-UWB Radiator Links.

| Antenna | Pulse Duration (ns) | Pulse Type | Peak Amplitude[a] |
|---|---|---|---|
| Vivaldi | 0.5 | Doublet | 0.80 |
| Ridged TEM Horn | 2.74 | Damped Sinusoid | 1.00 |
| LPD | 11 | Chirp | 0.21 |
| Archimedean Spiral | 13.5 | Chirp | 0.05 |
| Monopole (5 GHz) | 1.0 | Damped Sinusoid | 0.06 |
| Biconical Antenna | 0.2 | Damped Sinusoid[b] | 0.40 |

[a]Relative to ridged TEM horn.
[b]Damped sinusoid with cancellation effects (5.9 ns to 7.3 ns).

is an Antenna Research Associates model DRG-118/A with a range of 1.0–18.0 GHz; and the log-periodic toothed structure is an American Electronic Laboratories model APN-101B with a range of 1.0–12.4 GHz. The remaining antennas were built in-house and represent typical dimensions for such structures.

The comparison of these different classical broadband structures with regards to possible use in pulse UWB communications offers some interesting observations (see Table 4.2). The received peak amplitude in the far field obeys a 1/distance decrease in amplitude relative to the measured peak. The pulse shape, in general, will always stay the same and is not a function of distance, but it is dependent on the antenna geometry and angular orientation.

It can be seen that for directional applications the Vivaldi offers a good option, with a single-cycle sinusoid at the output, similar to the TEM horn. For omni-directional antennas, the monopole performs fairly well, though with a lower amplitude. Typical broadband structures, such as the log-periodic array and Archimedean spiral, suffer from an acute phase dispersion, rendering them undesirable for impulse UWB communication, but they can still be used for MC-UWB (multichannel) systems.

## 4.2.3    Frequency and Time Relationships

Observing the behavior of the UWB antennas in Table 4.1 we see a common trend in the transfer response. Basically the response has a resonance that splits the spectrum into two parts. In the region below resonance, the antennas act as constant effective-length structures, and the response is one of a capacitive mismatch with an $\omega^3$ link response due to the combination of the mismatch and the $\omega$ of the link.

Above the resonant frequency, $f_r$, the antennas generally act as constant gain antennas with a link response of $1/\omega$. These results are summarized in Figure 4.16.

The behavior in the vicinity of resonance is also important in the damping response of the system. If the maximum is peaked, the damping tends to be reduced, leading to a ringing of the response. For a smooth response, we have the form of the

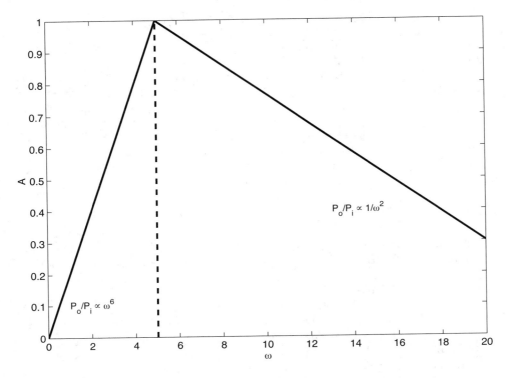

Figure 4.16: Simple Representation of the Link Response of a UWB Antenna.

system transmission, $s_{21}$, of Figure 4.17, and a corresponding result for a peaked response in Figure 4.18.

The latter represents the response typically obtained with a dipole. However, the constant gain conclusion for the dipole would be in error. As the frequency applied to a dipole increases, the gain actually tends to increase and focus to a narrower beam. The reason for the apparent constant-gain response is the result of the mismatch of the antenna. For the constant-gain observation to be valid the antenna must be well matched to the system.

### 4.2.4   Pattern Concept in Time

The concept of visualizing an antenna pattern in the time domain requires additional thought. In the frequency domain, we represent that pattern by plots similar to those of Figure 4.3. If the antenna behaves the same way over the bandwidth, the radiation pattern for a single frequency can be used. However, this frequency domain concept does not generalize easily to the time domain. There are two common methods for representing the pattern of an antenna in the time domain. Peak amplitude is the critical component for threshold detection in pulse systems. However, this peak

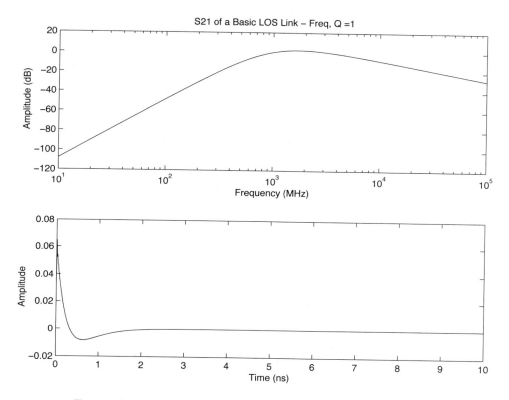

Figure 4.17: Response for a Slow Variation About Resonance.

information is limited, and it is common to represent the pattern by a collection of waveforms versus angle. Both of these representations are shown in Figure 4.19.

## 4.2.5  Time Domain Modeling (A Minimal Approach)

A pole-residue model provides a minimal representation for the antenna that is particularly useful for simulation. It also provides information for the frequency domain pattern creation from this simple representation. This type of characterization provides the following advantages:

1. Complete frequency and time domain characterization (after the pole and residues of an antenna system are obtained, it is easy to compute the time domain waveforms or frequency domain patterns in any arbitrary direction).

2. Minimal model to represent an antenna system.

3. Useful antenna model for channel models to predict the received time/frequency waveform (the effects of different antennas on received waveforms can be easily obtained when the model exists).

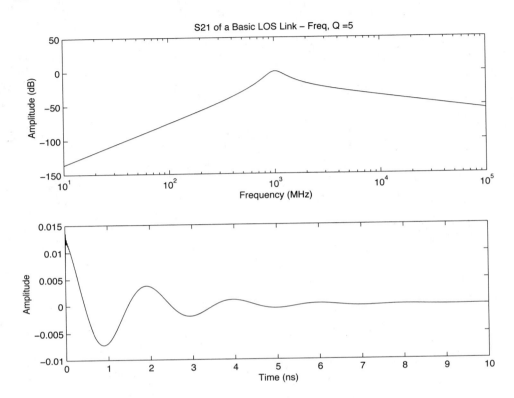

Figure 4.18: Response for a Peaked Resonance.

We present simple models for several antennas, providing insight into the operation of the antennas for UWB applications. Only seven poles are typically required
to model these antennas. Pole modeling of the monopole antenna is also demonstrated to show the results for a narrowband structure. The monopole requires six
to 39 poles to accurately model the antenna, depending on the region of interest.

The most complete model is presented for the Vivaldi antenna. The poles and
residues are obtained for the azimuth plane (Vivaldi antenna E-plane) for all 360
degrees in five-degree increments. The Vivaldi antenna frequency-domain patterns
were obtained using a 17-pole model from the transient responses at different azimuth angles. The frequency-domain patterns obtained agree well with the measurements obtained using a near-field scanner.

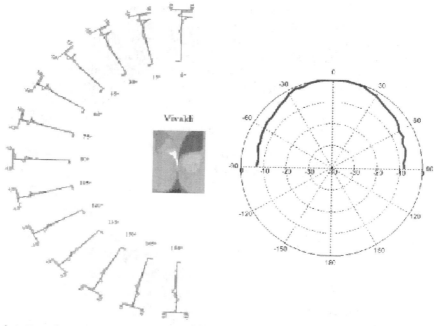

a) A Transient Plot Collection Versus Angle        b) A Peak Signal Response

Figure 4.19: Pattern Concepts for UWB.

SOURCE: S. Licul, "Unified Frequency and Time-Domain Antenna Modeling and Characterization," Ph.D. dissertation [7]. © S. Licul, 2004. Used by permission.

## Pole Modeling of Antenna Systems

The modeling is based on the measurements using the vector network analyzer, HP8510, as shown in Figure 4.5. The basic concept uses the form of (4.44) given by

$$b_r(t) = \frac{\mu}{2\sqrt{R_r R_t}} \frac{\partial}{\partial t} \left[ \mathbf{h}_{r_R}(\theta_r, \phi_r; t) \mathbin{\dot{*}} \frac{1}{2\pi r} \mathbf{h}_{t_R}(\theta_t, \phi_t; t - r/c) * a_t(t) \right] \quad (4.59)$$

The realized effective length includes the impedance mismatch not normally included in effective-length definitions. The antennas are assumed to be line-of-sight (LOS) with multipath components in the response deleted using time gating.

The link measurements combine the effects of the transmit/receive antennas and the channel as indicated in (4.59). The time domain waveform of the single-antenna, realized effective length can be obtained by using two identical antennas or a known reference antenna. After the realized effective length of the single antenna is obtained, one can obtain a pole and residue approximation using Prony's method

[16] or an equivalent pole and residue extraction technique, such as the matrix pencil method [17].

The pole-residue model was introduced with a theoretical electromagnetic basis by Carl Baum [18] and was used to develop approaches to target identification and related modeling issues in the 1970s and 1980s. This modeling approach uses the poles of the antenna to describe the realized effective length and the residues to represent the angular information. The model is useful in both time and frequency domains with the same set of parameters. The representation of the realized effective length in terms of poles and residues is given in the frequency domain as

$$\mathbf{h}_R(\theta, \phi; \omega) = \sum_{\alpha=1}^{N} \frac{\mathbf{R}_\alpha(\theta, \phi)}{j\omega - s_\alpha} \tag{4.60}$$

and in the time domain as

$$\mathbf{h}_R(\theta, \phi; t) = \sum_{\alpha=1}^{N} \mathbf{R}_\alpha(\theta, \phi) e^{s_\alpha t} \tag{4.61}$$

Representations (4.60) and (4.61) indicate that after the poles $s_\alpha$ and residues $\mathbf{R}_\alpha(\theta, \phi)$ are determined, it is straightforward to obtain the realized effective length in both time and frequency for any arbitrary direction. The factor $N$ represents the total number of poles used. The poles come in complex-conjugate pairs or negative-real values in order to represent a real function. It is important to note that the poles are independent of the direction, with all the directional information contained in the residues. Thus, the same set of poles is present in all the directions. In order to obtain the antenna frequency-domain pattern, one only needs to sum all the residue contributions as in (4.60). Likewise, the time-domain pattern information is obtained from (4.61).

This type of modeling provides a convenient way of describing the antenna system because only pole and residues are required to have an antenna characterization, providing a minimal form of representation for the antenna. The reflection response can also be computed in terms of the poles.

The poles and residues of the transient waveform can be obtained by variations of Prony's method [16]. The expression of (4.61) does not take into account the early transient response of the antenna during pulse excitation, but only the damped, characteristic impulse response of the antenna. The transient response of the antenna consists of an early time portion that represents the excitation and a late time transient that represents the antenna characteristics.

For the antennas measured, the early time duration was less than 50 ps. This time is usually associated with the finite frequency window of the measurement system. The Vivaldi antenna appears to have an early time delay of about 533 ps. However, the 533 ps corresponds to the time that is required for the excitation pulse to reach the phase center of the antenna from the reference plane with additional radiation from the feed region of the antenna.

Another issue with Prony's method is the possible existence of a Nyquist pole that appears as a result of the pole-residue extraction in a sampled data system.

A Nyquist pole has an imaginary part corresponding to the Nyquist sampling frequency of the data. The Nyquist pole should be split with a conjugate mate for the corresponding residue value in the model. For the TEM horn and biconical antenna models, the complex conjugate of the Nyquist pole is added to the model with a corresponding split in the residue values.

## Biconical Antenna Model

Two biconical antennas were measured in a boresight, line-of-sight link. The distance between the reference planes was 1.7700 meters. The distance response of the LOS link was removed in (4.45) to obtain the square of the effective length. A square root gives the effective length after some care in unwrapping the phase. From the single response of the biconical antenna, Prony's method was applied to obtain the poles and residues of the antenna system. Figure 4.20 shows the phase and amplitude response of the single biconical antenna (dashed) and the Prony's method approximation (solid). The time domain window used to obtain the response of the single biconical antenna was 10.0 ns. The window used in Prony's method was 0.45 ns.

Figure 4.21 shows the pole structure and the realized effective length of the biconical antenna in the time domain for a six-pole model. The main pulse shape is preserved with the six-pole model.

## Monopole Antenna Model

The response of the single monopole was obtained by using the TEM horn antenna of Figure 4.13 as a reference. The measurements were done by using the 4 x 4 ft ground plane in order to eliminate problems caused by a balun structure and effects caused by an edge of the finite ground plane required for monopole type structures. The distance between the reference planes of the monopole and the TEM horn antenna was 1.3500 meters. Figure 4.22 shows the amplitude and phase of the 39-pole model of the monopole antenna compared to measured, frequency-domain data.

The 39-pole model is in good agreement with the measured data. The monopole antenna exhibits many resonant frequencies, as the amplitude response shows. Thus, a monopole antenna requires a large number of poles for a complete characterization, indicating that resonant-antenna system models require a larger number of poles compared to wideband antenna systems.

The pole structure and the time-domain, realized effective length of the monopole are shown in Figure 4.23. The poles appear structured in three layers, as has been found previously by several authors [19]. It is expected that the first pole layer has the most dominant effect on the antenna system. The realized effective length in the time domain is in good agreement with the 39-pole model.

It is also interesting to consider the results obtained from reducing the number of poles used to represent such a narrowband antenna. Figure 4.24 shows the frequency and time responses of the antenna with only six poles. The data provides a good fit in the time domain, which is dominated by the low-frequency pole, but does

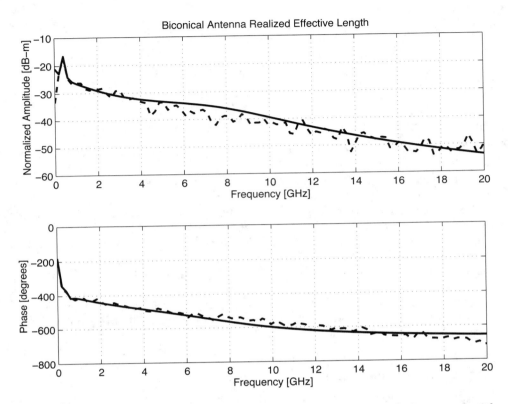

Figure 4.20: The Six-Pole Model of the Biconical Antenna (Solid) Compared with Measured Frequency Domain Data Using a 2.5 ns Time Domain Window (Dashed).

SOURCE: S. Licul, "Unified Frequency and Time-Domain Antenna Modeling and Characterization," Ph.D. dissertation [7]. © S. Licul, 2004. Used by permission.

not match the higher frequencies well. Indeed, the poles shown in Figure 4.25 are basically the dominant poles of the 39-pole model.

## Vivaldi Antenna Model and Frequency-Domain Antenna Patterns

It is possible to obtain the frequency domain patterns from the pole-residue antenna model. The Vivaldi antenna was used to prove this concept. The results show that after the residues and poles are known, the frequency domain pattern can be obtained for any arbitrary frequency. In order to obtain the frequency domain pattern, one needs to obtain residues for different antenna orientations.

The Vivaldi antennas were measured using the setup shown in Figure 4.26. The distance between the reference planes of the measured Vivaldi antennas was 2.0625 meters. The antennas were measured in a VTAG laboratory room using the vector

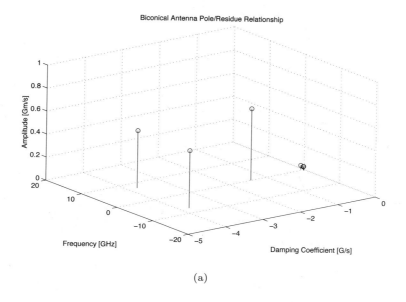

(a)

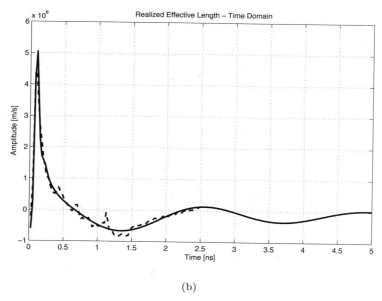

(b)

Figure 4.21: The Six-Pole Model Structure and Resulting Time Domain Waveform for the Biconical Antenna.

SOURCE: S. Licul, "Unified Frequency and Time-Domain Antenna Modeling and Characterization," Ph.D. dissertation [7]. © S. Licul, 2004. Used by permission.

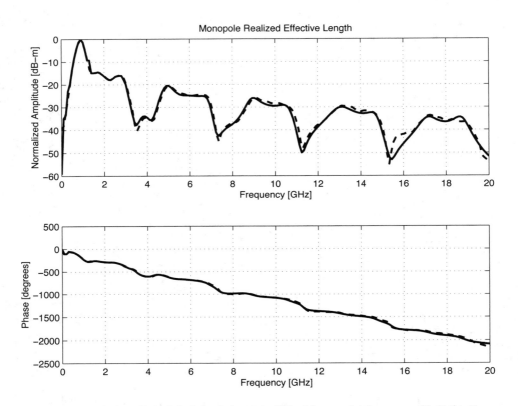

Figure 4.22: The 39-Pole Model of the 1.0 GHz Monopole Antenna (Solid), Compared to Measured Frequency Domain Data Using a 2.5 ns Time Domain Window (Dashed).

SOURCE: S. Licul, "Unified Frequency and Time-Domain Antenna Modeling and Characterization," Ph.D. dissertation [7]. © S. Licul, 2004. Used by permission.

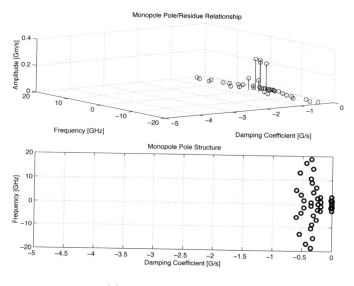

(a) 39-Pole Model Structure

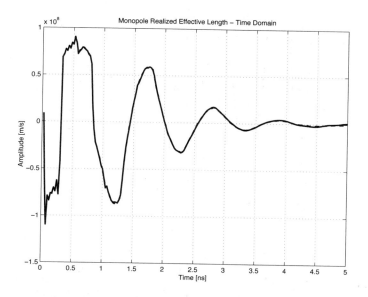

(b) Resulting Time Domain Waveform for the 1.0 GHz Monopole Antenna

Figure 4.23: SOURCE: S. Licul, "Unified Frequency and Time-Domain Antenna Modeling and Characterization," Ph.D. dissertation [7]. © S. Licul, 2004. Used by permission.

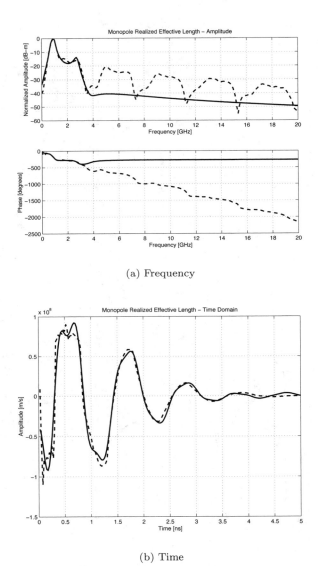

(a) Frequency

(b) Time

Figure 4.24: The Six-Pole Model of the 1.0 GHz Monopole Antenna (Solid) Compared to Measured Frequency and Time Data Using a 2.5 ns Time Domain Window (Dashed).

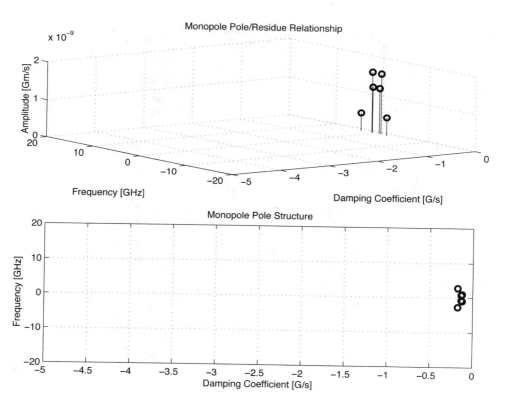

Figure 4.25: The Poles for the Six-Pole Model of the 1.0 GHz Monopole.

SOURCE: S. Licul, "Unified Frequency and Time-Domain Antenna Modeling and Characterization," Ph.D. dissertation [7]. © S. Licul, 2004. Used by permission.

Figure 4.26: The Measurement Setup.

network analyzer. One antenna was stationary while the other antenna was rotated. The measurements were taken in the azimuth plane (Vivaldi antenna E plane) at 5-degree increments for all 360 degrees. For complete characterization one needs to measure a complete sphere surrounding the antenna, but generally an azimuthal measurement is adequate. The concept is demonstrated by a measurement in a single plane.

The accurate description of the Vivaldi antenna requires a 17-pole model. However, the 17-pole model indicates that eight frequencies dominate the antenna response. Figure 4.27 shows the amplitude and phase of the measured (dashed) and 17-pole model approximation (solid) of the Vivaldi antenna response. The model agrees very well with the measured results. The Vivaldi antenna shows a "hybrid" type of operation compared to the monopole and TEM horn antenna models.

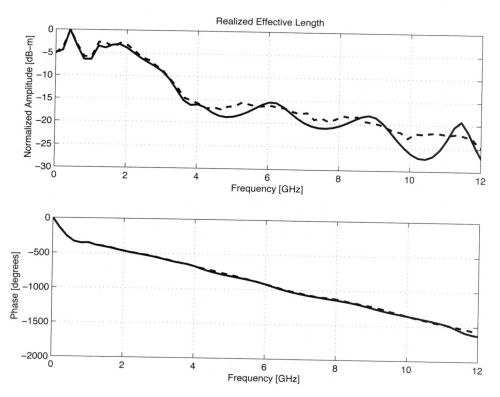

Figure 4.27: The 17-Pole Model of the Vivaldi Antenna (Solid) Compared to Measured Frequency Domain Data Using a 2.5 ns Time Domain Window (Dashed).

SOURCE: S. Licul, "Unified Frequency and Time-Domain Antenna Modeling and Characterization," Ph.D. dissertation [7]. © S. Licul, 2004. Used by permission.

A resonance occurs at low frequencies, yet the response moves toward a constant gain response at higher frequencies.

Figure 4.28a shows the pole structure of the Vivaldi antenna. The pole structure shows that it is possible to reduce the number of poles by eliminating poles with high damping coefficients. However, one needs to take the residues associated with a particular pole into consideration. Large amplitude residues can have a pronounced impact on the model even if the damping coefficients are high. The pole reduction has been pursued and a substantial reduction in the number of required poles is possible. The time-domain match of Figure 4.28b shows that the model approximates the early part of the waveform (<0.5 ns) very accurately.

After all the residues for the azimuth plane are obtained, one can reconstruct frequency domain patterns for any arbitrary frequency based on (4.60). The pattern is the combination of all the residues and poles. The poles used in the pattern calculations in Figure 4.29 are the poles obtained from the boresight direction. It may be necessary to evaluate the poles for different antenna directions in order to ensure a correct set of poles to represent the antenna system. The association of the poles from multiple directions can be used to reduce experimental error in the pole extraction. As an alternative, the poles obtained from the boresight can be used for all directions, with care taken to not miss some poles that cannot be represented adequately at the boresight. The poles obtained from the boresight measurement of the Vivaldi antenna provide a good choice of poles.

The field patterns, shown in black, were obtained by using the antenna in an anechoic chamber. The antenna was mounted on a positioner consisting of a metal mast surrounded by absorber. The measurements partially reflect from the positioner and are blocked by the positioner when the antenna is directed away from the test probe.

Indeed, it is possible to develop a useful minimal model representation for an antenna system. Pole modeling of antenna systems provides complete frequency and time domain antenna characterization with a minimal representation (minimal data set required to represent an antenna). The corresponding model can be used with channel models in order to predict the received time-domain or frequency responses. The antenna pole structure is the same regardless of the antenna orientation. The residues are obtained from the transient responses at different angles.

## 4.2.6  Transient Responses of Scatterers

The singularity expansion method (SEM) [20] allows characterization of canonical objects in terms of poles and residues from a measured time domain system response. The data collected through transient measurements were postprocessed using Prony's algorithm to obtain a pole response. The models considered include canonical structures of a sphere, cylinder, and straight edge of different dimensions. The data were recorded for the scatter data from these canonical objects based on equivalent time-domain measurements using an HP8510.

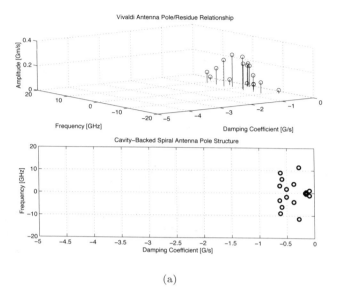

(a)

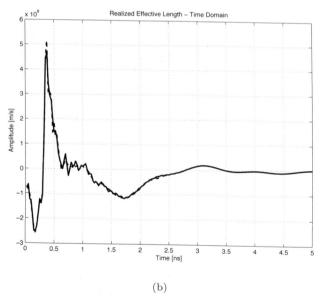

(b)

Figure 4.28: The (a) 17-pole Model Structure and (b) Resulting Time Domain Wave form for the Vivaldi Antenna.

SOURCE: S. Licul, "Unified Frequency and Time-Domain Antenna Modeling a Characterization," Ph.D. dissertation [7]. © S. Licul, 2004. Used by permission

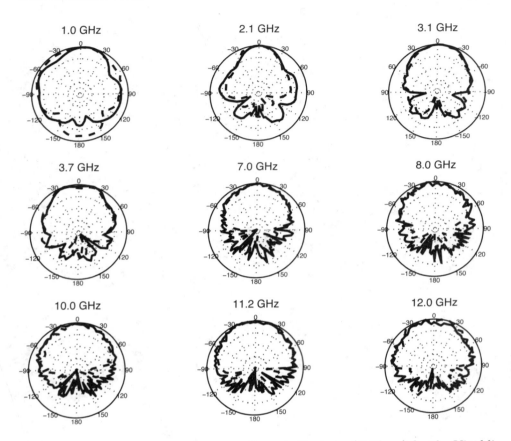

 gure 4.29: The Normalized Frequency Domain Patterns (E Plane) for the Vivaldi
enna (Dashed: Model; Solid: Near-Field Measurements).

CE: S. Licul, "Unified Frequency and Time-Domain Antenna Modeling and
cterization," Ph.D. dissertation [7]. © S. Licul, 2004. Used by permission.

## ment Setup Using Planar TEM Horn Antennas

measurements were conducted to characterize canonical objects, one
TEM horn antennas with a ground plane and one using wideband Vi-
without a ground plane. A 4 x 4 feet square ground plane was used
r TEM horn antennas mounted at diagonal corners of the ground
l unobstructed line-of-sight measurement was conducted with the
each other, which was used to deconvolve the effects of the anten-
tennas were oriented toward a third corner, as shown in Figure
data was recorded with the antennas pointed toward the third
t. This reference response was used to subtract the effect of the

201

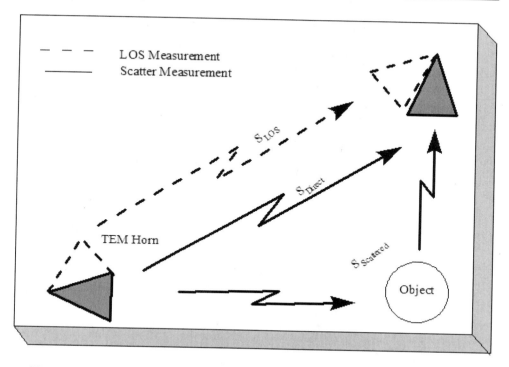

Figure 4.30: Scatter Measurement Setup Using Planar TEM Horn Antennas.

surrounding environment and the direct path. The objects were placed at a distance of 28 inches from the antenna feed point. The edge of the ground plane was covered with absorber to prevent diffraction from the edge, as well as to prevent scattering from the surrounding environment.

The scattered data was postprocessed to subtract the effect of the surrounding region and the direct path, as well as to deconvolve the effects of the antennas used for the measurement. The first canonical object used was a hemisphere of six inches in diameter. The image from the ground plane allowed the hemisphere to behave like a sphere in free space. The transient response from the hemisphere is shown in Figure 4.31. The response consists of a specular component followed by an apparent resonant component. The specular reflection is inverted for positive pulse excitation.

The frequency response of the hemisphere is shown in Figure 4.32. Periodic nulls occur with a staggered stair-like response. The periodic frequency of the ringing observed in Figure 4.31 seems evident from the peak observed at 700 MHz. The nearly flat high frequency response illustrates the smooth specular component. The extracted complex poles for the hemisphere are shown in Figure 4.33. A total of 25 pairs of complex-conjugate poles were extracted and illustrate the absence of double

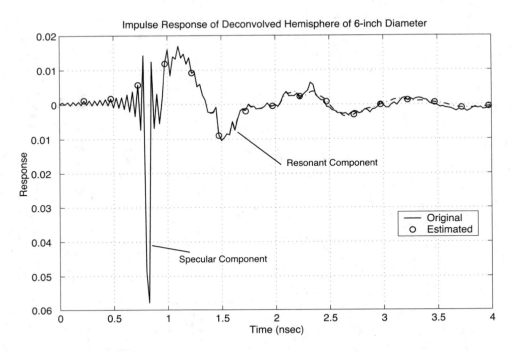

Figure 4.31: Transient Scatter Response of a Hemisphere of Six-Inch Diameter.

poles, with the two poles of about 700 MHz dominating the resonate part of the response.

Additional plots for other canonical objects are also shown in Figures 4.34–4.36. These plots include a cylinder four inches in diameter, the edge of a cube of side length 90 mm, and the edge of a 4 feet x 4 feet conducting surface. The study of such canonical shapes is expected to provide additional information into the performance of multipath scattering in a UWB environment. Most of the objects considered demonstrate both a resonant and a specular response. A large edge, such as might be expected in a doorway or large building, is not expected to have the same resonant response, but a response characteristic of diffraction of an edge or corner. Though this research is in its infancy, it provides insight into the variations of the responses that may be needed for appropriate modeling of transient and frequency multipath scattering.

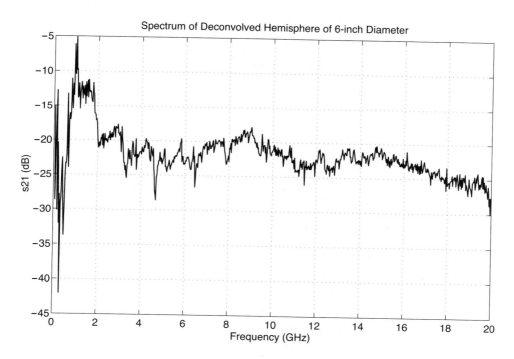

Figure 4.32: Frequency Scatter Response of a Hemisphere of Six-Inch Diameter Using Planar TEM Horn Antenna.

SOURCE: R. M. Buehrer, W. A. Davis, A. Safaai-Jazi, and D. Sweeney, "Ultra-wideband Propagation Measurements and Modeling," [22]. © Mobile and Portable Radio Research Group, 2004. Used by permission.

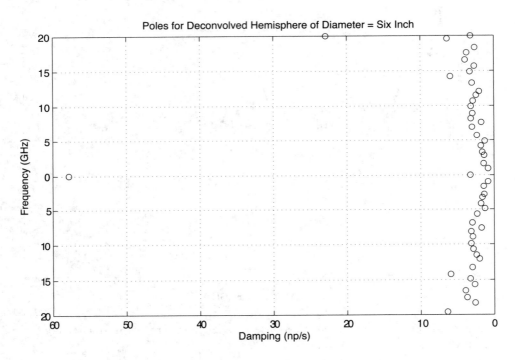

Figure 4.33: Pole Plot of a Hemisphere of Six-Inch Diameter.

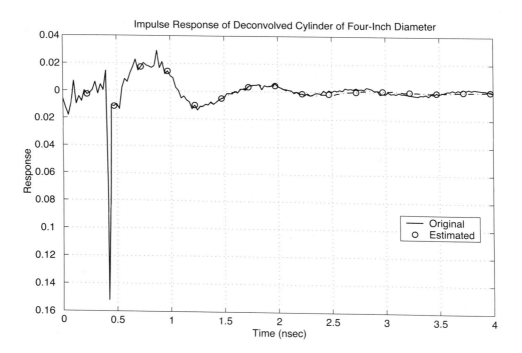

Figure 4.34: Transient Scatter Response of a Cylinder of Four-Inch Diameter.

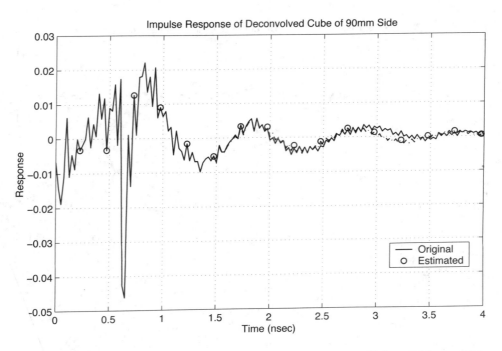

4.35: Transient Scatter Response of an Edge from a Cube of 90 mm Side

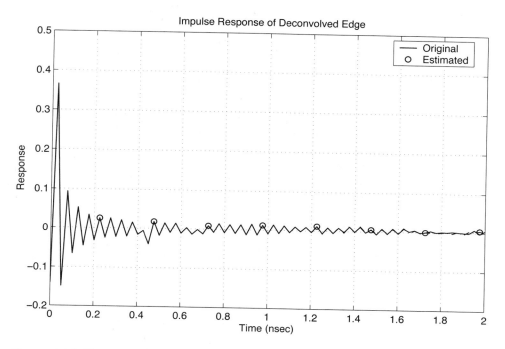

Figure 4.36: Transient Scatter Response of an Edge (Four Feet).

## 4.3    Summary

In this chapter, the reciprocity theorem has been used along with basic circuit relationships to evaluate the form of the various parameters of antennas, including effective aperture, effective length, mismatch loss, efficiency, polarization match, and related information. The advantage of the approach presented in this chapter is the formal process through which all the aspects of the antennas and environment can be integrated theoretically. A connection to a link budget form for particular transmit energy spectra or waveforms was also considered briefly. The link budget is expanded in concept in Chapter 2.

The Impulse (I-UWB) implementation of UWB uses impulses, approximated by nanosecond-wide Gaussian pulses, to communicate. For MC-UWB systems, each carrier band can be considered as a separate, narrowband signal. In the latter case, classic link analysis using antenna gain is sufficient. However, the wideband characteristics of the impulse response can be applied to both cases, and still identifies the basic features of antenna transmission needed for the reliability and capacity of UWB communications.

Measurements of different antenna types with both frequency and time domain data were presented to provide insight into the performance of the antenna and the link. In order to provide a better evaluation of the antennas, and also to provide a simple model for link simulation, the singularity expansion method (SEM) was used to characterize the line-of-sight link, specifically the antennas. The theory of the SEM is based on Maxwell's equations and observations of measured data for many pulsed electromagnetic systems. It was found that typical finite sized objects and antennas can be characterized as a set of poles, with residues that describe the angular performance of the structure. This representation leads to a minimal form of modeling that provides both the minimal property and insight into the performance of the antenna systems.

A measurement system was developed for determining the scattering properties of several canonical objects. The scattering for finite-sized objects was described in terms of poles and residues from the measured time-domain system response. A sphere and a cylinder were two of the canonical objects used for characterization. The data collected through transient measurements were postprocessed to obtain the poles and residues and to represent the frequency and time domain nature of the canonical structures.

# Bibliography

[1] W. L. Stutzman and G.A. Thiele, *Antenna Theory and Design*, 2nd ed., New York: John Wiley & Sons, 1998.

[2] W. L. Stutzman and W. A. Davis, "Antenna Theory," *Wiley Encyclopedia of Electrical and Electronics Engineering*, vol. 1, Webster, John G., ed., New York: John Wiley & Sons, 1999.

[3] K. Kurokawa, "Power waves and the scattering matrix," *IEEE Transactions on Microwave Theory and Techniques*, vol. MTT-13, pp. 194-202, 1965.

[4] Antenna Standards Committee of the IEEE Antennas and Propagation Society, IEEE Standard Definitions of Terms for Antennas, IEEE Std 145-1993, IEEE, New York, 1993.

[5] E. G. Farr and C. E. Baum, "Time Domain Characterization of Antennas with TEM Feeds," *Sensor and Simulation Notes*, Note 426, Air Force Research Laboratory, October 1998.

[6] L. J. Chu, "Physical limitations on omni-directional antennas," *Journal of Applied Physics*, vol. 19, pp. 1163-1175, December 1948.

[7] S. Licul, "Unified Frequency and Time-Domain Antenna Modeling and Characterization," Ph.D. dissertation, to be published, Virginia Tech.

[8] S. N. Samaddar and E. L. Mokole, "Some Basic Properties of Antennas Associated With Ultrawideband Radiation," *Ultra-Wideband, Short-Pulse Electromagnetics 3*, Baum et al., eds., New York: Plenum Press, 1997.

[9] C. C. Bantin, "Radiation from a Pulse-Excited Thin Wire Monopole," *IEEE Antennas and Propagation Magazine*, vol. 43, no. 3, June 2001.

[10] C. C. Bantin, "Pulsed Communication Link Between Two Dipoles," *IEEE Antennas and Propagation Magazine*, vol. 44, no. 5, October 2002.

[11] G. S. Smith, "On the Interpretation for Radiation from Simple Current Distributions," *IEEE Antennas and Propagation Magazine*, vol. 40, no. 4, August 1998.

[12] I. M. Besieris, "Far-Zone Radiation of Pulsed Thin-Wire Antennas," Personal Notes.

[13] L. B. Felsen, ed., *Transient Electromagnetic Fields*, Berlin: Springer-Verlag, 1976.

[14] C. E. Baum, L. Carin, and A. P. Stone, eds., *Ultra-wideband, Short-Pulse Electromagnetics 3*, New York: Plenum Press, 1997.

[15] M. Kanda, "The effects of resistive loading of 'TEM' Horns," *IEEE Transactions on Electromagnetic Compatibility*, vol. EMC-24, pp. 245-255, May 1982.

[16] M. L. Van Blaricum and R. Mittra, "Problems and Solutions Associated with Prony's Method for Processing Transient Data," *IEEE Transactions on Antennas and Propagation*, vol. 26, no. 1, pp. 174-182, January 1978.

[17] Y. Hua and T. K. Sarkar, "Matrix Pencil Method for Estimating Parameters of Exponentially Damped/Undamped Sinusoids in Noise," *IEEE Transactions on Acoustics, Speech, and Signal Processing*, vol. 38, no. 5, pp. 814-824, May 1990.

[18] C. E. Baum, "On the Singularity Expansion Method for the Solution of Electromagnetic Interaction Problems," *Interaction Notes*, Note 88, Air Force Research Laboratory, December 1971.

[19] F. M. Tesche, "On the Analysis of Scattering and Antenna Problems Using the Singularity Expansion Technique," *IEEE Transactions on Antennas and Propagation*, vol. AP-21, no. 1, pp. 53-62, January 1973.

[20] C. E. Baum, "The singularity expansion method," *Transient Electromagnetic Fields*, L. B. Felsen, ed., pp. 129-179.

[21] G. Joshi, "Ultra-Wideband Channel Dispersion: Characterization and Modeling," Ph.D. Preliminary, Virginia Tech, 2004.

[22] R. M. Buehrer, W. A. Davis, A. Safaai-Jazi, and D. Sweeney, "Ultra-wideband Propagation Measurements and Modeling," DARPA NETEX Program Final Report, January 31, 2004.

# Chapter 5

# TRANSMITTER DESIGN

Dennis Sweeney,

Dong S. Ha,

Annamalai Annamalai,

and Sridharan Muthuswamy

In this chapter, we focus on two themes. We first describe some of the widely used signal generation techniques unique to UWB communication systems. This is followed by a comprehensive review of different modulation/signaling schemes for UWB transmission. Transmitters for UWB systems fall into two broad categories: impulse-based communication (I-UWB) and multicarrier-based modulation (MC-UWB). While their construction is quite different, they do share problems common to any broadband system, such as signal fidelity during wideband amplification.

The organization of this chapter is as follows. In Sections 5.1 and 5.2 we describe several signal generators and modulators for I-UWB. Section 5.3 illustrates different modulation formats for the transmission of I-UWB signals. Section 5.4 describes multicarrier (MC) modulation for UWB systems. Section 5.5 examines transmitter structure for spectrally encoded UWB systems and discusses its merits and implementation issues. In Section 5.6 we provide some concluding remarks.

## 5.1  I-UWB Signal Generators

The heart of any I-UWB system is some type of fast-rise time step or pulse generator. These pulse generators are used for both transmitting and receiving. I-UWB transmitters convert data bits directly to fast-rise time pulses. Matched filter correlation receivers must generate a template pulse that matches the incoming waveform.

Conventional transmitters, including UWB transmitters, employing Orthogonal Frequency Division Multiplexing (OFDM) typically generate a continuous wave (CW) signal and then modulate this signal in some fashion. The modulation modifies the signal amplitude, phase, and/or frequency. While this modulation can be characterized in both the time and frequency domain, it is common to do the analysis in the frequency domain. CW transmitter circuitry employs familiar frequency synthesizers, mixers, modulators, and amplifiers.

The circuits and techniques described in this section are typical of transmitters used for I-UWB communications and ranging. I-UWB transmitters generally employ quite different circuitry from that used in continuous wave transmitters. Circuits that generate fast-rise time pulses are common. The modulation tends to be related to pulse timing so it lends itself more to time domain analysis. It is difficult to amplify I-UWB pulses due to the peak powers involved. High-voltage pulser circuits are attractive. In addition, the limited bandwidth and interface mismatch of filters, amplifiers, and mixers distort fast-rise time pulses. The pulse generator is often part of or connected directly to the radiating element in an I-UWB transmitter.

## 5.1.1   Avalanche Pulse Generators

A transistor driven into avalanche breakdown can produce a very fast-rise time pulse. Early UWB communications systems employed avalanche transistors for both the transmitter and receiver [1]. A complete description of the avalanche process is beyond the scope of this book. However, some understanding of the avalanche process is important. The process begins when the electric field across a reverse biased junction is great enough to pull electrons off the semiconductor atoms [2]. These electrons (and their corresponding holes) are accelerated through the semiconductor crystal and collide with other atoms, creating additional electron-hole pairs. These electron-hole pairs then produce additional collisions, producing a positive feedback where the number of charge carriers grows rapidly. The current through the junction quickly rises to a value limited only by external circuit restraints. Uncontrolled avalanche usually destroys the junction, but if the external current is limited an extremely fast-rise time step can be produced.

Generally, avalanche is to be avoided, so relatively little about it appears in textbooks. Information about avalanche for pulse generation can be found in [3]. Figure 5.1 shows the conditions for operating in the avalanche breakdown region. The safe operating active region (SOAR) defined by C-D-E in Figure 5.1 is the normal active region for bipolar junction transistors (BJT). $BV_{CEO}$ is the collector-base breakdown voltage with the emitter open, and $BV_{CEX}$ is the collector emitter breakdown voltage with the base reversed biased [2]. For avalanche breakdown, the device is biased somewhere between $BV_{CEO}$ and $BV_{CEX}$ with the base biased at zero volts or a small negative voltage. Figure 5.2 is a circuit that can be used to produce an avalanche pulse.

In Figure 5.2, the base is driven positive with the trigger pulse and the device begins to conduct. The instant conduction begins, the device goes into avalanche breakdown. The collector emitter voltage follows the dynamic load line from point A to B as shown in Figure 5.1. The dynamic load line defined by $C_O$ and $R_L$ is entirely outside the SOAR, and $V_{CE}$ rapidly collapses to point B because the entire load line is in the avalanche region. The DC load line that traverses the active region in Figure 5.1 is defined by $R_C$. As the device avalanches, the charge in $C_O$ is quickly dumped through the device into $R_L$; the result is a fast-rising edge. The rise

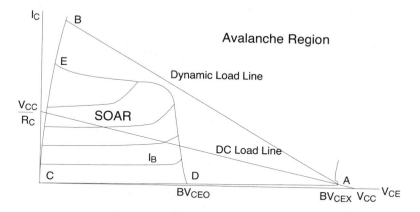

Figure 5.1: Plot of $I_C$ Versus $V_{CE}$ Defining the Avalanche and Normal Operating Region for a BJT.

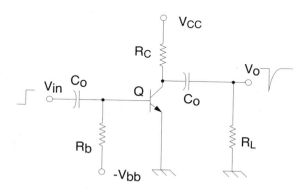

Figure 5.2: Avalanche Pulse Generator Circuit.

time is effectively limited by the inductance in the loop formed by the transistor, $C_O$, and $R_L$. It is possible to obtain subnanosecond rise time edges with microwave construction techniques.

As $C_O$ is discharged, the collector operating point passes from point B in Figure 5.1 to point C. The device turns off and returns to point A to await retriggering; $C_O$ is recharged through $R_C$. A major limitation on avalanche pulse generators is the rate at which the device can be retriggered. The values of $C_O$ and $R_C$ control the current through the device to limit its dissipation. The value for $R_C$ must be large enough to prevent destruction of the device, but if $R_C$ is larger it will take longer to recharge $C_O$.

Figures 5.3 and 5.4  show two variations on the avalanche circuit. In Figure 5.3, the load is in the emitter of the transistor and a positive going pulse is obtained. The circuit in Figure 5.4 replaces the $C_O$ storage capacitor with a section of open circuited transmission line. When the transistor goes into avalanche breakdown, it effectively becomes a short circuit, and the energy stored in the transmission line begins to be discharged into the load. The discharge current propagates to end of the transmission line where the open circuit boundary condition causes an inverted reflection current to propagate down the line in the opposite direction. When this current reaches the collector, the sum of the reflection current and the discharge current is zero. The transistor collector current falls to zero, and the transistor returns to the nonconducting state. The result is a square pulse whose duration is twice the propagation time down the transmission line.

Avalanche breakdown is difficult to characterize, and it is temperature- and process-dependent [2,3]. The same transistor part from different manufacturers may exhibit different avalanche characteristics. It is often necessary to select devices. The literature [1,4,5] shows that garden-variety silicon transistors, such as the 2N3904, 2N2222, and 2N918, seem to work well. The high voltages required for avalanche breakdown make battery operation difficult.

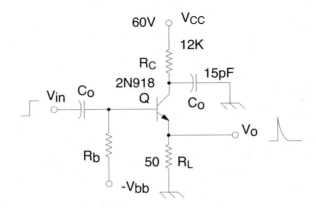

Figure 5.3: Positive Going Avalanche Pulse Generator.

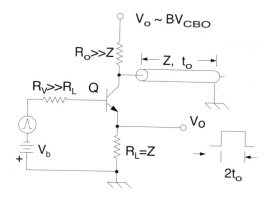

Figure 5.4: Square Pulse Avalanche Pulse Generator.

SOURCE: H.-M. Rein, "Subnanosecond-Pulse Generator with Variable Pulsewidth Using Avalanche Transistors," *IEEE Electronics Letters* [4]. © IEEE, 1975. Used by permission.

## 5.1.2   Step Recovery Diode Pulse Generators

Step Recovery Diodes (SRD) or "snap off" diodes can be used to make very fast-rise time pulses. SRDs were first commercially introduced by Hewlett Packard (now Agilent) for use as harmonic generators and as pulse generators in instruments that employed sampling. The Hewlett Packard Application Note 918 is an excellent introduction to pulse shaping and forming with the SRD [6].

Conventional diodes conduct when forward biased and shut off when reverse biased. SRDs have a P-I-N structure, and charge is stored in the intrinsic layer when the SRD is forward biased. This allows the SRD to continue to conduct when the device is reverse biased. This reverse conduction continues until the charge is swept out of the intrinsic layer; with the charge gone, the diode abruptly stops conducting and "snaps off." The SRD presents low impedance during the forward/reverse conduction and transitions to high impedance when it snaps off.

Figure 5.5 plots charge versus time when a current is applied. During the forward bias condition, minority carriers are injected into the junction. Due to the intrinsic layer, recombination requires a finite amount of time so charge is stored in the junction.

Let $i_d$ be the instantaneous diode current, $Q$ the charge stored, and $\tau$ the carrier lifetime. We have the following

$$i_d = \frac{dQ}{dt} + \frac{Q}{\tau} \tag{5.1}$$

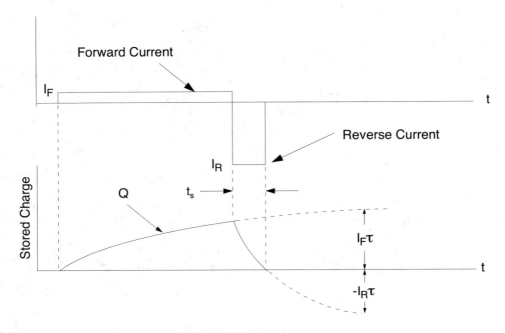

Figure 5.5: SRD Current and Stored Charge.

SOURCE: Application Note 918, Pulse and Waveform Generation with Step Recovery Diodes [6]. © Agilent Technologies Archives, 1984. Used by permission.

When a constant forward bias current is applied to the diode, charge is stored in the junction. The charge for a forward constant bias current is given by

$$Q_f = I_F \tau \left( 1 - e^{-t_F/\tau} \right) \tag{5.2}$$

where $Q_f$ is the charge stored due to forward current and $t_F$ is the time that forward current is applied. If the forward current is applied for a time significantly longer than the carrier lifetime, $\tau$, the stored charge can be approximated by

$$Q_f \approx I_F \tau \tag{5.3}$$

When the reverse current is applied, the stored charge is depleted. The depletion time is given by

$$\frac{t_s}{\tau} = \ln \left[ 1 + \frac{I_F \left( 1 - e^{-t_F/\tau} \right)}{I_R} \right] \approx \ln \left[ 1 + \frac{I_F}{I_R} \right] \quad t_F \gg \tau \tag{5.4}$$

where $t_s$ is the time to remove the stored charge and $I_R$ is the reverse current. Typical SRD circuits operate with $\frac{I_F}{I_R} \ll 1$, so (5.4) can be further simplified as

$$\frac{t_s}{\tau} \approx \frac{I_F}{I_R} \qquad (5.5)$$

The value of $\tau$ is specified by the diode manufacturer. Typical values of $\tau$ range from 10-100 nanoseconds, so with the proper choice of $I_F$ and $I_R$, it is possible to reduce the rise time of the transition applied to the SRD. Transition times range from 35-250 picoseconds. The practical limit on the minimum transition time is generally set by diode package parasitic capacitances. Figure 5.6 is a pulse sharpener circuit. $V_b$ and $R_b$ establish $I_F$ in the diode. The positive going input pulse has the effect of reducing the amplitude of $I_F$. As the input voltage becomes more positive, the diode current $i_d$ reverses and charge depletion begins. After the discharge period, $t_s$, the charge is totally depleted and the diode snaps off. During the time that the diode is conducting, it represents low impedance, and the voltage across it is low. When the diode snaps off, the voltage across the diode quickly rises as the diode transitions to a high impedance state. The diode acts as a charge controlled switch. The voltage waveforms are shown in Figure 5.7.

SRD pulse sharpeners can be cascaded for even faster rise time. The literature reports a 70 picosecond rise time using an avalanche transistor driving two SRD pulse sharpeners [7]. Additional pulse sharpener circuits can be found in [6]. With the addition of transmission line circuits, it is possible to obtain impulse and mono-cycle outputs with SRDs. Figure 5.8 shows an SRD square pulse generator. When the SRD snaps off, a step is propagated in both directions away from the diode. The step arrives at the shorted end of the transmission line stub and is reflected back inverted. The first step arrives at the output, and the inverted step arrives after the round trip delay in the stub. The sum of the two steps produces a square pulse with a width that is approximately the round trip delay.

The addition of an R-C differentiator to the circuit in Figure 5.8 produces a monocycle output, as shown in Figure 5.9. A more sophisticated version of this circuit that suppresses the inevitable ringing that occurs as result of the impedance mismatch caused by the differentiator is available in the literature [8,9].

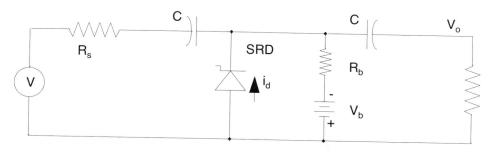

Figure 5.6: Pulse Sharpener Circuit Employing an SRD.

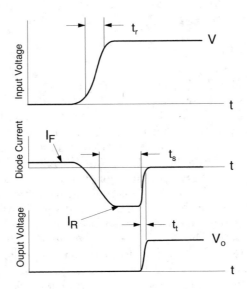

Figure 5.7: Voltage Waveforms in Pulse Sharpener Circuit.

SOURCE: Application Note 918, Pulse and Waveform Generation with Step Recovery Diodes [6]. © Agilent Technologies Archives, 1984. Used by permission.

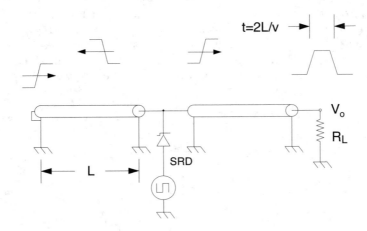

Figure 5.8: SRD Pulse Generator.

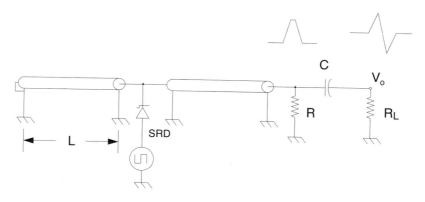

Figure 5.9: SRD Monocycle Generator.

SRDs can act as impulse generators as well as step generators [10,11]. A suitable circuit for an SRD impulse generator is shown in Figure 5.10. When the diode is forward biased, charge is stored in the diode. The forward conduction capacitance $C_{Fwd}$ is assumed to be very large, and the diode essentially acts as a short circuit. During forward conduction, energy is stored in the inductor $L$. Just prior to the transition from $C_{Fwd}$ to the reverse capacity $C_{VR}$, the energy stored in the inductor is given by $\left(\frac{1}{2}\right) I_p'^2 L$. At the transition, energy is transferred from the inductor to $C_{VR}$. This creates an impulse of amplitude $E_p'$ and width $t_p'$, as shown in Figure 5.11. It is assumed that the energy in the impulse is equal to the energy stored in the inductor just prior to the transition. This is the case if the loading from $R_L$ is assumed to be light.

$$\frac{1}{2}I_p'^2 L \approx \frac{1}{2}E_p'^2 C_{VR} \qquad (5.6)$$

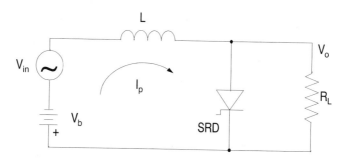

Figure 5.10: SRD Impulse Generator.

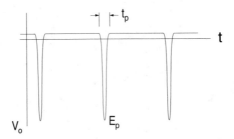

Figure 5.11: Output of SRD Impulse Generator.

and

$$E'_p \approx I'_p \sqrt{L/C_{VR}} \tag{5.7}$$

The pulse width $t'_p$ can be approximated by the resonant frequency of the circuit formed by $L$ and $C_{VR}$. This is given by

$$t'_p \approx \pi \sqrt{LC_{VR}} \tag{5.8}$$

Loading the circuit with $R_L$ tends to decrease the amplitude of $E'_p$ and increase the pulse width $t'_p$. The loaded amplitude and pulse width are given by

$$E_p = E'_p \exp\left(-\pi\zeta/2\sqrt{1-\zeta^2}\right) \tag{5.9}$$

$$t_p = \frac{\pi\sqrt{LC_{VR}}}{\sqrt{1-\zeta^2}} \tag{5.10}$$

Zeta is the loading factor which is controlled by $R_L$ and the other circuit parameters

$$\zeta = \frac{1}{2R_L}\sqrt{\frac{L}{C_{VR}}} \tag{5.11}$$

For light loading, $\zeta \ll 1$, the power in the impulse train can be calculated as

$$P_{impulse} = \frac{E_p^2 \pi \zeta C_{VR} f_i}{\sqrt{1-\zeta^2}} \approx E_p^2 \pi \zeta C_{VR} f_i \tag{5.12}$$

### 5.1.3  Tunnel Diode Pulsers

Tunnel diodes are PN junction devices that are capable of producing very fast transitions. A tunnel diode differs from a conventional diode in that the semiconductor doping is a factor of 1,000 or greater than that used in a conventional diode. Tunnel diodes are fabricated from germanium or gallium arsenide, and the resulting high

conductivity semiconductor material produces a diode with a very thin junction
(depletion region). Due to quantum effects it is possible for electrons to tunnel directly through the thin junction even when they do not have sufficient energy to
surmount the junction potential barrier.

Under reverse bias, no majority charge carriers are injected into the junction,
but the diode still conducts. This is due to the tunneling of valance electrons of
the semiconductor atoms close to the junction. With a small forward bias, valence
electrons continue to tunnel through the junction even though they do not have the
requisite energy to overcome the junction potential. As the forward bias is increased,
the energy of the free electrons in the N region becomes greater than valence electrons in the P region, and the tunneling current decreases. The free electrons still
do not have enough energy to cross the junction, so the total current decreases. The
result is a negative resistance region. As the forward bias is increased still further,
normal diode conduction ensues. Figure 5.12 shows the I-V characteristic of a tunnel
diode.

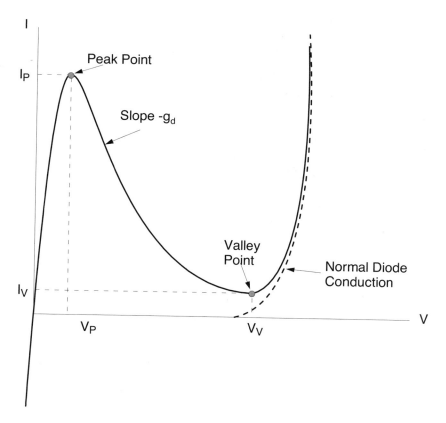

Figure 5.12: Tunnel Diode I-V Characteristic.

Tunnel diodes are typically specified in terms of the peak point, $I_P$ and $V_P$, the valley point, $I_V$ and $V_V$, and the slope of the negative resistance region, $-g_d$. An example is the 1N3716 (TD-3) with $I_P = 4.7$ mA, $V_P = 55$ mV, $I_V = 1.04$ mA, $V_V = 350$ mV, and $g_d = 0.04$ S. A bistable circuit that produces a fast-rise time step with a tunnel diode is shown in Figure 5.13. The diode is biased to point A just below $V_P$ with a current through the diode. A small trigger voltage is applied across the diode. When the sum of the trigger and the bias voltage exceeds $V_P$, the current source forces the diode over into the unstable negative resistance region. The diode voltage rises until it reaches the stable point, B. The result is a very fast rise in the voltage across the diode. The diode may be reset by momentarily forcing the diode voltage below $V_P$ or by interrupting the diode current. Because the tunnel diode produces a step, transmission line circuits similar to the ones used with the SRD can also employ tunnel diodes.

A major limitation on tunnel diode pulsers is the size of the step. The step size is limited to approximately 250–500 mV. In addition, the bias point is temperature sensitive, so a stable trigger point can be difficult to maintain. The current

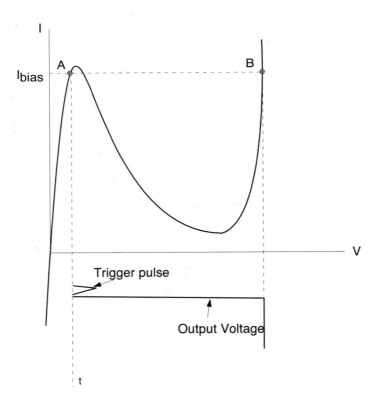

Figure 5.13: Tunnel Diode Bistable Operation.

applications for tunnel diode are limited to microwave detectors and pulse generator circuits. Because of the limited applications, tunnel diodes are expensive and can be difficult to obtain. The small step size limits the use of tunnel diodes in I-UWB transmitters, but the tunnel diode bistable circuit still has utility as a I-UWB receiver.

### 5.1.4   Pulse Circuits Suitable for Integrated Circuits

The UWB circuits shown thus far have limited utility in integrated configurations. It is highly desirable to design I-UWB transmitters that are compatible with CMOS and other IC technology. The following discusses several integrated pulse generator implementations.

**Scholtz Monocycle Generator**

An integrated implementation of the Scholtz monocycle generator was published in [12]. The circuit is shown in Figure 5.14. Q1–Q5 form a squaring circuit, and the

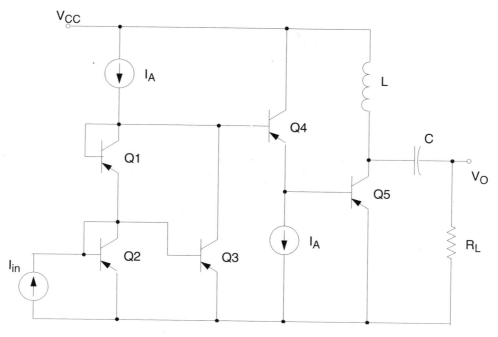

Figure 5.14: Scholtz Monocycle Generator Using BJT's.

SOURCE: H. Kim, D. Park, and Y. Joo, "Design of CMOS Scholtz's Monocycle Pulse Generator," *IEEE Conference on Ultra Wideband Systems and Technologies* [12]. © IEEE, 2003. Used by permission.

L and the C perform a double differentiation on Q5's collector current.

The circuit squares the input current, and a Gaussian pulse can be approximated by squaring a tanh(t) input current. As the circuit squares the input current, a Gaussian pulse can be approximated by a $\mathrm{sech}^2(t)$ function.

$$y(t) \propto e^{-t^2} = 1 - t^2 + \frac{t^4}{2!} - \frac{t^6}{3!} + \cdots \tag{5.13}$$

$$e^{-t^2} = \mathrm{sech}^2(t) = 1 - t^2 + \frac{16t^4}{24} - \frac{152t^6}{720} + \cdots \tag{5.14}$$

The input current is a tanh(t) function that can be obtained by the exponential function in the BJT. The collector current of Q5 in Figure 5.12 is the square of the input current

$$\begin{aligned} I_{CQ5} &= a - I_{in}^2 \\ &= a - \tanh^2(t) \\ &= a - \left(1 - \mathrm{sech}^2(t)\right) = b - \tanh^2(t) \end{aligned} \tag{5.15}$$

The collector current of Q5 is a Gaussian impulse. Due to the inductor in Q5's collector circuit, the collector voltage is proportional to the derivative of the collector current. The resulting collector voltage is a Gaussian monocycle. Due to the coupling capacitor, the voltage across the load is the derivative of the collector voltage. The output voltage is the second derivative of Q5's collector current. The resulting output is a Scholtz monocycle. Figure 5.15 is a graphic representation of the voltages and currents.

## Emitter Coupled Logic

Emitter Coupled Logic (ECL) is another technology that can be integrated. The 10E/100E series ECL logic can drive coaxial transmission lines, and they have approximately an 800 mV output swing and a 400 picosecond rise time. Figure 5.16 depicts a typical ECL OR/NOR gate. ECL logic is built around a differential amplifier. ECL may be fast, but its power consumption is high because all the devices are biased into the active region. Figure 5.17 is an input output voltage characteristic of a 10E series ECL. The voltage swing is approximately one volt, and ECL can drive 50 ohm loads.

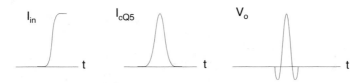

Figure 5.15: Input and Output Current and Output Voltage for the Monocycle Generator in Figure 5.14.

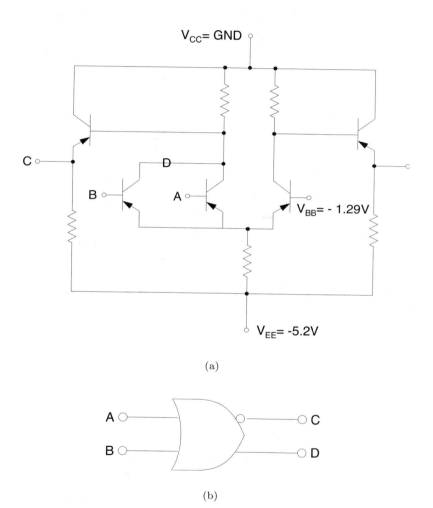

(a)

(b)

Figure 5.16: ECL Gate.

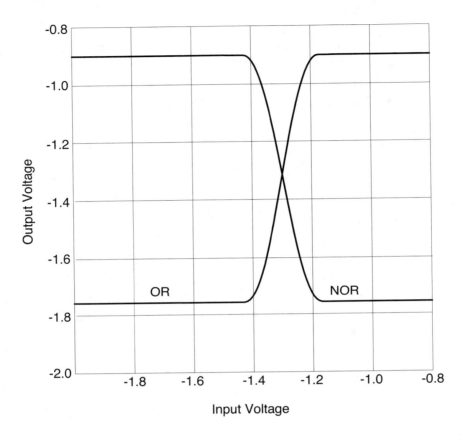

Figure 5.17: 10E Series ECL Input/Output Voltage Characteristic.

## Differential Circuits, The H Bridge

ECL gates are built around differential amplifiers. Differential drive is the basis
for an additional I-UWB transmitter that is suitable for integration. Figure 5.18
shows an H bridge circuit. The antenna forms the cross bar in the H, and a set of
complementary switches forms the vertical bars of the H. As point A is driven low,
point B is simultaneously driven high. The result is that point D goes low at the
same time C is driven high. The effect is to double the slew rate. It is possible to
drive A high while driving B low to generate a pulse of the opposite polarity.

The disadvantage of the H bridge is that neither side of the load (the antenna
in Figure 5.18) can be grounded. However, this is not a problem for current loop
antennas of the type used by Harmuth [13, 14]. Harmuth uses a BJT H bridge to
drive his current loop antennas. The gate and H bridge circuits lend themselves to
pulse shaping. The pulse length and timing can be controlled by the input signal.

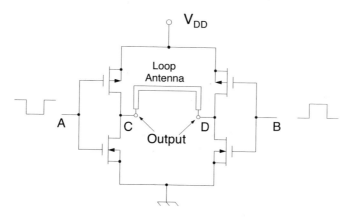

Figure 5.18: H Bridge Transmitter Output Stage.

A loop antenna and an H bridge circuit is employed in the transmitter of Aether Wire's location device [15]. With a loop antenna, the radiated energy is a function of the derivative of the current running through it. Gaussian impulses are radiated with an exponential rise or fall of current. This system lends itself to the transmission of doublets as shown in Figure 5.19. Strings of consecutive impulses of the same polarity will require ever increasing or decreasing antenna current. The doublets have zeros in their spectrum at DC and multiples of $1/t_o$.

The Aether Wire location device employs doublet-encoded sequences. Each doublet is a "chip" in conventional spread spectrum terminology. The radiated spectrum can be controlled by the doublet timing and the time between the doublets. An example of the transmitted pulses is shown in Figure 5.20. A "1" or a "0" can be transmitted by changing the polarity of the pulses.

Aether Wire employs a type of sliding correlator called a time integrating correlator [16] in the receiver. The polarity of the correlator in the receiver is controlled

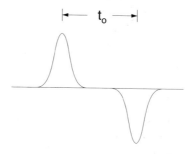

Figure 5.19: Gaussian Doublet Generated by H Bridge Transmitter.

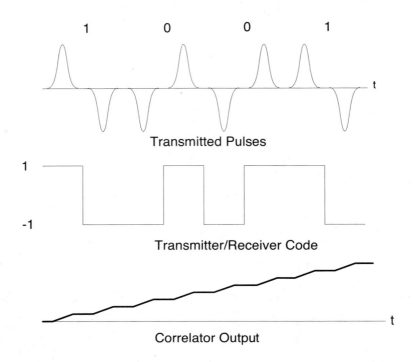

Figure 5.20: Gaussian Doublet and Code Used by Aether Wire Location Devices.

by the code. If the receiver code matches the transmitter code, the output of the correlator continues to grow as shown in Figure 5.20.

## Programmable CMOS Pulse Generator

A CMOS pulse generator circuit is proposed by Marsden, Lee, Ha, and Lee in [17]. Figure 5.21 shows a pulse generator circuit that is capable of synthesizing different waveforms by controlling the timing and sequence of transistor switching. Consider an example where devices A and D are switched off, and devices B and C are switched on (see Figure 5.21). This is the quiescent state, so no current flows through the load; however, DC current flows through the drain load inductor. Device B is now biased off. The current through the inductor cannot change instantaneously, so current flows into the coupling capacitor through the load, causing the load voltage to rise. At some later time, devices A and D are turned on. The current through A, B, and D now exceeds the current through L, so C discharges and the voltage across $R_L$ goes negative. Device C is now turned off, and the coupling capacitor, C, charges back to the quiescent condition. The resulting output is an approximation of a Gaussian monocycle. Figure 5.22 shows the timing. A range of pulses can be generated with this circuit. Figure 5.23 shows the state diagram and the resulting output pulse. The example described previously begins at the 0110 (A on, B and

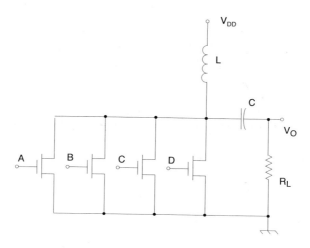

Figure 5.21: CMOS Switch Pulse Generator.

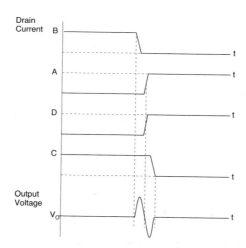

Figure 5.22: Pulse Generator Timing and Output Voltage.

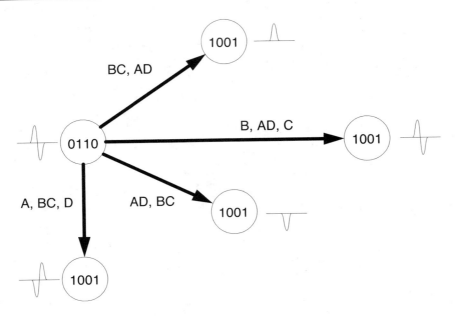

Figure 5.23: Pulse Generator Transition States.

C off, D on) state and transitions with B off, then A and D on, and then C off (B, AD, C) to the 1001 state on the right. Similar transition sequences can produce an opposite polarity monocycle or positive and negative going impulses.

## 5.2  Modulators

Modulation in the conventional sense of modifying some characteristic of a carrier wave is not employed in I-UWB transmitters. There is no carrier wave, but the characteristic of the pulses must be modified in some manner. The pulses must have some unique property in order to attach information. In carrier-based systems, there is usually some circuit that specifically performs the modulation. Pulse generator circuits in I-UWB transmitters perform a direct data bits to RF conversion so the modulation is often done by bit timing.

The simplest modulation is amplitude modulation or on/off keying (OOK). The pulser is turned on and off to represent data bits. While very simple, an OOK system is at a significant power disadvantage because, assuming equal number of ones and zeros, the transmitter is off half the time. More advantageous are pulse position modulation, (PPM), biphase pulse polarity modulation (antipodal PAM), and pulse width modulation (PWM). The Aether Wire transmitter previously described uses a type of antipodal PAM [15].

PPM can be accomplished with bit timing; both analog and digital modulation is possible. Avalanche transistor, SRD, and other pulser circuits simply put fast edges on pulses coming from a baseband signal processor. Analog PPM was used in early analog I-UWB systems.

Digital control of the time delay between pulses is possible. A dedicated timing IC is available [18]. Figure 5.24 is a simplified block diagram of the timing device. The VCO controlled by a phase-locked loop (PLL) runs at approximately 2.5 GHz with a 10 MHz reference. Each cycle of the 2.5 GHz VCO advances the 8-bit synchronous counter one count. Each count represents a 390 picosecond time advance. The value loaded into the comparator determines the number of 390 picosecond

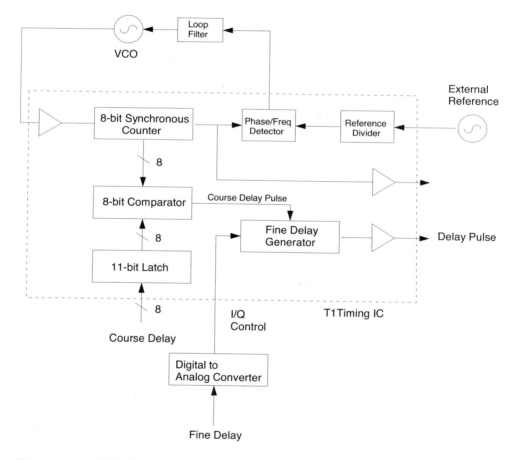

Figure 5.24: UWB Timing IC.

SOURCE: D. Rowe, B. Pollack, J. Pulver, W.Chon, P. Jett, L. Fullerton, and L.A. Larson, "A Si/SiGe HBT timing generator IC for high-bandwidth impulse radio applications," *Proceedings of the IEEE Custom Integrated Circuits Conference* [18]. © IEEE, 1999. Used by permission.

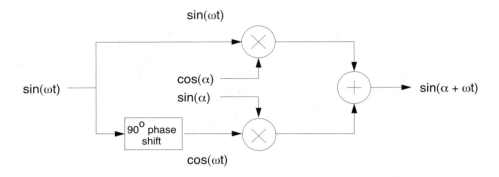

Figure 5.25: Fine Time Delay Generator Block Diagram.

steps that can occur before the output of a course delay pulse. The value loaded into the comparator can be changed dynamically so that the interval between output pulses can be controlled.

The fine delay generator is an analog circuit based on an I/Q vector modulator. Figure 5.25 is a block diagram of the fine delay generator. The fine delay generator is a voltage controlled phase shifter. The values for $\sin(\alpha)$ and $\cos(\alpha)$ are generated by an 8-bit digital to analog converter (DAC). With the 8-bit DAC it is possible to divide the 390 picosecond coarse delay pulse into 1.52 picosecond increments. The value in the comparator and the DAC are controlled by a baseband processor.

The output pulse from the timer is used to drive an SRD or other fast-rise time pulser. An improved version of this device is described in [19].

Digital PWM can be accomplished by combining a pair of pulsers. Two SRD pulsers of the type shown in Figure 5.8 with different pulse widths can be combined through a RF power combiner, and the SRD with the desired pulse width can be triggered at the desired time.

## 5.3    I-UWB Transmitters

In this section, signal models (and system models where needed) that capture all the different kinds of modulation techniques proposed to date for I-UWB are presented. These models include time-hopped pulse position modulation (TH-PPM), time-hopped antipodal pulse amplitude modulation (TH-A-PAM), optical orthogonal coded pulse position modulation (OOC-PPM), direct sequence spread spectrum modulation (DS), transmitted reference (TR), and pilot waveform assisted modulation (PWAM) .

### 5.3.1    TH-PPM and TH-(A-PAM) UWB Signals

A typical TH-PPM modulated UWB signal can be modeled as [20]

$$s^{(k)}_{TH-PPM}(t) = \sum_{j=-\infty}^{\infty} \sqrt{E_w} w \left( t - jT_f - c_j^{(k)} T_c - \delta \alpha^{(k)}_{\lfloor j/N_s \rfloor} \right) \tag{5.16}$$

The TH-(A-PAM) modulated UWB signal can be modeled as [20]

$$s^{(k)}_{TH-A-PAM}(t) = \sum_{j=-\infty}^{\infty} \sqrt{E_w} \beta^{(k)}_{\lfloor j/N_s \rfloor} w \left( t - jT_f - c_j^{(k)} T_c \right) \tag{5.17}$$

where $s^{(k)}(t)$ is the random process describing the signal transmitted by the $k^{th}$ user and $w(t)$ is the signal pulse normalized so that $\int_{-\infty}^{+\infty} |w(t)|^2 dt = 1$. The energy of the transmitted waveform is $E_w$. The pseudorandom time hopping code $\left\{ c_j^{(k)} \right\}$, $0 \le c_j^{(k)} \le N_h$ $\forall k$ provides an additional shift in order to reduce the effect of collisions in multiple access schemes. The binary information stream $\left\{ \alpha_j^{(k)} \right\}$ is transmitted using a PPM modulation format introducing an additional shift, $\delta$, that is used to distinguish between pulses carrying bit 0 and bit 1. $\beta_j^{(k)} = 1 - 2\alpha_j^{(k)}$ for TH-A-PAM. Each information bit is transmitted using $N_s$ consecutive pulses, and the resulting bit rate is $R_b = (N_s T_f)^{-1}$. $T_c$ is the chip duration.

## 5.3.2  OOC-PPM UWB Signals

Optical orthogonal codes are families of binary sequences typically employed to provide multiuser capabilities to "positive systems," which are systems in which the transmitted waveforms cannot be summed to obtain a zero. Refer to [20] for details. These codes are characterized by four parameters, $F$, $K$, $\lambda_a$, and $\lambda_c$ that symbolize respectively the code length, the number of '1's in the code words, and the maximum out of phase autocorrelation and cross correlation values. If $\left\{ A_i^{(k)} \right\}$, $0 \le A_j^{(k)} \le N_h$ $\forall k$ is the code word assigned to user $k$, the transmitted signal can be described as

$$s^{(k)}_{OOC-PPM}(t) = \sum_{j=-\infty}^{\infty} \sum_{i=0}^{N_h-1} A_i^{(k)} \sqrt{E_w} w \left( t - jT_f - iT_c - \delta \alpha_j^{(k)} \right) \tag{5.18}$$

## 5.3.3  DS-UWB Signals

A DS-UWB modulated signal can be modeled as

$$s^{(k)}_{DS-UWB}(t) = \sum_{j=-\infty}^{\infty} \sqrt{E_w} \alpha_j^{(k)} \sum_{i=0}^{N_h-1} c_i^{(k)} w \left( t - jT_f - iT_c \right) \tag{5.19}$$

where $\left\{ c_j^{(k)} \right\} \in \{-1, +1\}$ is the spread spectrum code assigned to user $k$, and $\left\{ \alpha_j^{(k)} \right\} \in \{-1, +1\}$ is the data bit of user $k$.

## 5.3.4   Transmitter Reference (TR) UWB System

TR systems were first proposed in the 1920s [21]. In a TR system, a pair of un-modulated and modulated signals is transmitted, and the former is employed to demodulate the latter. The receiver for this transmitter can capture the entire signal energy for a slowly varying channel without requiring channel estimation. Another potentially attractive feature of UWB autocorrelation receivers is their relative robustness to synchronization problems [22]. However, fundamental system weaknesses, such as bandwidth inefficiency and high noise vulnerability, coupled with the advent of stored reference and matched filter implementations in the 1950s and 1960s, largely diminished research interest in TR schemes [23]. However, research in UWB autocorrelation receivers has been relatively active in the last two years. A delay hopped, TR communications system was recently built by the research and development center of GE. Experiments show the viability of such a system in an indoor multipath environment [22, 24]. Giannakis, et al., [25, 26], introduced a general pilot waveform assisted modulation (PWAM) scheme, which subsumes TR as a special case. The values of the system's parameters were derived to minimize the channel's mean square error (MSE) and maximize the average capacity. The circumstances under which the UWB autocorrelation-TR system is optimal were also analyzed. We briefly describe the TR and the PWAM I-UWB transmitters here.

### TR System Model

The TR system described in [21] and [27] employs binary pulse position modulation (PPM). The transmitted signals consist of $N_p$ UWB pulses, $p(t)$, of duration $T_p$ and energy $E_p$. The waveforms are comprised of $\frac{N_p}{2}$ unmodulated pulses interleaved with $\frac{N_p}{2}$ PPM-modulated pulses. Time hopping is not considered in this model. The signals $s_0(t)$ and $s_1(t)$ are equally likely, and are transmitted on $t \in [0, N_p, T_f]$, where $T_f$ is the pulse repetition period. The duty cycle is assumed to be very low ($T_f \gg T_p$). The signals can be written as

$$s_j(t) = \sum_{i=0}^{\frac{N_p}{2}-1} \sqrt{E_p}\left[p(t - 2iT_f) + p(t - (2i+1)T_f - \varepsilon_{j,i}\tau_p)\right], \quad j = 0, 1 \quad (5.20)$$

where $\tau_p$ is the delay associated with PPM. Moreover, the authors of [21] and [27] assume $\varepsilon_{0,i} = i \bmod 2$ and $\varepsilon_{1,i} = 1 - \varepsilon_{0,i}$.

### PWAM Optimization

In [25] and [26], Giannakis, et al., described a general PWAM scheme and derived conditions for which the scheme operates optimally. In these works, minimum channel MSE and maximum average capacity are the two optimality measures. The system model can be viewed as a generalization of the model proposed by Stark, et al., [21], [27]. Every binary symbol is shaped by $p(t)$ and is transmitted repeatedly over $N_f$ consecutive frames, each of duration $T_f$ seconds. The channel is assumed to be static over a burst of duration $\overline{N}T_f$ seconds. Each burst includes up to $N = \frac{\overline{N}}{N_f}$

symbols that are either pilot or information bearing. Pilot waveforms are used at the receiver to form a channel estimate. During each burst, $N_s$ distinct information symbols are sent, corresponding to $\overline{N}_s = N_s N_f$ waveforms. The number of pilot waveforms is thus $\overline{N}_p = \overline{N} - \overline{N}_s$. The power of the $n_p^{th}$ pilot waveform is denoted by $P(n_p)$; likewise, the power of the $n_d^{th}$ data waveform is denoted by $P(n_d)$.

The following four constraints must be observed to ensure that the system is optimal:

1. Given the total number of pilot waveforms per burst, $\overline{N}_p$, and the total energy assigned to pilot waveforms, $\varepsilon_p$, equi-probable pilot power waveforms minimize the channel MSE.

2. Given the total data energy, $\varepsilon_d$, and the total pilot energy, $\varepsilon_p$, equi-powered information symbols maximize the average capacity, $C$.

3. Given the total data energy, $\varepsilon_d$, the total pilot energy, $\varepsilon_p$, and the number of waveforms per burst $\overline{N}$, the number of pilot waveforms that maximizes the average capacity is given by $\overline{N}_p^*$, defined by: $\overline{N}_p^* = (N - N_s^*) N_f$, where

$$N_s^* = \frac{\overline{N}}{N_f} - 1 \text{ if } \frac{\overline{N}}{N_f} \text{ is an integer}$$

$$N_s^* = \left\lfloor \frac{\overline{N}}{N_f} \text{ otherwise} \right\rfloor$$

4. With fixed burst size $N$, number of information symbols per burst $N_s$, and total transmission energy $\varepsilon$, the optimal energy allocation factor, $\alpha = \frac{\varepsilon_s}{\varepsilon}$, that maximizes the average capacity is

$$\alpha = \left\{ \begin{array}{ll} \frac{1}{2}, & \frac{\varepsilon}{N\sigma^2} \ll \\ \frac{N_s}{1+N_s}, & \frac{\varepsilon}{\sigma^2} \ll \end{array} \right. \tag{5.21}$$

For a proof of these claims, the reader is referred to [26]. Note that for $N = 2$, we get $N_p = N_s = 1$ and $\alpha = \frac{1}{2}$ for optimal PWAM. This corresponds to a system where the transmitted waveforms are split evenly between pilot and data waveforms of equal power. This resulting PWAM system turns out to be equivalent to the TR autocorrelation system described by Stark. (Note that one of Stark's TR models assumes the channel to be time invariant over $2N_pT_f$ seconds, which corresponds to $N = 2$.)

## 5.4   MC-UWB Transmitters

In multicarrier systems, a single data stream is split into multiple parallel data streams of reduced rate, with each stream transmitted on a separate frequency (subcarrier). Each carrier is modulated at a low enough rate to minimize inter-symbol interference (ISI). Subcarriers must be properly spaced so that they do not interfere. For a $N$ subcarrier system, each subchannel is tolerant of $N$ times as much dispersion as the original single carrier system. Multicarrier UWB (MC-UWB) communication systems use orthogonal UWB pulse trains and multiple subchannels to achieve reliable high bit rate transmission and spectral efficiency [28]. Some of the advantages of MC-UWB systems are better time resolution, which leads to better performance in multipath fading channels; better spectrum utilization, which leads to higher bit rate communications; and simple decoupled system design, which leads to a simple transmitter implementation. We briefly describe the transmitter for MC-UWB system here.

The binary MC-UWB transmitted signal can be represented as

$$x(t) = \beta \sum_r \sum_{k=0}^{K-1} b_k^r p(t - rT_p) \exp\left(j2\pi k f_0 (t - rT_p)\right) \tag{5.22}$$

where $b_k^r$ is the symbol that is transmitted in the $r^{th}$ transmission interval over the $k^{th}$ subcarrier, $K$ is the number of subcarriers, and $\beta$ is a constant that controls the transmitted power spectral density (PSD) and determines the energy per bit. The fundamental frequency is $f_0 = \frac{1}{T_p}$. In the next two sections, we describe three interesting MC-UWB transmitter models: CI-UWB, FH-UWB, and OFDM-UWB.

## 5.4.1   CI-UWB Signals

In [29] the use of a multicarrier Carrier Interferometry (CI) waveform as a means to enable higher throughput/improved multiaccess in UWB (without significant performance degradation) is proposed. Interferometry [30], a classical method in experimental physics, refers to the study of interference patterns resulting from the superposition of waves. The presence of distinct peaks and nulls in the interference patterns has motivated the widespread use of interferometry. The ideas underlying interferometry lend themselves naturally to multiple access applications in telecommunications. For example, in antenna arrays supporting space division multiple access, electromagnetic (EM) waves are emitted simultaneously from multiple antenna elements and initial phases are chosen to ensure that interference patterns create a peak at the desired user location and nulls at the position of other users. CI methods have initially been studied in [29], [31], and [32]. The CI pulse waveform in UWB corresponds to the superpositioning of $N$ orthogonal carriers. The idea is that with carefully chosen phase offsets (spreading codes) that ensure a periodic main lobe in the time domain (with side lobe activity at intermediate times) we could null out ISI, MAI, or any other form of interference. At the receiver side, the received UWB CI signal is decomposed into carrier components and recombined to exploit frequency diversity and minimize ISI/MAI from other information bearing pulses. We briefly describe the TH-UWB, CI-UWB transmitted signal.

The transmitted signal for user $k$, $k = 0, 2, \ldots, K - 1$ in a multiaccess scenario (or the $k^{th}$ bit per frame in a high bit rate scenario) is

$$s^k(t) = \sum_{m=-\infty}^{\infty} s_m^k(t) \tag{5.23}$$

where $m$ is the index indicating the frame number (each frame is of duration $T_f$), the sum over $m$ is the sum over all frames of duration $T_f$, and $s_m^k(t)$ is user $k$'s transmitted bit in the $m^{th}$ frame

$$s_m^k(t) = b_m^k p\left(t - mT_f - c_m^k T_c\right) \tag{5.24}$$

where $b_m^k \in \{-1, 1\}$ denotes user $k$'s information bit in the $m^{th}$ frame, $p(t)$ is the pulse waveform with main lobe duration $T_c$, $c_m^k \in \{0, 1, \ldots, N-1\}$ is the time hopping sequence assigned to user $k$, and $N = \frac{T_f}{T_c}$. Considering the $0^{th}$ transmitted frame, the total transmitted signal is given by

$$s(t) = \sum_{k=0}^{K-1} s_0^k(t) = \sum_{k=0}^{K-1} b_0^k p\left(t - c_0^k T_c\right) \tag{5.25}$$

CI is an alternative for UWB pulse shaping and corresponds to the superposition of carriers equally spaced in frequency by $\Delta f = \frac{1}{NT_c}$. That is

$$p(t) = \text{Re} \left\{ \sum_{n=0}^{N-1} A \exp\left(j2\pi nt\Delta f\right) \right\} \tag{5.26}$$

The CI signal given above is constructed via an $N$-point IFFT. Specifically, we have

$$p\left(\frac{kT_f}{N}\right) = \text{Re} \left\{ \sum_{n=0}^{N-1} A \exp\left(\frac{j2\pi nk}{N}\right) \right\} \tag{5.27}$$

This corresponds to the $k$ value in the $N$-point IFFT shown in Figure 5.26.

The frequency based processing of CI pulses in UWB systems enables high throughputs while maintaining excellent BER performance. This idea has been extended to CI-DS-UWB [33].

## 5.4.2    FH-UWB System

In [34], a residue number system (RNS) assisted multistage frequency hopped spread spectrum multiaccess scheme is studied. We refer to this system as FH-UWB. This FH-UWB scheme is capable of efficiently dividing a huge number of users into a number of reduced-size user groups. Note that multiuser interference only affects users within the same group. Because the number of users within the same group is only a small fraction of the total number of users supported by the synchronous FH-UWB system, advanced multiuser detection algorithms can be employed to achieve near single user performance at an acceptable complexity. The following are some of the advantages of FH-UWB systems over DS-UWB systems:

1.  Because the hop rate of FH-UWB is typically lower than the chip rate of DS-UWB systems, synchronization requirements for FH-UWB are less stringent than those of DS-UWB systems.

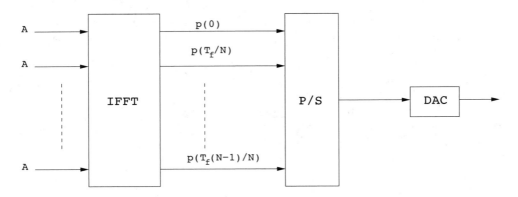

Figure 5.26: CI-UWB Transmitter.

SOURCE: Z. Wu, F. Zhu, and C. R. Nassar, "Ultra Wideband Time Hopping Systems: Performance and Throughput Enhancement via Frequency Domain Processing," *36th Asilomar Conference on Signals, Systems and Computers* [48]. © IEEE, 2002. Used by permission.

2. For each hop, the FH-UWB system is analogous to that of a narrowband system because detection has to be performed in each hop and the demodulation operation is significantly less demanding than demodulating a signal during a chip interval. This results in considerably lower power consumption.

3. FH systems are in general less susceptible to the near-far problem than DS-UWB systems.

4. I-UWB has spectrum overlapping with existing narrowband systems; consequently, it has to cope with intersymbol interference. The frequency bands used by FH-UWB need not be contiguous; hence, we can avoid using the frequency bands occupied by other systems. As a consequence, FH-UWB is more robust in an interference limited environment.

In Figure 5.27, the transmitter block diagram of a FH-UWB system employing MFSK modulation is shown. Readers are referred to [32] for details and a description of the transmission mechanism. We present the figure in this section for the sake of completeness.

### 5.4.3 OFDM-UWB System

OFDM is a special case of multicarrier transmission that permits subcarriers to overlap in frequency without mutual interference, resulting in increased spectral efficiency. OFDM exploits signal-processing technology to obtain cost-effective means of implementation. Multiple users can be supported by allocating each user a group of subcarriers. OFDM-UWB is a novel system that has been proposed as a physical

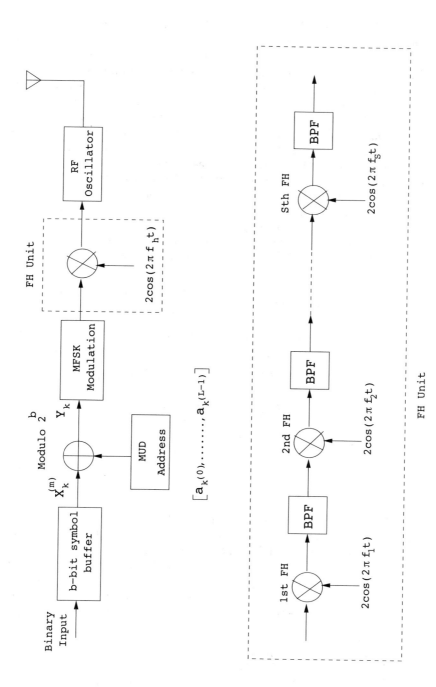

Figure 5.27: FH-UWB Transmitter.

241

layer for high bit rate, short-range (10 m–20 m) communications networks in high performance computing clusters. The OFDM-UWB transmitter relies on splitting orthogonal subcarriers into a train of short pulses, sending them over a channel, and reassembling them at the receiver to get orthogonality and to recover each sub-carrier separately [35], [36]. This new system offers more flexibility in shaping the transmitted spectrum because it has more degrees of freedom than a single carrier system. OFDM-UWB provides more multipath resolution than single carrier UWB. Unlike narrowband OFDM, a given tone in OFDM-UWB is transmitted only during parts of the transmission interval. Reliable communication results from integrating several pulses and high throughput from transmitting frequencies in parallel. Unlike narrowband OFDM, the OFDM-UWB spectrum can have gaps between subcarriers. Another major difference between OFDM-UWB and narrowband OFDM is their spectral shapes. OFDM-UWB is being proposed as the physical layer standard in 802.15.3a Wireless Personal Area Networks [37]. See Appendix 10.B for a detailed description and block diagram of this physical layer standard.

OFDM-UWB is a MC-UWB system that uses a frequency coded pulse train as a shaping signal. The frequency coded pulse train is defined as follows

$$p(t) = \sum_{n=0}^{N-1} s(t - nT) \exp\left(-j2\pi c(n)\frac{t}{T_c}\right) \tag{5.28}$$

where $s(t)$ is an elementary pulse with unit energy and duration $T_s < T$. $p(t)$ has duration $T_p = NT$. Each pulse is modulated with a frequency $f_n = \frac{c(n)}{T_c}$, where $c(n)$ is a permutation of the integers $\{0, 1, \ldots, N - 1\}$. It can be easily verified that

$$\int_0^{T_p} (p(t) \exp(j2\pi k f_0 t)) (p^*(t) \exp(-j2\pi m f_0 t)) \, dt$$
$$= \exp(j\pi (k - m) f_0 T_s) \operatorname{sinc}((k - m) f_0 T_s) \tag{5.29}$$

Because $f_0 = \frac{1}{NT}$, the signals

$$p_k(t) = p(t) \exp(j2\pi k f_0 t) \tag{5.30}$$

are orthogonal for $k = 0, 1, \ldots, N - 1$. Further, if $T_s = T$, the signals given by (5.30) are orthogonal for any $k$. Costas showed that if $c(n)$ is a Costas sequence, the $p_k(t)$'s remain near-orthogonal under different multipath delays. Further, (5.30) can be efficiently implemented by taking samples of the signal $p(t)$, which are IFFT samples, and passing them through a DAC. The most challenging part of this structure is designing the DAC. In these structures, DACs typically operate around hundreds of MHz. In the next section, we describe OFDM-UWB using sigma-delta modulators that trade off the high operating rate of DACs for lower resolution.

## $N$-Tone Sigma-Delta OFDM-UWB

A new method for generating and detecting the OFDM-UWB signal using a modified sigma-delta modulator is proposed in [38], which is an improvement of the existing OFDM-UWB system discussed in Section 5.4.3 for high data rate applications. Unlike narrowband OFDM, IFFT/FFT cannot be directly used to generate and receive OFDM-UWB signals because of the high bit rates involved. To solve this problem, a procedure to move the bulk of the processing load from analog devices to the digital base band section is described in [38]. Sigma-Delta A/D and D/A converters are a good choice for high bit rate wireless communications [39]. Traditional versions of the sigma-delta modulators cannot be used in OFDM-UWB transceivers because they would require prohibitively high sampling rates. A sigma-delta modulator modified to fit the characteristics of OFDM-UWB signals is proposed that enables most of the processing to be done digitally in both the transmitter and receiver. This approach efficiently uses the IFFT/FFT algorithm to generate and demodulate OFDM-UWB signals and also gets rid of the high peak-to-average ratio (PAR) problem that occurs with OFDM systems. The modified sigma-delta modulator, called the $N$-tone sigma-delta modulator, introduces $N$ zeros at the frequencies in the quantization noise spectrum. These zeros match the locations of frequencies used by the OFDM system, and the quantization noise spectrum fills the gaps in the spectrum of the OFDM-UWB signal. In fact, this new structure can be used in other UWB systems any time there are gaps in the spectrum of the transmitted signal. However, these advantages come at the expense of lower spectral efficiency. The $N$-tone sigma-delta quantizer structure is shown in Figure 5.28. A transmitter structure for generating UWB-OFDM signals using the sigma-delta modulator is shown in Figure 5.29 [40].

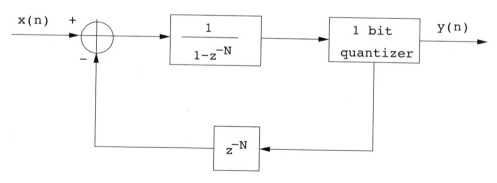

Figure 5.28: *N-tone Sigma-Delta Modulator.*

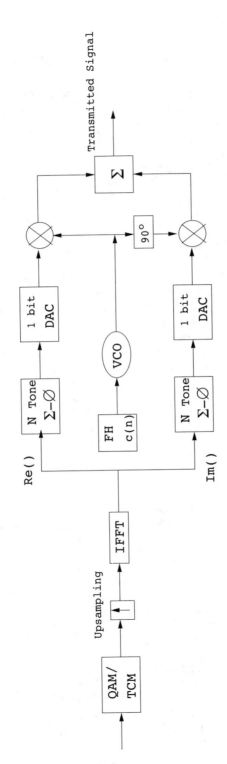

Figure 5.29: OFDM-UWB Transmitter with $N$-tone Sigma-Delta Modulators.

## 5.5   Spectral Encoded UWB Communication System

In this section, we describe a UWB communications system that uses spectral encoding as a multiple access scheme with interference suppression capability [41]. The main advantage of this technique is that transmitted signal spectrum can be conveniently shaped to suppress narrowband interference. The concept of spectral encoding is not new. It has been proposed both for an optical communication system [42, 43] and for wireless communications scenarios [44, 45]. In this technique, a message signal is multiplied by a spreading sequence in the frequency domain, as opposed to conventional direct sequence code division multiaccess systems in which the encoding is in the time domain. As a result, the signal will be spread in time. Multiaccess capability is achieved by assigning distinct spreading sequences to different users. When narrowband interference suppression is desired, a spreading sequence is used that has a spectral null where the narrowband interference is located. Moreover, the narrowband interference is "notched out" at the receiver when the received signal spectrum is multiplied by the conjugate of the transmitted signal spectrum (correlation receiver). Details can be found in [46]. The Fourier Transform of the data signal is performed via a well-established implementation that corresponds to the multiplication and convolution of the desired signal by chirp (linear FM) waveforms. This method of spectrally encoding the UWB signal can suppress narrowband interference and can create spectral nulls where desired, simplifying the requirement to meet the FCC mask.

A schematic of the transmitter is shown in Figure 5.30. The modulated signal, $f(t)$, is non-zero only in $t \in \left[-\frac{T}{2}, \frac{T}{2}\right]$. The waveform $f_1(t)$ is given by

$$
\begin{aligned}
f_1(t) = {} & \frac{1}{2} F_R(t) \cos \left( \omega_a \left( t - \frac{T}{2} \right) + \beta \left( t - \frac{T}{2} \right)^2 \right) \\
& - \frac{1}{2} F_I(t) \sin \left( \omega_a \left( t - \frac{T}{2} \right) + \beta \left( t - \frac{T}{2} \right)^2 \right)
\end{aligned}
\tag{5.31}
$$

where $\omega_a$ and $\beta$ are respectively the center frequency and the half slope of the chirp signals, and $F_R(t)$ and $F_I(t)$ are respectively the real and imaginary components of the Fourier Transform of the modulated signal, $f(t)$, and are given by

$$
F_R(t) = 2 \int_{\frac{T}{2}}^{\frac{T}{2}} f(u) \cos (\omega_c u) \cos \left( 2\beta \left( t - \frac{T}{2} \right) u \right) du
$$

$$
F_I(t) = -2 \int_{\frac{T}{2}}^{\frac{T}{2}} f(u) \cos (\omega_c u) \sin \left( 2\beta \left( t - \frac{T}{2} \right) u \right) du
\tag{5.32}
$$

where $\omega_c$ is the carrier frequency. $f_1(t)$ is valid only in $t \in [T, T_1]$, where $T_1$ is the duration of the impulse response of the filter labeled as $\cos \left( \omega_a t + \beta t^2 \right)$. The signal $f_1(t)$ is spread to $f_2(t)$ by a spreading sequence such that its spectrum is given by

$$
PN(\omega) = PN_R(\omega) + j PN_I(\omega)
\tag{5.33}
$$

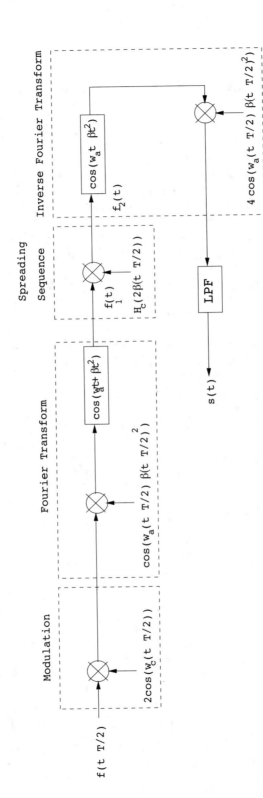

Figure 5.30: Transmitter Schematic of Spectrally Encoded UWB.

where $PN(\omega)$ is in general a complex waveform bandlimited to $|f| \leq \frac{W}{2}$. $PN_R(\omega)$ and $PN_I(\omega)$ are real waveforms that consist of a sequence of 1s or -1s (and 0s when interference suppression is considered). The function $H_c\left(2\beta\left(t - \frac{T}{2}\right)\right)$ in Figure 5.4 relates to the desired spreading sequence by

$$H_c(\omega) = \frac{8\beta}{\pi}\left(PN_R(\omega)\cos(\omega T_1) + PN_I(\omega)\sin(\omega T_1)\right) \tag{5.34}$$

The output transmitted signal $s(t)$ is the inverse Fourier Transform of the signal $f_2(t)$ and is equal to

$$s(t) = \frac{1}{\pi}\int_{\omega_c - \frac{W}{2}}^{\omega_c + \frac{W}{2}}$$

$$\cdot\left(R(\omega)\cos\left(\omega\left(t - \frac{T}{2} - T_1\right)\right) - I(\omega)\sin\left(\omega\left(t - \frac{T}{2} - T_1\right)\right)\right)d\omega \tag{5.35}$$

where $R(\omega)$ and $I(\omega)$ are defined as $R(\omega) = F(\omega)PN_R(\omega)$ and $I(\omega) = F(\omega) \cdot PN_I(\omega)$. $W$ is the bandwidth of $f(t)$. The transmitted signal is valid only in $t \in [T_1, T_1 + T]$. The receiver structure for this mechanism is shown in Chapter 6, Section 6.3.

## 5.6   Summary

In this chapter, we have discussed signal generation techniques for impulse-based UWB communication systems. A comprehensive study of the different popularly used modulation formats and also some of the recent developments in UWB modulators are presented. The receiver designs for some of the major types of transmitters described here are discussed in detail in the next chapter. We have not talked about pulse shaping issues in UWB transmitter design, and assume that the pulses satisfy the FCC spectral mask. Further, channel coding, interleaving issues, and other advanced developments at the transmitter, such as space time coding and transmitter beamforming, are natural extensions for UWB transmitter systems. Appendix 10.B gives the IEEE 802.13.3a standard, which considers DS-UWB–based and MC-UWB–based transmitters.

# Bibliography

[1] L. W. Fullerton, Spread Spectrum Radio Transmission System, U.S. Patent 4,641,317, February 1987.

[2] Application Note 1628D, Understanding Power Transistor Breakdown Parameters, On Semiconductor. Available: http://onsemi.com.

[3] W. B. Mitchell, "Avalanche transistors give fast pulses," *Electronic Design*, pp. 202-209, March 1968.

[4] H. -M. Rein, "Subnanosecond-Pulse Generator with Variable Pulsewidth Using Avalanche Transistors," *IEEE Electronics Letters*, vol. 11, no. 1, pp. 21-23, January 1975.

[5] E. K. Miller, ed., *Time-Domain Measurements in Electromagnetics*, Chapter 4, New York: Van Norstrand Reinhold, 1986.

[6] Application Note 918, Pulse and Waveform Generation with Step Recovery Diodes, Hewlett Packard, October 1984.

[7] R. Tielert, "Subnanosecond-Pulse Generator Employing 2-Stage Pulse Step Sharpener," *IEEE Electronics Letters*, vol. 12, no. 3, pp. 84-54, February 1976.

[8] J. S. Lee, C. Nguyen, and T. Scullion, "New Uniplanar Subnanosecond Monocycle Pulse Generator and Transformer for Time-Domain Microwave Applications," *IEEE Transactions on Microwave Theory and Techniques*, vol. 49, no. 6, pp. 1126-1129, June 2001.

[9] J. Han, and C. Nguyen, "A New Ultra-wideband, Ultra-Short Monocycle Pulse Generator with Reduced Ringing," *IEEE Microwave and Wireless Components Letters*, vol. 12, no. 6, pp. 206-208, June 2002.

[10] S. Hamilton and R. Hall, "Shunt-Mode Harmonic Generation Using Step Recovery Diodes," *Microwave Journal*, pp. 69-78, April 1967.

[11] Application Note 920, Harmonic Generation Using Step Recovery Diodes and SRD Modules, Hewlett Packard, 1988.

[12] H. Kim, D. Park, and Y. Joo, "Design of CMOS Scholtz's Monocycle Pulse Generator," *IEEE Conference on Ultra Wideband Systems and Technologies*, Reston, VA, November 2003.

[13] H. F. Harmuth, "Antennas and Waveguides for Nonsinusodial Waves," *Advances in Electronics and Electron Physics*, Supplement 15, Orlando: Academic Press, 1984.

[14] H. F. Harmuth, Efficient Driving Circuit for Current Radiator, U.S. Patent 5365240, November 15, 1994.

[15] U.S. Patents 5748891, May 1998; 6002708, December 1999; 6385268, May 2002; and 64000754, June 2002. Available: http://www.aetherwire.com/.

[16] B. E. Burke and D. L. Smythe, Jr., "ACCD Time-Integrating Correlators," *IEEE Journal of Solid State Circuits*, vol. SC-18, no. 6, pp. 736-744, December 1983.

[17] K. Marsden, H. -J. Lee, D. S. Ha, and H. -S. Lee, "Low Power CMOS Reprogrammable Pulse Generator for UWB Systems," *IEEE Conference on Ultra Wideband Systems and Technologies*, Reston, VA, pp. 443-447, November 2003.

[18] D. Rowe, B. Pollack, J. Pulver, W.Chon, P. Jett, L. Fullerton, and L.A. Larson, "A Si/SiGe HBT timing generator IC for high-bandwidth impulse radio applications," *Proceedings of the IEEE Custom Integrated Circuits Conference*, pp. 221-224, May 1999.

[19] D. Kelly, R. S. Reinhardt, M. Einhorn, "PulsON Second Generation Timing Chip; Enabling UWB Through Precise Timing," *IEEE Conference on Ultra Wideband Systems and Technologies*, Baltimore, MD, May 2002.

[20] G. Durisi and S. Benedetto, "Performance Evaluation and Comparison of Different Modulation Schemes for UWB Multiaccess Systems," *IEEE International Conference on Communications*, Anchorage, AK, 2003.

[21] J. D. Choi, and W. E. Stark, "Performance of ultra-wideband communications with suboptimal receivers in multipath channels," *IEEE Journal on Selected Areas in Communications*, vol. 20, no. 9, December 2002.

[22] R. T. Hoctor and H. W. Tomlinson, "An Overview of Delay-Hopped, Transmitted-Reference RF Communications," *Technical Information Series*, G.E. Research and Development Center, pp. 1-29, January 2002.

[23] M. K. Simon, J. K. Omura, R. A. Sholtz, and B. K Levitt, "Spread Spectrum Communications, Volume-I," Computer Science Press, 1985.

[24] R. T. Hoctor and H. W. Tomlinson et al., "Delay Hopped Transmitted Reference Experimental Results," *Technical Information Series*, G. E. Research and Development Center, 2002GRC099, April 2002.

[25] L. Yang and G. B. Giannakis, "Optimal pilot waveform assisted modulation for ultra-wideband communications," *Conference Record Thirty-Sixth Asilomar Conference on Signals, Systems and Computers*, vol. 1, pp. 733-737, November 2002.

[26] L. Yang and G. B. Giannakis, "Optimal pilot waveform assisted modulation for ultra-wideband communications," *IEEE Transactions on Wireless Communications*, 2004 (to appear).

[27] J. D. Choi, and W. E. Stark, "Performance of autocorrelation receivers for ultra-wideband communications with PPM in multipath channels," *Digest of Papers IEEE Conference on Ultra Wideband Systems and Technologies*, May 2002.

[28] E. Saberinia and A. H. Tewfik, "Single and multicarrier UWB Communications," *Proc. 7th International Symposium on Signal Processing and its Applications*, vol. 2, pp. 343-346, July 2003.

[29] S. S. Kolenchery, J. K. Townsend, and J. A. Freebersyser, "A Novel Impulse Radio Network for Tactical Military Wireless Communications," *IEEE Military Communications Conference*, vol. 1, pp. 59-65, Boston, MA, October 1998.

[30] W. H. Steel, *Interferometry*, 1st ed., Cambridge UK: Cambridge University Press, 1967.

[31] B. Natarajan, C. R. Nassar, and S. Shattil, "Innovative Pulse Shaping for High Performance Wireless TDMA," *IEEE Communications Letters*, vol. 5, no. 9, pp. 372-374, September 2001.

[32] B. Natarajan, C. R. Nassar, and S. Shattil, "Throughput Enhancements in TDMA Through Carrier Interferometry Pulse Shaping," *IEEE Vehicular Technology Conference*, pp. 1799-803, Boston, MA, September 2000.

[33] C. R. Nassar, F. Zhu, and Z. Wu, "Direct Sequence Spreading UWB Systems: Frequency Domain Processing for Enhanced Performance and Throughput," *IEEE International Conference on Communications*, vol. 3, pp. 2180-2186, May 2003.

[34] L. -L. Yang and L. Hanzo, "Residue Number System Assisted Fast Frequency Hopped Synchronous Ultra-Wideband Spread-Spectrum Multiple-Access: A Design Alternative to Impulse Radio," *IEEE Journal on Selected Areas in Communications*, vol. 20, no. 9, pp. 1652-1663, December 2002.

[35] E. Saberinia and A. H. Tewfik, "Multicarrier UWB," IEEE P802.15 Wireless Personal Area Networks, P802.15-02/03147r1-TG3a, November 2002.

[36] E. Saberinia and A. H. Tewfik, "Synchronous UWB-OFDM," *IEEE International Symposium on Wearable Computers*, pp. 41-42, September 2002.

[37] E. Saberinia and A. H. Tewfik, "High Bit Rate UWB-OFDM," *IEEE Global Communications Conference*, vol. 3, pp. 2260-2264, November 2002.

[38] A. Batra, et al.,"Multi-band OFDM Physical Layer Proposal for IEEE 802.15 Task Group 3a," IEEE P802.15 Wireless Personal Area Networks, P802.15-03/268r1, September 2002.

[39] E. Saberinia and A. H. Tewfik, "N-Tone Sigma-Delta UWB-OFDM Transmitter and Receiver," *IEEE International Conference on Acoustics, Speech and Signal Processing*, vol. 4, pp. 129-132, April 2003.

[40] J. C. de Mateo Garcia and A. Garcia Armada, "Effects of Bandpass Sigma-Delta Modulation on OFDM Signals," *IEEE Transactions on Consumer Electronics*, vol. 45, no. 2, pp. 119-124, May 1999.

[41] E. Saberinia and A. H. Tewfik, "Generating UWB-OFDM Signal Using Sigma-Delta Modulator," *57th IEEE Vehicular Technology Conference*, vol. 2, pp. 1425-1429, April 2003.

[42] C. R.C.M. da Silva and L. B. Milstein, "Spectral-Encoded UWB Communication Systems," *IEEE Conference on Ultra-Wideband Systems and Technologies*, November 2003.

[43] J. A. Salehi, A. M. Weiner, and J. P. Heritage, "Coherent Ultrashort Light Pulse Code-Division Multiple Access Communication Systems," *Journal of Lightwave Technology*, vol. 8, pp. 478-491, March 1990.

[44] K. S. Kim, D. M. Marom, L. B. Milstein, and Y. Fainman, "Hybrid Pulse Position Modulation/Ultrashort Light Pulse Code-Division Multiple Access Systems—Part I: Fundamental Analysis," *IEEE Transactions on Communications*, vol. 50, pp. 2018-2031, December 2002.

[45] P. Crespo, M. Honig, and J. Salehi, "Spread-Time Code-Division Multiple Access," *IEEE Transactions on Communications*, vol. 43, pp. 2139-2148, June 1995.

[46] M.G. Shayesteh, J. A. Salehi, and M. Nasiri-Kenari, "Spread-Time CDMA Resistance in Fading Channels," *IEEE Transactions on Wireless Communications*, vol. 2, pp. 446-458, May 2003.

[47] L. B. Milstein and P. K. Das, "An Analysis of a Real-Time Transform Domain Filtering Digital Communication System—Part I: Narrow Band Interference Rejection," *IEEE Transactions on Communications*, vol. 28, pp. 816-824, June 1980.

[48] Z. Wu, F. Zhu, and C. R. Nassar, "Ultra Wideband Time Hopping Systems: Performance and Throughput Enhancement via Frequency Domain Processing," *36th Asilomar Conference on Signals, Systems and Computers*, vol. 1, pp. 722-727, November 2002.

# Chapter 6

# RECEIVER DESIGN PRINCIPLES

Annamalai Annamalai,
Sridharan Muthuswamy,
Dennis Sweeney,
R. Michael Buehrer,
Jihad Ibrahim,
and Dong S. Ha

This chapter provides a comprehensive review of various UWB receiver architectures designed for different modulation formats and signaling schemes. Rather than being overly concerned with the effect of pulse shaping, antenna design and channel impairments, this chapter focuses on different mechanisms of demodulating the received UWB signal optimally in the sense of minimizing the probability of bit error. In practice, however, receiver complexity and power consumption are important design concerns. Thus, we shall also highlight suboptimal receivers that are good candidates for different UWB applications. Figure 6.1 shows an overview of the UWB receiver architectures discussed in this chapter. We will employ an L-tap multipath channel model for a simplified analysis of various receivers. However, it may be more appropriate to apply the channel model proposed in [1] by the IEEE 802.15.3a working group (which is a modification of the Saleh-Valenzuela channel model described in [2]) that takes into account the clustering phenomenon observed in several UWB channel measurements [3].

In Section 6.1, we describe several receiver structures suitable for I-UWB receivers. I-UWB offers many potential advantages, such as high resolution in multipath reducing fading margins in link budget analysis, allowing for low transmit

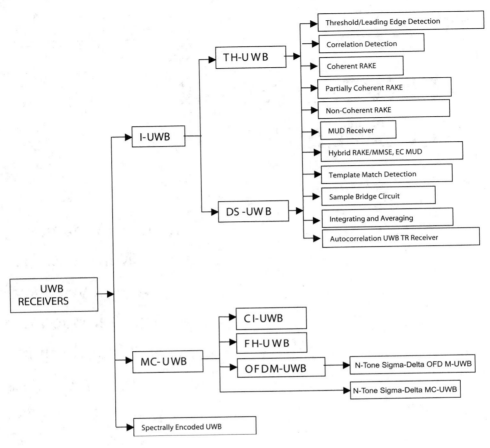

Figure 6.1: Overview of Different UWB Receiver Architectures.

powers and low complexity implementation. However, ultra-fine time resolution requires long synchronization intervals, and I-UWB may require additional correlators to capture adequate signal energy for demodulation. Further, an all-digital receiver requires highly sophisticated receiver processors. For example, assuming a pulse width of 500 picoseconds, Nyquist sampling of 4 samples/pulse (for both in-phase and quadrature-phase) requires a sampling rate of 8 Giga/samples per second. Using 6 bits per sample, the receiver must process a data stream of 48 Gbps. Processing huge amounts of data requires large memory and high processing speeds, which makes it expensive to implement, and thus the goal of a low-cost, simple, ubiquitous communication system is defeated. A number of methods have been developed for pulse detection and reception using a hybrid analog and digital receiver or, alternatively, a fully analog receiver.

RAKE receiver designs for UWB systems are much different from that of narrowband/wideband systems. Consider the following example that demonstrates some of the challenges of RAKE receivers for I-UWB systems. The delay spread $\sigma_T$ of a realistic indoor channel is approximately 5 to 14 ns [1]. Because the pulse duration is much smaller than the delay spread, the channel is frequency selective. Let us assume a delay spread of 10 ns. If the bandwidth of the transmitted UWB signal is $W = 7.26$ GHz, the number of possible multipath components is given by $L = \lfloor W\sigma_T \rfloor = 72$, where $\lfloor x \rfloor$ denotes the closest integer smaller than $x$. That is, we need a RAKE receiver with 72 fingers. Designing a 72-finger RAKE receiver is a challenging task. For outdoor channels where the delay spread is of the order of microseconds, the number of RAKE fingers needed could be on the order of 1,000. This is practically impossible to design.

We will discuss many different forms of I-UWB RAKE receivers with maximum ratio combining and also several multi-user detector-based optimum combining schemes. Further channel estimation is an important task in RAKE reception. Estimating a channel at low energy-to-noise ratio becomes difficult. Imperfect channel estimates further degrade the RAKE receiver performance.

To summarize, I-UWB RAKE receivers suffer from two major drawbacks. First, the energy capture is relatively low for a moderate number of fingers when Gaussian pulses are used. As we have seen, a typical NLOS channel can have up to ∼70 resolvable dominant specular components. Even if a RAKE receiver with this many fingers is realizable, it would only be able to capture part of the signal energy [4]. Second, each multipath undergoes a different channel, which causes distortion in the received pulse shape and makes the use of a single LOS path signal suboptimal.

Synchronization and timing are extremely important in I-UWB design. Timing offsets will have a significant impact on the quality of the signal output by all of the receiver techniques (with the exception of the leading edge detector receiver). Particularly, correlation detector-based receivers will see a dramatic drop in the SNR at the receiver output due to imperfect match or correlation with the receiver waveform. Additionally, if PPM is used and the cumulative timing offset becomes large enough, bit errors will occur. In addition to maintaining precise timing at the receiver, acquiring and synchronizing to the transmitted signal is essential for UWB communication systems. Previous studies [5,6] have shown that the performance of I-UWB is very sensitive to timing jitter and tracking over a range of pulse interference levels.

Section 6.2 explores different variations of UWB receiver architectures for MC-UWB. MC-UWB communication systems use orthogonal UWB pulse trains and multiple subchannels to achieve reliable high bit rate transmission and spectral efficiency [7]. Some of the advantages of MC-UWB systems are better time resolution, which leads to better performance in multipath fading channels; better spectrum utilization, which leads to higher data rates; simple decoupled system design; and more freedom in signal design, which leads to a simple transmitter implementation. MC-UWB systems suffer from interchannel interference (ICI) in a multipath fading channel because the received subcarriers arriving via different paths are no longer orthogonal. Hence synchronization is extremely important [7,8]. Multicarrier systems also have the problem of the peak-to-average power ratio (PAPR) problem, which must be taken care of in order to maintain the proper linear operating region of the receiver.

Section 6.3 discusses spectrally encoded UWB transceivers systems. In Section 6.4, we present a case study on RAKE receivers for improving the range of communication in UWB systems.

## 6.1  I-UWB Radio Receivers

I-UWB radio communicates with pulses of very short duration, typically on the order of a few nanoseconds, spreading the energy of the radio signal from near 0 to several GHz. Thus, I-UWB can accommodate a large number of simultaneous users [9, 10]. Regulatory considerations over such a broadband limit the radiated power and impose stringent pulse shaping requirements to comply with the FCC mask. Impulse radios, operating in the highly populated lower range of the frequency band below 3.1 GHz, must coexist with a variety of interfering signals, including self interference from multipaths and multiuser interference. Further, UWB signals must not interfere with narrowband radio systems operating in dedicated bands (legacy systems). To meet these requirements, spread spectrum techniques are often used. A simple means for spreading the spectrum of low duty cycle pulse trains is time hopping, with pulse position modulation for data modulation at the rate of many pulses per data bit. This signaling scheme is often described as time hopped pulse position modulation, or TH-UWB. Another technique of spreading the spectrum is direct sequence spread spectrum modulation, or DS-UWB. Recently, other schemes have been considered in order to obtain better performance in particular multiuser scenarios. In particular, time hopped phase shift keying (TH-PSK) and a novel technique based on pulse position modulation and optical orthogonal codes (OOC-PPM) are considered in [11, 12]. In Section 6.1.1, we describe a generic system model that covers all the different kinds of spread spectrum multiaccess techniques proposed so far for I-UWB systems.

Receivers for impulse I-UWB can be broadly categorized as threshold or leading edge detectors (LED), correlation detectors (CD), and RAKE receivers. Multi-user detectors (MUD) and hybrid RAKE/MUD-UWB receivers for robust narrowband interference suppression are becoming popular. We describe in depth LED, CD, and RAKE receivers for UWB systems in Sections 6.1.2, 6.1.3, and 6.1.4, respectively. MUD-UWB is discussed briefly in Section 6.1.5. Hybrid RAKE/MUD-UWB receivers are analyzed in sufficient depth in Section 6.1.6. Some of the other less popular methods of I-UWB signal reception and detection available in the literature are the sampling bridge circuit, integration and averaging, and template match detection [13–15], which we briefly describe in Section 6.1.7. An alternative approach to UWB signal demodulation that has recently gained significant attention is the autocorrelation transmitted reference (TR-UWB) receiver [16]. We provide a detailed analysis of TR-UWB in Section 6.1.8. Synchronization and timing issues are examined in Section 6.1.9. A brief introduction to the concept of fully digital I-UWB receivers is described in Section 6.1.10. Section 6.1.11 describes Intel's proposal for impulse-based UWB communication.

## 6.1.1  System Model

In this section, a general system model that captures all proposed spread spectrum multiaccess techniques for I-UWB is presented and the principle characteristics of the system are described [3]. The time axis is divided into symbol intervals of $T_s$ seconds, each one further subdivided into smaller intervals called chips of duration $T_c$ seconds. The signature sequence assigned to each user for multiple access purposes is a periodic signal with a period $T$ seconds. For simplicity, we assume the baud duration $T = N_s T_s$ and $T_s = N_c T_c$, where $N_s$ and $N_c$ may denote any positive integer values. The period $T$ is longer than the symbol time $T_s$, whenever the multiaccess code spans more than a single symbol.

The signal transmitted by user $k$ can be represented as follows

$$s^{(k)}(t) = \sum_i p_i^{(k)}(t - iT) \tag{6.1}$$

where

$$p_i^{(k)}(t) = \sum_{j=0}^{N_s-1} \sum_{h=0}^{N_c-1} c_{j,h}^{(k)} g\left(t - hT_c - jT_s; a_{iN_s+j}^{(k)}\right) \tag{6.2}$$

and $a_l^{(k)}$ is the $l^{th}$ symbol transmitted by user $k$, chosen independently and with uniform distribution from an alphabet $A_a$. The notation $c_{j,h}^{(k)}$ in (6.2) indicates the value assumed by the spread spectrum code assigned to user $k$ in the chip $h$ for the $j^{th}$ symbol interval in each period $T$. We will assume $c_{j,h}^{(k)} \epsilon \{0, \pm 1\}$. In this way, all proposed spread spectrum techniques, such as TH, DS, and OOC, for I-UWB are included. The signal $g(t;a)$ in (6.2) depends on the modulation format and the UWB pulse used in the communication. In order to be as general as possible, let us assume that the transmitted signal is constituted by a sequence of basic waveforms that can be both pulse amplitude and pulse position modulated. In particular, assume that the possible PPM positions are $M$, and each of them is identified by its PPM delay $\tau_l$, where $l = 0, 1, \ldots, M - 1,$ . It is possible to associate each symbol $a_l^{(k)}$ with a vector $d_i^{(k)} = \left[d_{i,0}^{(k)}, \ldots, d_{i,M-1}^{(k)}\right]$, containing the amplitude of the pulses transmitted in each PPM position. If an $N$ level amplitude modulation is employed, then in the most general case $d_{i,j}^{(k)}$ belongs to a subset of $A_a = \{2p - 1 - N\}_{p=1}^N \cup \{0\}$.

Let the UWB basic pulse waveform be denoted by $w(t)$ with duration $T_w$ chosen such that $T_w + \max_l (\tau_l) \leq T_c$ and with unitary energy. We then have the following relation

$$g\left(t; a_i^{(k)}\right) = \sqrt{E_w} \sum_{l=0}^{M-1} d_{i,l}^{(k)} w(t - \tau_l) \tag{6.3}$$

where $w(t) = \widetilde{w}(t)/\sqrt{E_w}$, and $E_w$ is the energy of the unnormalized pulse, $w(t)$, defined as

$$E_w = \int_{-\infty}^{\infty} |\widetilde{w}(t)|^2 \, dt \qquad (6.4)$$

The UWB channel model proposed by the IEEE 802.15.3a working group [1], is a modification of the Saleh-Valenzuela channel model [2]. This model takes into account the clustering phenomenon observed in several UWB channel measurements [3]. According to [2] the channel impulse response for user $k$ can be modeled as

$$h^{(k)}(t) = \sum_{l=0}^{L} \sum_{h=0}^{H} \alpha_{l,h}^{(k)} \delta\left(t - T_l^{(k)} - \tau_{l,h}^{(k)} - \tau_a^{(k)}\right) \qquad (6.5)$$

where $\left\{\alpha_{l,h}^{(k)}\right\}$ are the multipath gain coefficients, and $\left\{T_l^{(k)}\right\}$ and $\left\{\tau_{l,h}^{(k)}\right\}$ represent the delay of the $l^{th}$ cluster and of the $k^{th}$ multipath relative to the $l^{th}$ cluster arrival time, $\tau_a^{(k)}$ is a random variable characterizing the delay due to asynchronism between users, and is assumed to be uniformly distributed over the interval $T$. The channel impulse response $h^{(k)}(t)$ is assumed to be time-invariant and assumed to have a maximum delay of $\tau_{\max}$ seconds.

Assuming there are $N_u$ active users, the received signal can then be written as

$$r(t) = \sum_{k=1}^{N_u} \sqrt{A_k} s^{(k)}(t) * h^{(k)}(t) + \sqrt{A_b} n_b(t) + n(t) \qquad (6.6)$$

where $*$ denotes the convolution operation, $A_k$ and $A_b$ represent the attenuations due to path loss (which are a function of the transmitter receiver distance), $n(t)$ is the white Gaussian noise process with two-sided power spectral density $\frac{N_0}{2}$, and $n_b(t)$ is the narrowband interference modeled as an ergodic, zero mean, Gaussian random process. Equation (6.6) is the most general form of the received signal that captures all the channel impairments and interference in an I-UWB communication system. In the previous discussion we ignore nonlinear antenna effects. However, (6.6) in its most general form is not analytically tractable as such. We make assumptions, when necessary, to simply (6.6) and to design suboptimal receivers based on our assumptions. We will start with the most simple and basic threshold detection-based receiver and then study other advanced receivers.

## 6.1.2   Threshold/Leading Edge Detection

Threshold detectors, also known as leading edge detection (LED) receivers, were some of the earliest and probably the simplest of all I-UWB receivers [17, 18]. The LED receiver sets a threshold at the receiver, and any incoming pulse that crosses the threshold is detected and demodulated. The problem with the threshold reception technique is that noise spikes which happen to cross the threshold will also be erroneously detected as a data pulse (known as a "false alarm" or "false detection"). To mitigate the problem of false detections, the receiver must continuously monitor the input noise signal and adaptively set a threshold such that only a small percentage of false detections will occur, similar to the constant false alarm rate (CFAR) used in radar [19]. To operate properly, the LED receiver must use a device that is capable of responding to a very sharp change in received voltage in a very short time span. A tunnel diode, with its bistable operating point, would seem to be an ideal candidate for the leading edge detection receiver. However, its bias point varies with temperature, and a temperature compensation network must be designed into the receiver. The LED receiver is advantageous in that it is simple to implement and may be used for the case where only one pulse per data bit is transmitted. The LED receiver has some disadvantages. This receiver is susceptible to noise spikes and is incapable of taking advantage of multipath signals in the channel. Additionally, the receiver front-end, must, by necessity, have a very large bandwidth; any interfering signals (or intentional jammers) will either be sufficiently strong to cross the threshold and trigger numerous false detections or will cause the receiver to increase its threshold and reduce its range. We further study the threshold detector in detail next.

The threshold receiver consists of a circuit fast enough to detect the presence of the I-UWB pulse, and some type of pulse stretcher or one-shot monostable circuit that outputs a pulse that is long enough for slower logic or analog circuits to process. Figure 6.2 is the basic block diagram of a threshold receiver. Note that the pulse stretcher can be viewed as a power gain because the energy in the output pulse is much greater than in the input pulse.

Threshold receivers tend to be simple circuits. An important characteristic of threshold receivers is that they are capable of detecting a single pulse. But, because they respond to a minimum amplitude, threshold receivers cannot take advantage of collecting energy from multipath components, and they are sensitive to noise and interference.

The earliest threshold receivers employed tunnel diode bistable circuits. Figure 6.3 is a tunnel diode bistable circuit suitable for a threshold detector. A tunnel diode is connected to a broadband antenna, and the bias current through the diode is adjusted to point A in Figure 6.4. The voltage from the antenna is applied across the diode, and when the sum of the antenna voltage and the voltage at point A

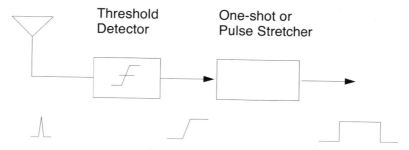

Figure 6.2: Block Diagram of a Threshold Receiver.

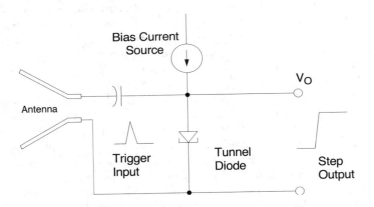

Figure 6.3: Tunnel Diode Peak Threshold Circuit Suitable for an I-UWB Receiver.

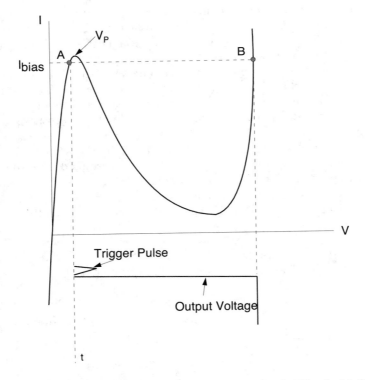

Figure 6.4: I-V Characteristic of a Tunnel Diode Peak Threshold Circuit.

exceeds the peak voltage, $V_P$, the diode is pushed over into the negative resistance region. The voltage across the diode rapidly rises to the stable bias point B. The diode must be reset at some point by interrupting the diode current or by forcing the diode voltage below $V_P$. The external reset circuit can be conventional slow speed circuitry and integrated with the pulse stretcher.

A complete tunnel diode threshold detector circuit is shown in Figure 6.5 [17]. The diode current is set by R1. When the threshold is exceeded by a signal from the antenna, the voltage across D1 rapidly becomes more negative, and Q1 begins to conduct turning on D3 and D4. The result is that the current through R1 that normally biases the tunnel diode D1 is now shunted through Q1. This forms a regenerative loop because, as the current through the diode falls, the diode voltage drops below $V_P$, cutting off Q1. The tunnel diode can now be retriggered. The conduction of Q1, D3, and D4 is relatively slow compared to the tunnel diode transition time. When Q1 conducts, it also drives the base of the avalanche transistor Q2 positive, starting the avalanche breakdown. R2 controls the avalanche point. When Q2 avalanches, it discharges the 680 pF capacitor. The output voltage is taken across the capacitor.

Avalanche transistors can be used for single hit threshold I-UWB receivers. Figure 6.6 is an avalanche transistor detector [18]. Q is biased for avalanche operation. The incoming pulse triggers the avalanche, and the energy stored in the transmission line begins to discharge into the load resistor, $R_L$. The length of the output pulse is twice the propagation delay of the transmission line. The transmission line acts as the pulse stretcher. The pulse across $R_L$ triggers a one-shot monostable. During the one-shot time-out period, the input of the avalanche transistor is held low. This prevents triggering of the avalanche transistor while the transmission line recharges. If the time-out period is set just less than the expected interpulse period, the transistor cannot be retriggered on noise during the majority of the interpulse period. This simple time gating reduces the receiver's sensitivity to noise and interference.

The main disadvantage of the threshold receiver is that any signal that exceeds the threshold will produce an output. This makes it sensitive to noise and interference. The threshold can be raised to reduce these effects at the cost of receiver sensitivity. An additional problem with the receivers in Figures 6.4 and 6.5 is that their threshold is sensitive to temperature variations. Temperature sensitivity can be reduced by applying feedback. The threshold receiver in Figure 6.7 employs a super regeneration technique [18] to reduce temperature sensitivity. The variable constant current source in Figure 6.7 charges capacitor C, resulting in a linear increase in the tunnel diode bias current. This change in bias is shown in Figure 6.8. The current through the diode increases until point 3 on the diode I-V curve in Figure 6.8 is reached. The tunnel diode then switches to its high voltage state. This transition triggers the first avalanche detector and the monostable pulse stretcher.

The monostable output turns on Q2 in Figure 6.7. This discharges the capacitor, and the bias current through the emitter following Q1 drops to a low level. This resets the tunnel diode to its low voltage state. After the monostable times out, the capacitor charges again, producing a sawtooth current waveform through the diode, as shown in Figure 6.8. The pulse stretcher output is also fed to a phase

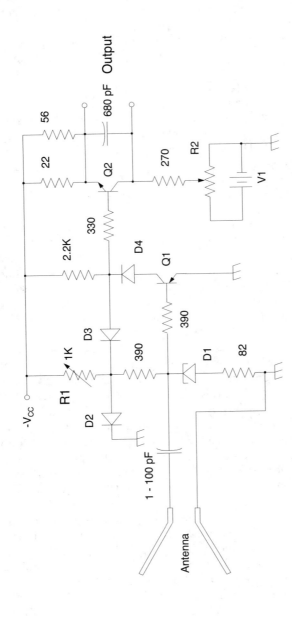

Figure 6.5: Tunnel Diode I-UWB Threshold Receiver.

SOURCE: United States Patent 3,662,316, Short Base-Band Pulse Receiver, Kenneth W. Robbins, May 9, 1972 [17].

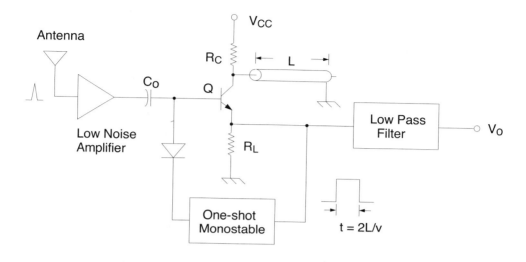

Figure 6.6: Avalanche Transistor I-UWB Detector.

detector. The phase error between the 10 kHz (100 sec) reference and the pulse stretcher output controls the capacitor charging current. If temperature causes the voltage at point 3 in Figure 6.8 to rise, it will take longer than 100 sec for the diode to switch. The resulting phase detector error voltage will increase the capacitor charging current, and the period of the sawtooth will be decreased. The result is that the most sensitive point on the diode's I-V curve is always reached at the same time.

The receiver described in Figure 6.7 is used as part of the radar system. The transmitter is triggered 1 $\mu$sec (99 $\mu$sec after the reset) before the diode switches. The return signal finds the receiver in its most sensitive state just before transition. The output pulse from the first avalanche detector triggers the second avalanche detector only when the returned signal corresponds to the delay determined by the range gate.

The feedback ensures that the receiver operates at the most sensitive point on the tunnel diode I-V characteristic independent of temperature, and the range gate activates the receiver only in the desired time window. This improves the receiver's stability and sensitivity at the same time, reducing its sensitivity to interference.

The circuit in Figure 6.7 addresses the problem of temperature stability but it does not fully address the problem of the optimum threshold. The problem of a false output or a false alarm in threshold detectors is well known in radar systems [20]. The number of false alarms can rise dramatically with only a small change in the threshold. The sensitivity of threshold receivers can be increased by using the CFAR metric to adjust the threshold.

The threshold of a CFAR receiver is constantly readjusted for a preset number of errors. Figure 6.9 is an example of a CFAR receiver [4]. The transmitter output only

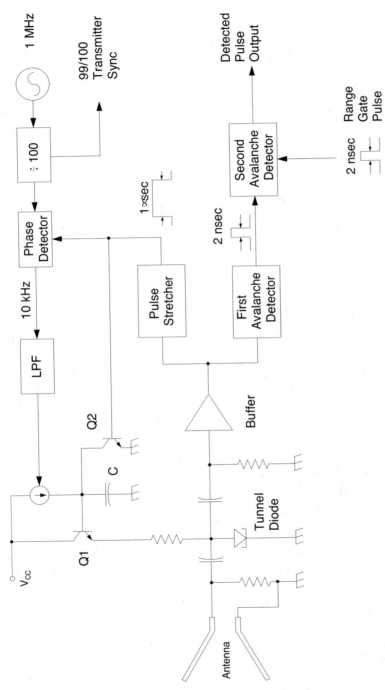

Figure 6.7: Super Regenerative I-UWB Receiver.

SOURCE: C. Bennett and G. F. Ross, "Time Domain Electromagnetics and Its Applications," *Proc. of the IEEE* [4]. © IEEE, 1978. Used by permission.

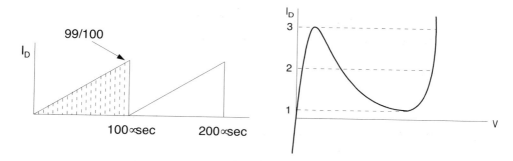

Figure 6.8: Bias Timing and Diode I-V Characteristics for the Receiver in Figure 6.7.

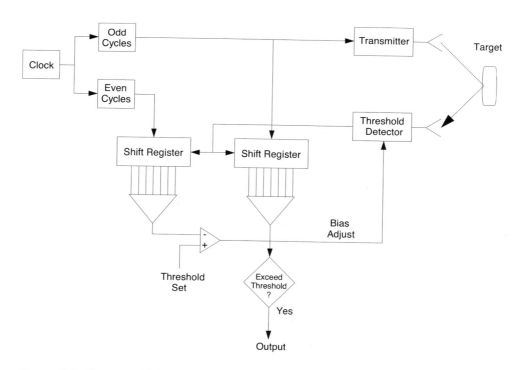

Figure 6.9: Constant False Alarm Rate (CFAR) Receiver.

SOURCE: C. Bennett and G. F. Ross, "Time Domain Electromagnetics and Its Applications," *Proc. of the IEEE* [4]. © IEEE, 1978. Used by permission.

occurs on the odd cycles of the clock. On the even cycles, there is no transmitter output, so any output from the receiver is a spurious or false alarm output. Every time there is a spurious output, a "1" is clocked into the even cycle shift register, and when there is no output, a "0" is clocked in. The output of this shift register is summed and compared with the threshold set. This voltage can be adjusted to produce a given number of "1" hits in the shift register during any given time. The odd cycle shift register is also loaded with a "1" each time the threshold is exceeded. Because a pulse is transmitted during the odd cycles, it is more likely that the threshold will be exceeded by the signal. If the threshold is exceeded a preset number of times, a signal present output is generated. It should be noted that this CFAR receiver is no longer a single pulse detector. The effect of the shift register is to average either the noise, in the case of the even shift registers, or the signal in the odd shift registers.

An analysis of threshold receiver performance is given by Harmuth [21]. Estimates of the probability of detection, $P_D$, and the probability of false alarm, $P_{fa}$, are necessary in order to estimate threshold receiver performance. The noise and the signal pulse noise distributions are assumed to be Gaussian. The zero mean noise distribution is given by

$$W_N\left(x\right) = \frac{1}{\sigma_G\sqrt{2\pi}}\exp\left(-\frac{x^2}{2\sigma_G^2}\right) \qquad (6.7)$$

where

$$\sigma_G^2 = \frac{\text{Average Noise Density During Signal Duration } T}{\text{Received Signal Energy}} \qquad (6.8)$$

and

$$x = \text{normalized noise voltage } = \frac{n}{V_{peak}} \qquad (6.9)$$

Because the received signal is the sum of the noise voltage and the signal voltage, the normalized signal plus noise voltage is given by

$$\left(n + V_{peak}\right)/V_{peak} = x + 1 \qquad (6.10)$$

Thus, the sum of the noise voltage and the signal voltage is also Gaussian distributed whose mean value is shifted by the signal. This results in a pdf for the signal, plus the superimposed noise given by

$$W_{N+S}\left(x\right) = \frac{1}{\sigma_G\sqrt{2\pi}}\exp\left(-\frac{(x-1)^2}{2\sigma_G^2}\right) \qquad (6.11)$$

The normalized pdfs for $W_N\left(x\right)$ and $W_{N+S}\left(x\right)$ are plotted in Figure 6.10. Suppose the threshold is given by $F$ times $\sigma_G$ or $F\sigma_G$. Any value of the noise pdf that exceeds the threshold is a false alarm, and any value of the signal plus noise pdf below the threshold is an error. By integrating the pdfs over the proper

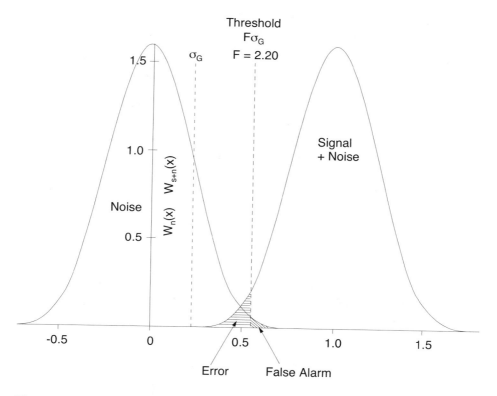

Figure 6.10: Threshold Detection in Gaussian Noise.

SOURCE: H. F. Harmuth, "Radar Equation for Nonsinusodial Waves," *IEEE Transactions on Electromagnetic Compatibility* [21]. © IEEE, 1989. Used by permission.

regions, the probability of detection, $P_D$, and probability of false alarm, $P_{fa}$, can be computed as given by the following equations

$$P_D = \frac{1}{2} \left( 1 + \mathrm{erf} \left[ \frac{1 - F\sigma_G}{\sigma_G \sqrt{2}} \right] \right) \tag{6.12}$$

$$P_{fa} = \frac{1}{2} \left( 1 - \mathrm{erf} \left[ \frac{1 - F\sigma_G}{\sigma_G \sqrt{2}} \right] \right) \tag{6.13}$$

A 15 dB signal-to-noise ratio (SNR) is required for a 0.99 probability of detection with a $10^{-8}$ false alarm rate. Increasing the threshold decreases the false alarm rate but decreases the probability of detection for a given signal-to-noise ratio.

## 6.1.3   Correlation Detection (CD) Receivers

The correlation detection (CD) receiver is also known as a matched filter receiver and has been used in narrowband communication systems for several decades. A block diagram of a simple correlation receiver is shown in Figure 6.11. The incoming pulse is multiplied by a template waveform. The output from the multiplier is a function of how well the template waveform generator matches the incoming waveform in time and shape. The correlation receiver is a matched filter system, and as such, it can provide the optimum detection SNR if the template waveform exactly matches the time and shape of the incoming waveform.

Next, we describe optimal matched filter and simple matched filter performance in multipath fading channels for A-PAM and PPM modulation schemes. We consider the case of a single-user system. For A-PAM modulation, (6.1) can be simplified and the transmitted signal can be written as

$$s\left(t\right) = \sum_j a_j w\left(t - jT\right) \tag{6.14}$$

where $a_j = \pm 1$ are the data bits. We assume that the symbol or baud duration $T$ is long enough to cancel any ISI effects. We also assume a simple $L$-tap multipath channel with impulse response given by

$$h\left(t\right) = \sum_{l=0}^{L-1} \alpha_l \delta\left(t - \tau_l\right) \tag{6.15}$$

where $\alpha_l$ and $\tau_l$ are the magnitude and delay of the $l^{th}$ path, respectively. Now the received signal may be written as

$$r\left(t\right) = \sum_j \sum_{l=0}^{L-1} \alpha_l a_j w\left(t - jT - \tau_l\right) + n\left(t\right) \tag{6.16}$$

Similarly for PPM, the transmitted signal can be written as

$$s\left(t\right) = \sum_j w\left(t - jT - \delta a_j\right) \tag{6.17}$$

where $a_j \in \{1, 0\}$ are the data bits, and $\delta$ is the modulation index that is used to optimize BER performance. Now the received signal can be written as

$$r\left(t\right) = \sum_j \sum_{l=0}^{L-1} \alpha_l w\left(t - jT - \delta a_j - \tau_l\right) + n\left(t\right) \tag{6.18}$$

### Optimal Matched Filter Performance

In this section, we derive an expression for the BER performance of the optimal matched filter. It is assumed that the filter template is exactly matched to the

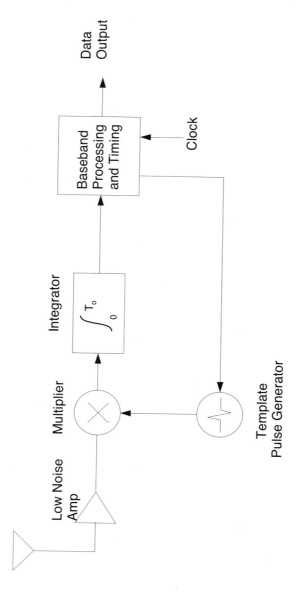

Figure 6.11: Basic Correlation Receiver.

received signal. Such a receiver is studied in [22] for PPM modulation. By averaging over a large number of channel realizations, it is shown that the fading margin of this system is only 1.5 dB due to the ability of the UWB pulse to resolve multipaths. The optimal receiver provides a benchmark against which we can test performance of other suboptimal receivers.

**A-PAM Modulation**    Assume that we have perfect knowledge of the channel. The template used in the matched filter receiver is given by

$$y_{A-PAM}(t) = \sum_{l=0}^{L-1} \alpha_l w(t - \tau_l) = h(t) \tag{6.19}$$

Hence, the decision statistic for the $j^{th}$ bit is given by

$$z = \int_0^T \left( a_j \sum_{l=0}^{L-1} \alpha_l w(t - \tau_l) + n(t) \right) y_{A-PAM}(t)\, dt \tag{6.20}$$

Let $u_{\pm 1}$ and $\sigma_{\pm 1}^2$ be the mean and the variance of $z$ conditioned on $a_j = \pm 1$ respectively. The following is obtained

$$u_1 = -u_{-1} = \int_0^T y_{A-PAM}^2(t)\, dt \tag{6.21}$$

$$\sigma_1^2 = \sigma_{-1}^2 = \frac{N_0}{2} \int_0^T y_{A-PAM}^2(t)\, dt \tag{6.22}$$

$$\int_0^T y_{A-PAM}^2(t)\, dt = E_w \left[ \sum_{l=0}^{L-1} \alpha_l^2 + \sum_{i=0}^{L-1} \sum_{j=0, j \neq i}^{L-1} \alpha_i \alpha_j R(\tau_i - \tau_j) \right] \tag{6.23}$$

where $R(\tau)$ denotes the normalized autocorrelation of pulse $w(t)$, and is defined as

$$R(\tau) = \frac{\int_{-\infty}^{\infty} w(t) w(t - \tau)\, dt}{E_w} \tag{6.24}$$

Thus, the BER is given by

$$P_e = Q \left( \sqrt{\frac{2}{N_0} \int_0^T y_{A-PAM}^2} \right)$$

$$= Q \left( \sqrt{\frac{2E_w}{N_0} \left[ 1 + \sum_{i=0}^{L-1} \sum_{j=0, j \neq i}^{L-1} \alpha_i \alpha_j R(\tau_i - \tau_j) \right]} \right) \tag{6.25}$$

where $Q(.)$ denotes the Gaussian probability integral, and the channel impulse response energy is assumed to be normalized to 1, $\sum_{l=0}^{L-1} \alpha_l^2 = 1$. Recognizing the expected value of $\sum_{i=0}^{L-1} \sum_{j=0,j\neq i}^{L-1} \alpha_i \alpha_j R(\tau_i - \tau_j) = 0$, the BER reduces to

$$P_e = Q\left(\sqrt{\frac{2E_w}{N_0}}\right) \tag{6.26}$$

upon averaging (6.25) over all possible channel realizations.

**PPM Modulation**   For PPM modulation, the template is given by

$$y_{PPM}(t) = \sum_{l=0}^{L-1} \alpha_l w(t - \tau_l) - \sum_{l=0}^{L-1} \alpha_l w(t - \tau_l - \delta) \tag{6.27}$$

Now the decision statistic for the $j^{th}$ bit is given by

$$z = \int_0^T \left(\sum_{l=0}^{L-1} \alpha_l w(t - \delta a_j - \tau_l) + n(t)\right) y_{PPM}(t)\, dt \tag{6.28}$$

Let $u_0 = E(g| a_j = 0)$, $u_1 = E(g| a_j = 1)$ and $\sigma^2$ be the variance of $z$. We have

$$u_1 = -u_0 = E_w \sum_{i=0}^{L-1} \sum_{j=0}^{L-1} \alpha_i \alpha_j \left[R(\tau_i - \tau_j - \delta) - R(\tau_i - \tau_j)\right]$$

$$\sigma^2 = N_0 E_w \sum_{i=0}^{L-1} \sum_{j=0}^{L-1} \alpha_i \alpha_j \left[R(\tau_i - \tau_j) - R(\tau_i - \tau_j - \delta)\right] = N_0 u_0 \tag{6.29}$$

Hence the BER is given by

$$P_e = \frac{1}{2}\left[Q\left(\frac{u_0}{\sigma}\right) + Q\left(\frac{-u_1}{\sigma}\right)\right] \tag{6.30}$$

## Simple Matched Filter Performance

In this section, we look into the simple matched filter receiver, which assumes that the pulse shape was not modified by the channel. The performance of such a receiver in an AWGN channel for A-PAM and PPM modulation is analyzed in [23]. A-PAM performance is identical to BPSK modulation in AWGN. PPM performance depends on the autocorrelation properties of the UWB pulse. An expression for the BER with PPM in a multipath channel is derived in [24].

**A-PAM Modulation**   The decision statistic for the $j^{th}$ bit at the output of this receiver is given by

$$z = \int_0^T r(t + jT)\, w(t)\, dt \tag{6.31}$$

The decision rule is as follows

$$
\begin{aligned}
z \geq 0 &\rightarrow a_j = 1 \\
z < 0 &\rightarrow a_j = -1
\end{aligned}
\qquad (6.32)
$$

For a specific channel realization, because the noise is a Gaussian random variable, $r(t)$ also follows a Gaussian distribution, and $z$ is therefore also a Gaussian random variable. We derive an expression for the BER by first finding the mean and the variance of $z$ conditioned on a bit, and then applying the $Q$ function for the decision statistics. The mean of $z$ conditioned on $a_j = 1$ is given by

$$
u_1 = E\left[\int_0^T \left(\sum_{l=0}^{L-1} \alpha_l w(t - \tau_l) + n(t)\right) w(t)\, dt\right] = E_w \sum_{l=0}^{L-1} \alpha_l R(\tau_l) \qquad (6.33)
$$

Similarly, the variance of $z$ conditioned on $a_j = 1$ is given by

$$
\sigma_1^2 = \frac{N_0}{2} E_w \qquad (6.34)
$$

By symmetry, we have $u_{-1} = -u_1$ and $\sigma_{-1}^2 = \sigma_1^2$. The BER is thus given by

$$
P_e = Q\left(\sqrt{\frac{2E_w}{N_0}\left(\sum_{l=0}^{L-1} \alpha_l R(\tau_l)\right)^2}\right) \qquad (6.35)
$$

Note that the BER performance is limited by the factor $\left(\sum_{l=0}^{L-1} \alpha_l R(\tau_l)\right)^2$, which depends on the autocorrelation properties of the pulse $w(t)$, as well as on the channel characteristics. In fact, for any delay $\tau_l$ larger than the pulse width, $R(\tau_1) = 0$, and the energy in path $l$ is lost. Because the channel delay spread is typically much larger than the pulse width, we expect the performance of this simple matched filter to be highly suboptimal.

**PPM Modulation**    The performance of a simple matched filter using PPM modulation was derived in [24]. The analysis is similar to the one discussed in the previous section. In this case the matched filter statistic is given by

$$
z = \int_0^T r(t + jT)\left[w(t) - w(t - \delta)\right] dt \qquad (6.36)
$$

Thus, the BER can be computed as

$$
P_e = \frac{1}{2}\left[Q\left(\frac{u_0}{\sigma}\right) + Q\left(\frac{-u_1}{\sigma}\right)\right] \qquad (6.37)
$$

where

$$
u_0 = E_w \sum_{l=0}^{L-1} \alpha_l\left[R(\tau_l) - R(\tau_l - \delta)\right] \qquad (6.38)
$$

$$u_1 = E_w \sum_{l=0}^{L-1} \alpha_l \left[ R\left(\tau_l + \delta\right) - R\left(\tau_l\right) \right] \tag{6.39}$$

$$\sigma = N_0 E_w \left(1 - R\left(\delta\right)\right) \tag{6.40}$$

Notice that, in general, $u_0$ and $-u_1$ are not equal.

**Performance of Matched Filter Based on Pilots**    We now look at a receiver where the matched filter template is obtained by averaging $N$ received pilot signals. Pilots are symbols known to the receiver and are primarily used in synchronization and channel estimation. This system can be thought of as a generalization of an auto-correlation or transmitted reference (TR) receiver discussed in Section 6.1.8. In a typical TR system, a pair of unmodulated and modulated signals is transmitted, and the former is employed to demodulate the latter. This receiver can capture the entire signal energy for slowly varying channels without requiring channel estimation. However, it suffers from the use of noisy received signals as a template for demodulation. In the subsequent sections we derive the BER performance of the pilot-based matched filters for A-PAM and PPM modulation schemes.

**A-PAM Modulation**    Suppose we use $N$ pilots (modulated as ones) to estimate the correlator template. The estimated template can be written as

$$\widehat{y}_{A-PAM}\left(t\right) = \frac{1}{N} \sum_{i=0}^{N-1} \left[ \sum_{l=0}^{L-1} \alpha_l w\left(t - \tau_l\right) + n_i\left(t\right) \right] \tag{6.41}$$

or equivalently as

$$\widehat{y}_{A-PAM}\left(t\right) = \sum_{l=0}^{L-1} \alpha_l w\left(t - \tau_l\right) + \frac{1}{N} \sum_{i=0}^{L-1} n_i\left(t\right) \tag{6.42}$$

The decision statistic for the $j^{th}$ bit is given by

$$z = \int_0^T \left( a_j \sum_{l=0}^{L-1} \alpha_l w\left(t - \tau_l\right) n\left(t\right) \right) \widehat{y}_{A-PAM}\left(t\right) dt \tag{6.43}$$

We first show the need for a bandpass filter at the output of the receiver. Assuming a '1' was transmitted, the decision statistic can be written as

$$z = \int_0^T \left[ s\left(t\right) + n\left(t\right) \right] \left[ s\left(t\right) + n_a\left(t\right) \right] dt \tag{6.44}$$

where

$$n_a\left(t\right) = \frac{1}{N} \sum_{i=0}^{N-1} n_i\left(t\right) \tag{6.45}$$

The problematic term in (6.44) is

$$z_n = \int_0^T n(t)\, n_a(t)\, dt \tag{6.46}$$

We know that $E[z_n] = 0$. The second moment of the $z_n$ is given by

$$E[z_n^2] = \int_0^T \int_0^T R_n(t-\lambda)\, R_{n_a}(t-\lambda)\, dt d\lambda \tag{6.47}$$

where $R_n(\tau)$ and $R_{n_a}(\tau)$ are the autocorrelation functions of $n(t)$ and $n_a(t)$ respectively. Moreover, we have

$$R_n(\tau) = \frac{N_0}{2}\delta(\tau) \tag{6.48}$$

$$R_{n_a}(\tau) = \frac{N_0}{2N}\delta(\tau) \tag{6.49}$$

Consequently, we have

$$E[z_n^2] = \frac{N_0^2}{4N}\int_0^T \int_0^T \delta(t-\lambda)\,\delta(t-\lambda)\, dt d\lambda = \infty \tag{6.50}$$

Equation (6.50) shows that the noise power at the output of the receiver is not limited. Thus, we need to band limit the signal first. The same conclusion applies to PPM systems.

Assume that the signal is processed by a bandpass filter of bandwidth $W$ and center frequency $f_c$ before passing through the "imperfect" matched filter. We assume that $W$ is large enough such that ISI and interpulse interference are negligible. The new autocorrelation functions are given by

$$R_n'(\tau) = N_0 W \operatorname{sinc}(W\tau)\cos(2\pi f_c \tau) \tag{6.51}$$

$$Rn_a'(\tau) = \frac{N_0}{N} W \operatorname{sinc}(W\tau)\cos(2\pi f_c \tau) \tag{6.52}$$

Therefore

$$E[z_n^2] = \frac{(WN_0)^2}{N}\int_0^T \int_0^T \operatorname{sinc}^2(W(t-\lambda))\cos^2(2\pi f_c(t-\lambda))\, dt d\lambda \tag{6.53}$$

The other two variance terms are

$$\sigma_1^2 = WN_0 \int_0^T \int_0^T s'(t)\, s'(\lambda)\operatorname{sinc}(W(t-\lambda))\cos(2\pi f_c(t-\lambda))\, dt d\lambda \tag{6.54}$$

$$\sigma_2^2 = \frac{W N_0}{N} \int_0^T \int_0^T s'(t) s'(\lambda) \operatorname{sinc}(W(t - \lambda)) \cos(2\pi f_c(t - \lambda)) \, dt d\lambda \qquad (6.55)$$

where $s'(t)$ is the filtered version of $s(t)$. The total noise power is given by

$$\sigma^2 = \sigma_1^2 + \sigma_1^2 + E\left[z_n^2\right] \qquad (6.56)$$

To obtain a BER in terms of $\frac{E_w}{N_0}$ we make the following transformation

$$s'(t) = \sqrt{E_w} s_0'(t) \qquad (6.57)$$

Thus, the total noise power can be restated as

$$\sigma^2 = E_w N_0 \left(1 + \frac{1}{N}\right) X_2 + \frac{N_0^2}{N} X_1 \qquad (6.58)$$

where

$$X_1 = W^2 \int_0^T \int_0^T \operatorname{sinc}^2(W(t - \lambda)) \cos^2(2\pi f_c(t - \lambda)) \, dt d\lambda \qquad (6.59)$$

$$X_2 = W \int_0^T \int_0^T s_0'(t) s_0'(\lambda) \operatorname{sinc}(W(t - \lambda)) \cos(2\pi f_c(t - \lambda)) \, dt d\lambda \qquad (6.60)$$

Now the BER can be calculated as

$$P_e = Q\left(\sqrt{\frac{E_w}{N_0} \frac{\left[\int_0^T s_0'^2(t) \, dt\right]^2}{\left(1 + \frac{1}{N}\right) X_2 + \frac{X_1}{N}\left(\frac{E_w}{N_0}\right)^{-1}}}\right) \qquad (6.61)$$

Note that the noise term becomes negligible as $\frac{E_w}{N_0}$ increases. Also, decreasing the bandwidth of the bandpass filter limits the noise power at the expense of increased pulse distortion. Increasing the integration time increases energy capture at the expense of more noise. Figures 6.12 and 6.13 show simulation studies for varying the number of pilot bits and also varying the integration time. In the simulation, a bandpass filter of bandwidth 7 GHz with center frequency 3.05 GHz is implemented. This roughly corresponds to the -20 dB bandwidth of the line-of-sight pulse. It is observed that increasing the integration time improves performance up to a certain point, after which the energy capture saturates and only noise is added. When 250 pilot bits are used, the performance is only 1 dB worse than the theoretical lower bound. The power loss incurred by used pilots is not taken into consideration because it depends on the data block size.

**PPM Modulation** A similar approach can be used to derive the BER expression for the pilot-based performance of PPM modulation. The decision statistic now is given by

$$z = \int_0^T (s'(t) + n'(t))(s'(t) - s'(t - \delta) + n_a'(t)) \, dt \qquad (6.62)$$

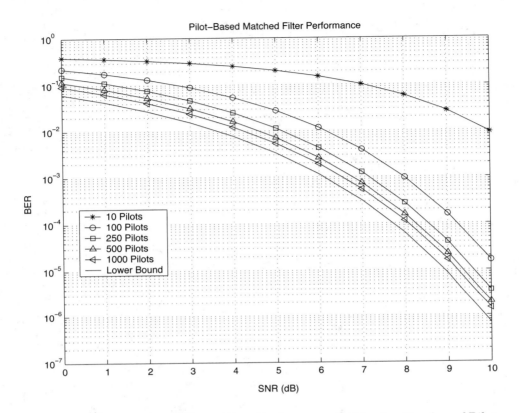

Figure 6.12: Pilot-Based Matched Filter Performance for Varying Number of Pilots ($T = 50$ns).

where $s'(t)$, $n'(t)$, and $n'_a(t)$ are the filtered versions of $s(t)$, $n(t)$, and $n_a(t)$ respectively. The total noise power is contributed by three components as before, and is given by

$$\sigma^2 = E_w N_0 Y_1 + \frac{E_w N_0}{N} Y_2 + \frac{N_0^2}{N} Y_3 \tag{6.63}$$

where

$$Y_1 = W \int_0^T \int_0^T [s'_0(t) - s'_0(t - \delta)] [s'_0(\lambda) - s'_0(\lambda - \delta)]$$
$$\cdot \operatorname{sinc}(W(t - \lambda)) \cos(2\pi f_c (t - \lambda)) \, dt d\lambda \tag{6.64}$$

$$Y_2 = W \int_0^T \int_0^T s'_0(t) \, s'_0(\lambda) \operatorname{sinc}(W(t - \lambda)) \cos(2\pi f_c (t - \lambda)) \, dt d\lambda \tag{6.65}$$

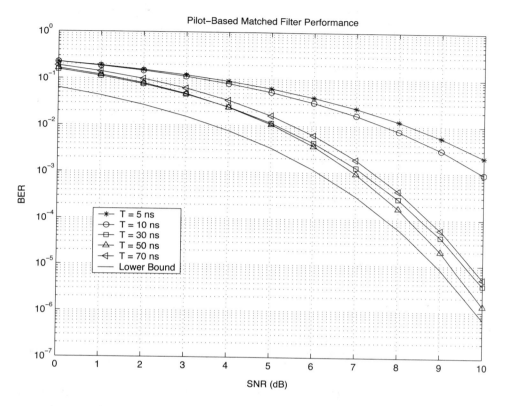

Figure 6.13: Pilot-Based Matched Filter Performance for Varying Integration Time (Pilots = 250).

$$Y_3 = W^2 \int_0^T \int_0^T \text{sinc}^2 \left( W \left( t - \lambda \right) \right) \cos^2 \left( 2\pi f_c \left( t - \lambda \right) \right) dt d\lambda \qquad (6.66)$$

Hence, the BER is given by

$$P_e = Q \left( \sqrt{ \frac{E_w}{N_0} \frac{ \left( \int_0^T s_0'(t) \, s_0'(t - \delta) \, dt \right)^2 }{ Y_1 + \frac{Y_2}{N} + \frac{Y_3}{N} \left( \frac{E_w}{N_0} \right)^{-1} } } \right) \qquad (6.67)$$

Similar approaches can be adopted for the analysis of TH-PSK, OOC-PPM, and DS-PSK multiuser systems. BER performance curves are presented in Figures 6.14 and 6.15 [12].

The CD receiver is advantageous in that the correlation operation can be done either in analog or digital circuits. The receiver can also take advantage of multipath signals by creating a bank of correlators, one for each multipath signal. The primary

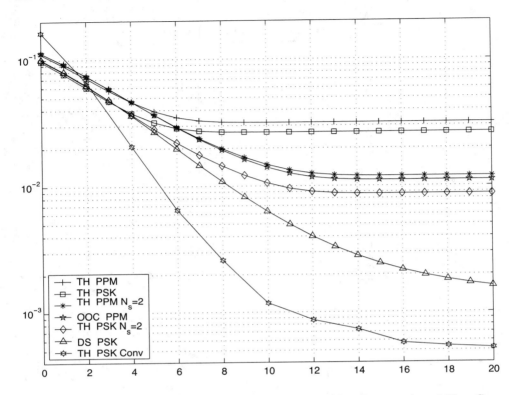

Figure 6.14: Performance Comparison of Different I-UWB Systems in a 5 User System. (The BER of convolution encoded TH-PSK was obtained by simulation.)

SOURCE: G. Durisi and S. Benedetto, "Performance Evaluation and Comparison of Different Modulation Schemes for UWB Multiaccess Systems," *ICC 2003* [12]. © IEEE, 2003. Used by permission.

disadvantage of the CD receiver is imperfect correlations resulting from distorted input pulses. To correct this problem, an adaptive equalizer may be added or the CD receiver may be combined with the template match detection technique and perform a matched filter operation with a series of template waveforms, assuming the resulting increase in digital complexity is allowable.

The LED receiver is sensitive to peak power while the CD receiver is sensitive to total received power. The SNR for the CD receiver is maximized by capturing all the energy in the incoming waveform. The SNR is degraded by a mismatch between the template waveform and the incoming waveform as well as timing errors. Figure 6.16 shows the effect of timing error on a correlation receiver.

The basic building block of this receiver is the correlator. Therefore, the template that is correlated with the incoming signal is very important. In [25], the sinusoidal

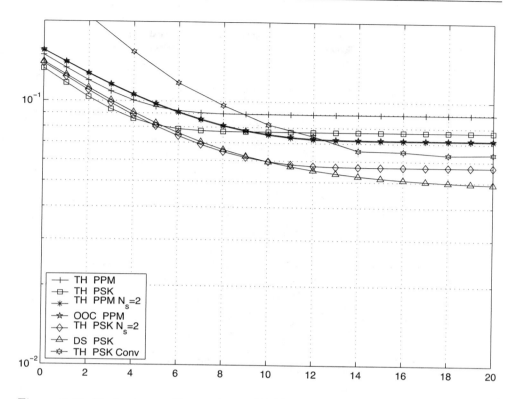

Figure 6.15: Performance Comparison of Different I-UWB Systems in a 14 User System. (The BER of convolution encoded TH-PSK was obtained by simulation.)

SOURCE: G. Durisi and S. Benedetto, "Performance Evaluation and Comparison of Different Modulation Schemes for UWB Multiaccess Systems" *ICC 2003* [12]. © IEEE, 2003. Used by permission.

template for this receiver is compared with the ideal template. The optimal receiver matches the received waveform to the locally generated copy of the received signal. Assuming there is no intersymbol interference, and that a transmitted pulse only suffers from attenuation and delay (neglecting diffraction effects), the received signal would be a delayed and scaled version of the transmitted signal, which is a second derivative of a Gaussian pulse. This second order Gaussian pulse is difficult to generate using electronic circuits. Current system designs typically use a rectangular gate on the central peak of the receiver signal as a template. The sinusoidal template studied in [25] can simplify the UWB receiver structure compared to using the second order Gaussian derivative pulse template. A basic analog phase locked loop (APLL)-type UWB system is shown in Figure 6.17.

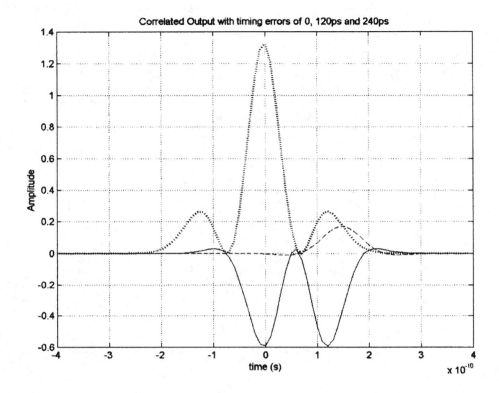

Figure 6.16: Correlator Receiver Output with Time Errors. The Temple Pulse Matches the Input Pulse. The Maximum Correlation Peak Occurs for Zero Time Offset. The Double Negative Peaks Are for 120 ps. Offset, and the Dotted Line Is the Output for 240 ps. Offset.

The switch between the multiplier and the integrator, and the sample and hold (S/H) block, controls the correlation time with respect to the hopping sequence. When the switch is on, the incoming signal correlates with the sinusoidal template. After the correlation, when the switch is off, the integrator works as the signal holder. The S/H block holds the correlated signal before the correlation output is reset. The VCO keeps generating template signals with the same pulse repetition rate that occurs when the switch is on. Therefore, the correlation output signal locks on the optimal template pulse repetition rate at the output of the VCO. Another advantage of this system is that the VCO provides the optimal clock signal to the microprocessor, which controls the time hopping sequence and integration time. The autocorrelation loop at the bottom half of Figure 6.17 ensures that the output of this system keeps the optimal SNR. The primary implementation advantage is that most of the hardware components used can be imported from the current

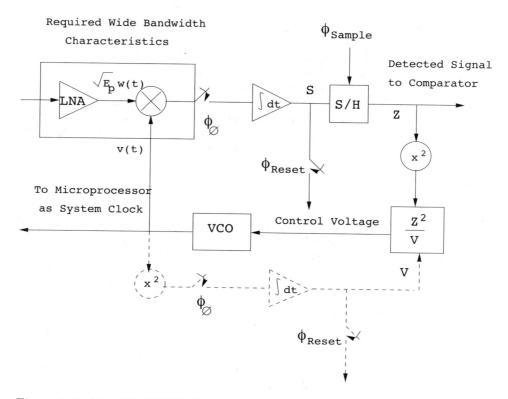

Figure 6.17: Simplified UWB Correlation Receiver Using a Modified Analog PLL.

SOURCE: S. Lee, "Design and Analysis of Ultra-Wide Bandwidth Impulse Radio Receiver," Ph.D. Dissertation [25]. © S. Lee, 2002. Used by permission.

technology, except for the low noise amplifier (LNA) and the mixer that is part of the correlator. Wide bandwidth LNA can be found on board-level designs that are used in basestations and radar systems, but it is hard to find them on chip-level designs. Design of the LNA and mixer for the LD receiver architecture is critical [25].

## 6.1.4   RAKE Receivers

UWB systems development for high-speed indoor communications requires an analysis to establish which type of multiaccess and modulation format, and which receiver structure, offers the best compromise between complexity and robustness against multipath, multi-user and narrowband interference. As shown in [12], DS can better deal with multi-user interference than TH on AWGN channels. For a list of comprehensive literature on spread spectrum systems, see [26]. In this section, we present some of the receiver structures and design issues for DS-UWB systems.

The RAKE receiver is used in any kind of spread spectrum communication system to accumulate the energy in the significant multipath components. A RAKE receiver is a bank of correlators. Each finger of the RAKE is synchronized to a multipath component. The output of each finger is coherently combined using MRC (maximum ratio combining). Channel estimation is required in the combining scheme. Synchronization is critical in this form of receiver. RAKE receivers ideally reject out self-interference due to multipath effects, in addition to other forms of interference like multiaccess interference and adjacent channel interference, by using long, almost orthogonal spreading sequences like the PN or Gold sequence. RAKE receivers for CDMA systems are described in [26].

I-UWB RAKE receivers suffer from two major drawbacks. First, the energy capture is relatively low for a moderate number of fingers when Gaussian pulses are used. As we have seen, a typical NLOS channel can have up to approximately ~70 resolvable dominant specular components. Even if a RAKE receiver with so many fingers is possible, it would only be able to capture part of the signal energy. Second, each multipath undergoes a different channel, which causes distortion in the received pulse shape, and makes the use of a single LOS path signal as a suboptimal template [4]. Another issue is channel estimation. The energy-to-noise ratio available per path is given by

$$ENR = \frac{P_{av}\tau_c}{N_o W \sigma_T} \sim \frac{1}{W} \tag{6.68}$$

where $P_{av}$ is the average transmitted power, $\tau_c$ is the coherence of the UWB channel, $N_o$ is the one-sided power spectral density of the background receiver additive white Gaussian noise. It is seen that the energy-to-noise ratio per path is inversely related to the signal bandwidth; hence, it becomes increasingly difficult to estimate the channel coefficients at 7.26GHz. Channel estimation is critical in RAKE receivers for MRC, so degradation in performance is observed due to imperfect channel estimates. Synchronization (acquisition and tracking) is another major problem for pulses of subnanosecond duration. Also, the design of an all-digital RAKE receiver (similar to conventional RAKE for narrowband and wideband systems) is highly impractical. To summarize, UWB RAKE receivers suffer from two major drawbacks. First, the energy capture is relatively low for a moderate number of fingers when Gaussian pulses are used. It has been shown that a typical NLOS channel may have up to 70 resolvable dominant specular components. Even if a RAKE receiver with so many fingers is possible, it would only be able to capture part of the signal energy. Second, each multipath undergoes a different channel, which causes distortion in the received pulse shape and makes the use of a single LOS path signal as a suboptimal template [4].

Several papers investigate impulse radio systems using the AWGN channel or other simplified multipath channels where taps are spaced at an integer multiple of one pulse period [27, 28]. In general, several multipath components arrive during the duration of one pulse, which generates intrapulse interference. A fractionally spaced (FS) receiver can compensate for channel distortion due to intrapulse interference by sampling at least as fast as the Nyquist rate [29, 30]. An optimal maximum likelihood (ML) detector is far too complex to be implemented for a realistic system, so a suboptimal coherent RAKE receiver to reduce the complexity and still obtain diversity is presented in [31]. The coherent receiver estimates the delay, amplitude, and phase of the channel taps, but does not attempt to account for the correlation between received versions of the same signal. It is possible to reduce complexity even more by using noncoherent receivers that apply nonlinear operations on the received signal to eliminate the necessity of full channel estimation [31]. Such solutions incur additional performance losses.

We briefly discuss coherent and partially/noncoherent RAKE receivers, and consider the problem of digital signaling over a frequency selective fading channel [24]. We consider a single-user system, and model the channel by a tapped delay line with statistically independent tap weights $\{\alpha_l\}_{l=0}^{L-1}$ where $L$ is the number of delays in the delay line. The dispersive channel response $h(t)$ is given by (6.15). Equation (6.15) is considerably different from the comprehensive UWB channel model given by (6.5). But for simpler analysis, and to highlight the benefit of RAKE receivers, we use (6.15) in subsequent analysis. It is apparent that this model provides us with $L$ replicas of the same transmitted signal at the receiver. Hence, a receiver that processes the received signal in an optimum manner will achieve the performance of an equivalent $L^{th}$ order diversity system. Let us consider binary signaling with two equal energy lowpass signals $s_{l,i}(t)$ and $i\epsilon\{1,2\}$, which are either antipodal or orthogonal, and assume that the bandwidth of the signal exceeds the coherence time of the channel. Then the received signal can be expressed as

$$r_l(t) = \sum_{k=0}^{L-1} c_k(t) s_{l,i}\left(t - \frac{k}{W}\right) + n(t) = v_i(t) + n(t) \qquad (6.69)$$

where the time-variant tap weights $\{c_k(t)\}$ are complex-valued stationary random processes. Assuming that the channel tap weights are known, the optimum demodulator consists of two filters matched to $\nu_1(t)$ and $\nu_2(t)$. The demodulator is sampled at the symbol rate, and the samples are passed to a decision circuit that selects the signal corresponding to the largest output. An equivalent optimum demodulator uses cross-correlation instead of matched filtering. In either case, the decision variables for coherent detection can be expressed as

$$z = \mathrm{Re}\left[\int_0^T r_l(t) \nu_m^*(t) dt\right] \qquad (6.70)$$

Figure 6.18[1] illustrates the operations involved in the computation of the decision variables. In effect, the tapped delay line demodulator attempts to collect the signal energy from all the received signal paths that fall within the span of the delay line and carry the same information.

Channel estimation (estimating the tap weights) is crucial in RAKE demodulation and leads to complex receiver structures. If we choose not to estimate the channel, we may either use differential phase-shift keying (DPSK) signaling (partially

---

[1]Figures 6.18–6.20 are modified from Figures 14.5-3, 14.5-6, and 14.5-7 in [30].

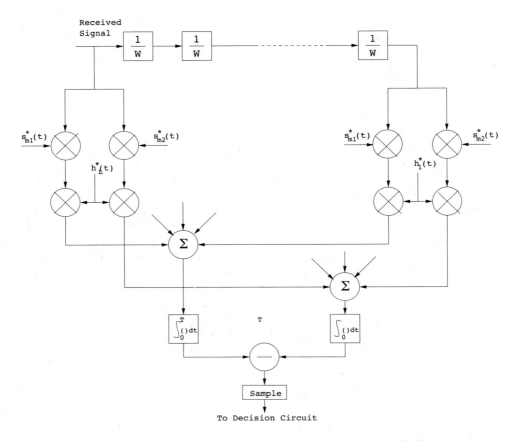

Figure 6.18: Coherent RAKE Demodulator for Binary UWB Signals.

coherent) or noncoherently detected orthogonal signaling. The RAKE demodulator structure for DPSK is shown in Figure 6.19. When the fading is rapid enough to preclude a good estimate of the channel tap weights or the cost of implementing the channel estimators is high, noncoherent RAKE with orthogonal signaling and square-law detection at the receiver can be used. The receiver structure of a square-law-based receiver is shown in Figure 6.20. The RAKE demodulator for orthogonal signaling is assumed to contain a signal component at each delay. If this is not the case, the performance will be degraded because some of the tap correlators will only contribute noise. The RAKE demodulator presented here can be generalized to multilevel signaling. In the following subsections, we further expand on the analysis of RAKE receivers for A-PAM and PPM modulation schemes with resolvable and nonresolvable paths and also analyze a realistic RAKE receiver.

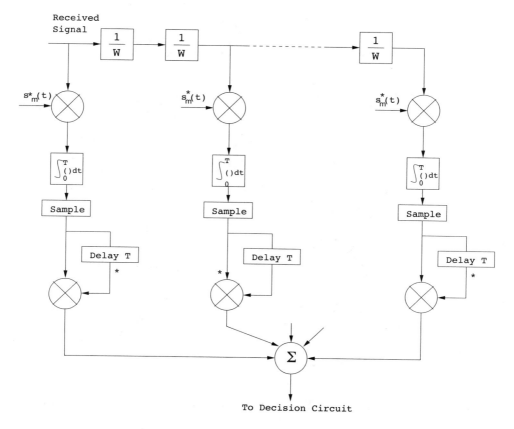

Figure 6.19: Partially Coherent RAKE Demodulator for DPSK-UWB Signals.

## A-PAM Modulation

**Resolvable Paths**    Assuming all paths are resolvable, that is, the minimum time between any two paths is larger than the pulse width, the template created by an $F$-finger RAKE receiver is given by

$$y_{A-PAM}(t) = \sum_{i=1}^{F} \alpha_{fi} w\left(t - \tau_{fi}\right) \qquad (6.71)$$

where $\alpha_{fi}$ and $\tau_{fi}$ denote the amplitude and delay of the $i^{th}$ strongest resolvable paths respectively. Following the development leading to (6.19), the BER in this case is given by

$$P_e = Q\left(\sqrt{\frac{2E_w}{N_0} \sum_{i=1}^{F} \alpha_{fi}^2}\right) \qquad (6.72)$$

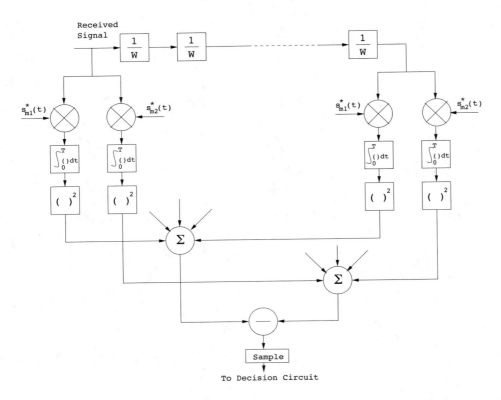

Figure 6.20: Noncoherent RAKE Demodulator for Square-Law Combining of Orthogonal Signals.

where the term $\sum_{i=1}^{F} \alpha_{fi}^2$ represents the energy capture. When all the paths are resolvable, $\sum_{i=1}^{F} \alpha_{fi}^2 = 1$ for $F = L$, we get $P_e = Q\left(\sqrt{\frac{2E_w}{N_0}}\right)$ as expected.

**Nonresolvable Paths**    Let us now assume that the two paths may be less than a pulse width apart. The autocorrelation function of the wave is now evident. We assume that the template constructed by the RAKE is still given by (6.71). Then the decision statistic is given by

$$z = \int_0^T \left[\sum_{l=0}^{L-1} \alpha_l w\left(t - \tau_l\right) + n\left(t\right)\right]\left[\sum_{i=1}^{F} \alpha_{fi} w\left(t - \tau_{fi}\right)\right] dt \qquad (6.73)$$

The mean and variance of the decision statistic can be readily shown as

$$E\left[z\right] = E_w\left[\sum_{i=1}^{F} \alpha_{fi}^2 + \sum_{i=1}^{F}\sum_{l=0,l\neq fi}^{L-1} \alpha_l\alpha_{fi}R\left(\tau_l - \tau_{fi}\right)\right] \qquad (6.74)$$

$$\sigma^2 = E_w \frac{N_0}{2} \left[ \sum_{i=1}^{F} \sum_{j=1}^{F} \alpha_{fi} \alpha_{fj} R\left(\tau_i - \tau_j\right) \right] \tag{6.75}$$

The BER is then given by

$$P_e = Q \left( \sqrt{\frac{2E_w}{N_0} \frac{\left[ \sum_{i=1}^{F} \alpha_{fi}^2 + \sum_{i=1}^{F} \sum_{l=0,l\neq fi}^{L-1} \alpha_l \alpha_{fi} R\left(\tau_l - \tau_{fi}\right) \right]^2}{\left[ \sum_{i=1}^{F} \sum_{j=1}^{F} \alpha_{fi} \alpha_{fj} R\left(\tau_i - \tau_j\right) \right]}} \right) \tag{6.76}$$

where $\sum_{i=1}^{F} \sum_{l=0,l\neq fi}^{L-1} \alpha_l \alpha_{fi} R\left(\tau_l - \tau_{fi}\right)$ represents the additional normalized term due to path correlations. Because the delays and amplitudes of the RAKE fingers will be shifted when path correlations are nonzero, the development of (6.76) is a simplified analysis. Figure 6.21 shows that in the case of perfect channel estimation, the theoretical and simulated performances are very close. In the case of noisy

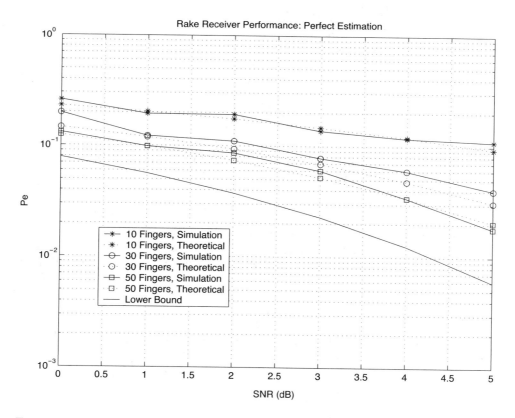

Figure 6.21: RAKE Performance for A-PAM Modulation and Perfect Channel Estimation. 500 ps Pulse Is Assumed.

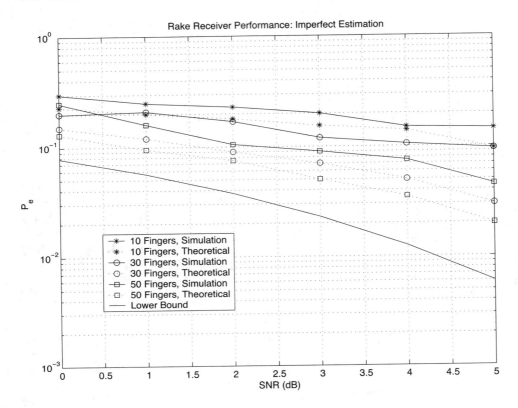

Figure 6.22: RAKE Performance for A-PAM Modulation and Imperfect Channel Estimation. 500 ps Pulse Is Assumed.

channel estimation (Figure 6.22), the theoretical expression leads to more significant error. This is because the RAKE receiver is no longer able to select the $F$ strongest paths. However, Figure 6.21 shows that the estimation error caused by multipath is negligible.

## PPM Modulation

**Resolvable Paths**  Assuming all paths are resolvable, the template created by an $F$-finger RAKE receiver for PPM modulation is given by

$$y_{PPM}(t) = \sum_{i=1}^{F} \alpha_{fi} w(t - \tau_{fi}) - \sum_{i=0}^{F} \alpha_{fi} w(t - \tau_{fi} - \delta) \qquad (6.77)$$

Notice that all the fingers are shifted by $\delta$ when a '1' is sent. Let us assume for now that $\delta$ is greater than the pulse width. Because the mean and the variance of the decision statistic are given by

$$E[z] = E_w \sum_{i=1}^{F} \alpha_{fi}^2 \tag{6.78}$$

$$\sigma^2 = E_w N_0 \sum_{i=1}^{F} \alpha_{fi}^2 \tag{6.79}$$

we obtain an expression for the BER as

$$P_e = Q\left(\sqrt{\frac{E_w}{N_0} \sum_{i=1}^{F} \alpha_{fi}^2}\right) \tag{6.80}$$

**Nonresolvable Paths**   We now look at the more realistic case where the correlation among the paths is nonzero. The template reconstructed by the RAKE is still given by (6.77). The decision statistic is now given by

$$z = E_w \left[ \sum_{i=1}^{F} \alpha_{fi}^2 + \sum_{i=1}^{F} \sum_{l=0,l\neq fi}^{L-1} \left[ \alpha_l \alpha_{fi} R\left(\tau_l - \tau_{fi}\right) - \alpha_l \alpha_{fi} R\left(\tau_l - \tau_{fi} - \delta\right) \right] \right]$$
$$+ n'(t) \tag{6.81}$$

The mean and the variance of the decision statistic are given by

$$E[z] = E_w$$
$$\cdot \left[ \sum_{i=1}^{F} \alpha_{fi}^2 + \sum_{i=1}^{F} \sum_{l=0,l\neq fi}^{L-1} \left[ \alpha_l \alpha_{fi} R\left(\tau_l - \tau_{fi}\right) - \alpha_l \alpha_{fi} R\left(\tau_l - \tau_{fi} - \delta\right) \right] \right] \tag{6.82}$$

$$\sigma^2 = E_w N_0 \sum_{i=1}^{F} \sum_{j=1}^{F} \alpha_{fi} \alpha_{fj} \left[ R\left(\tau_i - \tau_j\right) - R\left(\tau_i - \tau_j - \delta\right) \right] \tag{6.83}$$

The BER is then given by

$$P_e = Q\left(\sqrt{\frac{E_w}{N_0} \frac{\left(\sum_{i=1}^{F} \alpha_{fi}^2 + \sum_{i=1}^{F} \sum_{l=0,l\neq fi}^{L-1}[\alpha_l \alpha_{fi} R(\tau_l - \tau_{fi}) - \alpha_l \alpha_{fi} R(\tau_l - \tau_{fi} - \delta)]\right)^2}{\sum_{i=1}^{F} \sum_{j=1}^{F} \alpha_{fi} \alpha_{fj}[R(\tau_i - \tau_j) - R(\tau_i - \tau_j - \delta)]}}\right) \tag{6.84}$$

## Realistic RAKE Receiver

In this section, we derive the BER expression for a more realistic RAKE finger. The main assumption is that the RAKE receiver "knows" the location of the $F$ strongest resolvable paths. However, the RAKE receiver has to estimate the MRC weights. Assume that the RAKE attempts to estimate the weight at a multipath with known delay $\tau_f$. The correlator output is given by

$$X = \int_0^T \left[ \sum_{l=0}^{L-1} \alpha_l w\left(t - \tau_l\right) + n\left(t\right) \right] w\left(t - \tau_f\right) dt \tag{6.85}$$

$$X = \alpha_f E_w + E_w \sum_{l=0, l \neq f}^{L-1} \alpha_l R\left(\tau_l - \tau_f\right) + \int_0^T n\left(t\right) w\left(t - \tau_f\right) dt \tag{6.86}$$

The term $\int_0^T n\left(t\right) w\left(t - \tau_f\right) dt$ is Gaussian with zero mean and variance $\frac{N_0 E_w}{2}$. Assuming an $N$-pilot training sequence, an estimate of $\alpha_f$ is therefore

$$\widehat{\alpha}_f = \alpha_f + \sum_{l=0, l \neq f}^{L-1} \alpha_l R\left(\tau_l - \tau_f\right) + n_f = \alpha'_f + n_f \tag{6.87}$$

where $n_f$ is a zero mean Gaussian random variable with variance $\frac{N_0}{2N E_w}$. The decision statistic is then given by

$$z = \int_0^T \left[ \sum_{l=0}^{L-1} \alpha_l w\left(t - \tau_l\right) + n\left(t\right) \right] \left[ \sum_{i=0}^{F-1} \left(\alpha'_{fi} + n_{fi}\right) w\left(t - \tau_{fi}\right) \right] dt \tag{6.88}$$

We can write $z$ as

$$z = U + X_1 + X_2 + X_3 \tag{6.89}$$

where

$$U = E\left[z\right] = E_w \sum_{l=0}^{F-1} \alpha'^2_{fl} + E_w \sum_{i=0}^{F-1} \sum_{j=0, j \neq i}^{L-1} \alpha_j \alpha_{fi} R\left(\tau_j - \tau_{fi}\right) \tag{6.90}$$

$$X_1 = \int_0^T n\left(t\right) \sum_{i=0}^{F-1} \alpha'_{fi} w\left(t - \tau_{fi}\right) dt \tag{6.91}$$

$$X_2 = \int_0^T n\left(t\right) \sum_{i=0}^{F-1} n_{fi} w\left(t - \tau_{fi}\right) dt \tag{6.92}$$

$$X_3 = \int_0^T \sum_{l=0}^{L-1} \alpha_l w\left(t - \tau_l\right) \sum_{i=0}^{F-1} n_{fi} w\left(t - \tau_{fi}\right) dt \tag{6.93}$$

Because the variance of $z$ is given by $E\left[X_1^2\right] + E\left[X_2^2\right] + E\left[X_3^2\right]$, the BER can be calculated as

$$Pe = Q\left(\frac{U}{\sqrt{\frac{N_0}{2} E_w Y_1 + \frac{F E_w N_0^2}{4N} + \frac{E_w N_0}{2N} Y_2}}\right) \tag{6.94}$$

where

$$Y_1 = \sum_{i=0}^{F-1} \alpha_{fi}'^2 + \sum_{i=0}^{F-1} \sum_{l=0,l\neq i}^{F-1} \alpha_{fl}' \alpha_{fi}' R\left(\tau_{fi} - \tau_{fl}\right) \tag{6.95}$$

$$Y_2 = \sum_{i=0}^{L-1} \sum_{j=0}^{L-1} \sum_{k=0}^{F-1} \alpha_i \alpha_j R\left(\tau_i - \tau_{fk}\right) R\left(\tau_j - \tau_{fk}\right) \tag{6.96}$$

Equation (6.94) can be written compactly as

$$P_e = Q\left(\sqrt{\frac{2E_w}{N_0} \frac{\left[\sum_{l=0}^{F-1} \alpha_{fl}'^2 + \sum_{i=1}^{F-1} \sum_{j=0,j\neq i}^{L-1} \alpha_j \alpha_{fi}' R\left(\tau_j - \tau_{fi}\right)\right]^2}{Y_1 + \frac{Y_2}{N} + \frac{F}{2N}\left(\frac{E_w}{N_0}\right)^{-1}}}\right) \tag{6.97}$$

Figures 6.23 and 6.24 test the validity of this theoretical expression. When evaluating the theoretical expression, it is assumed that the $F$-finger RAKE receiver selects the $F$ largest $\widehat{\alpha}_f$ evaluated in the absence of noise. Notice that the theoretical expression becomes more precise for larger number of pilots.

The performance of DS-UWB and TH-UWB for a single user link is compared for a data rate of 100 Mbps using the newly proposed realistic UWB channel model from IEEE P802.15 [32]. The IEEE UWB channel characteristic for four different scenarios is summarized in Table 6.1. The performance of suboptimal fractionally spaced (FS) coherent RAKE receivers with a pulse matched filter is evaluated for different channel estimation algorithms suitable for coherent detection. The FS-RAKE receiver is compared with pulse and symbol spaced RAKE receivers and an optimal maximum likelihood detector. The FS-RAKE receiver suffers 1 dB loss compared with the ML detector assuming perfect channel estimates, and the best evaluated channel estimation algorithm has a fractional loss compared with the perfect estimated channel. DS-UWB outperforms the TH-UWB by 2 dB when the transmitter-receiver separation is within 4 meters. Therefore, DS-UWB is more suitable for high-speed indoor links than TH-UWB systems. BER performance curves are shown in Figures 6.25 to 6.29.

The RAKE receiver using maximum ratio combining (MRC) is optimal only when the disturbance to the desired signal is sourced by additive white Gaussian noise. Wireless personal area networks (WPANs), including those with a UWB physical layer, will be typically required to operate in proximity to other wireless networks. In the presence of narrowband interference, a UWB receiver with a

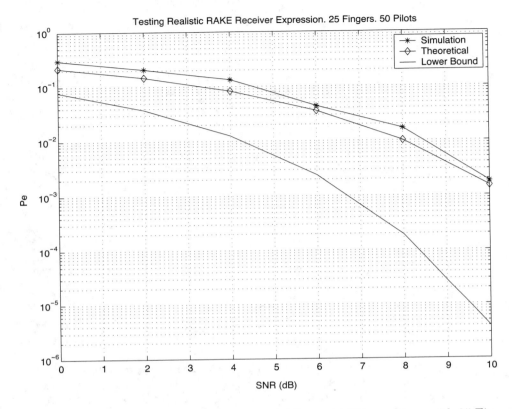

Figure 6.23: Testing Realistic RAKE Receiver Theoretical Expressions with 25 Fingers and 50 Pilots.

conventional RAKE combiner will exhibit an error floor dependant on the signal-to-interference plus noise ratio (SINR). A more suitable diversity scheme to employ in this case is optimum combining (OC), whereby the received signals are weighted and combined to maximize the output SINR [34]. The minimum mean square error (MMSE) RAKE and eigen analysis-based eigen canceller (EC) are possible implementations of OC [35, 36]. The next sections discuss these receiver structures. The transmitter model assumed for the receivers in Sections 6.1.4 and 6.1.5 is shown in Figure 6.30.

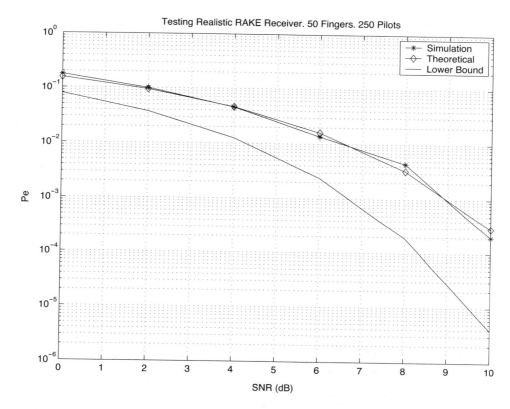

Figure 6.24: Testing Realistic RAKE Receiver Theoretical Expressions with 50 Fingers and 250 Pilots.

Table 6.1: IEEE UWB Channel Characteristics for Four Different Scenarios.

SOURCE: B. Mielczarek, M. -Ol. Wessman, and A. Svensson, "Performance of Coherent UWB Rake Receivers with Channel Estimators," *Proc. IEEE Vehicular Technology Conference* [33]. © IEEE, 2003. Used by permission.

| Target Channel Characteristic | CM1 | CM2 | CM3 | CM4 |
|---|---|---|---|---|
| Distance (m) | 0–4 | 0–4 | 4–10 | |
| (Non-) Line of Sight | LOS | NLOS | NLOS | NLOS |
| Mean excess delay $\tau_m$ (ns) | 5.05 | 10.38 | 14.18 | |
| RMS delay spread $\tau_{rms}$ (ns) | 5.28 | 8.03 | 14.28 | 25 |
| $NP_{10dB}$ | | | 35 | |
| NP (85%) | 24 | 36.1 | 61.64 | |

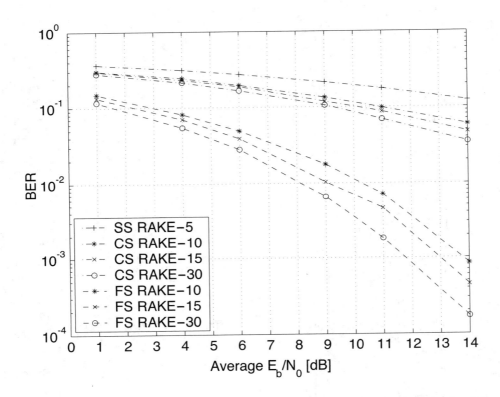

Figure 6.25: Different Detection Method with Perfect Channel Estimation for DS-UWB on CM1.

SOURCE: B. Mielczarek, M.-Ol. Wessman, and A. Svensson, "Performance of Coherent UWB Rake Receivers with Channel Estimators," *Proc. IEEE Vehicular Technology Conference* [33]. © IEEE, 2003. Used by permission.

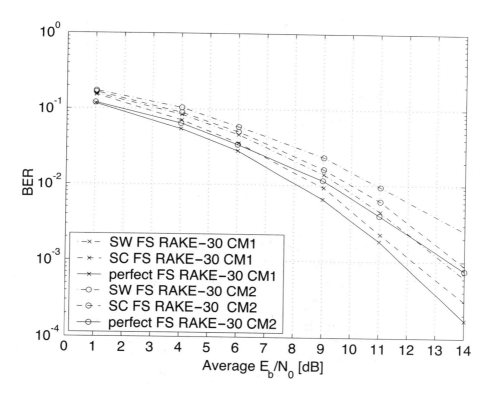

Figure 6.26: Different Channel Estimation Methods for DS-UWB on CM1 and CM2 Assuming 100 Pilot Symbols.

SOURCE: B. Mielczarek, M. -Ol. Wessman, and A. Svensson, "Performance of Coherent UWB Rake Receivers with Channel Estimators," *Proc. IEEE Vehicular Technology Conference* [33]. © IEEE, 2003. Used by permission.

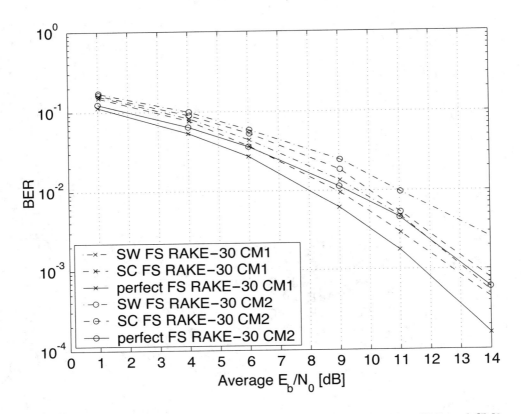

Figure 6.27: Different Channel Estimation Methods for TH-UWB on CM1 and CM2 Assuming 100 Pilot Symbols.

SOURCE: B. Mielczarek, M. -Ol. Wessman, and A. Svensson, "Performance of Coherent UWB Rake Receivers with Channel Estimators," *Proc. IEEE Vehicular Technology Conference* [33]. © IEEE, 2003. Used by permission.

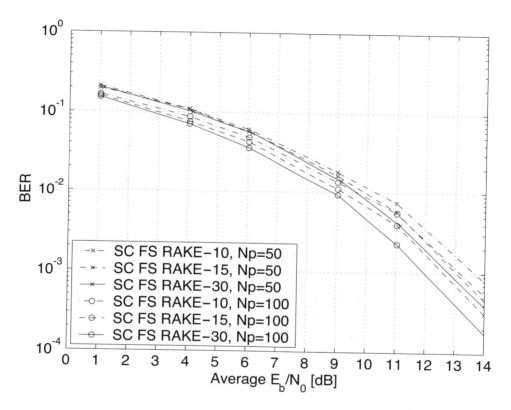

Figure 6.28: Different RAKE Detectors for DS-UWB on CM1.

SOURCE: B. Mielczarek, M. -Ol. Wessman, and A. Svensson, "Performance of Coherent UWB Rake Receivers with Channel Estimators," *Proc. IEEE Vehicular Technology Conference* [33]. © IEEE, 2003. Used by permission.

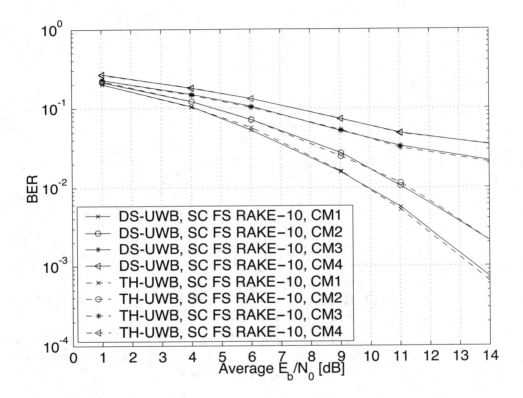

Figure 6.29: TH-UWB and DS-UWB with Successive Channel Estimation Algorithm Using 100 Pilot Symbols.

SOURCE: B. Mielczarek, M.-Ol. Wessman, and A. Svensson, "Performance of Coherent UWB Rake Receivers with Channel Estimators," *Proc. IEEE Vehicular Technology Conference* [33]. © IEEE, 2003. Used by permission.

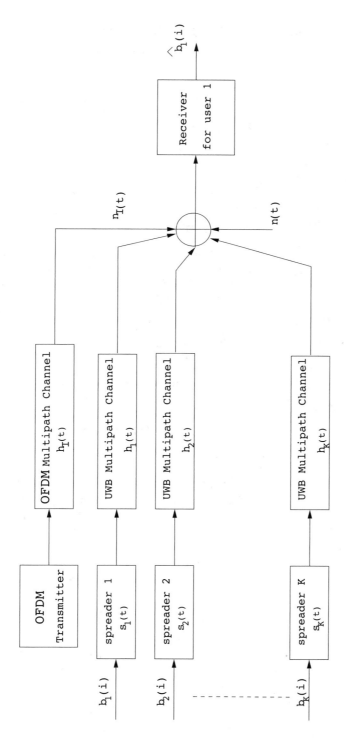

Figure 6.30: Narrowband Interference Model for DS-UWB.

SOURCE: Q. Li and L. A. Rusch, "Muti-user Receivers for DS-CDMA UWB," *IEEE Conference on Ultra Wideband Systems and Technologies* [37]. © IEEE, 2003. Used by permission.

## 6.1.5   Multi-User Detection (MUD) UWB Receivers

The effectiveness of MUD for an UWB pulse-based direct sequence spread spectrum system is described in [37]. It is shown in this chapter that adaptive minimum mean square error (MMSE) MUD receivers are able to gather the multipath energy and reject intersymbol and interchip interference for these channels to a much greater extent than conventional RAKE receivers. Adaptive MMSE is able to reject an IEEE 802.11a OFDM interferer even for SIR as high as $-30$ dB. The adaptive MMSE receiver exhibits only a 6 dB penalty in a 16 user case relative to the single user case. Practical RAKE receivers are incapable of effectively rejecting either strong narrowband interference or heavily loaded wideband interference. This chapter shows that even moderate levels of interference causes significant performance degradation of RAKE receivers. Figure 6.31 shows this MMSE MUD UWB receiver architecture. The BER performance plots to illustrate this are shown in Figure 6.32 and 6.33 [38].

This receiver has high complexity, as the number of taps is equal to at least four times the length of the spreading code (for an observation window of two symbols and two samples per chip). The sampling rate of the analog to digital (A/D) converter is twice the chip rate, which for high bit rate systems leads to very high A/D sample rates.

## 6.1.6   Hybrid RAKE/MUD UWB Receivers

In this section, we describe a hybrid RAKE MUD UWB receiver that is an enhancement of RAKE reception, replacing MRC combining with optimum combining based on MMSE or eigen analysis [hereafter referred to as the eigen canceller (EC)].

### Minimum Mean Square Error (MMSE) MUD UWB RAKE Receiver

The effectiveness of combining multiple arm RAKE reception with adaptive MMSE combining to combat both narrowband and wideband interference in a UWB pulse-based spread spectrum system is described in [39]. This receiver takes an $N$ finger RAKE and replaces the MRC with an adaptive MMSE combiner. The proposed receiver offers robust suppression of narrowband interference for both a single user and a multi-user DS-UWB system as compared to the RAKE only receiver. The receiver provides much greater rejection of multiaccess interference (wideband interference), avoiding an error floor for moderate system load. RAKE reception alone exhibits an error floor at light system load. The proposed hybrid is considered as a reduced-complexity alternative to adaptive MMSE MUD-UWB.

The new receiver gathers multipath energy and rejects intersymbol and interchip interference for the 2–8 GHz indoor channel to a much greater extent than simple RAKE receivers. It is also demonstrated that the adaptive combiner is able to reject the IEEE 802.11a OFDM interferer, even for signal-to-interference ratios (SIR) as severe as $-30$ dB. The adaptive combiner offers improved multiaccess interference rejection, with no visible error floor at $10^{-3}$ bit error rate. Simple RAKE receivers are incapable of effectively rejecting either the strong narrowband interference or the wideband interference.

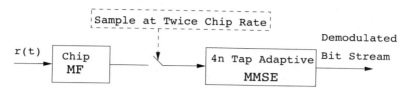

Figure 6.31: MMSE MUD-UWB Receiver.

SOURCE: Q. Li and L. A. Rusch, "Hybrid RAKE / Multiuser Receivers," *Proc. Radio and Wireless Conference (RAWCON '03)* [39]. © IEEE, 2003. Used by permission.

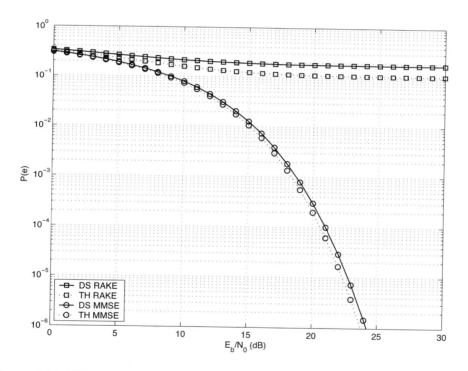

Figure 6.32: TH and DS Performance in Presence of Strong Narrowband Interference. Six Asynchronous Users Are Active. BER Curves Are Depicted for Both RAKE and MMSE Receivers.

SOURCE: G. Durisi, J. Romm, and S. Benedetto, "Performance of TH and DS UWB Multiaccess Systems in Presence of Multipath Channel and Narrowband interference," *IWUWBS 2003* [38]. © IEEE, 2003. Used by permission.

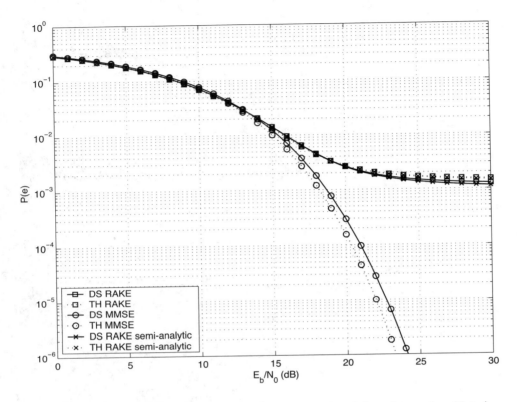

Figure 6.33: TH and DS Performance When Narrowband Interferers Are Not Active. BER Curves Are Depicted for Both RAKE and MMSE Receivers.

SOURCE: G. Durisi, J. Romm, and S. Benedetto, "Performance of TH and DS UWB Multiaccess Systems in Presence of Multipath Channel and Narrowband interference," *IWUWBS 2003* [38]. © IEEE, 2003. Used by permission.

Figure 6.34 gives the block diagram for this receiver. The receiver essentially consists of a chip-matched filter (or suboptimally a bandpass filter) and an adaptive filter. The chip-matched filter suppresses noise and interference out of a signal band. The adaptive filter is an FIR filter that essentially acts as a correlator. The adaptive filter would perform minimum mean square error combining of the RAKE fingers using, for instance, the least mean squares (LMS) or recursive least squares (RLS) algorithm. The adaptation would be first to a known training sequence and later to data directed adaptation. While MRC is optimal in Rayleigh fading channels with additive white Gaussian noise, the adaptive MMSE filter is effective against narrowband and wideband interferers. The BER performance plots are presented in Figures 6.35 and 6.36 [39].

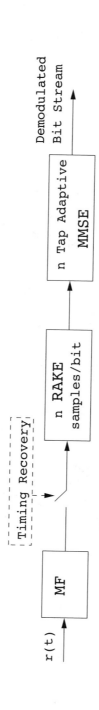

Figure 6.34: Hybrid RAKE/MMSE-MUD UWB Receiver.

SOURCE: Q. Li and L. A. Rusch, "Hybrid RAKE / Multiuser Receivers," *Proc. Radio and Wireless Conference (RAWCON '03)* [39]. © IEEE, 2003. Used by permission.

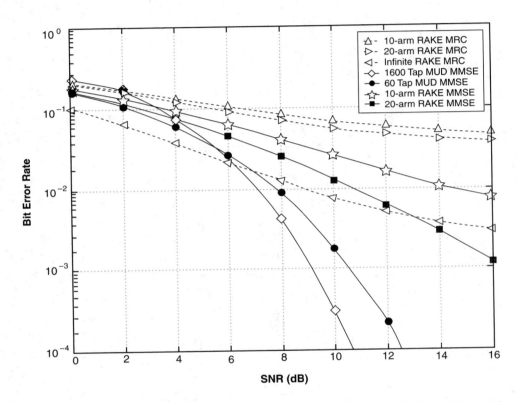

Figure 6.35: BER Performance Comparison of RAKE MRC, MMSE MUD, and Hybrid RAKE/MMSE MUD-UWB in NLOS UWB Channels in the Presence of 5 UWB Interferers All with the Same Received Power.

SOURCE: Q. Li and L. A. Rusch, "Hybrid RAKE / Multiuser Receivers," *Proc. Radio and Wireless Conference (RAWCON '03)* [39]. © IEEE, 2003. Used by permission.

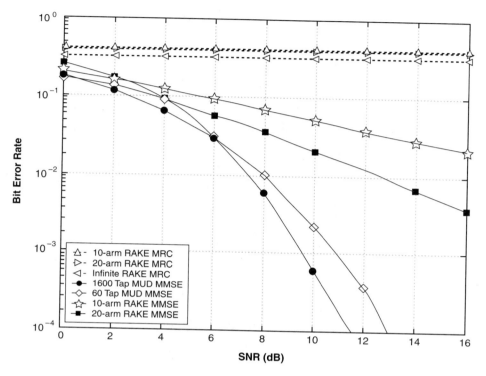

Figure 6.36: BER Performance Comparison of RAKE MRC, MMSE MUD, and Hybrid RAKE/MMSE MUD-UWB in NLOS UWB Channels in the Presence of 5 UWB Interferers All with the Same Received Power and One OFDM Interferer with SIR=-30 dB.

SOURCE: Q. Li and L. A. Rusch, "Hybrid RAKE / Multiuser Receivers," *Proc. Radio and Wireless Conference (RAWCON '03)* [39]. © IEEE, 2003. Used by permission.

## Eigen Canceller (EC) MUD-UWB RAKE Receiver

The MMSE scheme is optimal in the sense that it achieves the maximum likelihood solution for Gaussian interference plus noise if the correlation of the received signal (aggregate of transmitted signals, interference, and noise) is known. When this correlation matrix has to be estimated, the MMSE solution is affected by measurement noise and is not optimal anymore. An eigen analysis-based OC scheme referred to as eigen canceller (EC) has been suggested for various applications, among them suppression of narrowband interference in direct sequence spread spectrum [40]. The EC exploits the inherent low rank property of the narrowband interference correlation matrix. It is designed as a weight vector orthogonal to the interference subspace. The interference subspace is defined as the signal space spanned by the eigen vectors associated with the dominant eigen values. For the method to be effective, the dominant eigen values need to be contributed mainly by the interference. This is the case for low SINR. The EC is motivated by the observation that the correlation matrix of the received signal consists of a limited number of large eigen values contributed by the narrowband interference and a large number of small and almost equal eigen values contributed by the desired signal and AWGN. A tap weight vector orthogonal to the interference eigen vectors effectively cancels the interference, leaving most of the data untouched. The EC is computed from relatively few, stable eigen vectors spanning the interference subspace. Thus even with a short data record it can obtain a high degree of interference cancellation. The next section compares MMSE and EC-based UWB RAKE receivers.

## MMSE Versus EC MUD UWB RAKE Receiver

We briefly compare an MMSE versus EC MUD UWB RAKE over dispersive channels for TH-UWB for A-PAM and binary PPM modulation formats [41]. These ideas can be extended to DS-UWB.

Consider a binary stream $\{d_k\} \in \{\pm 1\}$ transmitted over a multipath channel where each data bit is represented by a sequence of $N_p$ time delayed pulses. The basic pulse $w(t)$ is chosen to meet the limits imposed by the FCC. The basic waveform representing a data bit is given by

$$q(t) = \sum_{j=1}^{N_p} w(t - jT_f - c_j T_c) \tag{6.98}$$

where $T_f$ denotes the pulse repetition period.

Polarity reversals can eliminate the spectral lines and reduce the peak-to-average power ratio. Two types of modulation, A-PAM and binary PPM, are considered. Hence, the following model is used for the received signal at time epoch $k$, namely,

$$r_k(t) = \begin{cases} d_k q(t - kT_s) * h(t) + I(t) + n(t) & \text{A-PAM} \\ q(t - kT_s - d_k \tau_p) * h(t) + I(t) + n(t) & \text{binary PPM} \end{cases} \tag{6.99}$$

where the subscript $k$ represents the bit index, $T_s = T_f N_p$ is the symbol duration, $\tau_p$ is a PPM shift that ensures orthogonality between the two symbols of the modulation, $I(t)$ denotes the narrowband interference and residual ISI, and the channel response $h(t)$ is as given in (6.15).

An optimum combining RAKE receiver is composed of $L$ correlators followed by a linear combiner, as shown in Figure 6.37. The RAKE receiver samples the received signals at the symbol rate and correlates them with suitably delayed references given by

$$\nu(t) = \begin{cases} q(t) & \text{A-PAM} \\ q(t - \tau_p) - q(t + \tau_p) & \text{binary PPM} \end{cases} \tag{6.100}$$

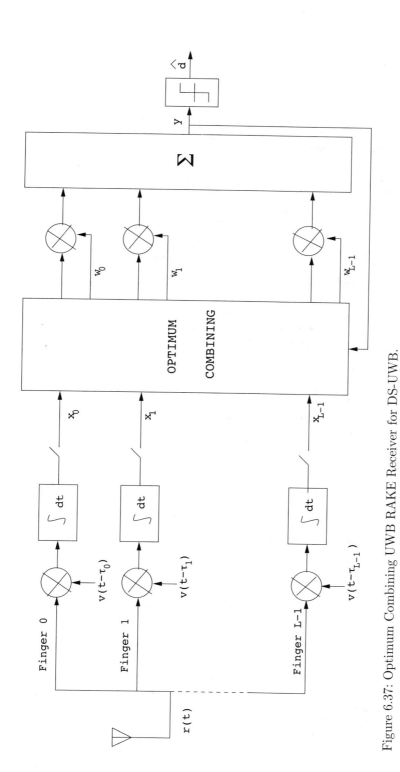

Figure 6.37: Optimum Combining UWB RAKE Receiver for DS-UWB.

The received signal at the output of the correlator for the $l^{th}$ finger of the RAKE receiver can be written as

$$x_{l,k} = \int_{-\infty}^{\infty} r_k(t)\, \nu(t - kT_s - \tau_l)\, dt = d_k h_l + I_{l,k} + n_{l,k} \qquad (6.101)$$

while the received signal samples at the input to the optimum combiner (in a vector form) are given by

$$\mathbf{x}_k = d_k \mathbf{h} + \mathbf{I}_k + \mathbf{n}_k \qquad (6.102)$$

where $\mathbf{x}_k = [x_{0,k}, \ldots, x_{L-1,k}]^T$, $\mathbf{h} = [h_0, \ldots, h_{L-1}]^T$, $\mathbf{I}_k [I_{0,k}, \ldots I_{L-l,k}]^T$, and $\mathbf{n}_k = [n_{0,k} \ldots, n_{L-l,k}]^T$. A bit decision is made at the output of the combiner $\widehat{d}_k = sgn(\mathbf{w}^T \mathbf{x}_k)$, where $\mathbf{w} = [w_0, \ldots, w_{L-1}]^T$ is the combiner weight.

**Concept of MMSE Receiver Combining**   The MMSE filter parameters are varied such that the mean squared error between the desired signal and the actual output is minimized. The minimum mean squared error (MMSE) criterion intends to find a weight vector that will minimize the mean squared error (MSE) between the combined signal and some desired (or reference) signal. The error signal can be defined as

$$e_k = a_k - \mathbf{w}^H \mathbf{x}_k \qquad (6.103)$$

where $a_k$ is the reference signal. The weight vectors are estimated from a training sequence that is known to the receiver. The MSE is given by

$$J = E\left[|e_k|^2\right] = E\left[\left|a_k - \mathbf{w}^H \mathbf{x}_k\right|^2\right] = |a_k|^2 - \mathbf{w}^H \mathbf{r}_{xa} - \mathbf{r}_{xa}^H \mathbf{w} + \mathbf{w}^H \mathbf{R}_{xx} \mathbf{w} \quad (6.104)$$

where $\mathbf{R}_{xx} = E\left[\mathbf{x}_k \mathbf{x}_k^H\right]$ is the covariance matrix of the received signal, and $\mathbf{r}_{xa} = E\left[\mathbf{x}_k a_k^*\right]$ is the cross-correlation vector between the received signal vector and the reference signal. The MSE $J$ is minimized when the gradient vector defined by

$$\nabla(J) = \frac{\partial J}{\partial w^*} \qquad (6.105)$$

is equal to the null vector. We have the optimum weight defined as $\underline{w}_{MMSE}$, and the following relation using (6.104) and (6.105)

$$\nabla(J)|\, w_{MMSE} = \underline{0}$$
$$\mathbf{R}_{xx} \mathbf{w}_{MMSE} = \mathbf{r}_{xa} \qquad (6.106)$$

Equation (6.106) is the well-known Weiner-Hopf equation, and the solution for optimal weights known as the Weiner solution is given by

$$\mathbf{w}_{MMSE} = \mathbf{R}_{xx}^{-1} \mathbf{r}_{xa} \qquad (6.107)$$

In practice, the matrix $\mathbf{R}_{xx}$ is estimated from a block of training symbols. The maximum likelihood estimate is given by the sample covariance matrix

$$\widehat{\mathbf{R}}_{xx} = \frac{1}{N} \sum_{k=0}^{N-1} \mathbf{x}_k \mathbf{x}_k^T \tag{6.108}$$

where $N$ is the block size. Because training is an overhead function that consumes resources, it is of interest to develop techniques that can work with short training sets. It is well-known that the number of vector samples required in estimating a $L \times L$ correlation matrix within 3 dB of its true value is $2L$ [42]. For a dispersive channel resulting in a large number of nonzero taps $L$, a large number of samples is required to train MMSE.

**Concept of EC Receiver Combining**  In this section, a reduced rank optimum combiner based on an eigen analysis approach that is more robust to errors caused by short training sets is presented [41]. By eigen decomposition, the covariance matrix is expressed as

$$\mathbf{R}_{xx} = \mathbf{Q}_I \Lambda_I \mathbf{Q}_I^H + \mathbf{Q}_n \Lambda_n \mathbf{Q}_n^H \tag{6.109}$$

where the columns of $\mathbf{Q}_1$ and $\mathbf{Q}_n$ consist of the interference eigen vectors and noise eigen vectors respectively. The matrices $\Lambda_I$ and $\Lambda_n$ are diagonal, and contain the interference and noise eigen values respectively. For an interference with bandwidth considerably smaller than the signal bandwidth, the eigen analysis of the interference-plus-noise matrix reveals a few large eigen values and a large number of small eigen values. The eigen vectors associated with large eigen values span the interference subspace. Because the interference subspace is orthogonal to the noise subspace, a tap weight vector residing in the noise subspace will effectively cancel the interference, leaving most of the information data unaffected. The tap weight of the EC is designed to minimize the norm of the weight vector while maintaining linear and eigen vector constraints given by

$$\min \left( \mathbf{w}^H \mathbf{w} \right) \text{ subject to } \mathbf{w}^H \mathbf{h} = c \text{ (constant), } \mathbf{Q}_I^H \mathbf{w} = 0 \tag{6.110}$$

Using the method of Lagrange multipliers, we define the objective function

$$\mathbf{J} = \mathbf{w}^H \mathbf{w} - \left( \mathbf{w}^H \mathbf{h} - c \right) \lambda - \mathbf{w}^H \mathbf{Q}_I \mu \tag{6.111}$$

The derivative of the objective function with respect to $\underline{w}^H$ is given as

$$\nabla \left( \mathbf{J} \right) = \mathbf{w} - \mathbf{h}\lambda - \mathbf{Q}_I \mu \tag{6.112}$$

We have the optimum weight defined as $\underline{w}_{EC}$ and the following relation

$$\nabla \left( \mathbf{J} \right) \big|_{w_{EC}} = 0$$
$$\mathbf{w}_{EC} - \mathbf{h}\lambda - \mathbf{Q}_I \mu = 0$$
$$\mathbf{w}_{EC} = \mathbf{h}\lambda + \mathbf{Q}_I \mu \tag{6.113}$$

Using the null constraint in (6.110) and the result of (6.113) we have the following result

$$\mathbf{Q}_I^H \mathbf{w}_{EC} = 0$$
$$\mathbf{Q}_I^H \left(\mathbf{h}\lambda + \mathbf{Q}_I \mu\right) = 0$$
$$\mu = -\mathbf{Q}_I^H h\lambda \tag{6.114}$$

Using (6.113) and (6.114) we have the optimum weights for the eigen canceller as

$$\mathbf{w}_{EC} = \lambda \left(\mathbf{I} - \mathbf{Q}_I \mathbf{Q}_I^N\right) \mathbf{h} \tag{6.115}$$

Equation (6.115) shows that the weight vector of the EC is constructed from the stable eigen vectors of the largest eigen values of the received signal. It is unaffected by fluctuations in the noise eigen values. This leads to a high degree of interference cancellation. Figures 6.38 to 6.40 compare the performance of MMSE and EC-based receiver combining.

### 6.1.7   Other I-UWB Receivers

In this section, we briefly describe three other less popular I-UWB receivers: template match detectors (TMD), the sampling bridge circuit (SBC), and integration and averaging (IA). In practice, researchers and companies use a combination of these and other techniques discussed in earlier sections. For example, O'Donnell has implemented a design based on a combination of the sampling bridge circuit and correlation detection that is capable of demodulating either pulse amplitude modulation (PAM) or PPM [43]. Time Domain Corporation has also implemented a similar design that demodulates PPM signals [44]. Immoreev uses a leading edge detection receiver with on-off keying [45]. Cellonics uses an integration and averaging receiver [46].

### Template Match Detection

TMD receivers essentially take an input waveform and perform a cross-correlation operation with a bank of stored waveform templates to see which template generates a match. A basic block diagram of such a system is shown in Figure 6.41.

TMD reception would be useful if the receiver is expected to operate in a variety of environments that will impart a significant amount of pulse distortion, or environments that will try to jam the receiver by transmitting UWB-like pulses. The TMD technique, however, adds a significant amount of complexity to the receiver, thus limiting its practicality.

### Sampling Bridge Circuit

The SBC operates similar to a sampling oscilloscope by sampling the UWB signal at a very low rate (such as 200 Ksps) and combining multiple snapshots into a single received waveform. Typically, snapshots are sampled at random, interleaved intervals to ensure that the combined waveform is faithfully reproduced, a process that is illustrated in Figure 6.42.

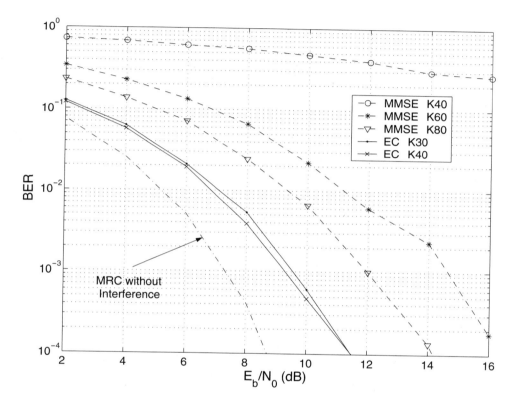

Figure 6.38: BER Performance Comparison of MMSE and EC Combiner with Selected $L = 40$ Largest Fingers over the 802.15.3a LOS Channel in the Presence of 802.11a Interference.

SOURCE: H. Sheng, A.M. Haimovich, A.F. Molisch, and J. Zhang, "Optimum Combining For Time Hopping Impulse Radio UWB Rake Receivers," *Proc. IEEE Conference on Ultra Wideband Systems and Technologies (UWBST '03)* [41]. © IEEE, 2003. Used by permission.

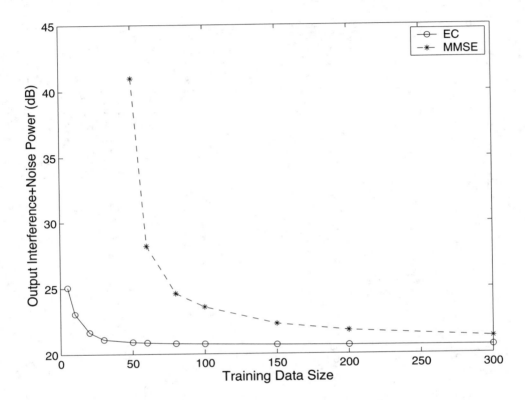

Figure 6.39: Comparison of the Output Power of the Interference Plus Noise as a Function of Training Data Size Between the MMSE and EC RAKE Receiver ($L = 50$ Taps are used).

SOURCE: H. Sheng, A.M. Haimovich, A.F. Molisch, and J. Zhang, "Optimum Combining For Time Hopping Impulse Radio UWB Rake Receivers," *Proc. IEEE Conference on Ultra Wideband Systems and Technologies (UWBST '03)* [41]. © IEEE, 2003. Used by permission.

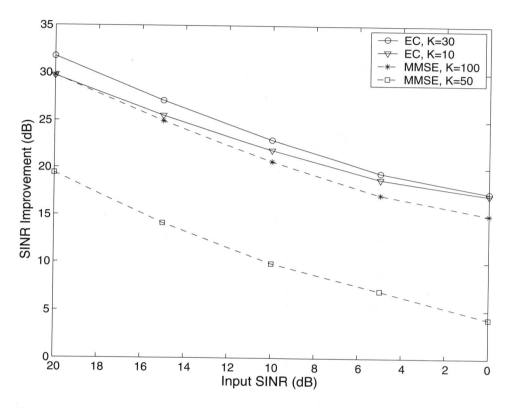

Figure 6.40: SINR Improvement as a Function of Input SINR in the Presence of 802.11a OFDM Interference over the 802.15.3a LOS Channel ($L = 50$ taps are used).

SOURCE: H. Sheng, A.M. Haimovich, A.F. Molisch, and J. Zhang, "Optimum Combining For Time Hopping Impulse Radio UWB Rake Receivers," *Proc. IEEE Conference on Ultra Wideband Systems and Technologies (UWBST '03)* [41]. © IEEE, 2003. Used by permission.

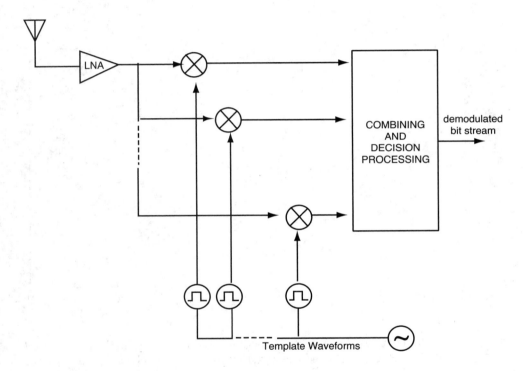

Figure 6.41: Template Match Detection.

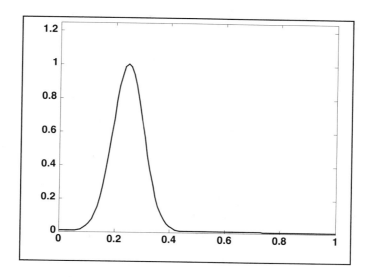

(a) Incoming Pulse Train (Only Single Pulse Shown)

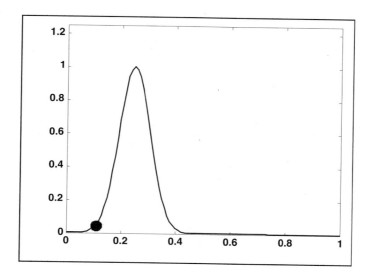

(b) First Sampled Pulse

Figure 6.42: A Sampling Bridge Circuit Samples an Input UWB Pulse Train at Interleaved Intervals and then Combines the Various Snapshots into a Single Received Waveform (continued).

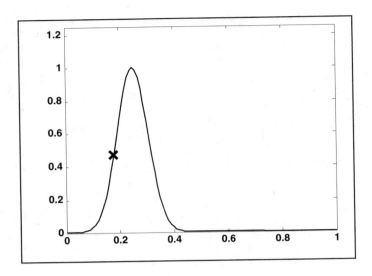

(c) Second Sampled Pulse

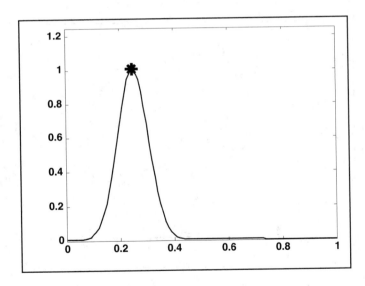

(d) Third Sampled Pulse

Figure 6.42: (cont.) A Sampling Bridge Circuit Samples an Input UWB Pulse Train at Interleaved Intervals and then Combines the Various Snapshots into a Single Received Waveform (continued).

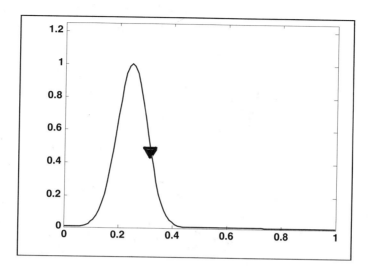

(e) Fourth Sampled Pulse

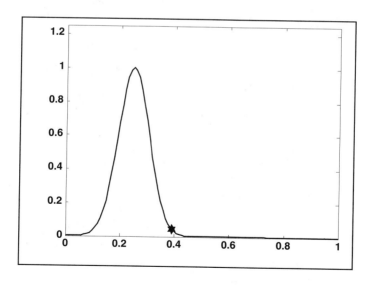

(f) Fifth Sampled Pulse

Figure 6.42: (cont.) A Sampling Bridge Circuit Samples an Input UWB Pulse Train at Interleaved Intervals and then Combines the Various Snapshots into a Single Received Waveform (continued).

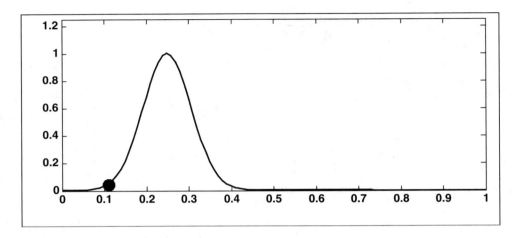

(g) Reconstructed Waveform

Figure 6.42: (cont.) A Sampling Bridge Circuit Samples an Input UWB Pulse Train at Interleaved Intervals and then Combines the Various Snapshots into a Single Received Waveform.

The SBC is most useful when a large number of pulses per data bit are being used by the UWB communication system, as it will automatically combine all of the transmitted pulses into a single received pulse. The SBC is an interesting approach in that by sampling a series of snapshots, the equivalent sampling rate may be very high, but the resulting data rate output is significantly reduced, thereby easing the requirements on the receiver hardware. The SBC also allows receivers to take advantage of multipath signals that may be present in the environment. A primary disadvantage of the SBC receiver is that it will only sample each input pulse a single time, with the individual samples combined together to produce a single pulse output. Thus, the SBC receiver does not fully realize the spreading gain that could be achieved if the total energy in each pulse were added together.

### Integration and Averaging

The Integrating and Averaging (IA) receiver is essentially a classical pulse compression receiver, where some number, $N$ of high-rate pulses are combined into a single time dilated pulse. The IA receiver may be implemented similar to a sliding correlator channel sounder [29, 47]. For sliding correlator implementation, the time dilation is related to the slide factor, namely,

$$k = \frac{T_{f_{TX}}}{T_{f_{TX}} - T_{f_{RX}}}$$

$$T_{d_{OUT}} = kT_{d_{IN}} \tag{6.116}$$

where $T_{f_{TX}}$ is the pulse repetition time for the transmitted pulses, $T_{f_{TX}}$ is the pulse repetition time for the receiver pulse generator, $k$ is the slide factor, $T_{d_{OUT}}$ is the time duration of the pulse output from the sliding correlator, and $T_{d_{IN}}$ is the time duration of the pulse input to the sliding correlator.

For example, consider a system transmitting pulses 1 nanosecond pulse width at a rate of 10 MHz ($T_{f_{TX}} = 10$ ns ). The receiver correlates the pulses with a locally generated reference pulse clocked at a slightly faster rate ($T_{f_{RX}} = 9.9$ ns). Thus, the slide factor $k = 100$, and the output pulse has a duration of 0.1 microseconds. The net result of this sliding correlation operation is to reduce the requirements on the digital portion of the receiver because the output pulse may be sampled at a significantly lower rate as compared to the original input pulse. The primary disadvantage of the IA receiver is that a number of pulses must be transmitted per data bit as the receiver trades off a reduction in power and complexity for a lower data rate.

## 6.1.8   Autocorrelation Transmitted Reference (TR) UWB Receiver

The fine time resolution of UWB signals results in the channel being extremely frequency selective. This results in a significant number of resolvable multipath components at the receiver. Thus, the need arises for a receiver structure tailored to achieve maximum energy capture. Research has mainly concentrated on the analysis of RAKE receivers. However, RAKE receivers applied to UWB systems suffer from two major drawbacks: First, the energy capture is relatively low for a moderate number of fingers when Gaussian pulses are used; second, each multipath undergoes a different channel, which causes distortion in the received pulse shape and makes the use of a single LOS path signal as a template suboptimal [16].

An alternative approach is the use of an autocorrelation receiver that correlates the received signal with a previously received signal [16, 48–50]. More precisely, a transmitted reference (TR) system is used, where a pair of unmodulated and modulated signals is transmitted, and the former is employed to demodulate the latter. This receiver can capture the entire signal energy for a slowly varying channel without requiring channel estimation. A potentially attractive feature of UWB autocorrelation receivers is their relative robustness to synchronization problems [49]. However, it suffers from the use of a noisy received signal as the template for demodulation.

An analytical characterization of the performance of a UWB autocorrelation TR system can be found in [16]. Experimental results comparing the TR receiver with RAKE receiver structures can be found in [48]. It is shown that the TR receiver performs slightly better than a single finger RAKE receiver with coherent reception.

The TR system described in [16] and [48] employs binary PPM. The transmitted signals consist of $N_p$ UWB pulses $w(t)$ of duration $T_w$ and energy $E_w$. The waveforms are comprised of $\frac{N_p}{2}$ unmodulated pulses interleaved with $\frac{N_p}{2}$ PPM-modulated pulses. The signals $s_0(t)$ and $s_1(t)$ are equally likely, and transmitted on $t \epsilon [0, N_p T_f]$, where $T_f$ is the pulse repetition period. The duty cycle is assumed to be very low (i.e, $T_f \gg T_w$). The signals can be written as

$$s_j(t) = \sum_{i=0}^{\frac{N_p}{2}-1} \sqrt{E_w} \left[ w(t - 2iT_f) + p(t - (2i+1)T_f - \varepsilon_{j,i}\tau_w) \right], j = 0, 1 \quad (6.117)$$

where $\tau_w$ is the delay associated with PPM. Moreover, the authors take $\varepsilon_{0,i} = i \bmod 2$ and $\varepsilon_{1,i} = 1 - \varepsilon_{0,i}$.

Suppose the indoor multipath channel is modeled as a linear, randomly time varying filter that is time-invariant over $2N_p T_f$, with impulse response $h(t)$, and maximum delay spread $\tau_{\max}$. The pulse repetition time is also assumed to be long enough such that interpulse and intersymbol interference can be ignored. Then the received signal, assuming $s_0(t)$ is transmitted, can be written as

$$r(t) = \sum_{i=0}^{\frac{N_p}{2}-1} \sqrt{E_g \cdot E_w} \left[ \widehat{g}(t - 2iT_f) + \widehat{g}(t - (2i+1)T_f - \varepsilon_{0,i}\tau_w) \right] + n(t),$$

$$0 \le t \le N_p T_f \qquad (6.118)$$

where $g(t) = w(t) * h(t)$, $E_g = \int_0^{T_{\max}+T_w} g^2(t) \, dt$, $\widehat{g}(t) = \frac{g(t)}{\sqrt{E_g}}$, and $n(t)$ is zero mean additive white Gaussian noise (AWGN) with power spectral density $\frac{N_0}{2}$.

Before demodulation, the autocorrelation receiver passes the signal through a bandpass filter of center frequency $f_c$ and bandwidth $W$. Note that $W$ is an important design parameter in the system performance. In general, increasing $W$ increases the energy capture at the expense of increasing the noise present in the demodulation stage. Because the UWB pulse occupies a wide frequency spectrum, the filter's bandwidth should be sufficiently wide so that no interpulse or intersymbol interference occurs. After filtering, the received signal is

$$r'(t) = \sum_{i=0}^{\frac{N_p}{2}-1} \sqrt{E_g \cdot E_w}$$
$$\cdot \left[ \widehat{g}'(t - 2iT_f) + \widehat{g}'(t - (2i+1)T_f - \varepsilon_{0,i}\tau_w) \right] + n'(t) \qquad (6.119)$$

where $\widehat{g}'(t)$ and $n'(t)$ are the filtered version of $\widehat{g}(t)$ and $n(t)$, respectively.

To demodulate the $i^{th}$ data pulse, the receiver first multiplies the received signal during the $(2i+1)^{th}$ time frame with an appropriately delayed average of the $N(1 \le N \le \frac{N_p}{2})$ previously received unmodulated pulses and then integrates the product over a time duration $T_d$, where $(0 \le T_d \le \tau_{\max} + T_w)$. Increasing $T_d$ increases the energy capture, but also integrates more noise. The average of the $N$ unmodulated pulses can be written as

$$f(t) = \sqrt{E_g \cdot E_w} \widehat{g}'(t) + \frac{1}{N} \sum_{k=0}^{N-1} n'(t + 2(i-k)T_f) \qquad (6.120)$$

The autocorrelator system is illustrated in Figure 6.43. The autocorrelator outputs corresponding to the two possible received signals are

$$Z_{j,i} = \int_0^{T_d} r\left(t + (2i+1)\,T_f + \varepsilon_{j,i}\tau_w\right).f\left(t\right)dt,\ j \in \{0,1\} \tag{6.121}$$

The difference between the two outputs is calculated as

$$Z_{\Delta,i} = Z_{0,1} - Z_{1,i} = E_w E_g X_{1,i} + \sqrt{E_w E_g}\,X_{2,i}$$
$$+ \frac{\sqrt{E_w E_g}}{N} \sum_{k=0}^{N-1} X_{3,i,k} + \frac{1}{N} \sum_{k=0}^{N-1} X_{4,i,k} \tag{6.122}$$

where

$$X_{1,i} = \int_0^{T_d} \left[\widehat{g}\,'\left(t\right) - \widehat{g}\,'\left(t + (\varepsilon_{1,i} - \varepsilon_{0,i})\,\tau_w\right)\right]\widehat{g}\,'\left(t\right)dt \tag{6.123}$$

$$X_{2,i}$$
$$= \int_0^{T_d} \left[n'\left(t + (2i+1)\,T_f + \varepsilon_{0,i}\tau_w\right) - n'\left(t + (2i+1)\,T_f + \varepsilon_{1,i}\tau_w\right)\right]$$
$$\widehat{g}\,'\left(t\right)dt \tag{6.124}$$

$$X_{3,i,k} = \int_0^{T_d} \left[\widehat{g}\,'\left(t\right) - \widehat{g}\,'\left(t + (\varepsilon_{1,i} - \varepsilon_{0,i})\,\tau_w\right)\right]n'\left(t + 2\left(i-k\right)T_f\right)dt \tag{6.125}$$

$$X_{4,i,k}$$
$$= \int_0^{T_d} \left[n'\left(t + (2i+1)\,T_f + \varepsilon_{0,i}\tau_w\right) - n'\left(t + (2i+1)\,T_f + \varepsilon_{1,i}\tau_w\right)\right]$$
$$n'\left(t + 2\left(i-k\right)T_f\right)dt \tag{6.126}$$

The term $\frac{\sqrt{E_w E_g}}{N} \sum_{k=0}^{N-1} X_{3,i,k} + \frac{1}{N} \sum_{k=0}^{N-1} X_{4,i,k}$ reflects the degradation associated with correlating the received pulse with a noisy template. The decision statistic $Z_\Delta$ is obtained by summing all the differences $\{Z_{\Delta,i}\}$ between the correlator outputs and can be written as

$$Z_\Delta = Y_1 + Y_2 + Y_3 + Y_4 \tag{6.127}$$

where

$$Y_1 = E_w E_g \sum_{i=0}^{\frac{N_p}{2}-1} X_{1,i} \tag{6.128}$$

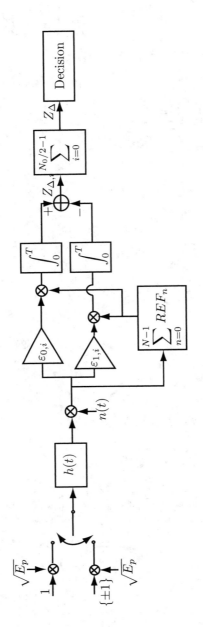

Figure 6.43: Autocorrelator System.

$$Y_2 = \sqrt{E_w E_g} \sum_{i=0}^{\frac{N_p}{2}-1} X_{2,i} \tag{6.129}$$

$$Y_3 = \frac{\sqrt{E_w E_g}}{N} \sum_{i=0}^{\frac{N_p}{2}-1} \sum_{k=0}^{N-1} X_{3,i,k} \tag{6.130}$$

$$Y_4 = \frac{1}{N} \sum_{i=0}^{\frac{N_p}{2}-1} \sum_{k=0}^{N-1} X_{4,i,k} \tag{6.131}$$

For a specific channel realization, the following assumptions hold:

1. $Y_1$ is deterministic.

2. $Y_2$, $Y_3$, and $Y_4$ are three independent Gaussian random variables.

The reader is referred to [16] for a justification of these assumptions. The decision statistic is therefore Gaussian with mean given by

$$E[Z_\Delta] = \frac{N_p E_w E_g}{2} \left[ \int_0^{T_d} \widehat{g}'^2(t)\, dt - \beta \right] \tag{6.132}$$

where $\beta = \frac{1}{2} \int_0^{T_d} [\widehat{g}'(t+\tau_w) + \widehat{g}'(t-\tau_w)] . \widehat{g}'(t)\, dt$.

The probability of error is thus given by

$$P_e = Q\left( \sqrt{ \frac{N_p E_w E_g \left[ \int_0^{T_d} \widehat{g}'^2(t)\, dt - \beta \right]^2}{2N_0 \left[ 2\zeta_2 + \frac{2}{3}\zeta_3 + \left( \frac{N_p E_w E_g}{4N_0 T_d W \zeta_4} \right)^{-1} \right]} } \right) \tag{6.133}$$

The terms $\zeta_2$, $\zeta_3$, and $\zeta_4$ in (6.133) are associated with the variances of $Y_2$, $Y_3$, and $Y_4$ of (6.129), (6.130), and (6.131) respectively. For an extended discussion of these terms, the reader is referred to [16]. A similar BER expression for on-off keying (OOK) modulation is also included. Note that the third term in the denominator in (6.133) corresponds to the variance of $Y_4$. This term becomes negligible at high SNR, while significantly degrading performance for low SNR or high $T_d W$.

In the following, we compare the autocorrelation receiver analyzed in [16] to a RAKE receiver. The analysis is performed using measured indoor channel data provided by Sholtz, et al [54]. Three different path selection schemes are investigated for the RAKE:

1. Scheme 1: The $L$ selected delays are optimally selected.

2. Scheme 2: The strongest path is tracked ($\tau_0$), while the remaining $L-1$ delays are placed in a suboptimal manner, such that $\tau_l = \tau_0 + lT_w$ for $l = 1, 2, \ldots, L-1$.

3. Scheme 3: The $L$ paths are detected regardless of their strength, such that $\tau_l = lT_w$, for $l = 0, 1, 2, \ldots, L - 1$.

Although $W$ (the width of the bandpass filter) is an important design parameter in the performance of the autocorrelation receiver, the filter bandwidth is not optimized in the experiment. The effect of the integration time $T_d$ on the autocorrelator's performance is studied. The BER performance is improved as $T_d$ is increased. The gains begin to diminish after a certain value of $T_d$ (20 ns for this experiment). When comparing the performance of the RAKE receiver and autocorrelation receiver, one must keep in mind that the autocorrelation receiver suffers a 3 dB penalty because half of the pulses are unmodulated. This 3 dB loss should be taken into account because although the RAKE receivers also require pilots for channel estimation, the data to pilot ratio is typically in the range of $10^{-1}$.

In spite of the 3 dB penalty, simulation results show that the autocorrelation receiver ($T_d = 20$ ns) performs slightly better than the first scheme RAKE receiver with one finger, better than the second scheme RAKE receiver with $L = 4$, and better than the third scheme RAKE receiver with $L = 16$. (Please compare Figures 6.44a and 6.44b [16].) Although the autocorrelation receiver captures a significant portion of the received energy, the performance is ultimately limited by the noise-on-noise term. Unless the effect of this term is limited, a RAKE receiver with multiple fingers will outperform the autocorrelation receiver.

Results from [16] and [48] suggest that the autocorrelation system's performance is severely limited by the use of noisy unmodulated pulses as templates at the receiver. In fact, although the autocorrelator can potentially capture the total signal energy, the BER behavior suffers from the noise-on-noise term. Therefore, noise suppression techniques must be applied in order to obtain robust performance. Time hopping the unmodulated-modulated pair coupled with averaging at the receiver might hinder the effect of the noise term. Also, choosing an optimal value for the bandwidth of the receiver's bandpass filter plays a significant role in limiting the noise. Finally, more efficient signaling schemes can be investigated.

In the design of I-UWB receivers it is important to take into consideration the effects of signal distortion due to diffraction by multiple frequency-dependent scatters [55]. Frequency-dependent dispersive effects are not a concern in narrowband systems because the characteristics of the receiver elements are essentially flat over small bandwidths. On the other hand, signal components at the low and high bounds of signal bandwidth will likely be affected by dispersion and loss. UWB generation affects receiver design because bandwidths depend on signal rise time instead of duration. The UWB signal may be viewed on a single pulse basis or as part of a string of pulses with specific intrapulse spacing. Each path of the multipath channel will have its own impulse response or frequency transfer characteristics. The frequency independence assumption is implied in the Turin model, which is widely adopted in tapped delay line models. This assumption is not true for I-UWB systems as in the case for radar target identification. Therefore, we should consider UWB signal distortion in the design of an optimum receiver that is coupled to the UWB channel propagation model. The physics behind the spatial-temporal resolution of scattering objects into multiple frequency-dependent scattering centers is described in [55].

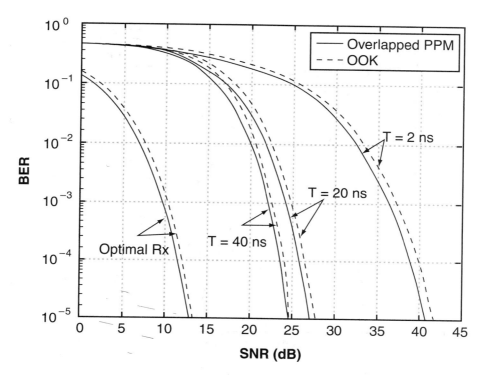

(a) Performance of TR System with Binary Block-Coded Overlapped PPM and OOK and Autocorrelation Receiver which Averages the Reference Pulses for $T = 2, 20, 40$ ns

Figure 6.44: Performance Comparison of TR System and MRC RAKE Receiver. Note that at $T = 40$ ns, the TR System Performs Slightly Better than the MRC RAKE Receiver (continued).

SOURCE: J. D. Choi and W. E. Stark, "Performance of ultra-wideband communications with suboptimal receivers in multipath channels," *IEEE Journal on Selected Areas on Communications* [16]. © IEEE, 2002. Used by permission.

UWB signal distortion due to multiple diffractions degrades the BER performance when a conventional correlation receiver is used. This is because the received UWB signal is a distorted version of the transmitted signal. Based on the matched filter principle, which is well-established in communications detection theory, optimum reception is obtained when the received signal is matched with a locally generated replica of the received signal. This is true for the narrowband case. Consequently, the correlation receiver is no longer optimum for UWB signal reception. A solution for this problem is to compensate for the diffraction by using a predistortion filter

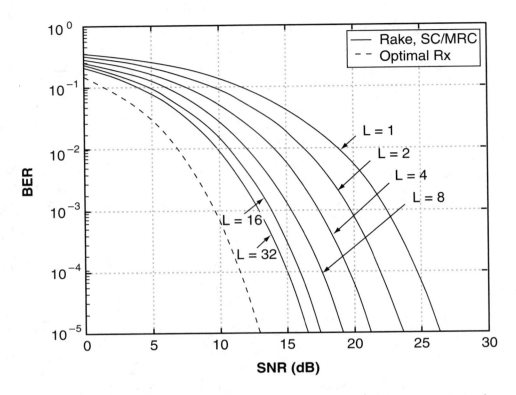

(b) Performance of Binary Block-Coded Overlapped PPM with RAKE Reception and SC/MRC for $(L = 1, 2, 4, 8, 16, 32)$

Figure 6.44: (cont.) Performance Comparison of TR System and MRC RAKE Receiver. Note that at $T = 40$ ns, the TR System Performs Slightly Better than the MRC RAKE Receiver.

SOURCE: J. D. Choi and W. E. Stark, "Performance of ultra-wideband communications with suboptimal receivers in multipath channels," *IEEE Journal on Selected Areas on Communications* [16]. © IEEE, 2002. Used by permission.

in the transmitter or by introducing a filter to restore the original shape of the received signal at the receiver.

## 6.1.9  Synchronization and Timing Issues

With the exception of the LED receiver, all the previous receiver techniques require highly accurate synchronization with the transmitter, as well as accurate oscillators to maintain that timing. Particularly with PPM, proper synchronization and timing is essential to correctly demodulate the received signals because information is conveyed in the time position of the pulse. Figure 6.45 shows the cumulative effect of a 50 picoseconds offset in the timing of the transmitter and receiver pulse train. Note that after only a few pulses the two waveforms are no longer aligned.

The timing offset will have a significant impact on the quality of the signal output by all the receiver techniques (with the exception of the LED receiver). Particularly, the TMD or CD receiver will see a dramatic drop in the SNR at the receiver output due to imperfect match or correlation with the receiver waveform. Additionally, if PPM is used and the cumulative timing offset becomes large enough, bit errors will occur.

In addition to maintaining precise timing at the receiver, acquiring and synchronizing to the transmitted signal is essential for UWB communication systems. Currently, very little has been published in the area of synchronization, as many in the industry consider synchronization algorithms to be proprietary information. One known technique is used by Immoreev and involves transmitting a long stream of regularly spaced impulses only until the receiver acquires synchronization [5]. Whenever the receiver loses synchronization, the transmitter will return to transmitting a stream of synchronization impulses. In [6], the effects of timing jitter and tracking on the performance of binary and 4-ary UWB communications is analyzed, and it is reported that the performance of the Impulse Radio is very sensitive to timing jitter and tracking over a range of pulse interference levels.

Another technique for acquisition and tracking is based on the sliding correlator principle. Bing, et al, has developed a "two-stage" technique that allows the receiver to acquire, capture, and track a UWB signal with a very simple hardware setup [57]. In their setup, the UWB transmitter produces pulses at a rate of 10 MHz. At the receiver, an identical copy of the transmitted pulse (the reference pulse) can be produced at rates of 10.01 MHz, 10.006 MHz, and 10.00 MHz. For coarse acquisition, the receiver first performs a sliding correlation operation with the reference pulse clocked at 10.01 MHz. After a correlation has been detected, the receiver switches to the 10.006 MHz clock signal in order to fine-tune the timing of the reference pulse. Finally, when the reference pulse is perfectly aligned with the received waveform, the receiver will switch to the 10.00 MHz clock signal, and begin demodulating data. The technique of Bing et al. is unique in that it does not require expensive hardware or signal processing techniques and can produce synchronization in a few milliseconds [57].

## 6.1.10 Digital I-UWB Implementation

The basic concept of a digital UWB receiver is to oversample the analog received signal (sample at equal to or greater than twice the Nyquist frequency) and then perform demodulation in the digital domain. Sampling the received signal may be accomplished using a Direct Sampling or a Time Interleaved Sampling approach.

Direct Sampling involves sampling the received signal at a very high rate of, say, 10–20 GHz. Thus, the overall waveform is essentially preserved, and demodulation in the digital domain may be performed. The major challenge in the Direct

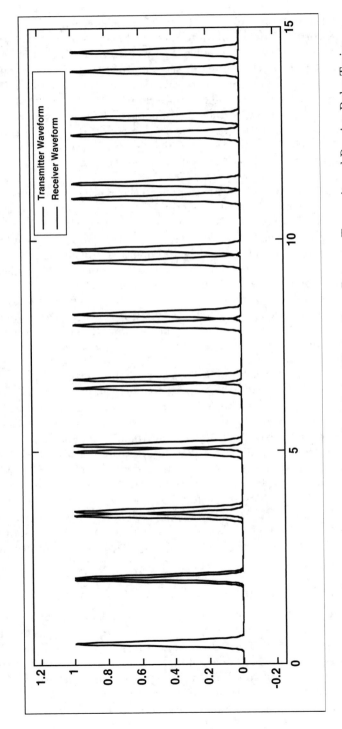

Figure 6.45: Effect of a 50 picosecond/pulse Cumulative Timing Error Between Transmit and Receive Pulse Trains.

328

Sampling approach is implementing a very high-speed data bus to transfer the samples from the ADC to the digital hardware. For the best performance, the ADC and digital demodulation hardware should be implemented on a single IC, where the designer has very tight control over signal propagation delay and skew. While implementing such a high-speed data bus on a PCB is possible, taking care of clock synchronization, distribution, and signal propagation delay is a major design challenge. Implementing very high-speed ADCs may be difficult and expensive. Some parallel (or time interleaved) sampling approaches have been developed especially for UWB systems [58]. The idea behind Time Interleaved Sampling is to relax the requirements on the data bus while still sampling the received signal at a very high rate. Time Interleaved sampling is a quite common technique for high-speed oscilloscope design. Essentially, the received waveform is sampled by a number of ADCs operating in parallel, with each ADC clock slightly offset from the others, as shown in Figure 6.46. Thus, each ADC samples a slightly different point of the time domain waveform, and the effective sampling rate is the individual ADC sample rate multiplied by the total number of ADCs, as illustrated in Figures 6.46 and 6.47 with a set of 4 ADCs.

An example of Time Interleaved Sampling is given in Figure 6.42 for a 5 ADC system. Typically, the signal is sampled at interleaved intervals by different ADCs to ensure that the combined waveform is faithfully reproduced. The time interval between the sampling of successive ADCs is given by $\tau$. That is, ADC #5 has a clock signal delayed by the same small interval $\tau$ as compared to ADC #4, $2\tau$ as compared to ADC #3, $3\tau$ as compared to ADC #2, and so forth. The data bus that connects each ADC to the digital hardware is able to run at a much lower rate. Additionally, the digital hardware can re-create the received signal as if it had been sampled by a single, high-speed ADC. The trade-off is in complexity; instead of having a single high-speed data bus, we now have several lower-speed busses. Additionally, the use of multiple ADCs increases the power consumption and chip area required to implement the receiver. Finally, small imperfections in the clock signal delay, as well as clock jitter and ADC aperture delay, lead to slight variations in the exact point at which each ADC samples—resulting in distortion of the UWB signal.

The *Effective Sampling Rate* of the Time Interleaved Sampling receiver may be represented mathematically by

$$f_{s_{effective}} = N_i f_s \qquad (6.134)$$

where $f_{s_{effective}}$ is the effective sampling frequency, $N_i$ is the number of ADCs, and $f_s$ is the sampling frequency of an individual ADC. The time delay, $\tau$, is simply the inverse of the effective sampling rate, or $\frac{1}{f_{s_{effective}}}$. For example, a system using 10 ADCs, each with a sampling frequency of 1 GHz, has an effective sampling frequency of 10 GHz, and the time delay between successive ADCs is therefore 100 picoseconds.

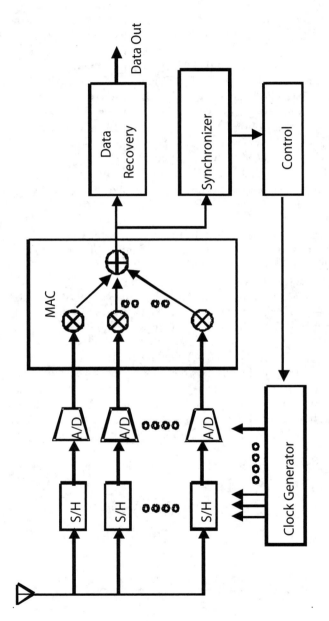

Figure 6.46: Digital UWB Receiver with Time Interleaving ADC.

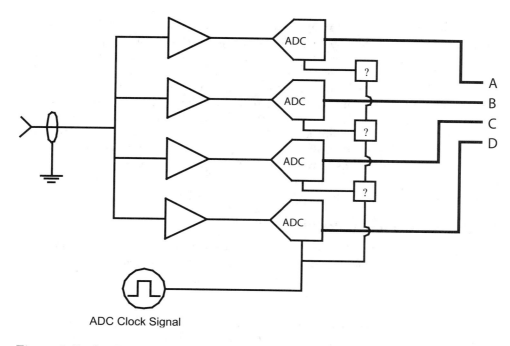

Figure 6.47: An Example Block Diagram of the Parallel Time Domain Sampling Technique Using 4 ADCs.

## 6.1.11  Example of IEEE Proposed Standards for PPM and DS-Based UWB Receivers

An overview of the IEEE 802.15.3a preliminary standard that describes the MAC and PHY layers of a wireless personal area network (WPAN) can be found in Appendix 10.A. At the time of publication, the PHY layer was not fully determined. We briefly describe the IEEE proposed standard for the PHY layer by Intel Corporation [59] for PPM and DS-based UWB to get some insight into the implementation details, although PPM-based UWB is not part of the 802.15.3a standard.

### Intel's Proposal for PPM-Based UWB

An impulse-based system with 7.26 GHz bandwidth is considered. For polarity modulation, 1 bit/pulse, polarity and 2-PPM modulation, 2 bits/pulse and polarity, and 4-PPM modulation, 4 bits/pulse is proposed. The PPM transceiver has direct RF sampling and a 20 GHz 1-bit ADC with 1,280 parallel digital matched filters. Tunable notch filters are used at the transmitter for pulse shaping and to create spectral holes for coexistence or regulatory issues (to comply with the FCC mask) and also at the receiver for rejecting out narrowband interference. The PPM receiver processing involves a matched filter bank with infinite fingers RAKE performance that captures multipath energy and is assumed to be insensitive to pulse shape. The channel estimator provides 50 picoseconds resolution, which is a simple increment or decrement operation, and complex math is not required (1-bit samples). The pulse repetition frequency (PRF) is 185 MHz and can be varied. Convolutional code is used for FEC. The studies by Intel Labs have shown that PPM offers excellent performance in the worst multipath environments. The link budget assumes a 7 dB system noise figure and 123 Mbps at 10 meters range with a margin of 6.8 dB. The estimated power consumption is 200 mW for the complete receiver which is 5 mm$^2$ in area. The channel estimator needs 175 K gates and is the dominant power consumer. Intel Labs have proposed reducing power consumption by MAC enhancements and also reducing the duty cycle to 10%. An all-digital scheme is also proposed because power will scale well according to Moore's law.

### Intel's Proposal for DS-UWB

The 7.26 GHz band is divided into three channels with center frequencies of 4 GHz, 7 GHz, and 9 GHz. The 7 GHz and 9 GHz channels are reserved for future use, perhaps for when the CMOS technology improves. This proposal avoids the 5 GHz WLAN bands worldwide. The emission bandwidth is restricted to 2 GHz. In the proposed DS-SS UWB system, the chip rate is 1 GHz and the modulation format is $\pi/2$-BPSK. Alternating 2 ns pulses on the in-phase and quadrature channels yields a constant envelope signal, thus easing CMOS implementation. The receiver processing uses a $4x$ oversampling rate (4 GHz). Four parallel despreaders are used for concurrent acquisition and channel estimation, which reduces the acquisition search time and also the RAKE multipath energy. The Viterbi decoder is not required at high SNR, thus reducing the power consumption using the software defined radio architecture. Individual or concatenated forward error correction (FEC) codes, convolution ($r = 1/2$, $K = 7$) and Reed Solomon codes (240, 224, 16), are used. The link budget assumes a 7 dB system noise figure, 4 GHz sampling frequency, and 115 Mbps at 10 meters range with margin of 7.7 dB. The link and the implementation margins are combined. The estimated power consumption is 150 mW for the complete receiver, which is 5–6 mm$^2$ in area. The 115 Mbps operation is in a $0.18\mu$ CMOS chip.

## 6.2   MC-UWB Receivers

In multicarrier systems, a single data stream is split into multiple parallel data streams of reduced rate, each of them transmitted on a separate frequency (subcarrier). Each carrier is modulated at a low enough rate that ISI is not a problem. Subcarriers must be properly spaced so that they do not interfere. For a $N$ subcarrier system each subchannel is tolerant of $N$ times as much dispersion as the original single carrier system. Multicarrier UWB (MC-UWB) communication systems may use orthogonal UWB pulse trains and multiple subchannels to achieve reliable high bit rate transmission and spectral efficiency [7]. Some of the advantages of MC-UWB are discussed in Section 5.4. MC-UWB systems, however, suffer from interchannel interference (ICI) in multipath fading channels because the received subcarriers arriving by different paths are no longer orthogonal. Hence, frequency synchronization is extremely important [7, 8]. Multicarrier systems also have the problem of the peak-to-average power ratio problem, which must be taken care of to maintain the proper linear operating region of the receiver.

The received MC-UWB signal is a multipath channel described in (6.15) over one transmission interval, and can be written as

$$r(t) = \beta \sum_{k=0}^{N-1} \sum_{l=0}^{L-1} b_k \alpha_l p(t - \tau_l) \exp(j2\pi k f_0(t - \tau_l)) + n(t) \tag{6.135}$$

where $b_k$ is the symbol that is transmitted in a transmission interval over the $k^{th}$ subcarrier, $N$ is the number of subcarriers, and $\beta$ is a constant that controls the transmitted power spectral density (PSD) and determines the energy per bit. The fundamental frequency is $f_0 = \frac{1}{T_p}$, where $T_p$ denotes the duration of waveform $p(t)$. To detect the information in subcarrier $m$, we examine the output of the filter matched to $p(t) \exp(j2\pi k f_0 t)$ given by

$$r_m(t) = \int r(u) p^*(u - t) \exp(j2\pi m f_0(t - u)) du$$

$$= \beta \exp(j2\pi m f_0 t) \sum_{k=0}^{N-1} \sum_{l=0}^{L-1} b_k \alpha_l \exp(-j2\pi m f_0 \tau_l) X(t - \tau_l, (m-k) f_0)$$

$$+ w_m(t) \tag{6.136}$$

In the previous equation

$$X(\tau, f) = \int_{-\infty}^{\infty} p(t) p^*(t - \tau) \exp(-j2\pi f t) dt \tag{6.137}$$

and $w_m(t)$ is the filtered complex noise. If we sample the output of the matched filter at times $t_i$, $i = 0, 1, \ldots, L - 1$, we get the following form

$$r_m(t_i) = \beta b_m \sum_{k=0}^{L-1} \alpha_l \exp(j2\pi m f_0(t_1 - \tau_l)) X(t_1 - \tau_l, 0)$$

$$+ \beta \sum_{k \neq m}^{N-1} \sum_{l=0}^{L-1} b_k \alpha_l \exp(j2\pi m f_0(t_i - \tau_l)) X(t_i - \tau_l, (m-k) f_0)$$

$$+ w_m(t_i) \tag{6.138}$$

The first part of this output is the multipath channel response to the $m^{th}$ subchannel. The second part is the ICI term. For best performance, we would like to separate the multipath contribution due to each subcarrier and combine them optimally to form the sufficient statistic that will be used in the detection process.

The optimum receiver for a multicarrier system is a multichannel maximum likelihood (MCML) detector that finds the data vector that maximizes the likelihood function over all subchannels. The complexity of this receiver increases exponentially with the number of subchannels. The optimum receiver is not practical when a large number of carriers are used. The study of the MCML detector is used to provide an upper bound on the performance of any other form of receiver for multicarrier systems. Suboptimal, less complex receiver structures also need channel parameter information, so channel estimation is an important block in the receiver design. The performance of a single channel RAKE receiver with channel information and that of a noncoherent receiver that uses no channel information is described in [8]. The single channel RAKE receiver treats ICI as additive Gaussian noise. The noncoherent receiver, dubbed peak detector, can be used with any orthogonal modulation schemes, such as on-off keying, but is inapplicable to more complex modulation schemes, such as quadrature amplitude modulation (QAM). The performance of these two receivers compared to the optimal MCML shows that the single carrier peak detector suffers the maximum in terms of BER performance. The single carrier RAKE is close to optimal performance at small SNR, but deviates from optimal performance at high SNR [8]. In the next two sections, we describe two interesting multicarrier UWB receiver architectures: CI-UWB and FH-UWB.

## 6.2.1   Carrier Interferometry (CI) UWB Receiver

The UWB nature of the signals in the frequency domain leads to ultra fine multipath resolution in the time domain (down to path differentials on the order of one nanosecond). The resolvable multipath components, regarded as independently faded delayed versions of the original signal, enable a path diversity gain (for example using RAKE reception as described in Section 6.1.3). In the originally proposed UWB system [60–62], data bits are modulated onto short pulses with a repetition period larger than the time delay spread of the multipath fading channel. The receiver implements a combining across the resolvable multipath components enabling high performance at low complexity. However, the data rate of such a system is low (relative to the total transmission bandwidth). In order to achieve higher data rates or to support multiple users, spread spectrum techniques have been proposed, as described in Section 6.1.3. When an attempt is made to improve the spectral efficiency (throughput) by co-locating multiple bits within the channel's delay spread, multipath effects lead to severe ISI, or MAI in the multiaccess case, and dramatic performance degradation with a small increase in the number of users. Therefore, such types of UWB may not be suitable for multiaccess and high throughput applications [63–65].

In [66], the use of the multicarrier carrier interferometry (CI) waveform as a means to enable higher throughput/improved multiaccess in UWB (without significant performance degradation) is proposed. The CI pulse waveform in UWB corresponds to the superpositioning of $N$ orthogonal carriers. The idea here is that with carefully chosen phase offsets (spreading codes) that ensure a periodic main lobe in the time domain (with side lobe activity at intermediate times), we could null out ISI, MAI, or any other forms of interference. At the receiver side, the received UWB CI signal is decomposed into carrier components and recombined to exploit frequency diversity and minimize ISI/MAI from other information-bearing pulses.

The CI-UWB transmitter is discussed in Section 5.4.1. Because the CI pulse shape can be decomposed into $N$ frequency components, the receiver can exploit the benefits of frequency diversity. Assuming a slow frequency-selective fading channel (typical of UWB transmission), frequency selectivity exists over the entire bandwidth. However, careful design of the CI waveform, i.e., proper selection of $N = \frac{T_f}{T_c}$ and $\Delta f = \frac{1}{N}T_c$, ensures that each of the $N$ individual carriers making up the CI pulse experience a flat fade. Note that $T_f$ denotes the frame duration, while $T_c$ corresponds to the duration of the main lobe of the CI pulse waveform, $p(t)$. In time-based receivers, the frequency-selective fade is usually modeled as a tapped-delay line with $L$ resolvable paths. In the frequency decomposable CI signal, the received signal is best characterized by the unique flat fade on each carrier component. Therefore, the received signal (assuming $0^{th}$ transmitted frames) can be shown as

$$r(t) = A \sum_{k=0}^{K-1} b_0^k \sum_{n=0}^{N-1} \alpha_n \cos(2\pi n\Delta ft - n\theta_k + \phi_n) + n(t) \qquad (6.139)$$

where $b_m^k \in \{-1, 1\}$ denotes user $K$'s information bit in the $m^{th}$ frame, $A$ is the amplitude of the carrier, $\alpha_n$ is the fading gain, and $\phi_n = 2\pi n\Delta f\tau_l$ is the phase offset of the $n^{th}$ carrier, $\tau_l$ is the delay of the $l^{th}$ multipath, and $n(t)$ is the AWGN noise. The receiver of Figure 6.48 is applied to detect bit $b_0^k$ where the angle $\theta_k$ in the figure corresponds to $\theta_k = 2\pi \frac{c_0^k}{N}$, while $c_m^k \in \{0, 1, \ldots, N-1\}$ is the time-hopping sequence assigned to user $k$. That is, the received signal is decomposed into its $N$ subcarriers, and crosscarrier combining is then performed to exploit available frequency diversity and to minimize ISI. The $m^{th}$ branch signal in Figure 6.50, $r_m^i$, is given by

$$
\begin{aligned}
r_m^i &= A \sum_{k=0}^{K-1} b_0^k \sum_{n=0}^{N-1} \alpha_n \int_0^{T_f} \cos(2\pi\Delta ft - n\theta_k + \phi_n)\cos(2\pi m\Delta ft - m\theta_i + \phi_m)\,dt \\
&\quad + \int_0^{T_f} n(t)\cos(2\pi m\Delta ft - m\theta_i + \phi_m)\,dt \\
&= \frac{AT_f}{2}\alpha_m \sum_{k=0}^{K-1} b_0^k \cos(m(\theta_k - \theta_i)) + n_m \\
&= \frac{AT_f}{2}b_0^i\alpha_m + \frac{AT_f}{2}\alpha_m \sum_{k\neq i}^{K-1} b_0^k \cos(m(\theta_k - \theta_i)) + n_m \qquad (6.140)
\end{aligned}
$$

The first term in the preceding equation represents the desired information, the second term represents the interchannel interference, and the third term represents the noise. When cross-carrier combining is applied (based on the MMSE criteria for minimizing ICI and noise while exploiting frequency diversity), the final decision

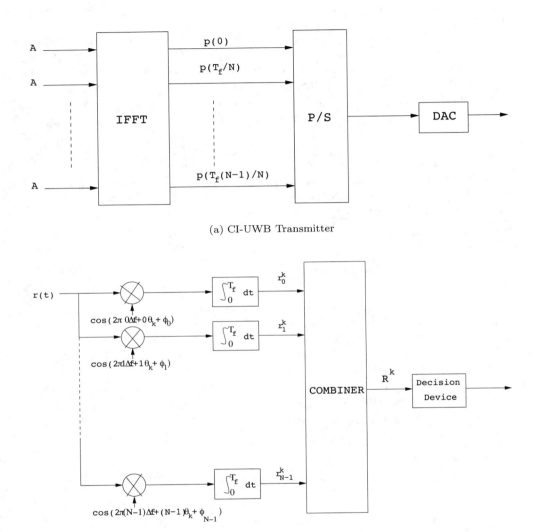

(a) CI-UWB Transmitter

(b) CI-UWB Receiver

Figure 6.48: Block Diagram of CI-UWB Transmitter and Receiver.

SOURCE: Z. Wu, F. Zhu, and C. R. Nassar, "Ultra Wideband Time Hopping Systems: Performance and Throughput Enhancement via Frequency Domain Processing," *36th Asilomar Conference on Signals, Systems and Computers* [63]. © IEEE, 2002. Used by permission.

variable, $R^i$ [63] can be shown to be

$$R^i = \sum_{m=0}^{n-1} \left( \frac{r_m^i \alpha_m}{KP_n \alpha_n^2 + \frac{N_0}{2}} \right) \tag{6.141}$$

where

$$P_n = \begin{cases} 1.0 & n = 0, \frac{N}{2} \\ 0.5 & otherwise \end{cases} \tag{6.142}$$

The frequency-based processing of CI pulses in UWB systems enables high throughputs while maintaining excellent BER performance. CI-TH-UWB gives better performance when compared to conventional RAKE processing [63]. The idea has been extended to CI-DS-UWB in [64].

BER performance plots for the comparison of time-based processing versus frequency-based processing for TH-UWB is presented in Figures 6.49 a and b [5], and for DS-UWB in Figures 6.50 a, b, and c [64].

## 6.2.2 Frequency Hopped (FH) UWB Receivers

In [70] a residue number system (RNS) assisted multistage frequency hopped spread spectrum multiaccess scheme is studied. We refer to this system as FH-UWB. This scheme is capable of efficiently dividing the huge number of users into a number of reduced size user groups; multi-user interference only affects the users within the same group. Because the number of users within the same group is only a small fraction of the total number of users supported by the FH-UWB system, advanced multi-user detection algorithms can be employed for achieving near single user performance at an acceptable complexity. Some of the advantages of the FH-UWB systems are discussed in Section 5.4.2.

Preliminary results [70] show that FH-UWB is capable of supporting an extremely large number of users while employing relatively simple quadratic receivers and multilevel frequency shift keying (MFSK). The readers are referred to [70] for a rigorous analysis and simulation results for BER performance. Figure 6.51a[2] depicts the transmitter block diagram. Figure 6.51b illustrates the receiver architecture of the FH-UWB system employing MFSK modulation.

A MC-UWB receiver architecture using sigma-delta modulators for the IEEE 802.15 standard has been proposed in [71]. In this design, the elementary pulse duration is chosen based on the peak power and pulse length limitations and desired side lobe behavior in the frequency domain. The interpulse separation and number of pulses is selected based on channel spread and average power limitation. This MC-UWB using $N$ tone sigma-delta modulators simplifies implementation. OFDM-UWB, a variation of MC-UWB, is increasingly more popular than conventional MC-UWB due to simple transmitter and receiver implementations via IFFT/FFT algorithms, respectively and is discussed next.

---

[2]Figures 6.51a and 6.51b are modified from Figures 1 and 3 of [70].

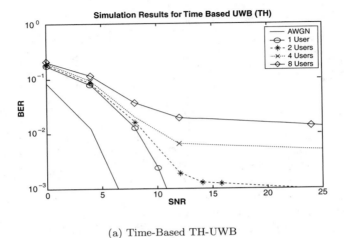

(a) Time-Based TH-UWB

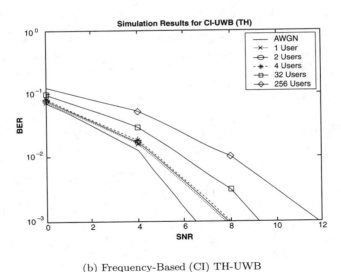

(b) Frequency-Based (CI) TH-UWB

Figure 6.49: Simulation Results.

SOURCE: Z. Wu, F. Zhu, and C. R. Nassar, "Ultra Wideband Time Hopping Systems: Performance and Throughput Enhancement via Frequency Domain Processing," *36th Asilomar Conference on Signals, Systems and Computers* [63]. © IEEE, 2002. Used by permission.

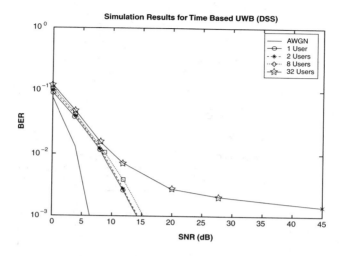

(a) Time-Based DS-UWB

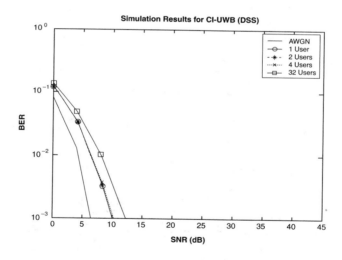

(b) Frequency-Based (CI) DS-UWB

Figure 6.50: Simulation Results (continued).

SOURCE: C. R. Nassar, F. Zhu, and Z. Wu, "Direct Sequence Spreading UWB Systems: Frequency Domain Processing for Enhanced Performance and Throughput," *IEEE International Conference on Communications* [64]. © IEEE, 2003. Used by permission.

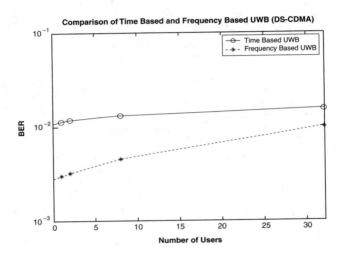

(c) Comparison of Time-Based and Frequency-Based DS-
UWB

Figure 6.50: (cont.) Simulation Results.

SOURCE: C. R. Nassar, F. Zhu, and Z. Wu, "Direct Sequence Spreading UWB Systems: Frequency Domain Processing for Enhanced Performance and Throughput," *IEEE International Conference on Communications* [64]. © IEEE, 2003. Used by permission.

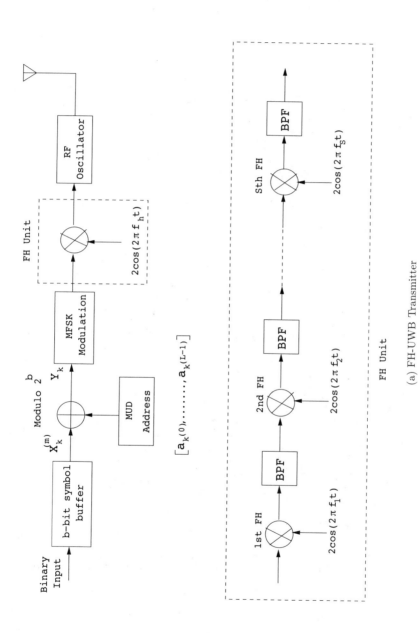

(a) FH-UWB Transmitter

Figure 6.51: Block Diagram of FH-UWB Transmitter and Receiver (continued).

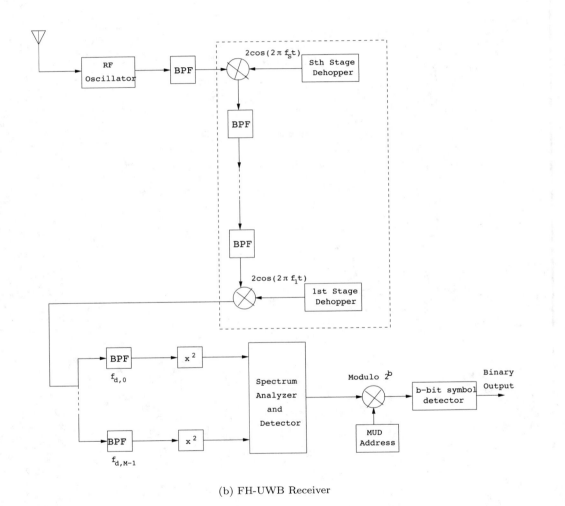

(b) FH-UWB Receiver

Figure 6.51: (cont.) Block Diagram of FH-UWB Transmitter and Receiver.

## 6.2.3   OFDM-UWB Receivers

OFDM is a special case of multicarrier transmission that permits subcarriers to overlap in frequency without mutual interference, resulting in increased spectral efficiency. OFDM exploits signal processing technology to obtain a cost-effective means of implementation. Multiple users can be supported by allocating a group of subcarriers to each user. OFDM-UWB is a novel system that has been proposed as a physical layer for a high bit rate, short-range (10 m–20 m) communication network in high performance computing clusters. Traditional UWB communications use PPM or PAM modulation schemes, different pulse generation methods, pulse rate and shape, and center frequency and bandwidth, as described in earlier sections. Earlier UWB systems were designed to be carrierless. In contrast, OFDM-UWB is a multicarrier UWB system that relies on splitting orthogonal subcarriers in a train of short pulses, sending them over the channel, and reassembling them at the receiver to get orthogonality and to recover each subcarrier separately [72, 73]. This new system offers more flexibility in shaping the transmitted spectrum because it has more degrees of freedom. OFDM-UWB provides more multipath resolution than single carrier UWB. A RAKE receiver can therefore achieve more multipath diversity gain and improve the overall performance. Unlike narrowband OFDM, a given tone in OFDM-UWB is transmitted only during parts of the transmission interval. Reliable communication results from integrating several pulses and high throughput from transmitting frequencies in parallel. OFDM-UWB is being proposed as the physical layer standard in 802.15.3a Wireless Personal Area Networks [74].

OFDM-UWB is a MC-UWB system that uses a frequency coded pulse train as a shaping signal. The frequency coded pulse train is defined as follows

$$p(t) = \sum_{n=0}^{N-1} s(t - nT) \exp\left(-j2\pi c(n) \frac{t}{T_c}\right) \tag{6.143}$$

where $s(t)$ is an elementary pulse with unit energy and duration $T_s < T$, and $p(t)$ has duration $T_p = NT$. Each pulse is modulated with a frequency $f_n = \frac{c(n)}{T_c}$, where $c(n)$ is a permutation of the integers $\{0, 1, \ldots, N-1\}$. It can be easily verified that

$$\int_0^{T_p} (p(t) \exp(j2\pi k f_0 t)) (p^*(t) \exp(-j2\pi m f_0 t)) \, dt$$

$$= \exp(j\pi (k - m) f_0 T_s) \operatorname{sinc}((k - m) f_0 T_s) \tag{6.144}$$

Because $f_0 = \frac{1}{NT}$, the signals

$$p_k(t) = p(t) \exp(j2\pi k f_0 t) \tag{6.145}$$

are orthogonal for $k = 0, 1, \ldots, N-1$. Further, if $T_s = T$, then the signals given by (6.145) are orthogonal for any $k$. Costas showed that if $c(n)$ is a Costas sequence, then the $p_k(t)$'s remain near-orthogonal under different delays. Further, (6.145) can be efficiently implemented by taking samples of the signal $p(t)$ which are IFFT samples and passing them through a DAC. The most challenging part of this structure is designing the DAC. These DACs typically operate around hundreds of MHz.

In the next section, we describe the receiver architecture for OFDM-UWB using sigma-delta quantizers which trade off the high operating rate of DACs for lower resolution.

### $N$-Tone Sigma-Delta OFDM-UWB Receiver

A new method for generating and detecting the OFDM-UWB signal using a modified sigma-delta modulator is proposed in [75]. Unlike narrowband OFDM, the OFDM-UWB spectrum can have gaps between subcarriers. Another major difference between OFDM-UWB and narrowband OFDM is their spectral shapes. Also unlike narrowband OFDM, IFFT/FFT cannot be used directly to generate and receive OFDM-UWB signals because of the high bit rates involved. To solve these problems, a procedure to move the bulk of the processing load from the analog devices to the digital baseband section is described in [75]. Sigma-Delta A/D and D/A converters are a good choice for high bit rate wireless communications [76]. Traditional versions of the sigma-delta modulators cannot be used in OFDM-UWB transceivers because they would require prohibitively high sampling rates. A modified sigma-delta modulator to fit the characteristics of OFDM-UWB signals is proposed that enables most of the processing to be done digitally in both the transmitter and receiver. This approach enables the efficient use of IFFT, FFT to generate and demodulate OFDM-UWB signals and also gets rid of the high peak to average ratio (PAR) problem that occurs with OFDM systems. The modified sigma-delta modulator, called the $N$-tone sigma-delta modulator introduces $N$ zeros at the frequencies in the quantization noise spectrum. These zeros match the locations of frequencies used by the OFDM system, and the quantization noise spectrum fills the gaps in the spectrum of the OFDM-UWB signal. In fact, this new structure can be used in other UWB systems anytime we have gaps in the spectrum of the transmitted signal. All these advantages come at the expense of a lower spectral efficiency unless one uses more complex multiband implementations. The $N$-tone sigma-delta quantizer structure is shown in Figure 6.52a. Generating UWB-OFDM signals using the sigma-delta modulator is shown in Figure 6.52b. The corresponding receiver structure is shown in Figure 6.52c [77].

## 6.2.4    Example on IEEE Proposed Standard for MC and OFDM-Based UWB Receivers

Details of the IEEE 802.15.3a preliminary standard describing the MAC and PHY layers of a wireless personal area network (WPAN) for MC-UWB and OFDM-UWB can be found in Appendix 10.A. We briefly describe the IEEE proposed standard for the PHY layer by Intel Corporation [59] for MC-UWB and OFDM-UWB to get some insight into the implementation details.

### Intel's Proposal for MC-UWB

The frequency band is subdivided into 13 smaller bands of 550 MHz, and time frequency codes are used for multiple access. The proposed system occupies the 3.25 GHz to 10.6 GHz range. Coexistence with other narrowband systems is achieved by

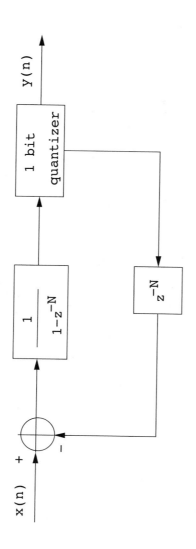

(a) *N*-Tone Sigma-Delta Modulator

Figure 6.52: Block Diagram of *N*-Tone Sigma-Delta Modulator, *N*-Tone Sigma-Delta OFDM-UWB Transmitter and Receiver (continued).

SOURCE: (a) E. Saberinia and A. H. Tewfik, *"N*-Tone Sigma-Delta UWB-OFDM Transmitter and Receiver," *IEEE ICASSP '03* [75]. © IEEE, 2003. Used by permission.

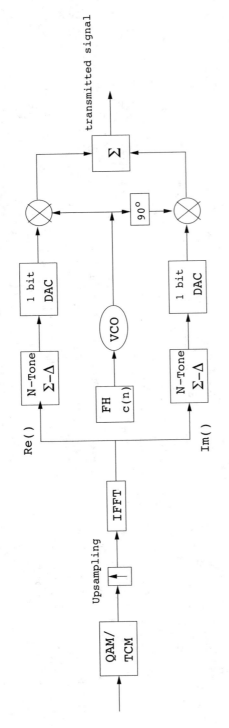

(b) *N*-Tone Sigma-Delta OFDM-UWB Transmitter

Figure 6.52: (cont.) Block Diagram of *N*-Tone Sigma-Delta Modulator, *N*-Tone Sigma-Delta OFDM-UWB Transmitter and Receiver.

SOURCE: (b) E. Saberinia and A. H. Tewfik, "Generating UWB-OFDM Signal Using Sigma-Delta Modulator," *57th IEEE Vehicular Technology Conference (VTC 2003)* [77]. © IEEE, 2003. Used by permission.

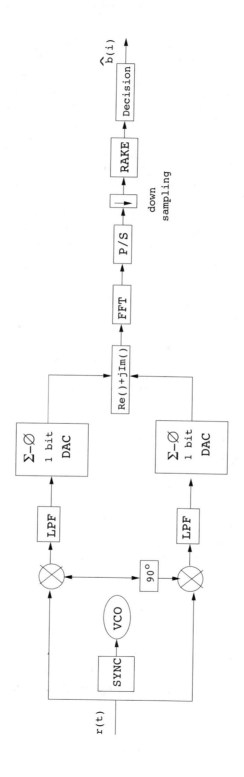

(c) *N*-Tone Sigma-Delta OFDM-UWB Receiver

Figure 6.52: (cont.) Block Diagram of *N*-Tone Sigma-Delta Modulator, *N*-Tone Sigma-Delta OFDM-UWB Transmitter and Receiver.

band dropping. Using pulsed transmission one band at a time yields low peak-to-average ratio (PAR). Reduced pulse repetition frequency reduces equalizer complexity. Pulses of duration 3 ns are used with QPSK modulation. The rectified cosine envelope yields a bandwidth of approximately 700 MHz, thus leading to the overlapping of bands. The time frequency coding scheme uses a single RF conversion chain and supports six coexisting piconetworks. Four symbols emitted per subband hop are used with interleaved $M$-ary Binary Orthogonal Keying (MBOK), which enables the use of the low complexity decision feedback equalizer. The Outer Reed-Solomon (221,255) code is used for error correction. The raw signaling rate for a symbol duration of 3 ns is 333 Mbps, and 2 bits per carrier (QPSK) is equivalent to 666 Mbps. The multiaccess performance in multipath shows that the proposed system can operate with interference 6 dB above the signal, is predicted to operate with four other piconets within a 5m radius, and improves in more dispersive channels. The link budget assumes a 7 dB noise figure with 108 Mbps at 10 meters with 9.3 dB margin. The link and implementation margin is combined. The estimated power consumption is 218 mW for the complete receiver. The 108 Mbps operation is in a 0.18 $\mu$ CMOS per chip.

### Intel's Proposal for OFDM-UWB

The frequency band is subdivided into smaller bands of 528 MHz. Three subbands occupy the 3.1 GHz to 4.6 GHz range. Symbols are pulsed on one subband at a time, which is a form of time interleaving. Higher frequencies are reserved for future enhancements, and the 5 GHz UNII band is avoided. OFDM uses parallel transmission using multiple carriers (tones) with 128 tones at 4.125 MHz, which gives a total bandwidth of 528 MHz. QPSK modulation is used at 2 bits per tone. 100 tones carry data, and 12 pilot tones are used for phase tracking and channel estimation. 10 tones are user-defined, and the remaining 6 tones including DC are null. The DC tone is null to simplify direct conversion receiver structures that have DC offset caused by on-chip LO leakage, and the use of the DC tone eliminates DC cancellation. Other null tones address world regulatory compliance, making spectral holes for radio astronomy, and so on. User-defined tones can be used for spectral or waveshaping. The symbol duration of 312.5 ns leads to a symbol rate of 32 MHz that includes 9.5 ns guard time for frequency hop. The raw signaling rate is 640 Mbps. The multipath performance of OFDM-UWB predicts that there is less than 0.1 dB loss of multipath energy, a 60.6 ns cyclic prefix adequate for expected channels, and there is no need for RAKE structures in the receiver. The link budget assumes a 7 dB noise figure with 110 Mbps at 10 meters with 8.1 dB margin. The link and implementation margin is combined. The estimated power consumption is 205 mW for the complete receiver, which has 3 mm$^2$ in analog and 3.8 mm$^2$ in digital. The 110 Mbps operation is in a 0.13 $\mu$ CMOS chip.

## 6.3   Spectral Encoded UWB Communication System

In this section, we describe a UWB communication system that uses spectral encoding as a multiple access scheme with interference suppression capability [78]. The main advantage of this technique is that the transmitted signal spectrum can be conveniently shaped to suppress narrowband interference. The concept of spectral coding has been proposed both in an optical communication system in [79, 80] and for a wireless communications scenario [81, 82]. In this technique, a message signal is multiplied by a spreading sequence in the frequency domain, as opposed to conventional direct sequence code division multiaccess systems in which the encoding is in the time domain. As a result, the signal will be spread in time. Multiaccess capability is achieved by assigning distinct spreading sequences to different users. When narrowband interference suppression is desired, a spreading sequence is used that has a spectral null where the narrowband interference is located. Moreover, the narrowband interference is "notched out" at the receiver when the received signal spectrum is multiplied by the conjugate of the transmitted signal spectrum (correlation receiver). See [83] for implementation details. Fourier Transform of the data signal is performed by a well-established implementation that corresponds to the multiplication and convolution of the desired signal by chirp (linear FM) waveforms using surface acoustic wave (SAW) devices. SAW devices have evolved to the GHz range, and currently submicron manufacturing and material techniques are improving greatly to frequencies up to 10 GHz.

A schematic of the transmitter is shown in Figure 6.53a.[3] The modulated signal $f(t)$ is a real and even function and is only nonzero in $t \in \left[-\frac{T}{2}, \frac{T}{2}\right]$. The waveform $f_1(t)$ is given by

$$
f_1(t) = \frac{1}{2} F_R(t) \cos\left(\omega_a \left(t - \frac{T}{2}\right) + \beta \left(t - \frac{T}{2}\right)^2\right)
$$

$$
- \frac{1}{2} F_I(t) \sin\left(\omega_a \left(t - \frac{T}{2}\right) + \beta \left(t - \frac{T}{2}\right)^2\right)
\tag{6.146}
$$

where $\omega_a$, $\beta$ are respectively the center frequency and half slope of the chirp signals, and $F_R(t)$, $F_I(t)$ are respectively the real and imaginary components of the Fourier Transform of the modulated signal $f(t)$, and are given by

$$
F_R(t) = 2 \int_{\frac{T}{2}}^{\frac{T}{2}} f(u) \cos(\omega_c u) \cos\left(2\beta \left(t - \frac{T}{2}\right) u\right) du
$$

$$
F_I(t) = -2 \int_{\frac{T}{2}}^{\frac{T}{2}} f(u) \cos(\omega_c u) \sin\left(2\beta \left(t - \frac{T}{2}\right) u\right) du
\tag{6.147}
$$

where $\omega_c$ is the carrier frequency. $f_1(t)$ is valid only in $t \in [T, T_1]$ where $T_1$ is the duration of the impulse response of the filter labeled $\left(\omega_a t + \beta t^2\right)$. The signal $f_1(t)$ is spread to $f_2(t)$ by a spreading sequence such that its spectrum is given by

$$
PN(\omega) = PN_R(\omega) + jPN_I(\omega)
\tag{6.148}
$$

---

[3]Figures 6.53a and 6.53b are modified from Figures 3, 4, and 5 of [78].

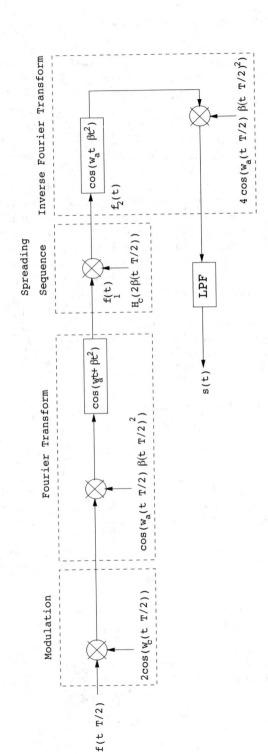

(a) Spectrally Encoded UWB Transmitter

Figure 6.53: Block Diagram of Spectrally Encoded UWB Transmitter and Receiver (continued).

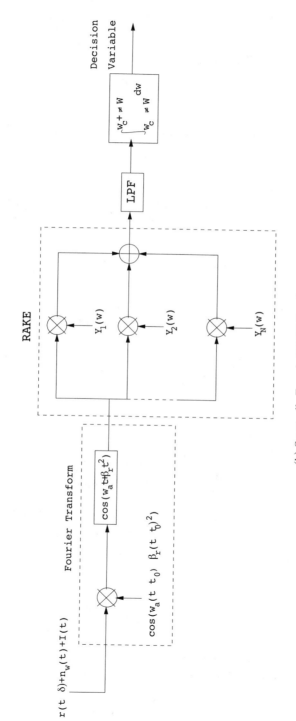

(b) Spectrally Encoded UWB Receiver

Figure 6.53: (cont.) Block Diagram of Spectrally Encoded UWB Transmitter and Receiver.

where $PN(\omega)$ is in general a complex waveform bandlimited to $|f| \leq \frac{W}{2}$. $PN_R(\omega)$, and $PN_I(\omega)$ are real waveforms and consist of a sequence of 1s or $-1$s (and 0s when interference suppression is considered). The function $H_c\left(2\beta\left(t - \frac{T}{2}\right)\right)$ in Figure 6.57a relates to the desired spreading sequence by

$$H_c(\omega) = \frac{8\beta}{\pi}\left(PN_R(\omega)\cos(\omega T_1) + PN_I(\omega)\sin(\omega T_1)\right) \tag{6.149}$$

The output transmitted signal $s(t)$ is the inverse Fourier Transform of the signal $f_2(t)$ and is equal to

$$s(t) = \frac{1}{\pi}\int_{\omega_c - \pi W}^{\omega_c + \pi W}$$
$$\cdot\left(R(\omega)\cos\left(\omega\left(t - \frac{T}{2} - T_1\right)\right) - I(\omega)\sin\left(\omega\left(t - \frac{T}{2} - T_1\right)\right)\right)d\omega \tag{6.150}$$

where $R(\omega)$ and $I(\omega)$ are defined as $R(\omega) = F_R(\omega)PN_R(\omega)$ and $I(\omega) = F_R(\omega)$ $\cdot PN_I(\omega)$. $W$ is the bandwidth of the signal $f(t)$. The transmitted signal is valid only in $t \in [T_1, T_1 + T]$. A receiver for the proposed system is shown in Figure 6.57b. The receiver presented has a RAKE stage due to the multipath channel. In addition to the desired signal, it is assumed that additive thermal noise $n_w(t)$ and interference $I(t)$ are present at the receiver's input. The channel is assumed to have a multipath spread, $T_m$, and the time window that the Fourier Transform window considers is $T + T_m$ long. The variable $t_0$ is defined as $t_0 = \delta + T_1 + \frac{T}{2}$, where a delay of $\delta$ was added in the transmitted signal as shown in Figure 6.57b. The RAKE stage has $Y_m(\omega)$ defined as

$$Y_m(\omega) = \frac{8\beta_r}{\pi}\alpha_m\left(\begin{array}{c} \cos\left(\omega_a(t - t_0) + \beta_r(t - t_0)^2\right)\Lambda_c(\omega, m) \\ -\sin\left(\omega_a(t - t_0) + \beta_r(t - t_0)^2\right)\Lambda_s(\omega, m) \end{array}\right) \tag{6.151}$$

where

$$\Lambda_c = (\omega, n) = R(\omega)\cos\left(\frac{\omega n}{W}\right) + I(\omega)\sin\left(\frac{\omega n}{W}\right)$$
$$\Lambda_s = (\omega, n) = I(\omega)\cos\left(\frac{\omega n}{W}\right) - R(\omega)\sin\left(\frac{\omega n}{W}\right) \tag{6.152}$$

for $m = 1, 2, \ldots, N$ and $\alpha_m$ is the channel gain.

A rigorous analysis of the receiver with numerical approximations and system performance curves is presented [78]. Figures 6.54 and 6.55 show the performance gains for this novel receiver structure. This method of spectrally encoding the UWB signal can suppress narrowband interference and create spectral nulls where desired, simplifying the requirement to meet the FCC mask.

## 6.4   Case Study: Improving Range of UWB Using RAKE Receivers

The fine time resolution of multipaths induced by the channel (typically in the order of 100–300 picoseconds) can be exploited using a RAKE receiver to capture a

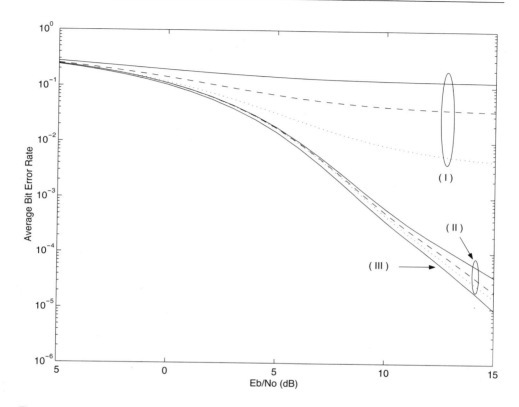

Figure 6.54: Average Bit Error Rate. (The decision is based on the first 10 received multipath components. The SIR is equal to $-5$ dB (dotted lines), $-10$ dB(dashed), and $-15$ dB(solid): (I) no interference suppresstion, (II) interference suppression, and (III) no interference).

SOURCE: C. R.C.M. da Silva and L. B. Milstein, "Spectral-Encoded UWB Communication Systems," *IEEE Conference on Ultra-Wideband Systems and Technologies* [78]. © IEEE, 2003. Used by permission.

significant amount of energy found in the multipath components, and to benefit from multipath diversity gain. Energy capture is important in UWB receiver design, and the number of detectable multipath components could be as large as a few hundred since UWB signals occupy bandwidth greater than 500 MHz or have a fractional bandwidth greater than 0.25. Clearly, classical RAKE receiver design solutions, such as MRC or selection combining, are no longer appropriate from the viewpoint of complexity and performance trade-offs. The realistic trade-off between the energy capture and RAKE diversity level in dense multipath for UWB systems have been

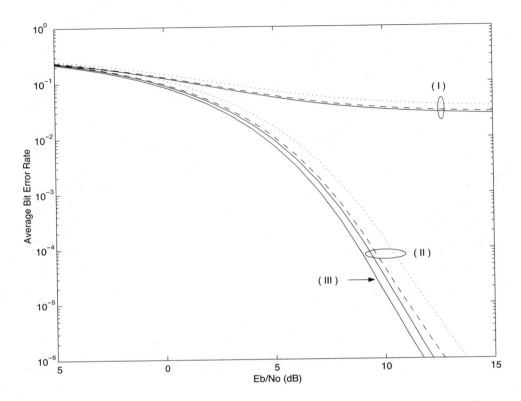

Figure 6.55: Average Bit Error Rate. (The decision is based on the first 20 (solid lines), 10 (dashed), and 5 (dotted) multipath components with largest absolute values, among the first 20 multipath components. The SIR is equal to −10 dB. (I) no interference suppression, (II) interference suppression, and (III) no interference with first 20 received multipath components considered).

SOURCE: C. R.C.M. da Silva and L. B. Milstein, "Spectral-Encoded UWB Communication Systems," *IEEE Conference on Ultra-Wideband Systems and Technologies* [78]. © IEEE, 2003. Used by permission.

quasi-analytically/experimentally presented in [54]. The experimental results are shown in Figure 6.56.

In this section, we examine the throughput range trade-off of UWB communication systems that employ two variants of RAKE receiver structures in multipath fading channels: (a) the generalized selection combining receiver (hereafter, referred to as GSC($N, L$)) that combines a subset of $N$ resolvable multipaths with the highest instantaneous SNR out of $L$ resolvable multipaths in a similar fashion to MRC; and (b) MRC with finite tap statistics or partial MRC (hereafter, referred to as

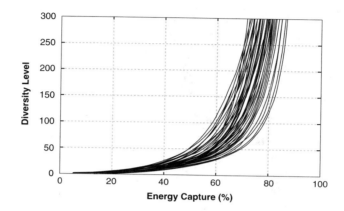

(a) "High SNR" Environment

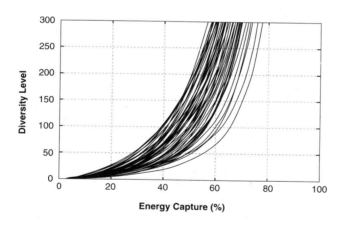

(b) "Low SNR" Environment

Figure 6.56: The Required Diversity Level, $L$, in a UWB RAKE Receiver as a Function of Percentage Energy Capture for Each of the 49 Received Waveforms in an Office Representing High, Low, and Extremely Low SNR Environments (continued).

SOURCE: M. Z. Win and R. A. Scholtz, "On the Energy Capture of Ultrawide Bandwidth Signals in Dense Multipath Environments," *IEEE Communications Letters* [54]. © IEEE, 1998. Used by permission.

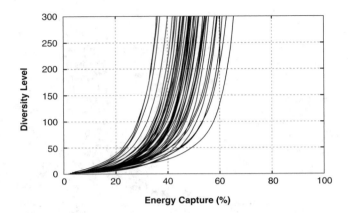

(c) "Extremely Low SNR" Environment

Figure 6.56: (cont.) The Required Diversity Level, $L$, in a UWB RAKE Receiver as a Function of Percentage Energy Capture for Each of the 49 Received Waveforms in an Office Representing High, Low, and Extremely Low SNR Environments.

SOURCE: M. Z. Win and R. A. Scholtz, "On the Energy Capture of Ultrawide Bandwidth Signals in Dense Multipath Environments," *IEEE Communications Letters* [54]. © IEEE, 1998. Used by permission.

PMRC($N, L$)) that combines $N$ multipaths with the largest mean SNR. Figures 6.57 a and b show the schematic for these receiver structures respectively.

Clearly, PMRC is simpler to implement compared to GSC owing to the simpler selection circuitry complexity (i.e., the need to select $N$ paths with the largest instantaneous SNR in the latter). The motivation for performance analyses of the previous two suboptimal receiver structures is based on the propagation measurements of UWB signals at Virginia Tech [84] that indicates the significant amount of energy that is found in a few dominant multipaths, particularly for the LOS case. For example, approximately 80% of the receiver energy could be captured by combining only 12 dominant multipaths. However, for the NLOS situation, approximately 75% of energy is captured by combining 30 dominant multipaths.

In [85], the performance comparisons between GSC and PMRC over Nakagami-$m$ channels have been studied in the context of spread spectrum communications but with only a few multipaths. In [86], the authors resorted to a semianalytical approach for analyzing the average bit error rate (ABER) performance of GSC($N, L$) and PMRC($N, L$) at a fixed distance. In contrast, we now examine the throughput range performance of UWB for two different reduced-complexity receiver structures

Received signal on $i$-th diversity path (at the demodulator output):

$$r_i = \sqrt{E_s}\,\alpha_i \exp[\,j(\phi_i + \theta_m)] + \eta_i$$

$\alpha_i$     random fading amplitude process

$\phi_i$     uniformly distributed random phase process

$\eta_i$     additive white Gaussian noise component

$\theta_m$     desired phase modulation

$E_s$     energy per information symbol

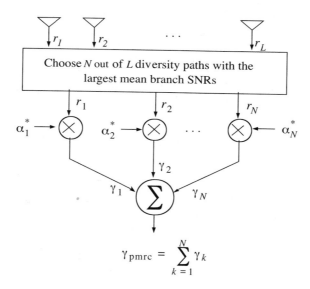

(a) PMRC $(N, L)$ RAKE Receiver that Combines $N$ Paths with Strongest Mean SNRs

Figure 6.57: Hybrid RAKE Receiver Structures (continued).

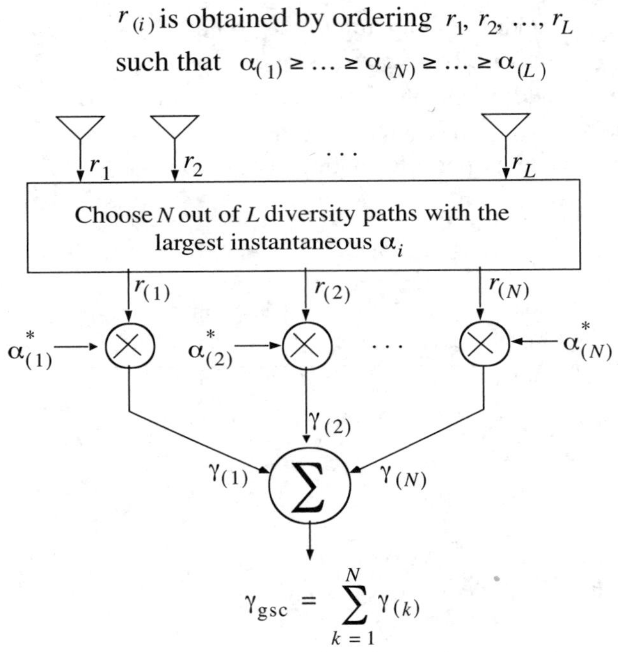

r (i) is obtained by ordering r1, r2, ..., rL
such that α(1) ≥ ... ≥ α(N) ≥ ... ≥ α(L)
r1
r2
rL
Choose N out of L diversity paths with the largest instantaneous αi
r(1)
r(2)
r(N)
α*(1)
α*(2)
α*(N)
γ(2)
γ(1)
γ(N)
Σ
γgsc = Σ γ(k), k = 1 to N

(b) GSC($N, L$) Receiver that Adaptively Combines $N$ Paths with Strongest Instantaneous SNRs

Figure 6.57: (cont.) Hybrid RAKE Receiver Structures.

over generalized fading channels using exact and approximate ABER expressions (rather than the semianalytical approach discussed in [86]).

We assume a UWB system operating in a Rayleigh or a Nakagami-$m$ fading environment and present analytical expressions for the throughput as a function of the distance for the GSC$(N, L)$ and PMRC$(N, L)$ receiver structures. The bit error probability for an $M$-ary PAM system is given by

$$
\begin{aligned}
P_b &= \frac{2\,(M-1)}{M} Q\left(\sqrt{\frac{6c\gamma}{M^2-1}}\right) \\
&= \frac{2\,(M-1)}{\pi M} \int_0^{\pi/2} \exp\left(-\frac{3c/\,(M^2-1)}{\cos^2\theta}\gamma\right) d\theta
\end{aligned}
\tag{6.153}
$$

where $M$ is the alphabet size and $c = \log_2 M$ denotes the number of information bits carried by each symbol. The instantaneous SNR/bit is $\gamma = (E_b/N_0)\,\alpha^2$ where $E_b$ denotes received energy per bit, $N_o$ is noise spectral density, and $\alpha$ denotes random fading amplitude. The average SNR per bit at the receiver input can be computed as $\overline{\gamma} = (P_r T_p/N_0)$, where $P_r$ is the average received signal strength and $T_b$ denotes the bit duration (time between transmitted pulses). The received power at a distance $d$ from the transmitter can be modeled as

$$
P_r = \frac{P_r G_t G_r}{L_s}\left(\frac{\lambda}{4\pi d}\right)^n
\tag{6.154}
$$

where $n$ denotes the path loss exponent, $L_s$ is the system loss factor not related to propagation $(L_s \geq 1)$, $\lambda$ is the carrier frequency used for computing distance loss, and $\{G_t, G_r\}$ denotes the transmitter and receiver antenna gains, respectively. Using those expressions, it can be shown that the maximum achievable throughput, $R_b$, at distance $d$, is given by

$$
R_b = c/T_b = \frac{P_t G_t G_r}{N_0 \overline{\gamma} L_s}\left(\frac{\lambda}{4\pi d}\right)^n
\tag{6.155}
$$

The noise spectral density is $N_0 = kTF$ where $k$ is Boltzman's constant, $T$ is system temperature, and $F$ denotes the noise figure. From (6.155) it is apparent that for fixed transmit power, throughput is a function of $d^{-n}$. The average SNR per bit required to evaluate (6.155) for a given bit error rate performance can be found by averaging (6.153) over the probability distribution function (PDF) of GSC$(N,L)$ or PMRC$(N,L)$.

## 6.4.1   GSC$(N, L)$ with Independent but Nonidentically Distributed Fading Statistics

### Exact Analysis of GSC$(N, L)$

Using the moment-generating function (MGF) approach, we obtain

$$
\overline{P_b} = \frac{2\,(M-1)}{\pi M} \int_0^{\pi/2} \phi_\gamma\left(\frac{3c/\,(M^2-1)}{\cos^2\theta}\right) d\theta
\tag{6.156}
$$

where $\phi_{\gamma(.)}$ denotes the MGF of the GSC($N$,L) output SNR, $\gamma_{gsc}$. The MGF expressions for the GSC($N$,L) output SNR in a myriad of fading environments has been developed in [87, 88], namely

$$\phi_{\gamma_{gsc}}(s) = \sum_{\sigma \epsilon T_{L,N}} \int_0^\infty e^{-sx} f_{\sigma(L-N+1)}(x) \left[ \prod_{t=1}^{L-N} F_{\sigma(t)}(x) \right]$$

$$\times \left[ \prod_{k=L-N+2}^{L} \phi_{\sigma(k)}(s,x) \right] dx,$$

$$1 \leq N \leq L \qquad (6.157)$$

where

$$\sum_{\sigma \epsilon T_{L,N}} = \sum_{\substack{\sigma \epsilon S_L, \sigma(1) < \sigma(2) < ... < \sigma(L-N) \\ \sigma(L-N+2) < ... < \sigma(L)}} \qquad (6.158)$$

while $S_L$ is the set of all permutations of integers $[1, 2, \ldots, L]$, and $\sigma \epsilon S_L$ denotes the specific function $\sigma = [\sigma(1), \sigma(2), \ldots, (L)]$ which permutes the integers $[1, 2, \ldots, L]$. In (6.157), $f_{\sigma(k)}(\cdot)$, $F_{\sigma(k)}(\cdot)$, and $\phi_{\sigma(k)}(\cdot)$ denote the PDF, cumulative distribution function (CDF), and the MGF of SNR of the $\sigma k^{th}$ multipath, respectively.

### Approximate Analysis of GSC($N, L$)

The throughput range performance of UWB systems employing GSC receivers can be analyzed using the previous equations. However, as the number of resolvable multipath components increases, it becomes computationally intensive to evaluate (6.156) along with (6.157), and turns out to be impractical for the case $L \geq 15$. As such, an approximate ABER formula is needed to predict the performance of GSC($N, L$) with independent but nonidentically distributed (i.n.d) multipath components. It seems that a very good approximation can be achieved by setting $\Omega = \frac{1}{L} \sum_{i=1}^{L} \overline{\gamma_i}$, $m \approx (K+1)^2 / (2K+1)$ and $m = \frac{1}{L} \sum_{i=1}^{L} m_i$ in the analytical expression for independent and identically distributed multipath components. This hypothesis is further validated by the numerical results presented in the following subsection.

### 6.4.2   PMRC($N, L$) with Independent but Nonidentically Distributed Fading Statistics

It is not difficult to show that the MGF of PRMC $(N, L)$ output SNR may be computed as

$$\phi_{\gamma_{pmrc}}(s) = \prod_{k=1}^{N} \phi_{\gamma_k}(s) \qquad (6.159)$$

where $\phi_{\gamma_k}(.)$ denotes the MGF of the $k^{th}$ strongest (mean SNR) diversity path. Hence, the ABER performance of PMRC may be analyzed using (6.156) and (6.159), and the maximum achievable throughput can be easily computed as before.

### 6.4.3    GSC($N, L$) with Equally Correlated Nakagami-$m$ Fading Statistics

In order to gain significantly from the use of diversity, there must be a sufficient level of statistical independence in the fading of the received signal in different diversity paths. In UWB communication systems with closely received resolvable multipath components, an assumption of independent multipath signals may lead to grossly overestimated diversity gains. As such, a study of the effect of multipath correlation on the performance of a UWB communication system is important. In this section, we will provide analytical expressions to evaluate the ABER performance of the GSC($N, L$) receiver structure in equally correlated Nakagami-$m$ multipath fading channels.

Let $\gamma_{(1)}, \ldots, \gamma_{(L)}$ denote the instantaneous SNRs in ascending order, namely $\gamma_{(1)} < \gamma_{(2)} < \ldots < \gamma_{(L)}$. Using the property of exchangeability, the joint PDF of the ordered set of instantaneous SNR can be expressed as

$$f_{\gamma_{(1)}, \ldots, \gamma_{(2)}}(x_1, \ldots, x_L) = \sum_{\sigma \epsilon S_L} f_{\gamma_1, \ldots, \gamma_L}(x_1, \ldots, x_L) \qquad (6.160)$$

The joint PDF of the ordered instantaneous SNRs in a Nakagami-$m$ channel can be obtained using Equation 21 of [90] and (6.160) as

$$f_{\gamma_{(1)}, \ldots, \gamma_{(2)}}(x_1, \ldots, x_L) = \xi^m \sum_{p=0}^{\infty} \frac{\Gamma(m+p)}{\Gamma(m)\, p!} \xi^p \sum_{\substack{0 \leq \ell_1, \ldots, \ell_L \leq p \\ \ell_1 + \ldots + \ell_L = p}} \frac{p!}{\ell_1! \ldots \ell_L!}$$

$$\times \sum_{\sigma \epsilon S_L} \prod_{k=1}^{L} g_{\sigma(k)}(x_k, \nu_k, \Lambda) \qquad (6.161)$$

where $m \geq \frac{1}{2}$ denotes the fading severity index, $\Omega$ is the average received SNR per bit per branch, $\rho$ denotes the power correlation coefficient $(0 < \rho < 1)$, $\xi = \left(\frac{1-\rho}{1+(L-1)\rho}\right)$, $\nu_k = \ell_k + m$, $\Lambda = \frac{m}{\Omega(1-\rho)}$, and $g_k(.,.,.)$ is defined as

$$g_k(x_k, \nu_k, \Lambda) = \frac{\Lambda^{\nu_k} x_k^{\nu_k - 1} e^{-\Lambda x_k}}{\Gamma(\nu_k)} \qquad (6.162)$$

It is apparent from (6.161) that we can compute the MGF $\phi_{\gamma_{gsc}}(.)$ in a similar fashion to GSC with i.n.d fading statistics [88], namely

$$
\phi_{\gamma_{gsc}}(s) = E\left[\exp\left(-s\sum_{i=L-N+1}^{L}\gamma_i\right)\right]
$$

$$
= \xi^m \sum_{p=0}^{\infty} \frac{\Gamma(m+p)}{\Gamma(m)\,p!}\xi^p \sum_{\substack{0\leq\ell_1\ldots\ell_L\leq p\\ \ell_1+\ldots+\ell_L=p}} \frac{p!}{\ell_1!\ldots\ell_L!} I(N,L) \qquad (6.163)
$$

where $I(.,.)$ is defined as

$$
I(N,L) = \sum_{\sigma\epsilon T_{L,N}} \int_0^{\infty} e^{-sx} g_{\sigma(L-N+1)}(x) \left[\prod_{t=1}^{L-N} G_{\sigma(t)}(x)\right]
$$

$$
\times \left[\prod_{k=L-N+2}^{L} \phi_{\sigma(k)}(s,x)\right] dx, 1 \leq N \leq L \qquad (6.164)
$$

where $G_k(y) = \int_0^y g_k(x)\,dx$, $\phi_k(s,x) = \int_x^{\infty} e^{-st} g_k(t)\,dt$. Using the identity in Equation 3.381.3 of [91], functions $G_k(.)$ and $\phi_k(.,.)$ may be computed in closed form as

$$
G_k(y) = 1 - \frac{\Gamma(\nu_k, \Lambda x)}{\Gamma(\nu_k)} \qquad (6.165)
$$

$$
\phi_k(s,x) = \left(\frac{\Lambda}{s+\Lambda}\right)^{\nu_k} \frac{\Gamma(\nu_k, (s+\Lambda)\,x)}{\Lambda(\nu_k)} \qquad (6.166)
$$

where $\Gamma(a,x) = \int_x^{\infty} t^{a-1} e^{-at} dt$ denotes the complementary incomplete Gamma function. It may be noted that (6.163) is much more concise, yet more general and numerically efficient, than Equation 36a of [90]. Now the throughput performance can now be easily analyzed using (6.155), (6.156), and (6.163).

The results discussed in this section use the assumptions in Table 6.2. Figure 6.58 compares the accuracy of "approximate ABER analysis" against the "exact ABER analysis" using (6.157) for a GSC($N$, 12) receiver with i.n.d diversity branches in Rayleigh fading channels. It is evident that the throughput curves corresponding to approximate ABER match closely with the exact throughput curves. Readers are referred to [92] for further details.

Figure 6.59 draws a comparison between the throughput performance of GSC($N$, 30) and PMRC($N$, 30) receiver structures in a Rayleigh environment with an exponentially decaying multipath intensity profile ($\delta = 0.2$). The average SNR of the $n^{th}$ diversity path is $\overline{\gamma_n} = Ce^{-n\delta}\overline{\gamma_b}$, where $\overline{\gamma_b}$ denotes the average SNR/bit, and the parameter $\Theta$ is chosen such that the constraint $\sum_{n=1}^{L}\overline{\gamma_n} = \overline{\gamma_b}$ is satisfied.

Table 6.2: Assumptions on the System Parameters for Generating Numerical Results.

SOURCE: S. Gaur and A. Annamalai, "Improving the Range of Ultra Wideband Transmission using RAKE Receivers," *Proc. IEEE Vehicular Technology Conference* [92]. © IEEE, 2003. Used by permission.

| Transmit Power spectral d | −41 dBm/Mhz |
|---|---|
| Transmitter Antenna Gain | 0 dBi |
| Receiver Antenna Gain, $G$ | 0 dBi |
| System Loss, $L_s$ | 5 dB |
| Noise Figure, $F$ | 6 dB |
| Path-loss exponent, $n$ | 1.6 (LOS) 2.7 (NLOS) |
| Operating Bandwidth, $B_s$ | 2.5 GHz |
| Center Frequency, $f_c$ | 3.75 GHz |

Solving for C yields

$$\overline{\gamma_n} = \frac{\left(1 - e^{-\delta}\right) e^{-\delta(n-1)}}{1 - e^{-L\delta}}\overline{\gamma_b} \tag{6.167}$$

As expected, the GSC receiver outperforms PMRC for a fixed number of combined paths, $N$. For instance, GSC(10,30) and PMRC(20,30) have identical throughput performances. From Figure 6.59 it is also apparent that as the number of combined paths $N$ are increased, the throughput performance of PMRC becomes virtually identical to GSC. Based on these observations, we can conclude that deployment of a GSC$(N, L)$ receiver is desirable for the cases in which $N << L$, whereas for $N \geq L/2$, a PMRC rake receiver is preferable because of its low complexity and virtually identical throughput performance when compared to GSC$(N, L)$. Figure 6.60 depicts the performance of the 5-tap PMRC$(5, 20)$ RAKE receiver in a Nakagami-$m$ fading channel for varying fade distributions. It is evident that improving channel conditions translates into a considerable improvement in the receiver throughput. Figure 6.61 presents the throughput analysis for a GSC$(N, 20)$ receiver for path loss exponent $n = 3.5$. A comparative study of Figures 6.60 and 6.61 suggests a major fallback in throughput with even the slightest increase in the path loss exponent $n$. The rate at which the throughput falls declines gradually as the range increases. Finally, Figure 6.62 illustrates the effect of multipath correlations on the throughput performance of the GSC receiver for different numbers of combined paths, $N$. The throughput performance degrades with increasing multipath correlation, as expected. Throughput falls by almost 50% for $\rho = 0.8$.

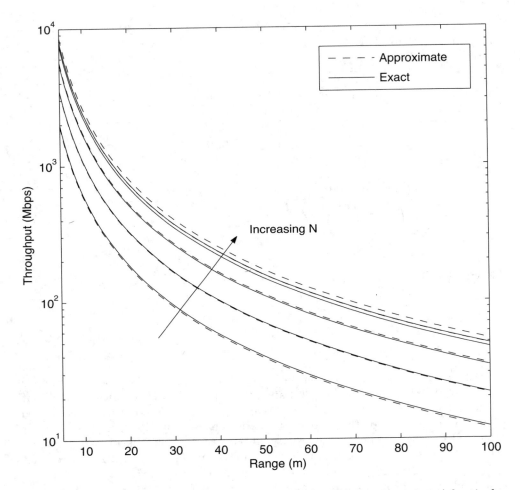

Figure 6.58: Throughput Performance of a GSC $(N, 12)$ Receiver that Adaptively Combines $N \in \{1, 2, 4, 8, 12\}$ Paths in a Rayleigh Channel with Nonidentical Fading Statistics, $\delta = 0.2$, and Path Loss Exponent, $n = 1.7$.

SOURCE: S. Gaur and A. Annamalai, "Improving the Range of Ultra Wideband Transmission using RAKE Receivers," *Proc. IEEE Vehicular Technology Conference* [92]. © IEEE, 2003. Used by permission.

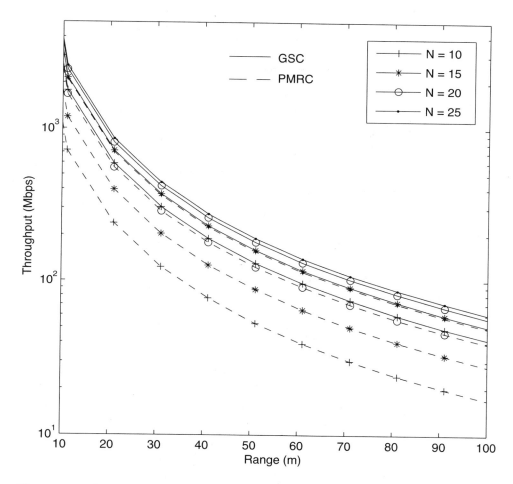

Figure 6.59: Comparison Between the Throughput Performances of a 2-ary PAM that Employs Either a GSC $(N, 30)$ Receiver, or a PMRC Receiver that Combines First $N$ Paths in an i.n.d Rayleigh Channel and Path Loss Exponent, $n = 1.7$.

Source: S. Gaur and A. Annamalai, "Improving the Range of Ultra Wideband Transmission using RAKE Receivers," *Proc. IEEE Vehicular Technology Conference* [92]. © IEEE, 2003. Used by permission.

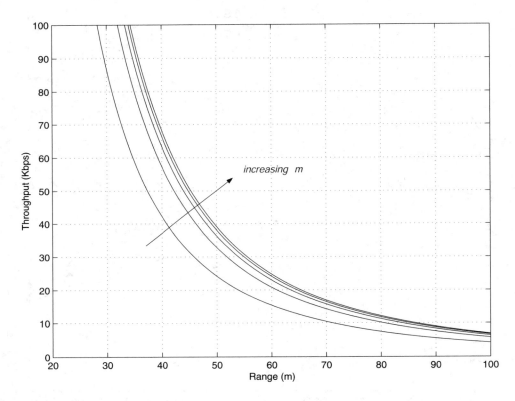

Figure 6.60: Throughput Performance of a 2-ary PAM in Conjunction with a PMRC(5,20) RAKE Receiver that Combines the First 5 Paths in a Nakagami-$m$ Fading Channel $m \epsilon \{1, 2, 3, 4, 5\}$, and Path-Loss Exponent $n = 2.7$.

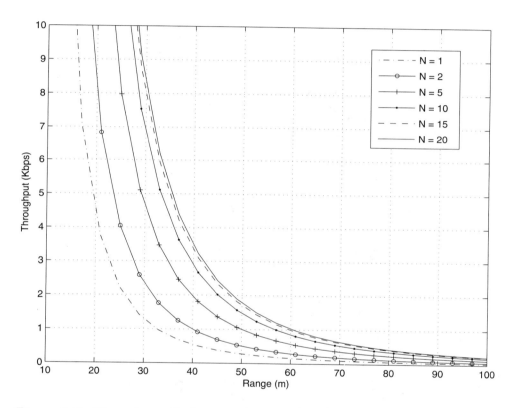

Figure 6.61: Throughput Performance of a 2-ary PAM in Conjunction with a GSC($N$, 20) Receiver in a Rayleigh Fading Channel and Path Loss Exponent $n = 3.5$.

SOURCE: S. Gaur and A. Annamalai, "Improving the Range of Ultra Wideband Transmission using RAKE Receivers," *Proc. IEEE Vehicular Technology Conference* [92]. © IEEE, 2003. Used by permission.

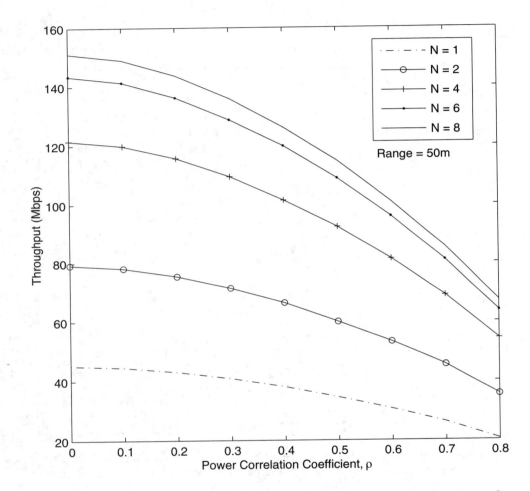

Figure 6.62: Effect of Multipath Correlation on the Throughput Performance of a 2-ary PAM in Conjunction with a GSC($N$, 8) Receiver in a Rayleigh Fading Channel and Path Loss Exponent $n = 1.7$.

SOURCE: S. Gaur and A. Annamalai, "Improving the Range of Ultra Wideband Transmission using RAKE Receivers," *Proc. IEEE Vehicular Technology Conference* [92]. © IEEE, 2003. Used by permission.

# 6.5  Summary

UWB communications opens new areas of research, as these systems are very different when compared to narrowband and wideband systems in regard to signal generation techniques, propagation characteristics, and receiver design. Receiver design is one of the most challenging aspects of the overall system design. This chapter has given the various receiver architectures for currently existing UWB systems and some proposed systems. Optimizing the UWB receiver performance is still an open, challenging task, and there are opportunities for more future work. Clearly, future developments in UWB receiver design will focus on lowering costs and reducing interference and multipath distortion to enable increased range. Developing low-power receivers will continue to be an important research area.

# Bibliography

[1] J. Foerster, et al., "Channel Modeling Sub-Committee Report Final," IEEE P802.15 Wireless Personal Area Networks, P802.15-02/490r1-SG3a, February 2003.

[2] A. Saleh and R. Valenzuela, "A Statistical Model for Indoor Multipath Propagation," *IEEE Journal on Selected Areas in Communications*, vol. 5, pp. 128-137, February 1987.

[3] J. Kunisch, and J. Pamp, "Measurement Results and Modeling Aspects for the UWB Radio Channel," *IEEE Conference on Ultra Wideband Systems and Technologies*, Digest of Papers, pp. 19-24, Baltimore, MD, 2002.

[4] C. Bennett and G. F. Ross, "Time Domain Electromagnetics and Its Applications," *Proc. of the IEEE*, vol. 66, no. 3, pp. 299-318, March 1978.

[5] I. J. Immoreev and A. A. Sudakov, "Ultra-Wideband (UWB) Interference Resistant System for Secure Radio Communication with High Data Rate," IEEE *1st International Conference on Circuits and Systems for Communications*, pp. 230-233, June 2002.

[6] W. M. Lovelace and J. K. Townsend, "The Effects of Timing Jitter and Tracking on the Performance of Impulse Radio," *IEEE Journal on Selected Areas in Communications*, vol. 20, no. 9, pp. 1646-1651, December 2002.

[7] E. Saberinia and A. H. Tewfik, "Single and Multi-Carrier UWB Communications," *Proc. 7th International Symposium on Signal Processing and Its Applications*, vol. 2, pp. 343-346, July 2003.

[8] E. Saberinia and A. H. Tewfik, "Receiver Structures for Multi-Carrier UWB Systems," *Proc. 7th International Symposium on Signal Processing and Its Applications*, vol. 1, pp. 313-316, July 2003.

[9] R. A. Scholtz and M. Z. Win, "Impulse Radio," Invited Paper, *IEEE PIMRC '97*, Helsinki, Finland.

[10] M. Z. Win and R. A. Scholtz, "Ultra-Wide Bandwidth Time-Hopping Spread-Spectrum Impulse Radio for Wireless Multiple-Access Communications," *IEEE Transactions on Communications*, vol. 48, no. 4, pp. 679-691, April 2000.

[11] G. Durisi, A. Tarable, J. Romme, and S. Benedetto, "A General Method for Error Probability Computation of UWB Systems for Indoor Multiuser Communications," *Journal of Communications and Networks*, vol. 5, no. 4, pp. 354-364, December 2003.

[12] G. Durisi and S. Benedetto, "Performance Evaluation and Comparision of Different Modulation Schemes for UWB Multiaccess Systems," *ICC 2003*, Anchorage, AK.

[13] T. W. Barrett, "History of Ultra Wideband Communications and Radar: Part 1, UWB Communications," *Microwave Journal*, pp. 22-56, January 2001.

[14] T. W. Barrett, "History of Ultra Wideband (UWB) Radar and Communications: Pioneers and Innovators," *Progress In Electromagnetic Symposium 2000 (PIERS 2000)*, Cambridge, MA, July 2000.

[15] C. R. Anderson, "Ultra Wideband Communication System Design Issues and Tradeoffs," Ph.D. Qualifier Report, Virginia Tech, October 2003.

[16] J.D. Choi and W. E. Stark, "Performance of Ultra Wideband Communications with Suboptimal Receivers in Multipath Channels," *IEEE Journal on Selected Areas in Communications*, vol. 20, no. 9, pp. 1754-1766, December 2002.

[17] United States Patent 3,662,316, Short Base-Band Pulse Receiver, Kenneth W. Robbins, May 9, 1972.

[18] United States Patent 4,641,317, Spread Spectrum Radio Transmission System, Larry W. Fullerton, February 3, 1987.

[19] M. I. Skolnik, *Introduction to Radar Systems*, 3rd ed., New York: McGraw-Hill, 2001.

[20] M. Skolnik, *Introduction to Radar Systems*, pp. 436-556, New York: McGraw-Hill, 1962.

[21] H. F. Harmuth, "Radar Equation for Nonsinusodial Waves," *IEEE Transactions on Electromagnetic Compatibility*, vol. 31, no. 2, pp. 138 -147, May 1989.

[22] F. Ramirez-Mireles, "On the Performance of Ultra Wideband Signals in Gaussian Noise and Dense Multipath," *IEEE Transactions on Vehicular Technology*, vol. 50, no. 1, pp. 244-249, January 2001.

[23] X. Huang and Y. Li, "Performance of Impulse Train Modulated UWB Systems," *IEEE International Conference on Communications*, vol. 2, pp. 758-762, April 28-May 2, 2002.

[24] L. G. Yue, and S. Affes, "On the BER Performance of Pulse-Position Modulation UWB Radio in Multipath Channels," *2002 IEEE Conference on Ultra Wideband Systems and Technologies*, Digest of Papers, pp. 231-234, May 2002.

[25] S. Lee, "Design and Analysis of Ultra-Wide Bandwidth Impulse Radio Receiver," Ph.D. Dissertation, EE Department, University of Southern California, August 2002.

[26] A. J. Viterbi, *CDMA Principles of Spread Spectrum Communication*, MA: Addison-Wesley, 1995.

[27] V. Lottici, A. D'Andrea, and U. Mengali, "Channel Estimation for Ultra-Wideband Communications," *IEEE Journal on Selected Areas in Communications*, vol. 20, no. 9, pp. 1638-1645, December 2002.

[28] M. Z. Win and R. A. Scholtz, "Characterization of Ultra-Wide Bandwidth Wireless Indoor Channels: A Communication-Theoretic View," *IEEE Journal on Selected Areas in Communications*, vol. 20, no. 9, pp. 1613-1627, December 2002.

[29] T. S. Rappaport, *Wireless Communications: Principles and Practice*, 2nd ed., New Jersey: Prentice Hall, 2002.

[30] J. G. Proakis, "Adaptive Equalization for TDMA Digital Mobile Radio," *IEEE Transactions on Vehicular Technology*, vol. 40, no. 2, pp. 333-341, May 1991.

[31] J. G. Proakis, *Digital Communications*, 4th ed., McGraw-Hill, 2001.

[32] B. Mielczarek, M. -Ol. Wessman, and A. Svensson, "Performance of Coherent UWB RAKE Receivers using different Channel Estimators," *Proc. IEEE International Workshop on Ultra Wideband Systems—UWB Cluster Day*, Oulu, Finland, June 2003.

[33] B. Mielczarek, M. -Ol. Wessman, and A. Svensson, "Performance of Coherent UWB Rake Receivers with Channel Estimators," *Proc. IEEE Vehicular Technology Conference*, vol. 3, pp. 1880-1884, Oct. 2003.

[34] J. H. Winters, "Optimum Combining in Digital Mobile Radio with Co-channel Interference," *IEEE Journal on Selected Areas in Communications*, vol. 2, no.4, pp. 528-539, July 1984.

[35] Q. Li and L. A. Rusch, "Multiuser Detection for DS-CDMA UWB in the Home Environment," *IEEE Journal on Selected Areas in Communications*, vol. 20, no. 9, pp. 1701-1711, December 2002.

[36] I. Bergel, E. Fishler, and H. Messer, "Narrow-band Interference Supression in Time Hopping Impulse Radio Systems," *Proc. IEEE Conference on Ultra-Wideband Systems and Technologies*, pp. 303-308, Baltimore, MD, December 2002.

[37] Q. Li and L. A. Rusch, "Muti-user Receivers for DS-CDMA UWB," *IEEE Conference on Ultra Wideband Systems and Technologies*, pp. 163-167, May 2002.

[38] G. Durisi, J. Romm, and S. Benedetto, "Performance of TH and DS UWB Multiaccess Systems in Presence of Multipath Channel and Narrowband interference," *IWUWBS 2003*, Oulu, Finland, June 2003.

[39] Q. Li and L. A. Rusch, "Hybrid RAKE / Multiuser Receivers," *Proc. Radio and Wireless Conference (RAWCON '03)*, pp. 203-206, August 2003.

[40] A. M. Haimovich and A. Vadhri, "Rejection of Narrow-Band Interference in PN Spread Spectrum Systems Using an Eigen Analysis Approach," *Proc. of IEEE Military Communications Conference (MILCOM '94)*, vol. 3, pp. 1002-1006, Fort Monmouth, NJ, October 1994.

[41] H. Sheng, A.M. Haimovich, A.F. Molisch, and J. Zhang, "Optimum Combining for Time Hopping Impulse Radio UWB Rake Receivers," *Proc. IEEE Conference on Ultra Wideband Systems and Technologies (UWBST '03)*, Reston, VA, November 2003.

[42] R. A. Monzingo and T. W. Miller, *Introduction to Adaptive Arrays*, New York: John Wiley and Sons, 1980.

[43] I. D. O'Donnell, M. S. W. Chen, S. B. T. Want, and R. W. Brodersen, "An Integrated, Low Power, Ultra-Wideband Transceiver Architecture for Low-Rate, Indoor Wireless Systems," *IEEE CAS Workshop on Wireless Communications and Networking*, September 2002.

[44] P. Withington, R. Reinhardt, and R. Stanley, "Preliminary Results of an Ultra-Wideband (Impulse) Scanning Receiver," *IEEE MILCOM*, vol. 2, pp. 1186-1190, Atlantic City, NJ, November 1999.

[45] I. J. Immoreev and A. A. Sudakov, "Ultra-wideband (UWB) Interference Resistant System for Secure Radio Communication with High Data Rate," *IEEE 1st International Conference on Circuits and Systems for Communications*, pp. 230-233, June 2002.

[46] —, Cellonics Technology Technical White Paper TWP-01, Application Note, Cellonics Incorporated Pvt. Ltd., 2003. Available: http://www.cellonics.com/downloads/files/TWP.pdf.

[47] C. R. Anderson, "Design and Implementation of an Ultra Broadband Millimeter-Wavelength Vector Sliding Correlator Channel Sounder and In-Building Measurements at 2.5 & 60 GHz," Master's thesis, Virginia Tech, May 2002. Available: http://scholar.lib.vt.edu/theses/index.html.

[48] J. D. Choi and W. E. Stark, "Performance of autocorrelation receivers for ultra-wideband communications with PPM in multipath channels," *2002 IEEE Conference on Ultra Wideband Systems and Technologies*, Digest of Papers, May 2002.

[49] R. T. Hoctor and H. W. Tomlinson, "An Overview of Delay-Hopped, Transmitted-Reference RF Communications," G.E. Research and Development Center, Technical Information Series, pp. 1-29, January 2002.

[50] R. T. Hoctor and H. W. Tomlinson, et al., "Delay Hopped Transmitted Reference Experimental Results," G.E. Research and Development Center, Technical Information Series, 2002GRC099, April 2002.

[51] M. K. Simon, J. K. Omura, R. A. Sholtz, and B. K Levitt, *Spread Spectrum Communications, Volume-I*, Computer Science Press, 1985.

[52] L. Yang and G. B. Giannakis, "Optimal pilot waveform assisted modulation for ultra-wideband communications," *Conference Record of the Thirty-Sixth*

*Asilomar Conference on Signals, Systems and Computers*, vol. 1, pp. 733-737, November 2002.

[53] L. Yang and G. B. Giannakis, "Optimal pilot waveform assisted modulation for ultra-wideband communications," *IEEE Transactions on Wireless Communications*, 2004 (to appear).

[54] M. Z. Win and R. A. Scholtz, "On the Energy Capture of Ultrawide Bandwidth Signals in Dense Multipath Environments," *IEEE Communications Letters*, vol. 2, no. 9, September 1998.

[55] R. C. Z. Qiu, "A Study of the Ultra-Wideband Wireless Propagation Channel and Optimum UWB Receiver Design," *IEEE Journal on Selected Areas in Communications*, vol. 20, no. 9, December 2002.

[56] I. D. O'Donnell, "Oscillator Design Issues for UWB," Poster Session, Berkeley Wireless Research Center Summer Retreat, Summer 2001.

[57] H. Bing, X. Hou, X. Yang, T. Yang, and C. Li, "A "Two-Step" Synchronous Sliding Method of Sub-Nanosecond Pulses for Ultra-Wideband (UWB) Radio," *IEEE International Conference on Communications, Circuits, and Systems and West Sino Expositions*, vol. 1, pp. 142-145, June 2002.

[58] C. R. Anderson, A. M. Orndorff, R. M. Buehrer, and J. H. Reed, "An Introduction and Overview of an Impulse-Radio Ultrawideband Communication System Design," MPRG Technical Report, Virginia Polytechnic Institute and State University, 2004.

[59] E. Green, "W241: System Architectures for High-Rate Ultra Wideband Communications," Intel Labs.

[60] M. Z. Win, X. Qui, R. A. Scholtz, and V. O. K. Li, "ATM-based TH-SSMA Network for Multimedia PCS," *IEEE Journal on Selected Areas in Communications*, vol. 17, no. 5, pp.824 - 836, May 1999.

[61] M. Z. Win and R. A. Scholtz, "Impulse Radio: How it Works," *IEEE Communication Letters*, vol. 2, pp. 36-38, February 1998.

[62] S. S. Kolenchery, J. K. Townsend, and J. A. Freebersyser, "A Novel Impulse Radio Network for Tactical Military Wireless Communications," *IEEE Military Communications Conference*, vol. 1, pp. 59-65, Boston, MA, October 1998.

[63] Z. Wu, F. Zhu, and C. R. Nassar, "Ultra Wideband Time Hopping Systems: Performance and Throughput Enhancement via Frequency Domain Processing," *36th Asilomar Conference on Signals, Systems and Computers*, vol. 1, pp. 722-727, November 2002.

[64] C. R. Nassar, F. Zhu, and Z. Wu, "Direct Sequence Spreading UWB Systems: Frequency Domain Processing for Enhanced Performance and Throughput," *IEEE International Conference on Communications*, vol. 3, pp. 2180-2186, May 2003.

[65] Z. Wu, F. Zhu, and C. R. Nassar, "Performance Comparison of Frequency Domain Processing in Ultra Wideband Links," *The 2002 45th Midwest Symposium on Circuits and Systems*, vol. 3, pp. III–591-596, August 2002.

[66] S. S. Kolenchery, J. K. Townsend, and J. A. Freebersyser, "A Novel Impulse Radio Network for Tactical Military Wireless Communications," *IEEE Military Communications Conference*, vol. 1, pp. 59-65, Boston MA, October 1998.

[67] W. H Steel, *Interferometry*, 1st ed., Cambridge UK: Cambridge University Press, 1967.

[68] B. Natarajan, C. R. Nassar, and S. Shattil, "Innovative Pulse Shaping for High Performance Wireless TDMA," *IEEE Communications Letters*, vol. 5, no. 9, pp. 372-374, September 2001.

[69] B. Natarajan, C. R. Nassar, and S. Shattil, "Throughput Enhancements in TDMA Through Carrier Interferometry Pulse Shaping," *IEEE Vehicular Technology Conference*, pp. 1799-1803, Boston, M, September 2000.

[70] L.-L. Yang and L. Hanzo, "Residue Number System Assisted Fast Frequency Hopped Synchronous Ultra-Wideband Spread-Spectrum Multiple-Access: A Design Alternative to Impulse Radio," *IEEE Journal on Selected Areas in Communications*, vol. 20, no. 9, pp. 1652-1663, December 2002.

[71] E. Saberinia and A. H. Tewfik, "Multi-carrier UWB," IEEE P802.15 Wireless Personal Area Networks, P802.15-02/03147r1-TG3a, November 2002.

[72] E. Saberinia and A. H. Tewfik, "Synchronous UWB-OFDM," *IEEE ISWC 2002*, pp. 41-42, September 2002.

[73] E. Saberinia and A. H. Tewfik, "High Bit Rate UWB-OFDM," *IEEE GLOBECOM '02*, vol. 3, pp. 2260-2264, November 2002.

[74] A. Batra, et al.,"Multi-band OFDM Physical Layer Proposal for IEEE 802.15 Task Group 3a," IEEE P802.15 Wireless Personal Area Networks, P802.15-03/268r1, September 2002.

[75] E. Saberinia and A. H. Tewfik, "$N$-Tone Sigma-Delta UWB-OFDM Transmitter and Receiver," *IEEE ICASSP '03*, vol. 4, pp. 129-132, April 2003.

[76] J. C. de Mateo Garcia and A. Garcia Armada, "Effects of Bandpass Sigma-Delta Modulation on OFDM Signals," *IEEE Transactions on Consumer Electronics*, vol. 45, no.2, pp. 119-124, May 1999.

[77] E. Saberinia and A. H. Tewfik, "Generating UWB-OFDM Signal Using Sigma-Delta Modulator," *57th IEEE Vehicular Technology Conference (VTC 2003)*, vol. 2, pp. 1425-1429, April 2003.

[78] C. R.C.M. da Silva and L. B. Milstein, "Spectral-Encoded UWB Communication Systems," *IEEE Conference on Ultra-Wideband Systems and Technologies*, November 2003.

[79] J. A. Salehi, A. M. Weiner, and J. P. Heritage, "Coherent Ultrashort Light Pulse Code-Division Multiple Access Communication Systems," *Journal of Lightwave Technology*, vol. 8, pp. 478-491, March 1990.

[80] K. S. Kim, D. M. Marom, L. B. Milstein, and Y. Fainman, "Hybrid Pulse Position Modulation/Ultrashort Light Pulse Code-Division Multiple Access Systems—Part I: Fundamental Analysis," *IEEE Transactions on Communications*, vol. 50, pp. 2018-2031, December 2002.

[81] P. Crespo, M. Honig, and J. Salehi, "Spread-Time Code-Division Multiple Access," *IEEE Transactions on Communications*, vol. 43, pp. 2139-2148, June 1995.

[82] M. G. Shayesteh, J. A. Salehi, and M. Nasiri-Kenari, "Spread-Time CDMA Resistance in Fading Channels," *IEEE Transactions on Wireless Communications*, vol. 2, pp. 446-458, May 2003.

[83] L. B. Milstein and P. K. Das, "An Analysis of a Real-Time Transform Domain Filtering Digital Communication System—Part I: Narrow Band Interference Rejection," *IEEE Transactions on Communications*, vol. 28, pp. 816-824, June 1980.

[84] A. Safaai-Jazi, A. Muqaibel, A. Attiya, A. Bayram, and S. Riad, "Ultra-Wideband Time-Domain Indoor Channel Measurements," Virginia Tech, February 2003.

[85] R. Wong, A. Annamalai, and V. K. Bhargava, "Evaluation of Pre-Detection Diversity Techniques for Rake Receivers," *Proc. IEEE PACRIM*, pp. 227-230, 1997.

[86] D. Cassioli, M. Z. Win, F. Vatalaro, and A. Molisch, "Performance of Low-Complexity Rake Reception in a Realistic UWB Channel," *IEEE International Conference on Communications*, vol. 2, pp. 763-767, 2002.

[87] A. Annamalai, G. Deora, and C. Tellambura, "Unified Error Probability Analysis for Generalized Selection Diversity in Rician Fading Channels," *Proc. IEEE Vehicular Technology Conference*, pp. 2042-2046, Birmingham, AL, May 2002.

[88] G. Deora, "Simulation and Mathematical Tools for Performance Analysis of Low-Complexity Receivers," Master's thesis, Virginia Tech, January 31, 2003. Available: http://scholar.lib.vt.edu/theses/available/etd-02042003-202146/.

[89] A. Annamalai, "Theoretical Diversity Improvement in $GSC(N, L)$ and T-$GSC(\mu, L)$ over Generalized Fading Channels," *Proc. IEEE ISWC 2002*, Victoria, September 2002.

[90] R. K. Mallik and M. Z. Win, "Analysis of Hybrid Selection/Maximal-Ratio Combining in Correlated Nakagami Fading," *IEEE Transactions on Communications*, vol. 50, no. 8, August 2002.

[91] I. S. Gradshteyn and I. M. Ryzhik, *Table of Integrals, Series and Products*, Academic Press, 1995.

[92] S. Gaur and A. Annamalai, "Improving the Range of Ultra Wideband Transmission using RAKE Receivers," *Proc. IEEE Vehicular Technology Conference*, Orlando, FL, Oct. 2003, pp. 597-601.

Chapter 7

# ON THE COEXISTENCE OF UWB AND NARROWBAND RADIO SYSTEMS

Ananthram Swami
and Brian M. Sadler

## 7.1 Introduction

UWB signals will encounter interference from many sources, primarily from relatively narrowband (NB) systems. In addition, UWB signals will also affect a large number of NB radios. Of critical importance is the potential interference with GPS and navigation bands, as well as cellular bands. There is a rich and growing literature on UWB radios. However, issues related to interference have only been partially addressed. Here, we assess the interference caused by UWB signals via analysis and simulations. Analytical results include the aggregate effect of spatially distributed UWB radios on an NB receiver, and theoretical BER expressions. The impact of NB interference on a UWB receiver is also studied. Some results on the UWB-to-UWB interference (the multiple access issue) are presented. Simulation results are included to verify the theory.

A wideband (WB) system, by its very nature, will interfere with existing NB services in the same bands. In turn, the NB signals will act as interferers to the WB system. The extent to which performance is degraded by the interference will clearly depend upon the number and distribution of the interferers, the relative powers, and the type of modulation used. Proposed applications for UWB radios include sensor networks, Wireless Personal Area Networking in the IEEE 802.15 framework [1] for indoor and outdoor command posts, geolocation, RF tagging, LPD/LPI applications, automobile collision avoidance systems, inventory control, and so on. As such, the density of fielded UWB devices could be high. UWB is also being considered, along with other alternatives, in the 802.15 WPAN 4a alternative for low rate applications such as sensor networks [2]. The FCC specifications [3] and European standards [4] essentially limit the EIRP to US Part 15 limits, that is, −41 dbm/MHz in the range 3.1–10.6 GHz, and the UWB signal is required to occupy an instantaneous bandwidth of at least 500 MHz.[1] There are significant differences in the Part 15 levels for intentional, unintentional, and incidental radiators. Conforming to Part 15 levels does not imply that interference to other systems, such as GPS, is harmless. The intent of the FCC mask is to enable co-existence of UWB radio services with currently licensed services, both commercial cellular as well as critical applications, such as GPS and navigation systems. A UWB radio with the pulsed Gaussian monocycle has a bandwidth of about 1.6 GHz centered around 2.0 GHz. Prototype radios with bandwidths of 3–4 GHz, centered at frequencies of 3–4 GHz have been reported. It is important to have tools to assess the potential interference with GPS and navigation bands as well as cellular bands.

Recent measurement campaigns indicate that GPS and some radar systems may be adversely affected, particularly if the pulse repetition frequency (PRF) is high, even though the device itself might conform to Part 15 emission limits [5–17]. With the exception of [14–17], these reports deal primarily with the impact of UWB radios on GPS.

Although there is a rich and growing literature on UWB radios [18–20, 22–24], issues related to interference have only been partially addressed. The effects of UWB interference have been considered in [25], [26], and [27], where the unconditional BER is evaluated via simulations.

There is a body of work that studies the degradation in the performance of UWB radios due to narrowband interference (NBI), such as tone jammers [28] and partial band interference (PBI) [29]. If the frequency of the tone jammer is known, several techniques can be used to shape the transmit pulse so as to avoid the interference [30–32]. In [33], spreading codes are designed that combat NBI by using frequency-spreading. The performance of generalized RAKE receivers and MMSE receivers, in the presence of MAI and NBI, has been analyzed in [34–37]. These techniques assume that all relevant user codes, multipath parameters, and noise parameters are known. Performance in the presence of multiple users has been studied. For example, [38] provides exact BER expressions for an interference limited uncoded TH-PPM and approximate expressions for the coded/asynchronous case. A Chernoff bounding approach is taken in [39], where bounds are established taking into account the episodic[2] nature of the transmission, the variance of the additive white Gaussian noise (AWGN), and imperfect channel estimates used by a RAKE receiver. However, results on aggregate effects, that is, the impact of multiple UWB emitters on NB radios, is rather sparse: [40], [25], [8], and [41] provide both analysis and simulation results. A simulation study is also reported in [42].

---

[1]Please see Chapter 1 for details.

[2]In contrast with a DS-CDMA radio, in a pulsed UWB system, pulses are not transmitted continuously; instead, there is a (variable) off time between pulses. This duty-cycled nature of the UWB signal has been termed *episodic* in [39].

In order to develop techniques for UWB interference suppression (such as blankers and other non-linear pre-processors), it is important to develop analytical and empirical models of the interference. In this chapter, we consider the interference effects of UWB signals on typical NB radios. We also consider the effect of NB interferers on UWB radios. The analysis tools are not very different from those used in studying the impact of conventional direct-sequence spread-spectrum and other multiple access signals on classical NB systems.

The theoretical analysis takes into account the waveforms of the interferers (Gaussian monocycles for UWB, typically NB Gaussian processes for NB), the spatial distribution of the sources (density and distances), propagation losses, and receiver models. The interfering UWB signal structure will cause effects different from those due to thermal or broadband noise. In particular, the pulse repetition rate, duty cycles, burst times, and specifics of the waveforms play significant roles.

The chapter is organized as follows: In Section 7.2, we consider the impact of UWB radios on an NB radio. After reviewing the signal model, we derive an expression for the conditional BER in the presence of one or more interferers and AWGN. In Section 7.3, we provide a statistical description of the impact of many interferers. In Section 7.4, we quantify the effect of NBI on an UWB signal. Finally, in Section 7.5, we summarize some results on the UWB-on-UWB problem, that is, the (perhaps asynchronous) multiple access interference (MAI) issue. Numerous simulation results are included. Most of our analysis is at "baseband" for convenience but holds for the "passband" case with simple frequency translation.

## 7.2  Interference of UWB on NB: Waveform Analysis

Our aim in this section is to derive an expression for the degradation in BER due to the UWB emitters. If the channel is noise-limited, the effect of the interference is essentially to raise the noise floor. But when the channel is interference limited, a more detailed analysis is required. We will consider the effect of one or more UWB radios on an NB radio. We do this in two steps: (a) detailed waveform analysis, corresponding to a single interferer; and (b) gross power analysis (discussed in the next section).

At the end of the linear receiver processing, conceptually we have a simple $M$-ary detection problem:

$$z = s_k + g + I \tag{7.1}$$

where $z$ is the decision variable at the output of the coherent detector, $s_k$, $k \in \{1, ..., M\}$, is the unknown symbol drawn from an $M$-ary alphabet, $g$ denotes AWGN, and $I$ denotes the effective interference. However, we do not have an expression for the pdf of $I$. Given that the BER is limited by the worst case, we need to pay more attention to the received waveform itself. Before proceeding with the analysis, we will describe the UWB signal model.

## 7.2.1   UWB Pulse Model

A popular choice for the basic UWB pulse is the second-derivative of the Gaussian pulse, which takes into account the differentiation effects of the antennas. This also leads to a symmetric pulse with an effectively limited temporal duration. We will see that the details of the waveform $p(t)$ are not critical to the analysis. Let $p(t)$ denote the Gaussian monocycle, [45, 46][3]

$$p(t) = [1 - 2(\pi t f_o)^2] \exp(-(\pi t f_o)^2) \qquad (7.2)$$

whose Fourier transform (FT) is given by,

$$P(f) := \int_{-\infty}^{\infty} p(t)e^{-j2\pi ft} \, dt = \frac{2}{\sqrt{\pi f_o^2}} \left(\frac{f}{f_o}\right)^2 \exp\left(-\frac{f^2}{f_o^2}\right) \qquad (7.3)$$

which peaks at $f = f_o$. The energy in this pulse is $E_p = \int p^2(t) \, dt = 3/\sqrt{32\pi f_o^2}$. Conceptually, a pulse-train (which may be dithered to reduce spectral lines, to accommodate user codes, to represent data via PPM, and so on) is convolved with the pulse-shape, so that the power spectrum of the transmitted UWB signal is essentially given by $|P(f)|^2$ [47]. The effect of dithering makes the modulation zero-mean and removes periodicities in the data, so that the PSD has no spectral lines. The pulse $p(t)$ and its 'PSD' $|P(f)|^2$ are shown in Figure 7.1; notice that the PSD is skewed. The bandwidth of the transmitted signal is given in Table 7.1. To a first-order approximation, the bandwidth is $2f_o$, and is centered at $f_o$. Assuming that the peak of $P(f)$ is normalized so that the FCC EIRP restriction is met, we notice that this waveform does not meet the spectral mask. Of course, the FCC specs can be met by adequately lowering the power, but at the expense of significantly reduced spectral efficiency. Various techniques have been proposed to shape the monocycle to meet the FCC mask [43] [44]. However, as we shall see, the choice of the waveform is not critical. Hence, we will use the Gaussian monocycle as an illustrative waveform.

We note from Figure 7.1 that the UWB PSD will be essentially constant over the bandwidth of a typical NB signal. As specific examples, note that the channel spacing or signal bandwidth is 30 KHz in AMPS, IS-54 and IS-136, 200 KHz in the 900 MHz GSM, 2.36 MHz in IS-95, and 5 to 20 MHz in W-CDMA and cdma2000, which are in the proposed 3G IMT-2000 standards. AMPS/IS-54/136/95 operate in the 900 MHz and 1.9 GHz bands; IMT-2000 is also expected to operate in the 1.9 GHz band. As such, these systems will neither interfere nor suffer interference from UWB systems operating in the 3.1–10.6 GHz band. Interference will come predominantly from the 5 GHz UNII bands, 802.11 WLANs that will operate at 5.3 GHz, and from the proposed 4G systems that will operate above 5GHz. In Japan, there are also reserved radio astronomy bands in the 3260-67 MHz and 3332-39 and 3345.8-3352.5 MHz.

Even over a 20 MHz range, the UWB spectrum is essentially flat; this will certainly be true with more carefully designed pulse shapes such as those proposed

---

[3]Please see Chapter 1 for details.

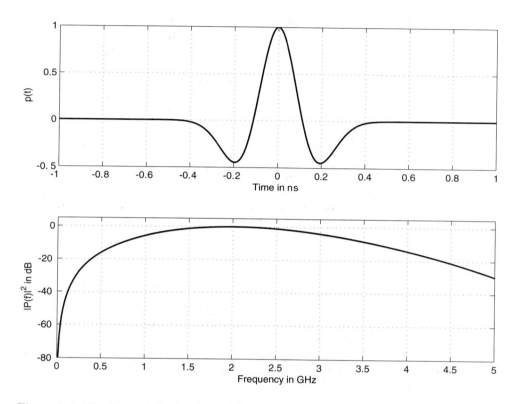

Figure 7.1: The Second-Derivative-of-Gaussian UWB Pulse, and its Spectrum, with $f_o = 2$ GHz.

Table 7.1: Bandwidth of UWB Signal using Gaussian Monocycle: Frequencies Normalized by $f_o$.

| Attenuation | Low $f$ | High $f$ | Bandwidth |
|---|---|---|---|
| 3 dB | 0.6169 | 1.4415 | 0.8246 |
| 20 dB | 0.1955 | 2.2113 | 2.0158 |
| 40 dB | 0.0607 | 2.7638 | 2.7031 |

in [43] [44]. Consequently, we expect that the effect of the UWB signal is essentially to raise the noise floor. If $P_R$ denotes the total received power from the UWB signal, then the noise floor will be raised by approximately $P_R B_{nb}/B_{uwb}$, where $B_{nb}$ and $B_{uwb}$ denote the bandwidths of the NB and UWB signals.

For a typical conventional wideband signal with a 20 MHz bandwidth, centered at $f_c = 6.85$ GHz, which is the center of the UWB band, and a UWB signal with the minimum 500 MHz bandwidth, we have $B_{nb}/B_{uwb} = 0.04$. For a more typical NB signal with a 30 KHz bandwidth, $B_{nb}/B_{uwb} = 6 \times 10^{-5} \ll 1$. As with any spread-spectrum type signal, its impact on an NB signal decreases as its bandwidth increases (assuming constant power).

As far as the signal waveform is concerned, the UWB pulse appears very impulsive if $B_{nb}$ is small; hence, we should expect to see the impulse response of the IF band-pass filter. Depending upon the precise ratios, some ringing may also be evident. Additionally, the power amplifier may be driven to saturation. We first consider the impact of a continuously pulsed UWB signal and then study the effects of an episodic pulse train (that is, a train of pulses with variable pulse spacing).

### 7.2.2  Effect of NB Receive Filter

In a typical NB radio, using digital modulation, the same symmetric baseband pulse-shape is used at both the transmitter and the receiver, namely the root-raised cosine filter (RRCF) with transfer function $H_{rrcf}(f)$ with nominal bandwidth $W$, roll-off factor or excess bandwidth parameter denoted by $\gamma$ ($0 \le \gamma \le 1$), and overall bandwidth $W(1 + \gamma)$ [48, p. 560].

Taking into account the fact that the front-end band-pass filter and the low-pass filter following the demodulator typically have larger bandwidth than $H_{rrcf}(f)$, we note the overall filter for the UWB impulse-train is given by,

$$
\begin{aligned}
G(f) &= \frac{1}{2}[P(f - f_c) + P(f + f_c)]H_{rrcf}(f) \\
&:= \bar{P}(f)H_{rrcf}(f)
\end{aligned}
\tag{7.4}
$$

where $P(f)$ is the FT of the Gaussian monocycle [c.f., eqn (7.3)], $\bar{P}(f) := [P(f - f_c) + P(f + f_c)]/2$, and $f_c$ is the carrier frequency for the NB system. If the bandwidth of $H_{rrcf}(f)$ is very small, then $P(f)$ is approximately constant over the range $[f_c - W(1 + \gamma), f_c + W(1 + \gamma)]$, so that $G(f) \approx P(f_c)H_{rrcf}(f)$.

### 7.2.3  BER Analysis

We first consider the effect of an isolated UWB pulse on the performance of an NB receiver. Later we consider the impact of a train of UWB pulses, including effects due to time-hopping.

Let $x(t)$ denote the received signal, after downconversion to baseband. Consider the usual matched filter (MF) plus threshold receiver for BPSK signaling, which is optimal for the AWGN channel. In the $k$-th symbol interval, the received signal is

$$
x(t) = \sqrt{E_b}b_k s(t) + w(t) + i(t)
$$

where $s(t)$ is the unit energy signal waveform with duration $T$ (corresponding to $H_{rrcf}(f)$), $E_b$ is the energy per bit, $b_k \in \{-1, 1\}$ is the unknown bit, $w(t)$ is AWGN with two-sided PSD $N_o/2$, and $i(t)$ is the interference. Applying the MF, we have

$$z = \int_0^T [\sqrt{E_b}b_k s(t) + w(t) + i(t)]s(t) \ dt$$

$$= \sqrt{E_b}b_k + \bar{w} + v(\epsilon) \tag{7.5}$$

Here $\bar{w}$ represents the 'noise' term and is zero-mean Gaussian, with variance $N_o/2$. The term $v(\epsilon)$ represents the interference. Assuming that the interfering pulse $i(t)$ is completely contained within the symbol period and has a relative delay of $\epsilon$, we have

$$v(\epsilon) = \int_0^T i(t)s(t)dt = \sqrt{E_p}\int_{-\infty}^{\infty} \bar{P}(-f)S(f)e^{j2\pi f\epsilon} \ df$$

where $\bar{P}(f)$ is defined in (7.4), and $E_p$ is the energy in the received UWB pulse. For an NB $S(f)$, $\bar{P}(f)$ is essentially constant over the bandwidth of $S(f)$, so that

$$v(\epsilon) \approx \sqrt{E_p}P(f_c)s(\epsilon) \tag{7.6}$$

We illustrate this via an example.

**Example 7.1.**   The NB modulation is BPSK; pulse shaping is with a RRCF with excess bandwidth parameter $\gamma = 0.5$, bit rate of 50 Kbps (symbol interval $T_s = 20\mu s$) and carrier frequency of 500 MHz. The UWB pulse is the Gaussian pulse shape $h(t) = \exp(-t^2/2\sigma^2)$, where $\sigma$ was chosen so that the 3-dB point is at 400 MHz. This waveform is somewhat different from that described in (7.2), but the detailed waveform is not important. Figure 7.2 shows the response of the RRCP matched filter to a single UWB pulse. The circles indicate the RRCP itself, indicating that the MF response to a single UWB pulse is a scaled version of the RRCP.  ∎

Define the SNR impairment factor,

$$\delta := \sqrt{\frac{E_p}{E_b}}P(f_c)s(\epsilon) \tag{7.7}$$

Conditioned on a given time offset $\epsilon$, the interference term, $v(\epsilon)$ in (7.6) acts as a fixed bias in the decision statistic of (7.5). Conditioned on the interference, the decision variable $Z$ in (7.5) is Gaussian with mean $\pm\sqrt{E_b} + v(\epsilon)$ and variance $N_o/2$; see Figure 7.3.

The MF receiver uses a threshold of zero and the bit-error rate (BER) is then given by

$$P_e(\delta) = \frac{1}{4}\mathrm{erfc}\left(\sqrt{\frac{E_b}{N_o}}(1 + \delta)\right) + \frac{1}{4}\mathrm{erfc}\left(\sqrt{\frac{E_b}{N_o}}(1 - \delta)\right) \tag{7.8}$$

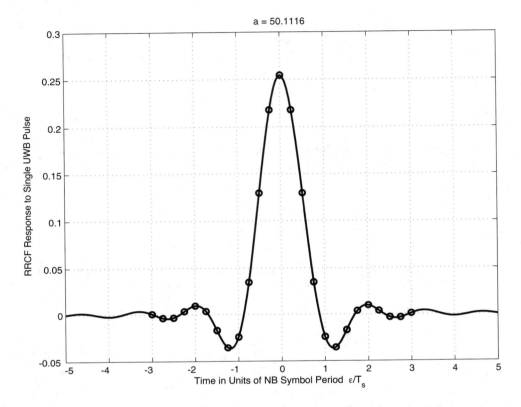

Figure 7.2: The Response of an NB System Matched Filter with UWB Pulse Input. The UWB Pulse Appears as an Impulse to the NB System, Whose Output Closely Approximates the NB System Impulse Response.

In deriving (7.8), we assumed NB BPSK modulation for simplicity, but clearly the result is readily generalized to other modulation formats such as MPSK and QAM. Figure 7.4 shows BER as a function of the impairment factor $\delta$ for nominal SNRs $(E_b/N_o)$ of 4, 6, and 8 dB. Figure 7.5 shows the extra SNR[4] required to keep the BER constant at target BERs of $10^{-2}$, $10^{-3}$, and $10^{-4}$. The performance loss is worse with higher SNR simply because the standard BER curve is steeper at higher SNR. But, for a fixed $\delta$, the required extra SNR does not change significantly with the target BER.

From (7.7) and Figure 7.4, we see that performance degrades as $\delta$ increases. For a RRCF $s(t)$, it is easy to verify that $\max_t |s(t)| = 1.61\sqrt{W}$, where $W$ is the nominal bandwidth. In Section 7.2.1, we saw that the Gaussian monocycle has an approximate bandwidth centered at $f = f_o$. To a first order approximation, we then have $P(f) \propto 1/\sqrt{(f_c)}$, and from (7.7), we obtain $\delta \propto \sqrt{W/f_o}$. For $W = 50\text{KHz}$,

---

[4]Extra SNR is defined as the difference between the SNR required to attain target BER in the presence of interference and that required in the pure AWGN case.

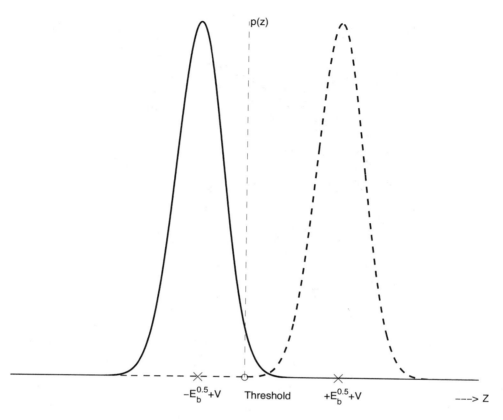

Figure 7.3: UWB Single Pulse Interference in a BPSK NB Receiver Acts to Create a Bias in the Bit Decision Variable.

$f_c = 2$ GHz, $f_o = 2$ GHz, $\delta \approx 5.8 10^{-3} \sqrt{E_p/E_b}$. To obtain $\delta = 0.1$, we need $E_p/E_b = 300$; from Figure 7.5, we see that a 3dB SNR impairment occurs when $\delta = 0.33$; this corresponds to a ratio $E_p/E_b \approx 3300$. As expected, significant degradation occurs only when the UWB pulse has large energy because only a small fraction of the UWB energy is passed by the IF filter. A more serious problem is saturation of the power amplifier (PA) due to the large peak power in the UWB pulse. Depending upon its hysteresis characteristics, the PA may need a long recovery time, during which period the received signal is distorted. This suggests the need for high-speed analog blanking circuits in order to cope with dynamic range issues with high-rate A/D convertors. Peak power issues also arise from low duty cycles, that is, when the off time between pulses is long. In this case, the effect of the interference is to wipe out occasional bits, which may be severe depending upon the strength of the coding used. In general, robust signal-processing techniques are required throughout the receiver chain.

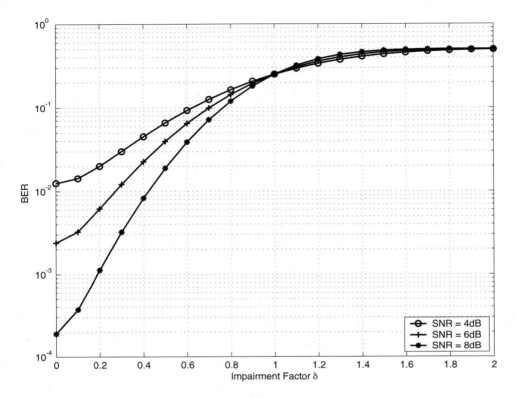

Figure 7.4: The BER Versus UWB Impairment Factor $\delta$, with Curves Parameterized by the SNR $= E_b/N_0$ (Bit Energy to Thermal Noise Ratio).

If the average UWB pulse rate is significantly less than the NB symbol rate (approximately the NB filter bandwidth), the filter response will settle down between individual pulses and the individual responses will be seen. In a typical NB receiver, RRCF filters will be used and the impulse response duration is approximately six symbol periods, so that a total of about seven symbols will be affected. This situation corresponds to impulsive or burst noise, and bit errors will occur in bursts. If the UWB pulse rate is large, that is, there are multiple pulses per bit, the responses will overlap. The effective interference then depends upon the pulse timings. In the extreme case, this could lead to a (DC) bias in the MF output. Example 7.2 illustrates this.

**Example 7.2.**    The NB system is similar to that used in Example 7.1, within a bit rate of 100 Kbps. The individual UWB pulse parameters are also the same, with a pulse-width of 2ns. The initial pulse has a random offset with respect to (wrt) the time origin. The next pulse occurs $U$ slots later, where $U$ is randomly picked from the set $[1, N_c]$ and the slot duration is the same as the pulse width, $T_p$. A total of $N_f$ random hop times were generated for each of the four curves shown in Figure 7.6.

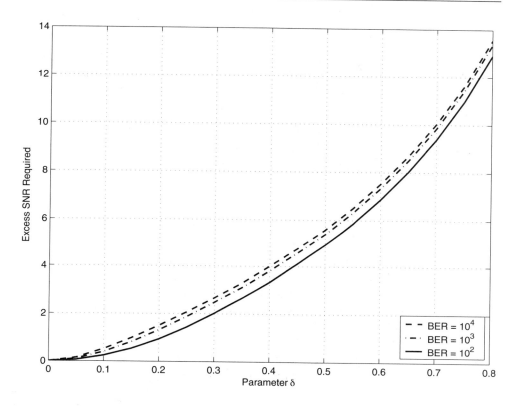

Figure 7.5: The Excess SNR $(E_b/N_0)$ Required to Maintain a Target BER as the UWB Interference Increases (that is, as the Impairment Factor $\delta$ Increases).

The duty cycle is $1/N_c$ so that the average pulse rate is $1/T_p N_c$, and the average number of pulses per NB symbol is $T_b/T_p N_c$, where $T_b$ is the average bit duration. Curves corresponding to 5, 50, and 500 pulses per NB symbol are shown in Figure 7.6. In the last case, shown by the '*' curve, the response is essentially constant. In the first case, depicted by the '+' and 'o' lines, the RRCF impulse response is recovered. The markers denote the nominal sampling points. In the intermediate case, shown by the 'x' curve, note that the sampling phase becomes critical. Except in rare cases, the sampling phase (or time offset) is a random variable, and the average BER must be computed by averaging across this phase. In this example, a separate spreading sequence was not applied to the time-hopped pulses, so that they all had the same polarity as the data (+1 here). This example is also representative of uncoded PPM. If the number of pulses is large and bipolar spreading codes are used, the interference would tend to look Gaussian.                                           ■

The BER in (7.8) is a function of the timing delay $\epsilon$; Figure 7.7 displays the average BER (averaged over $\epsilon$, or equivalently over $\delta$) as a function of the interference-to-signal ratio (ISR) defined by ISR $= \frac{E_p}{E_b} P^2(f_c)$. The curves are parameterized by

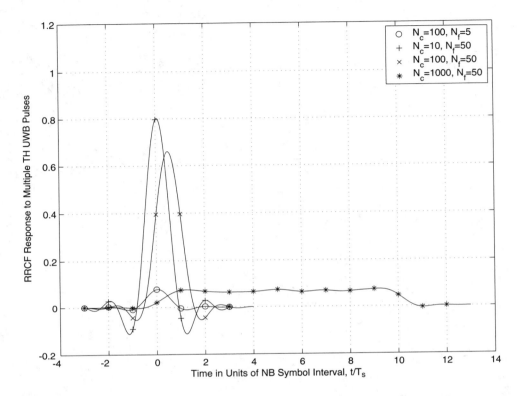

Figure 7.6: The Response of an NB System Matched Filter to a UWB Unipolar PPM Input, for Different Random UWB Pulse Rates.

nominal SNRs $(E_b/N_o)$ corresponding to target BER's of $10^{-2}$, $10^{-3}$, and $10^{-4}$. Note that as $\delta$ becomes large, the channel becomes interference limited. Because the basic receiver considered here ignores the interference, the BER goes to 0.5.

If the UWB pulse rate, $R_p$, is very small, not every NB symbol is affected by the interference. With $T_s$ denoting the NB symbol duration, the probability that interference is present is given by $p_c = T_s R_p$; the average BER is then given by

$$P_e = p_c E_\delta\{P_e(\delta)\} + (1 - p_c)\frac{1}{2}\text{erfc}\left(\sqrt{\frac{E_b}{N_o}}\right) \tag{7.9}$$

## 7.2.4   Time-Hopped Case

Many UWB radios use time-hopping to accommodate multiple users and randomize collisions in the asynchronous case. Time-hopping prevents complete collision of pulse trains when users are synchronized. Because the individual UWB pulse may have little energy, multiple pulses are transmitted for each data symbol. A DS-CDMA type spreading code may be used with these multiple pulses. A data symbol

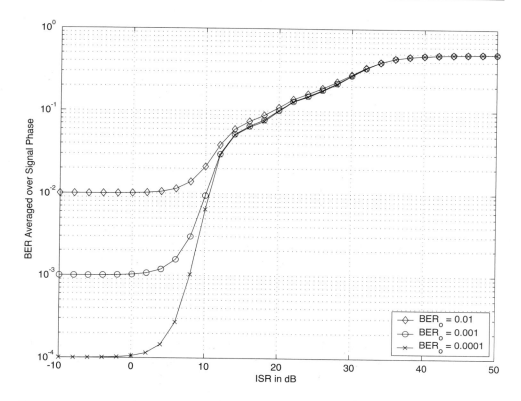

Figure 7.7: The BER of a Victim NB Receiver, as a Function of the Interference-to-Signal Ratio (ISR), and Averaged over Random Timing Offsets. The Curves are Parameterized by Nominal BER. For ISR Above 10 dB, the Receiver Becomes Interference Limited.

is conveyed by $N_f$ data modulated pulses, each with duration $T_p$. Each pulse is transmitted in a separate *frame* which has duration $T_f = N_c T_c$, where $T_c \geq T_p$ denotes the chip duration. The symbol duration is thus $T_{s,u} = N_f T_f = N_f N_c T_c$, so that the symbol rate is $R_s = 1/(N_f N_c T_c)$. In addition to facilitating multiple access, TH also provides some LPD capabilities. TH shifts the position of the UWB pulse from frame to frame. Let $c_k \in [0, N_c - 1]$ denote the chip position of the pulse in the $k$-th frame[5]. Let $a_i$, $0 \leq i < N_f$, denote a binary spreading code that may be used with or instead of the TH code, and $b_{j,u}$ denote the $j$-th binary bit in the UWB stream. Finally, let $\Delta$ denote the time shift if PPM is used, and $d_i := (b_{i,u} - 1)/2$. Then the transmitted waveform during the $j$-th UWB symbol interval is given by [45, 46]

---

[5]We have implicitly assumed symbol periodic TH code; this is not necessary, and the analysis carries over to long codes.

$$x_j(t) = b_{j,u}\sqrt{E_p} \sum_{k=0}^{N_f-1} a_k p(t - kT_f - c_k T_c - d_i \Delta) \qquad (7.10)$$

where $E_p$ denotes the energy per pulse, so that the energy per UWB bit is $N_f E_p$. The decision statistic can now be written as (c.f., eq. (7.5))

$$z = \sqrt{E_b} b_k + \bar{w} + v \qquad (7.11)$$

where the interference term is now

$$v = \sqrt{E_p} P(f_c) \sum_j \sum_{\ell=0}^{N_f-1} b_{j,u} a_\ell \; s(jT_{s,u} + \ell T_f + c_\ell T_c + d_i \Delta + \tau_d) \qquad (7.12)$$

where $\tau_d$ is propagation delay, and the sum over $(j, \ell)$ includes all UWB pulses seen within the $k$-th symbol period of the NB signal. The BER, conditioned upon the UWB parameters, is still given by (7.8), where the impairment factor $\delta$ is now

$$\delta := \sqrt{\frac{E_p}{E_b}} P(f_c) \sum_j \sum_{\ell=0}^{N_f-1} b_{ju} a_\ell s(jT_{s,u} + \ell T_f + c_\ell T_c + \Delta d_i + \tau_d) \qquad (7.13)$$

It is important to note that the BER can vary from symbol to symbol and that this BER must be averaged across the UWB symbols, $b_{j,u}$, the hop codes $c_\ell$, the spreading codes $a_i$ and the delay $\tau_d$ [26].

### 7.2.5   Simulation Results

The following examples demonstrate the effects of interference from multiple UWB emitters on an NB BPSK receiver.

**Example 7.3.**    The UWB pulse was modeled as a pure Gaussian pulse, $p(t) = \exp(-t^2/2\sigma^2)$. This waveform is somewhat different from that described in (7.2), but the detailed waveform is not important, as seen earlier. Parameter $\sigma \approx 1/f_{uwb}$ was adjusted so that the spectral energy is largely confined to the bandwidth $[-1, 1]$ GHz. The pulsewidth is about 2 ns. A sampling rate of $2\mu s$ was used to implement the NB receiver simulation.

The BPSK signal had a rate of 50 Kbps, that is, $T_b = 40$ $\mu$s, and carrier frequency $f_c = 500$ MHz. The receiver bandwidth is extremely narrowband with respect to the UWB pulse bandwidth, with an approximate fractional bandwidth of $(50 \times 10^3/10^9) \times 100 = 0.005\%$. A RRCF with $\gamma = 0.5$ (50% excess bandwidth) was used.

BER estimates, based on $10^5$ BPSK symbols, are shown in Figure 7.8. Three cases are shown: AWGN only, AWGN plus a single UWB interferer, and AWGN plus 10 UWB interferers. Each UWB interferer is assumed to have random non-overlapping pulse times, and the pulses are randomly bipolar. The UWB pulse rate is taken to be one fourth of the BPSK bit rate; thus, the average spacing between the pulses is $160\mu$s which corresponds to a very small duty cycle, $2 \times 10^{-9}/160 \times 10^{-6} = 1.25 \times 10^{-5}$. The pulse energy was set to $E_p = 10^4$, the energy per bit to $E_b = 1$,

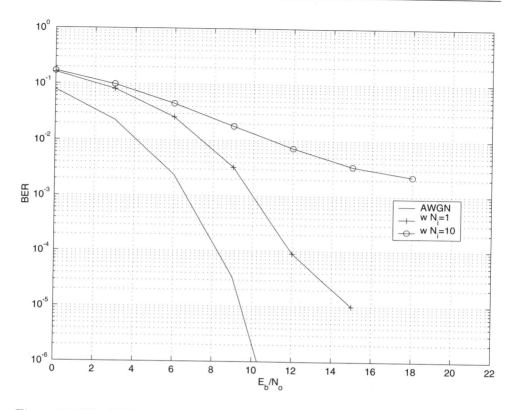

Figure 7.8: The BER Performance of a Victim NB Receiver in the Presence of One and Ten UWB Interferers. The Significant Level of Interference only Comes About Due to the Very High Power UWB Pulse Levels in This Example.

and the AWGN variance was varied to obtain the target SNR. Due to the very small amount of energy that passes the IF filter of the BPSK system, significantly large pulse energy is required to degrade performance. ∎

## 7.3 Aggregate UWB Interference Modeling

In the preceding, we derived detailed expressions for the BER of an NB receiver when subjected to interference from a single UWB emitter. In this section, we consider the aggregate effect due to multiple interferers. We derive a pdf for the asymptotic aggregate interference, which would be useful in a link budget analysis.

### 7.3.1 Received Power

Although narrowband propagation models do not readily apply to the UWB signal (see, for example, [49–51], the propagation channel is well-modeled as linear, and the narrowband propagation model can be applied because the NB receiver only sees a narrowband interfering signal.

A gross characterization of the geometric path loss is given by

$$P_R = P_o \left( \frac{d_o}{d} \right)^\beta \tag{7.14}$$

where $P_R$ denotes the received power at distance $d$ from the source, $\beta$ is the path loss exponent (typically between 2 and 6), and $P_o$ is the power measured at distance $d_o$ from the transmitter.

Consider a single UWB signal, with the Gaussian monocyle pulse, $p(t)$. As we saw earlier, the peak of the UWB signal spectrum is at $f = f_o$. Assuming $P(f)$ is approximately flat over the NB signal bandwidth, $[f_c - W f_c + W]$, with mean power $P^2(f_c)$ given by (7.3), the interference power due to the UWB pulse is easily found to be

$$P_I = P_R \frac{32}{3} \sqrt{\frac{2}{\pi}} \left( \frac{f_c}{f_o} \right)^5 \exp\left( -2\frac{f_c^2}{f_o^2} \right) \eta \tag{7.15}$$

where $P_R$ is the total received power from the UWB source, and $\eta = 2W/f_c \ll 1$ for the typical NB receiver. Figure 7.9 shows the normalized interference $\frac{P_I}{P_R}\eta$ versus the frequency ratio $f_c/f_o$. As expected, the interference is worst when the spectral peak of the UWB pulse matches the center frequency of the NB signal and decreases only slowly as the UWB peak shifts away from the center of the NB signal. Comparing the previous equation with (7.3), we note that this normalized interference is not the PSD of the UWB signal.

In order to assess the aggregate effect of a number of radios, we consider the following scenario. But it should be noted that the following analysis is not confined to UWB signals. We assume that the receiver uses an omni-directional antenna. The transmitters all have the same transmit power and are uniformly distributed, with density $\rho$ per square unit, over a concentric ring centered at the receiver with inner and outer radii $R_{\min}$ and $R_{\max}$. Figure 7.10 illustrates this. The uniform distribution implies that the number of transmitters in an area $A$ is Poisson distributed with rate parameter $\lambda = \rho A$. The inner ring represents a zone of exclusion and models the minimum separation that one might expect between the radios. The transmitted signals are assumed to be independent of each other. Additionally, it is assumed (although not strictly needed) that all transmitters use the same modulation format. The assumption of equal transmit powers is reasonable in a scheme where power control is difficult. The receiver's basic operation would be to project the received signals onto the signal space, which describes the signal of interest. For example, in a conventional $M$-ary narrow-band system, this projection would consist of a band-pass filter (BPF), followed by down-conversion to baseband, low-pass filtering to get rid of the double frequency terms, and matched filtering with a bank of filters corresponding to the $M$ waveforms.

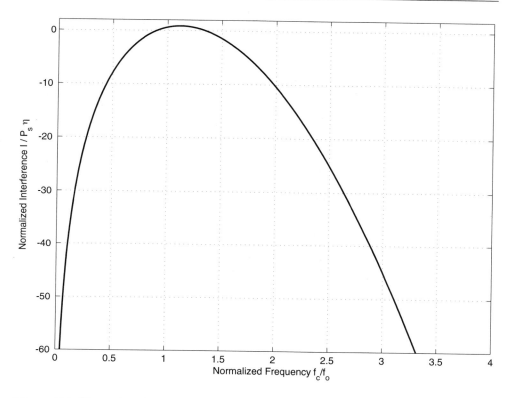

Figure 7.9: Normalized Interference Versus Normalized Frequency for a Single UWB Interferer into a Victim NB System. The Interference Is Worst When the UWB Spectral Peak Matches the NB Victim Center Frequency.

### Mean Received Power

The UWB emitters are randomly located and asynchronous. As such, their complex baseband signals (after front-end processing in the NB receiver) will not be in phase. Because these signals are independent, their powers will add at the receiver. The number of transmitters and their locations, are random; hence, the total received power is also a random variable. With $P_o$ denoting received power at reference distance $d_o$, as described previously, the expected total power received by the NB radio at the origin from all the transmitters within the annulus is given by

$$E\{P_R\} = P_o d_o^\beta \rho \int_{R_{\min}}^{R_{\max}} r^{-\beta}(2\pi r)\,dr$$

$$= P_o d_o^\beta \rho \frac{2\pi}{\beta - 2}[R_{\min}^{2-\beta} - R_{\max}^{2-\beta}]$$

$$\lim_{R_{\max} \to \infty} P_R = \left[P_o \left(\frac{d_o}{R_{\min}}\right)^\beta\right] \times \frac{2\pi R_{\min}^2 \rho}{\beta - 2} \qquad (7.16)$$

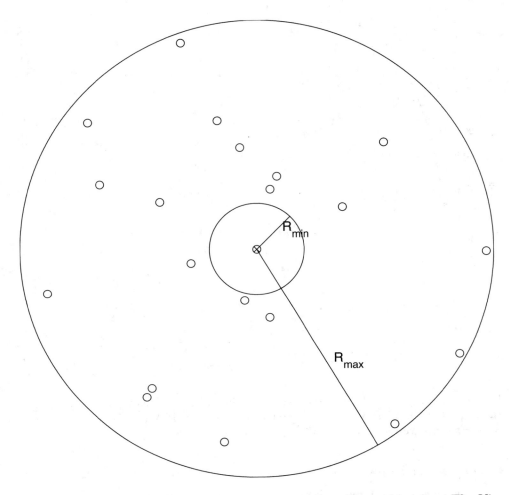

Figure 7.10: Physical Layout for UWB Aggregate Interference Modeling. The Victim Receiver Is at the Center of the Circle, with Interferers Uniformly Distributed in the Annulus with Inner Radius $R_{min}$ and Outer Radius $R_{max}$.

where we recall that $\beta > 2$. In the last equation, the first term describes the received power due to a transmitter at a distance $R_{\min}$ and the second term describes the aggregate effect due to *all* the transmitters. Note that because $\rho$ has dimensions of 1/area, the second term is dimensionless. The term $\pi R_{\min}^2 \rho$ represents the effective number of sources within a circle of radius $R_{\min}$; the denominator is related to the path loss exponent. Thus, the aggregate effect is represented by a single transmitter at the minimum distance, $R_{\min}$, whose received power at the reference point is $P'_o$, given by

$$P'_o = P_o \frac{2\pi R_{\min}^2 \rho}{\beta - 2} \qquad (7.17)$$

We next consider the moment generating function (MGF).

## MGF for Received Power

We just evaluated the expected value of the total received power, $P_R$. One way to characterize the random variable (r.v.) $P_R$ is to evaluate its moments. Following the steps leading to (7.16), we obtain for $\mu$ real,

$$
\begin{aligned}
E\{P_R^\mu\} &= [P_o d_o^\beta]^\mu \rho \int_{R_{\min}}^{R_{\max}} r^{-\beta\mu}(2\pi r)\ dr \\
&= [P_o d_o^\beta]^\mu \rho \frac{2\pi}{\beta\mu - 2}[R_{\min}^{2-\beta\mu} - R_{\max}^{2-\beta\mu}] \\
\lim_{R_{\max}\to\infty} P_R &= \left[P_o\left(\frac{d_o}{R_{\min}}\right)^\beta\right]^\mu \times \frac{2\pi R_{\min}^2 \rho}{\beta\mu - 2} \qquad (7.18)
\end{aligned}
$$

where we have assumed that $\beta\mu > 2$. When $\mu = 1$, we recover the result in (7.16). Note that the limiting moments do not exist if $\beta\mu < 2$, that is, the lower-order moments, $\mu < 2/\beta$ do not exist. This implies that the pdf is very peaked at the origin.

The variance of the received power is given by [6]

$$
\begin{aligned}
\text{Var}\{P_R\} &= \frac{2\pi\rho R_{\min}^2}{2(\beta - 1)(\beta - 2)^2} P_o^2 \left(\frac{d_o}{R_{\min}}\right)^{2\beta} \\
&\quad \times \left[\beta^2 - 4\beta(1 + \pi\rho R_{\min}^2) + 4(1 + \pi\rho R_{\min}^2)\right] \qquad (7.19)
\end{aligned}
$$

For $\beta \leq 2$ (corresponding to some reported indoor applications with LOS), some of the lower-order integer moments may not exist. For $\beta > 2$, all the integer moments exist, and we can formally write the MGF as

$$
\begin{aligned}
E\{e^{j\omega P_R/P_o}\} &= \sum_{k=0}^{\infty} \frac{j^k \omega^k}{k!} E\{P_R^k/P_o^k\} \\
&= \sum_{k=0}^{\infty} \frac{j^k \omega^k}{k!} \frac{2\pi R_{\min}^2 \rho}{k\beta - 2} \left(\frac{d_o}{R_{\min}}\right)^{k\beta} \qquad (7.20)
\end{aligned}
$$

Unfortunately this does not lead to a closed-form expression for the pdf.[7]

---

[6] An apparently convenient representation is obtained if the reference distance is also at the edge of the exclusion zone, that is, $d_o = R_{\min}$. In this case, we have

$$E\left\{\frac{P_R}{P_o}\right\}^\mu = \frac{2\pi R_{\min}^2 \rho}{\mu\beta - 2}$$

Unfortunately, this masks the effect of shrinking the exclusion zone, that is, this representation cannot be used in the limit $R_{\min} \to 0$. Hence, we will not pursue this approach.

[7] The pdf has $k$-th order moment, $1/(k\beta - 2)$.

## 7.3.2 Asymptotic Pdf of Aggregate Noise

We can conjecture that the aggregate received signal is heavy-tailed. The rationale is that the aggregate of impulsive processes must be impulsive. We establish this analytically.

In the preceding, we assumed that the transmitters are Poisson distributed over the plane. Let $\lambda$ denote the Poisson rate parameter and $n$ denote the dimension parameter. Thus $n = 2$ for sources in the plane, $n = 3$ for sources in a volume. If $\lambda$ denotes the rate of a n-D homogeneous point process (PP), with points located at distances $r_i$ from the origin, $\Gamma_i = r_i^n$ represent occurrence times on the line with rates $\pi\lambda$ (n=2), and $4\pi\lambda/3$ (n=3) [52].

**Theorem 1** *[53, Theorem 1.4.5]. Let $\{W_i\}$, and $\{\Gamma_i\}$ be independent sequences of random variables, where $\Gamma_i$ is the sequence of arrival times of a unit rate Poisson process. If the $W_i$ are positive valued with $E\{|W|^\alpha\} < \infty$, for some $\alpha \in (0,1)$, we can conclude as a special case of [53, Theorem 1.4.5] that the random variable*

$$z = \sum_{i=1}^{\infty} \Gamma_i^{-1/\alpha} W_i \qquad (7.21)$$

*converges almost surely to a positive-valued alpha-stable r.v., with index $\alpha$, and dispersion parameter $\sigma^\alpha = E|W|^\alpha \Gamma(2 - \alpha)\cos(\pi\alpha/2)/(1 - \alpha)$ where $\Gamma(\cdot)$ is the Gamma function.*

The second characteristic function of the r.v. $Z$ is ( [53, p. 7])

$$\Psi(\omega) := \ln\left(E\{\exp(j\omega Z)\}\right) = \sigma^\alpha\left[-|\omega|^\alpha + j\text{sign}(\omega)|\omega|^\alpha \tan\left(\frac{\pi\alpha}{2}\right)\right] \qquad (7.22)$$

With $r_i$ denoting the distances of the Poisson distributed interferers, the total normalized received power is

$$z = P_o d_o^\beta \sum_{i=1}^{\infty} W_i [r_i^n]^{-n/\beta} \qquad (7.23)$$

e positive-valued random variables $W_i$ represent the effects of shadowing and/or ing, assumed to be independent from transmitter to transmitter. The r.v.s $r_i^n$ Poisson distributed over $[0, \infty)$. Applying Theorem 1, we conclude that the alized aggregate received power converges to a positive stable distribution haracteristic exponent or index $\alpha = n/\beta$. Because $2 < \beta < 6$ typically, we /3 < \alpha < 1$ for sources in the plane[8]. Note that these pdfs are heavier-tailed e Cauchy.

ugh a closed-form expression (CFE) for the characteristic function of $Z$ n (7.22), a CFE for the pdf exists only when $\alpha = 1/2$, yielding the Levy n whose pdf is [53, p. 10]

$$p_Z(z) = \sqrt{\frac{\sigma}{2\pi}} z^{-3/2} \exp(-\sigma/2z)$$

---

in a volume, $n = 3$, we have $3/4 < \alpha < 3/2$. The r.v. $Z$ converges to a stable index $\alpha = 3/\beta$, and Theorem 1 has to be modified slightly.

and cdf is

$$\Pr(Z \leq z) = 2Q\left(\sqrt{\sigma}z\right)$$

where $Q(z)$ is the tail probability of the standard Gaussian random variable. Recall that if $u$ is a zero-mean unit variance Gaussian r.v., then $z$ can be generated as $\sigma/u^2$.

Although the pdf does not exist in closed form for other values of $\alpha$, a series expansion may be found in [54, Sec XVII.6],

$$p_Z(z) = \frac{1}{\pi z} \sum_{k=1}^{\infty} \frac{\Gamma(k\alpha+1)}{k!} (-1)^k z^{-k\alpha} \sin(k\pi(1-\alpha)/2) \qquad (7.24)$$

for $0 < \alpha < 1$ and $z > 0$.

From the series expansion, we note that the tails of the pdf decay algebraically as $1/z^{1+\alpha}$; as a consequence, $E\{Z^\mu\} < \infty$, only for $\mu < \alpha$. Because $\alpha < 1$, the mean is not defined, and neither is the variance.

The preceding analysis is an asymptotic one because we assumed that $R_{\min} = 0$ (that is, there is no exclusion zone) and that $R_{\max} = \infty$. In effect, the pdf becomes heavy-tailed because the interferer is allowed to come arbitrarily close to the victim receiver. If an exclusion zone $R_{\min}$ is prescribed, the analysis holds approximately, provided $R_{\min}$ is small. Simulations not shown here indicate that even with 10–20 interferers, the stable model is a very good approximation. Practically, the implication is that the aggregate signal is strongly non-Gaussian. We note that recent reports [14, 15] describe the signal due to a single UWB source at an NB receiver as being much heavier-tailed than the Gaussian. The index $\alpha$ quantifies this heavy-tailedness: As $\alpha$ decreases (that is, as decay exponent $\beta$ increases), the pdf becomes more heavy-tailed and more peaked at the origin. If the aggregate power is small enough, this can be lumped with the Gaussian thermal noise. Optimal detection in non-Gaussian noise involves non-linear pre-processing to eliminate the deleterious effects of the heavy-tailed noise. Robust algorithms for this are described in [55–57] and references therein.

It is of interest to compare the pdf of the aggregate interference with that of a single interferer. To this end, we derive the pdf of the location of the nearest interferer. We follow the approach in [41]. The probability that there is no interferer within distance $d$ of the receiver is also the probability that the nearest interferer is distance $d$ away. Because the sources are Poisson distributed with rate $\rho$, we have

$$\Pr(d_{nearest} < d) = e^{-\rho\pi(d^2 - R_{\min}^2)}$$

The received power from a source at distance $d$ is $P_R = P_o(d_o/d)^\beta$, where we recall that $P_o$ and $d_o$ are reference values. Thus $\Pr(P_R < z)$ is the probability that there are no interferers within distance $d = d_o[z/P_o]^{-1/\beta}$, yielding

$$\Pr(P_R < z) = \exp(-\rho\pi d_o^2 P_o^{2/\beta} z^{-2/\beta}) = \exp(-az^{-\alpha})$$

where we used $\alpha = 2/\beta$. For $z$ large, the pdf behaves as $1/z^{1+\alpha}$ and agrees with the alpha-stable pdf derived for the aggregate interference. In other words, if the scenario is interference limited, the pdf of the interference is adequately described by the nearest interferer that, under the equal transmitted power assumption, dominates the received power.

### 7.3.3 Amplitudes: Aggregate Pdf

We now derive the pdf for the complex amplitude due to the interferers. The received signal passes through the NB receiver's filters; it is converted to complex baseband, and correlated with a bank of $M$ matched filters. Let $\mathbf{I}_k$ denote the vector output of the MF bank due to the $k$-th interferer. Random vector $\mathbf{I}_k$ is well-modeled as circularly symmetric zero-mean complex Gaussian. The sources are assumed to be Poisson distributed. The development in [52] is applicable here. The decision variable is now a vector,

$$\mathbf{z} = \mathbf{s}_k + \mathbf{g} + \mathbf{I} \tag{7.25}$$

with $\mathbf{I}$ being spherically symmetric and alpha-stable, again with index $\alpha = 2/\beta$. Note that $\mathbf{z}$ is a M-dimensional complex vector due to the M-ary receiver. The joint second characteristic function of this $\mathbf{Z}$ is given by

$$\Psi_{\mathbf{Z}}(\eta_1, ..., \eta_M) = -\sigma^\alpha \left| \sum_{\ell=1}^{M} \eta_\ell^2 \right|^{\alpha/2} \tag{7.26}$$

### 7.3.4 Bernoulli and Poisson Models

In a TH-UWB system, a pulse occupies one of $T_h$ chip slots in each frame. The probability of slot occupancy in a given frame is $p = 1/T_h$. If we relax the requirement of one pulse per frame (see the discussion of episodic signaling in Section 7.5) and consider time in units of chip-slot widths, the sequence of pulses constitutes a Bernoulli random process, with probability of occurrence $p = 1/T_h$. In [25], we had considered this model briefly. If $p$ is small, the normalized kurtosis,

$$E|z/\sigma|^4 - 3 = (6p^2 - 6p + 1)/p(1 - p) \approx 1/p \tag{7.27}$$

indicating that the interference is strongly non-Gaussian. This non-Gaussianity can be exploited to estimate the frequency-selective channel encountered by the UWB signal. The $k$-th order normalized cumulants are roughly $p^{1-k/2}$.

A limiting case is the Poisson process model. Let $\tau_k$ denote the arrival times of the UWB pulses at a receiver. Given that the UWB signals are time-hopped, episodic, and asynchronous, it is reasonable to model the pulse times as the arrival times of a Poisson process. The effects of multipaths can be included in this model. To facilitate analysis, let us assume that the delays and amplitudes associated with each path are independent of one another, and identically distributed. Let $\lambda(t)$ denote the rate (pulses per second) of this inhomogeneous Poisson process (IPP), $\tau_k$ the arrival times, $a_k$ the amplitudes, and $p_\ell(t)$ the waveforms. The model is general enough because it does not require the different UWB sources to use identical pulse shapes, and it permits frequency-selectivity on a path-by-path basis. We can represent such a signal as

$$x(t) = \sum_k a_k p_k(t - \tau_k) \tag{7.28}$$

Such a model is called a marked point process (MPP) with multiplicative marks. The marks are the random variables, the $a_k$s. This is a special case of an MPP that we had considered in [58]. More recently, an unmarked homogeneous point process (HPP) model ($a_k \equiv 1$) was considered in [60].

In an NB receiver, the signal $x(t)$ would be filtered. If the bandwidth of the NB system is small relative to that of $x(t)$, it is reasonable to model the FT's of the individual pulses $p_k(t)$ to be flat over the NB system's band-pass region, $[f_c - W, f_c + W]$. Thus, the effective signal can be represented as

$$x_f(t) = \sum_k a_k h_{nb}(t - \tau_k) \tag{7.29}$$

where $h_{nb}(t)$ denotes the NB receive filter. The cumulants of this process are given by [58, eq. (9)] [59]

$$\text{cum}(x_f(t_1), \cdots, x_f(t_k)) = E\{a^k\} \int \lambda(\tau) \prod_{i=1}^{k} h(t_i - \tau) \, d\tau \tag{7.30}$$

For a HPP, $\lambda(t) \equiv \lambda_o$, the zero-lag cumulants are

$$\text{cum}_k(x_f(t), \cdots, x_f(t)) = \lambda_o E\{a^k\} \int h^k(\tau) \, d\tau \tag{7.31}$$

If $h(t)$ is band-pass, $H(0) = 0$, and $E\{x_f(t)\} = 0$. The variance is given by

$$\text{var}(x_f(t)) = \lambda_o E\{a^2\} \int h^2(t) \, dt = \lambda_o E\{a^2\} \tag{7.32}$$

for a receive filter with unit energy. The kurtosis is given by

$$\text{cum}_r(x_f(t)) = E\{x_f^4(t)\} - 3E^2\{x_f^2(t)\} = \lambda_o E\{a^4\} \int h^4(t) \, dt \tag{7.33}$$

This in turn yields the power at the output of an envelope detector

$$E\{x_f^4(t)\} = \lambda_o E\{a^4\} \int h^4(t) \, dt + 3\lambda_o^2 E^2\{a^2\}$$
$$= \lambda_o E\{a^4\} 2B + 3\lambda_o^2 E^2\{a^2\} \tag{7.34}$$

which follows readily from $h(t) = \sqrt{4B}\text{sinc}(2Bt)\cos(2\pi f_c t)$ for an ideal bandpass filter that has support over $|f_c - f| < B$. For an unmarked PP, according to [60], these results are validated by experimental results reported in [12]. The process

$x_f(t)$ is asymptotically Gaussian, and results on rates of convergence may be found in [61] [62]. When the pulse rate is small, the UWB interference appears as impulse noise and degrades the BER of the NB system. When the arrival rate is large, the Gaussian approximation holds and all NB bits are impacted about the same. Effectively, the noise floor is raised. If the modulation is not zero-mean (such as OOK or PPM without time-hopping), additional terms will appear on the right-hand side of (7.34) corresponding to lines in the spectrum of the interferer [41].

### 7.3.5   Simulation Examples

**Example 7.4.**     Here we consider the aggregate effect of Poisson distributed sources. The UWB source signal strength is adjusted to ensure a BER of $BER_u$ at the UWB receiver at a distance $d_u$ away. Here, $d_u$ represents the maximum required range of the UWB radios. BPSK modulation is used. $N_f$ pulses are transmitted per UWB bit; one pulse per frame. The pulse width is $T_p$, the frame length is $T_f = N_c T_p$. The pulses may be time-hopped by using symbol periodic or long-code hopping sequences. $P_R$ denotes the received power per pulse, so that the received energy per UWB bit is $E_b = P_R T_p N_f$. Assume that $P_R$ is such that the target BER is achieved in AWGN. The AWGN noise level is assumed to be identical at both the UWB and NB victim receiver (VR), this is a reasonable assumption (difference in noise power can be accommodated, but this facilitates the analysis). Let $N_o/2$ denote the two-sided noise power-spectral density. With $P_o$ denoting received power at reference distance $d_o$ from the UWB source, the received power at the UWB receiver is $P_{R,u} = P_o(d_o/d_u)^\beta$, where $\beta \geq 2$ is the path loss exponent. Let $d_n$ denote the distance between the UWB source and the NB receiver. The total UWB power at the NB receiver is $P_{R,n} = P_o(d_o/d_n)^\beta$. Because the UWB pulse shape does not play a critical role in the interference analysis, assume that it is rectangular; hence $E_b = T_p N_f A^2$, where $A$ is the received pulse amplitude, that is, $P_u = A^2$. Hence,[9] $E_b/N_o = \mathrm{erfcinv}^2(2BER_u)$ and $P_u = \frac{N_o}{N_f T_p}\mathrm{erfcinv}^2(2BER_u)$. So this is the received UWB power at distance $d_u$. Hence, at distance $d_n$, the received power is $P_{u,n} = P_u(d_u/d_n)^\beta$. The power due to the interference after the NB receive filtering is approximately given by $P_i = P_{u,n}B_u/B_n = P_u(d_u/d_n)^\beta B_u/B_n$, where $B_u$ and $B_n$ denote the bandwidths of the UWB and NB signals. Let the NB receiver be at the origin with UWB interferers Poisson distributed a minimum distance $R_{\min}$ beyond the victim receiver (VR). The squared distances from VR, $d_{u,k}^2$, are now Poisson arrival times with rate $\pi\rho$. The received powers are now $P_u(d_u/d_{u,k})^\beta$, before the NB receive filter.

We consider a scenario with UWB parameters $d_u = 20$m, $BER_u = 0.001$, $T_p = 2$ns, that is, $B_u = 1$ GHz, and $N_f = 100$, $N_c = 100$ (1% duty-cycle), and various values of $\rho = 10^{-k}$, $k = 2:6$. $\rho = 10^{-6}$ corresponds to a UWB density of 1 radio per square km. The NB-BPSK signal BW was fixed at 50 KHz, and the path loss parameter was $\beta = 3.5$. We consider two values of the exclusion radius, $d_{\min} = 20$, and $d_{\min} = 50$, with $d_{\max} = 1000$m. The aggregate interference can be well-modeled as Gaussian, taking into consideration the large number of interferers and

---

[9]In the following, x = erfcinv(y) denotes the unique inverse solution to y = erfc(x) .

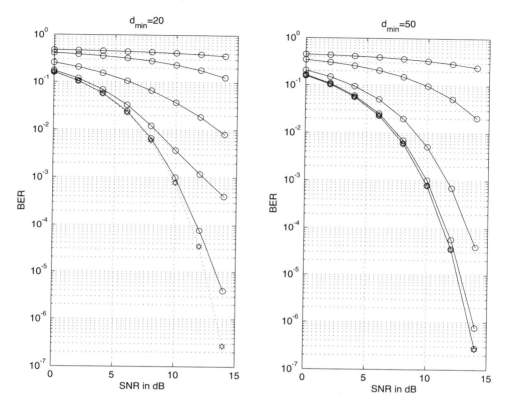

Figure 7.11: Effect of Aggregate UWB Interference on an NB Receiver, with Curves Parameterized by the Density of the UWB Interferers.

the effects of the NB front end filter. Figure 7.11 shows the BER curve for the victim receiver, for $d_{min} = 20$m (left panel), and $d_{min} = 50$m (right panel). The curves are parameterized by $\rho$. Also shown is the theoretical AWGN BER curve. As expected, the degradation is small if UWB density is low, the UWB duty cycle is low, the UWB radio range is short, or the exclusion radius is large.

Figure 7.11 shows BER versus SNR; the curves correspond to different values of the density of interferers. ∎

## 7.4   Interference Analysis: NB on UWB

In the frequency range above 3.1 GHz, the FCC mask restricts the radiated power to −41.25 dBm/MHz, with a −10 dB bandwidth of 7.5 GHz. In the 2.4 GHz ISM and 5 GHz NII bands, allowed emissions are 40+ dB higher per MHz. With a minimum required bandwidth of 500 MHz, the UWB receiver will see multiple NB waveforms. The range below 1 GHz, of course, is densely occupied by licensed NB systems (AM,

FM radios, TV stations, and so on). The wide-bandwidth of UWB systems permits large processing gains. However, the individual pulses have very little energy; as such, NBI signals, which may be expected to be 10–12 dB above the noise floor, may cause severe degradation in performance.

The impact of NBI on single-user systems is studied in [28], [29], [35], [37]. A DS-CDMA UWB system is shown to resist NBI from IEEE 802.11a OFDM signals [36], where the NBI was modeled as Gaussian noise, as was the MAI. Use of a maximal ratio combiner (MRC) can suppress MAI but not NBI [34]. BER expressions were derived in [38] for a multi-user TH system operating over a pure AWGN channel (that is, no NBI), which could provide a benchmark to study the performance of NBI suppression schemes.

Consider the time-hopped UWB waveform described in (7.10). Assume PAM modulation so that $\Delta = 0$. Then, the transmitted waveform during the $j$-th UWB symbol is

$$x_j(t) := x(t + jT_s)$$

$$= b_{j,u}\sqrt{E_p} \sum_{k=0}^{N_f-1} a_k p(t - kT_f - c_k T_c), \quad 0 \le t < T_s, \quad j = 0, 1...p \quad (7.35)$$

The received signal during the $j$-th UWB symbol interval is

$$y_j(t) = x_j(t) + w_j(t) + i_j(t) \quad (7.36)$$

where $w_j(t)$ denotes AWGN with two-sided psd $N_o/2$, and $i_j(t)$ is interference.

In AWGN, the optimal receiver is a MF, matched to the symbol waveform,

$$p_s(t) := \sum_{k=0}^{N_f-1} a_k p(t - kT_f - c_k T_c) \quad (7.37)$$

Equivalently, it does the following: it time gates the received waveform, followed by matched filtering with the pulse shape $p(t)$, and then despreading with the spreading code $\{a_k\}_{k=0}^{N_f-1}$; that is, the decision statistic is

$$Z_j = \sum_{k=0}^{N_f-1} a_k \int_0^{T_p} p(t) x_j(t + kN_c T_p) \, dt \quad (7.38)$$

$$= S_j + W_j + I_j \quad (7.39)$$

where $S_j$, $W_j$ and $I_j$ denote signal, noise, and interference components.

Because the pulse $p(t)$ has unit energy, and $\{a_k\}$ is a $\pm 1$ sequence, we readily obtain

$$S_j = b_{j,u}\sqrt{E_p} N_f \; .$$

The noise term $W_j$ is zero-mean Gaussian with variance $N_f N_o/2$.

The interference term $i_j(t)$ represents the aggregate effect of modulated or unmodulated narrow-band signals. Let $f_\ell$, $R_\ell$, and $P_\ell$ denote the carrier frequency, rate, and power of the $\ell$-th NBI. The received signal component due to the $\ell$-th interferer can be written as

$$i_{j,\ell}(t) = \sqrt{2P_\ell}\cos(2\pi f_\ell t + \theta_\ell) \sum_{n=-\infty}^{\infty} \bar{b}_{n,\ell} g_\ell(t - nT_\ell - \tau_\ell) \qquad (7.40)$$

where

- $\bar{b}_{n,\ell}$ is the $n$-th transmitted symbol of the $\ell$-th interferer; the symbol set is normalized so that $E\{|\bar{b}_{n,\ell}|^2\} = 1$

- $g_\ell(t)$ is the normalized NBI transmit pulse shape normalized with $\frac{1}{T_\ell}\int_0^{T_\ell} g^2(t)\,dt = 1$

- $\theta_\ell$ and $\tau_\ell$ denote the random carrier phase and delay. Phase $\theta_\ell \sim U[0, 2\pi)$ and $\tau_\ell \sim U[0, T_\ell)$.

Because the UWB signal is time-hopped, there is a random time-offset between successive pulses of the same bit. Hence, the starting phase of the NBI carrier changes randomly from the time associated with the beginning of one pulse to that of the next. Given a pulse duration $T_p$, and an NB symbol duration $T_\ell = 1/R_\ell$, the probability of a symbol transition within an UWB pulse is given by $T_p/T_\ell$. For a UWB pulse of duration 2 ns and a NB signal with a rate of 50 KHz, this probability is $2 \times 10^{-9} \times 50 \times 10^3 = 10^{-4}$. Even with a 20 MHz 'NBI', the probability is small, $2 \times 10^{-9} \times 20 \times 10^6 = 0.04$. To a first order, we can ignore the symbol transition. Note that the effect of a symbol transition is to make the integration less coherent. This introduces multiplicative noise, which broadens the spectrum; that is, the effective bandwidth of the interferer increases. We can then write the contribution of the $\ell$-th NBI to the decision statistic $z_j$ as

$$I_{j,\ell} = \sqrt{2P_\ell} \sum_{k=0}^{N_f-1} a_k \bar{b}_{j,k,\ell} \int_0^{T_p} p(t) g_\ell(t + \epsilon_{j,k,\ell}) \cos(2\pi f_\ell t + \phi_{j,k,\ell})\,dt \qquad (7.41)$$

where $\bar{b}_{j,k,\ell}$ denotes the symbol modulating the $\ell$th NBI during the $k$-th pulse of the $j$-th UWB bit, with $\epsilon_{j,k,\ell}$ denoting the corresponding delay. The phase term $\phi_{j,k,\ell}$ incorporates both the unknown NBI carrier phase as well as the random time shift between pulses due to time hopping. Given that $T_p \ll T_\ell$, $g_\ell(t + \epsilon_{j,k,l}) \approx g_{j,k,\ell}$, a constant over the duration of the pulse, but this amplitude varies randomly over $\{j, k, \ell\}$. For a rectangular pulse shape, $g_{j,k,\ell} = 1$. Recall that $T_p$ denotes the pulse duration[10] and $P(f)$ its FT; hence, we have

$$I_{j,\ell} = \sqrt{\frac{P_\ell}{2}} \sum_{k=0}^{N_f-1} a_k \bar{b}_{j,k,\ell}[P(f_\ell)e^{-j\phi_{j,k,\ell}} + P^*(f_\ell)e^{j\phi_{j,k,\ell}}] \qquad (7.42)$$

---

[10]For simplicity, we do not consider partial response signaling.

For a symmetric pulse shape, $P(f) = P(-f)$ is real-valued and

$$I_{j,\ell} = \sqrt{\frac{P_\ell}{2}} \sum_{k=0}^{N_f-1} a_k \bar{b}_{j,k,\ell} P(f_\ell) \cos(\pi f_c T_p + \psi_{j,k,\ell}) \qquad (7.43)$$

The phase variables, the $\phi$s, are random variables and are well modeled as being independent (across $j, k, \ell$) and uniform over $[0, 2\pi]$. If $\theta \sim U[0, 2\pi]$, we have for $m \neq 0$, $E\{e^{jm\theta}\} = 0$, and $\text{Var}\{e^{jm\theta}\} = 1 \ \forall m$. Assuming that the NBI bit $\bar{b}$ and phase $\phi$ are mutually independent, we have $E\{I_{j,\ell}\} = 0$ where the expectation is taken with respect to the $\phi$s and $b$s. Next, we compute the variance:

$$\begin{aligned} \text{var}\{I_{j,\ell}\} &= \frac{P_\ell}{2} \sum_{k=0}^{N_f-1} \sum_{m=0}^{N_f-1} a_k a_m E\{\bar{b}_{j,k,\ell} \bar{b}_{j,m,\ell}\} E\left\{[P(f_\ell)e^{-j\phi_{j,k,\ell}} + P^*(f_\ell)e^{j\phi_{j,k,\ell}}]\right. \\ &\quad \left. \times [P(f_\ell)e^{-j\phi_{j,m,\ell}} + P^*(f_\ell)e^{j\phi_{j,m,\ell}}]\right\} \\ &= \frac{P_\ell}{2} \sum_{k=0}^{N_f-1} a_k^2 E\{\bar{b}_{j,k,\ell}^2\} 2|P(f_\ell)|^2 \\ &= N_f P_\ell |P(f_\ell)|^2 \end{aligned}$$

If $N_f$ is large, we can approximate $I_{j,\ell}$ as a zero-mean Gaussian r.v. with variance $N_f P_\ell |P(f_\ell)|^2$. Because the interferers are modeled as independent, their variances will add.

Consequently, we can write the decision statistic as

$$z_j = b_{j,u} \sqrt{E_p} N_f + \eta_j \qquad (7.44)$$

where $\eta_j$ is zero-mean Gaussian with variance $N_f N_o/2 + N_f \sum_\ell P_\ell |P(f_\ell)|^2$. The BER is then given by

$$\begin{aligned} BER_{nbi} &= Q\left(\sqrt{\frac{E_p N_f^2}{N_f N_o/2 + N_f \sum_\ell P_\ell |P(f_\ell)|^2}}\right) \\ &= Q\left(\sqrt{\frac{N_f SNR_p}{1 + \sum_\ell SNR_\ell |P(f_\ell)|^2}}\right) \end{aligned} \qquad (7.45)$$

where $SNR_p := E_p/(N_o/2)$ is the per-pulse SNR, and $SNR_\ell := P_\ell/(N_o/2)$ is the interference-to-noise ratio.

We can derive (7.45) in a more heuristic, but insightful, way as follows. Assume that the UWB pulse has a flat FT so that $P(f) := 1/\sqrt{4B}$, $f_u - B < |f| < f_u + B$, and that the PSD of the NBI is flat over $f_c - W < |f| < f_c + W$, where $f_u$ and $f_c$ are the center frequencies, and $W, B$ are the bandwidths of the UWB and NBI signals. Assume that the support of the NBI PSD is contained within that of the UWB PSD. Then a simple calculation yields the interference power as $2 \times \frac{P_\ell}{4W} \times 2W \times \frac{1}{4B} = \frac{P_\ell}{4B} = P_\ell |P(f_\ell)|^2$. With the interference modeled as Gaussian, this leads to (7.45).

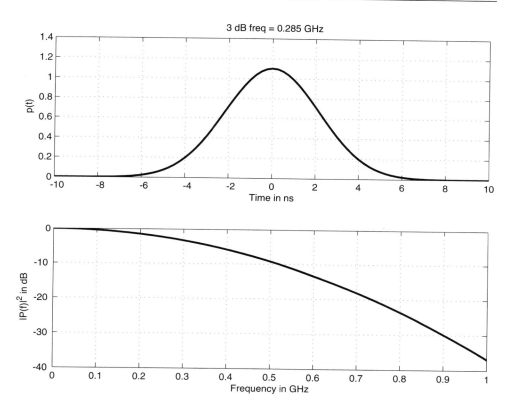

Figure 7.12: The Gaussian UWB Pulse, and its Power Spectrum.

As a specific example, consider the Gaussian pulse of Example 7.1, $p(t) = \exp(-t^2/2\sigma_u^2)$. The FT of this pulse peaks at $f = 0$; the FCC specifications require that the PSD be down by 34 dB at $f = 0.96$ GHz. To ensure this, we need $\sigma_u \geq (1/2)10^{-9}$. The time domain signal and its FT are shown in Figure 7.12 for $\sigma = (1/2)10^{-9}$. The pulse width is about 10–12 ns, and the 3 dB and 10 dB bandwidths are 0.28 GHz and 0.52 GHz.

Figure 7.13 shows signal spectra using the Gaussian pulse and the Raised Cosine Pulse (RCF) for different values of symbol period $T$ and excess bandwidth parameter $\beta$. These choices of the RCF meet the FCC specs at 0.96 GHz, while utilizing the spectrum more efficiently (ability to transmit more energy). Note also that the Gaussian pulse extends over 12–16 ns, the RCF pulse extends over typically 6–8 symbol periods, so that the RCF pulses have shorter duration. Finally, if the channel has no delay spread, the RCF pulse is desirable because it satisfies the Nyquist criterion.

**Example 7.5.**    We simulate a pulsed UWB system, with AWGN and narrow-band interference (NBI). The UWB signal is the baseband Gaussian pulse $p(t) = A \exp(-t^2/2\sigma^2)$, where $A$ is an amplitude factor. Parameter $\sigma$ was set to $2 \times 10^{-9}$ so

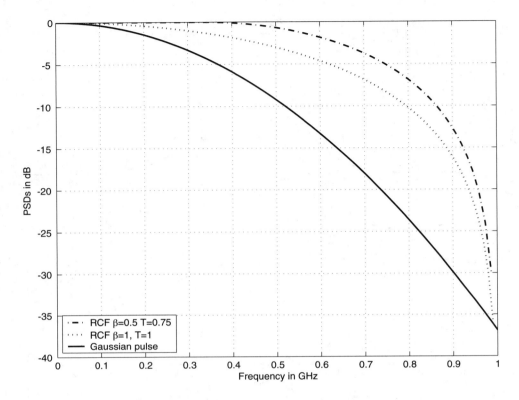

Figure 7.13: A Comparison of the Power Spectra of Gaussian and RCF Pulses.

that the pulse bandwidth is approximately 250 MHz; see Figure 7.12. The UWB processing gain, or $N_f$ the number of pulses per bit, is varied in the set $\{1, 32, 128, 512\}$. TH-UWB was assumed so that the NBI carrier phase changes randomly from pulse to pulse. The NBI interferer has a carrier frequency $f_c = 100$ MHz and is modulated by a BPSK sequence at rate 50 Kbps and has a constant interference-to-noise ratio of 10 dB. Figure 7.14 shows BER versus SNR with curves parameterized by $N_f$. The solid lines are the theoretical AWGN curves (without NBI) and the symbols denote simulation results. A degradation of about 1 dB is seen in the curves, indicating good resistance to NBI.                                                                    ■

**Example 7.6.**    We simulate a pulsed UWB system, with AWGN and narrowband interference (NBI).

The UWB signal is the baseband Gaussian pulse $p(t) = A \exp(-t^2/2\sigma^2)$, where $A$ is an amplitude factor. Parameter $\sigma$ was set to $\frac{1}{2}10^{-9}$ so that the pulse bandwidth is approximately 1 GHz; see Figure 7.12. The UWB processing gain, or $N_f$ the number of pulses per bit, is varied: $4^k$, $k = 0 : 1 : 4$. In TH-UWB, the pulse is duty-cycled $T_f \geq T_p$ and the $N_f$ pulses are time hopped, so that (7.10) models the transmitted waveform. The phase of the NBI carrier changes randomly from pulse

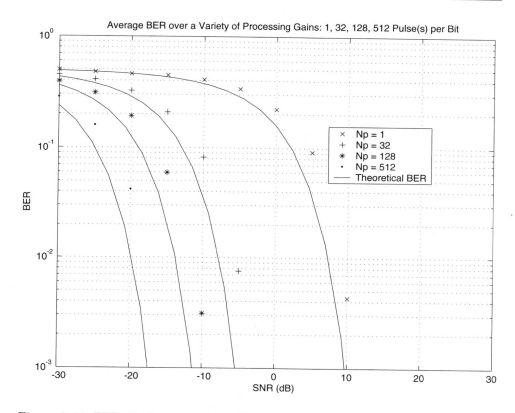

Figure 7.14: BER Performance of a Time-Hopped (TH) UWB System with $N_f$ Pulses per Bit, in the Presence of Modulated NB Interference (NBI).

to pulse. In DS-UWB, there is no off-time between pulses, that is, $T_f = T_c = T_p$, but an off-time is assumed between bits. The phase of the NBI carrier changes randomly from bit to bit, but not from pulse to pulse within a bit. The random phase-change from bit to bit is a reasonable model when the NBI and UWB rates are not harmonically coupled, even if there is no off-time between bits. Binary PAM was used so that $b_j = \pm 1$ and $\Delta = 0$. The MF receiver is time synchronized and perfect time gating is assumed so that the TH code and spreading code details are not relevant in the absence of NBI. Thus, without NBI, the system performance is equivalent to BPSK-DS, with processing gain $N_f$. In this example, note that SNR is defined per pulse and is not the SNR per bit, which is $N_f$ times larger. A sampling frequency of 2 GHz was used to digitally implement the MF receiver. This example consists of three parts described here.

(A) The NBI consisted of two pure tone interferers with frequencies 400 MHz and 600 MHz. In Figure 7.15, we plot BER versus SNR for both the TH-UWB (solid curves) and DS-UWB (dashed lines). The interference-to-noise

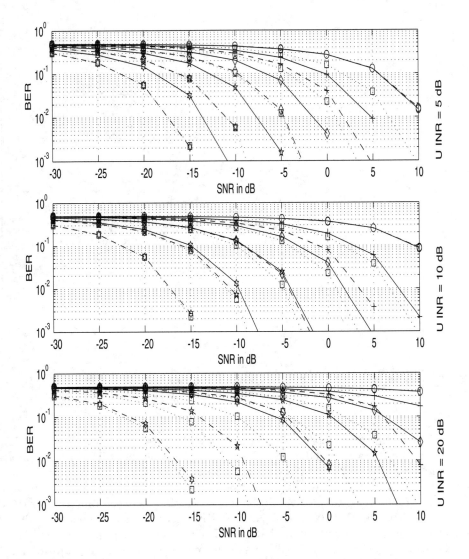

Figure 7.15: BER Performance of a Pulsed UWB System with AWGN and NB Interference Versus Per-Pulse SNR. The Panels Correspond to Different INR Ratios, with Curves Parameterized by Processing Gain $N_f$. Solid Curves Are for TH-UWB, and Dashed for DS-UWB. Example 7.6A Corresponds to Unmodulated NBI (that is, the NBI = Sinusoids).

ratio (INR) was varied: 5 dB, 10 dB, and 20 dB, and corresponds to the left, middle, and right panels of the figure. The curves are parameterized by processing gain, $N_f$, the number of pulses per bit, which increases from right to left, $2^k$, $k = 0, ..., 4$, corresponding to markers circles, plus sign, diamond, pentagram and hexagram. The square markers on the dotted lines denote the theoretical AWGN results in the absence of NBI. The UWB pulse duration was 21 ns, and we can verify that in the DS-UWB case, the NBI interferers did not have an integer number of cycles per bit duration; the fractional cycles were either 0.4 or 0.6 for the parameters used in this example. For DS-UWB, the impact of the NBI is minimal when the INR is small or the processing gain (PG) is large, as may be expected. In contrast, the performance of TH-UWB is not as good, but it does provide some NBI resistance (within 3 dB of AWGN bound in the presence of two 5-dB interferers). With TH-UWB, the phase of the NBI changes randomly from pulse to pulse. Equivalently, the unmodulated carrier is now modulated by a random phase sequence. This, in turn, spreads the effective bandwidth of the interferer and its impact on the UWB receiver. A conclusion from this example is that DS-UWB offers better resistance to unmodulated NBI than does TH-UWB.

(B)  The NBI was now modulated by a BPSK sequence with a bit rate of 50 Kbps. Because the NBI symbol duration of $20\mu s$ is much larger than that of the DS-symbol period with maximum PG, $21 \times 10^{-9} \times 256 = 5.4\mu s$, it was assumed that the NBI bit remained constant across each pulse in TH-UWB and across each bit in DS-UWB (we relax this assumption in part (C)). Notice that the performance of the two systems is now virtually identical. Both offer some resistance to modulated NBI, which increases with PG; see Figure 7.16. Note also that the performance of TH-UWB is about the same in both the modulated and unmodulated cases because the random phase change essentially incorporates phase changes due to the modulation.

(C)  The setting is now the same as in part (B), except that now the NBI symbol may change with probability $T_p/T_{nbi}$ over a TH-UWB pulse, or with probability $T_p N_f/T_{nbi}$ over a DS-UWB bit. Results are shown in Figure 7.17 and are similiar to those in Figure 7.16.

■

## 7.5   Interference Analysis: UWB on UWB

Under appropriate conditions, properly designed time-hopping and DS-CDMA type spreading codes can mitigate interference due to multiple users; however, in the presence of asynchrony, performance may degrade significantly. If the duty cycle is small or equivalently the load is light, that is, the number of active users $N_u \ll N_f$, random time hopping results in nearly orthogonal (or collision-free) low-rate multiple users.

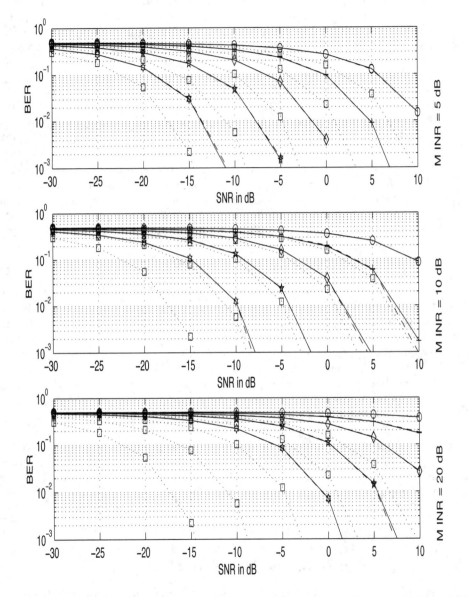

Figure 7.16: BER Performance of a Pulsed UWB System with AWGN and NB Interference Versus Per-Pulse SNR, as in Figure 7.15. For Example 7.6B, the NBI Is Now BPSK, whose Bit Boundaries Do Not Occur During a UWB Symbol.

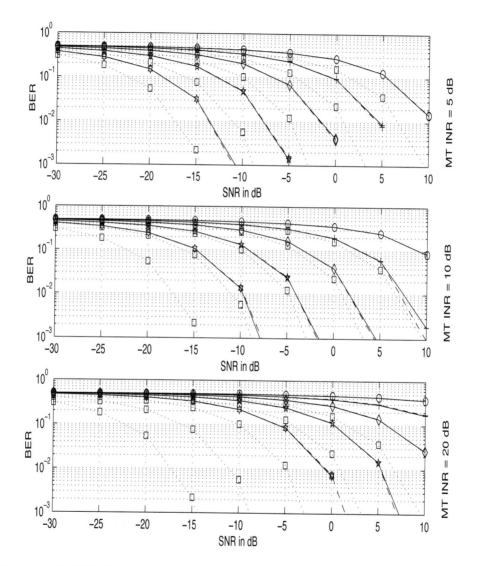

Figure 7.17: BER Performance of a Pulsed UWB System with AWGN and NB Interference Versus Per-Pulse SNR, as in Figures 7.15 and 7.16. For Example 7.6C, the NBI is BPSK, and Bit Boundaries May Randomly Occur During a UWB Symbol.

A transmitted symbol corresponding to $N_f$ pulses per bit occupies a total of $N_f N_c$ chip slots. In the traditional TH scenario, exactly one chip slot per frame (consisting of $N_c$ chip slots) is non-zero. Generalizing this, we let $a_i$, $i = 0, ..., N_f N_c - 1$, denote the length $N_f N_c$ spreading code. Rather than insisting that $N_f$ chips be non-zero, we will require that the expected number of non-zero chips be $\alpha N_f N_c$ per symbol. Thus, we model $a_i$ as i.i.d. random variables, taking on values $-1, 0, 1$ with probability mass function (PMF)

$$p_A(a) = (1 - 2\alpha)\delta(a) + \alpha \cdot \delta(a - 1) + \alpha \cdot \delta(a + 1) \tag{7.46}$$

where $0 < \alpha \le \frac{1}{2}$. The PMF is depicted in Figure 7.18. In a given chip slot, a chip (or pulse) occurs with probability $2\alpha$. If the chip occurs, it takes on values $\pm 1$ with

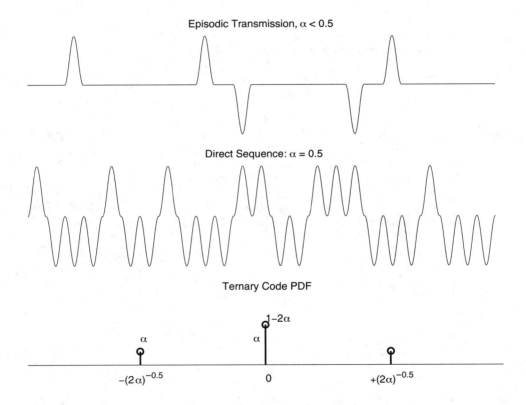

Figure 7.18: A Ternary Random Process Model (Pdf Shown at Bottom) May Be Used to Model Episodic Bipolar Pulsed Transmission. Cases with $\alpha < 0.5$ (Top) Correspond to Episodic Transmission (Off-Times Between Pulses), Whereas $\alpha = 0.5$ (Middle) Corresponds to a Conventional Binary Random Process Model (No Off-Times Between Pulses).

equal probability. The case $\alpha = 1/2$ corresponds to conventional non-episodic DS spreading. The duty cycle is $1/2\alpha$. If the processing gain $N_f$ is fixed (as is the usual case), the average symbol duration is $N_f/2\alpha$ chip slots or $N_f N_c T_c/2\alpha$ s.

This ternary PMF is very useful in capturing the effect of multiple access interference in a lightly loaded system. Specifically, if the user of interest is $k$, the receiver synchronizes itself to the $k$-th user's (random) hopping sequence, which it is assumed to know. Within any of the $N_f$ active chips in a symbol, the receiver will see interference from one or more users. With $N_u$ denoting the number of active users, the probability that any one given user causes a chip collision is $1/2\alpha$. The per-chip collision probability is then $(N_u - 1)/2\alpha$. The duty factor $1/2\alpha$ effectively reduces the interference seen by the receiver.

Let $k = 0$ denote the user of interest, with $N_u$ being the total number of users. A Chernoff bound on the BER was derived in [39], and is given by

$$P_e < \frac{1}{2} \exp\left( \frac{-N_f E_0}{2I_0} \right) \tag{7.47}$$

where $E_0$ is the desired user's energy per pulse, and $I_o$ denotes the effective interference,

$$I_0 = \sigma_n^2 + 2\alpha \sum_{k=1}^{N_u - 1} E_k \tag{7.48}$$

Note that the effective MUI is reduced by the duty cycle factor, $2\alpha$. In [39], this approach was used to obtain bounds on BER when training is used to obtain estimates of a random multipath UWB fading channel, which is then used in a RAKE receiver and in the presence of MUI.

**Example 7.7.**   [39] We simulated a 10-user  system, with processing gain $N_f = 128$. The desired user's SNR, $\text{SNR}_1 = 10 \log_{10}(E_1/\sigma_v^2)$ was varied, while the remaining 9 interfering users have SNRs fixed at 0 dB. We implemented a matched filter receiver with threshold at zero. The interfering bits were generated according to the ternary PMF to simulate the effects of random time hopping codes. Figure 7.19 shows simulation curves and the Chernoff bound for two values of $\alpha$. The bound becomes increasingly tight for meaningful BER's ($\leq 10^{-2}$), typically within 1 to 2 dB. For $\alpha = 0.003$, the performance is essentially that of the MUI-free case. In AWGN, the single user theoretical BER is given by $\frac{1}{2}\text{erfc}(\sqrt{N_f E_1/2\sigma_v^2})$. This bound is also shown in Figure 7.19, virtually coinciding with the $\alpha = 0.003$ simulation. ■

Here, the time-hopping pattern was considered to be random and the modulation A-PAM. With the hop pattern modeled as deterministic, BER expressions have been derived for PPM in [45] assuming that the MUI is Gaussian, and in [38] without this assumption.

## 7.6   Summary

UWB systems are inherently overlay systems, so that coexistence is a fundamental issue. Coexistence requires the interference question be addressed from both the

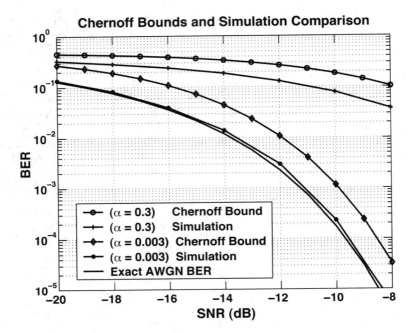

Figure 7.19: Chernoff Upper Bound on the BER of a UWB System Employing a Single-User RAKE Receiver, in the Presence of Multi-User Interference, and Compared with Simulations. All Users Have Processing Gain $n = 128$ Pulses/Bit. Interfering Users Have SNR = 0 dB, while the Desired User Per-Pulse SNR is Varied over $[-20, -8]$ dB. The Bounds are Increasingly Tight as the SNR Increases.

narrowband and UWB point of view. Although the FCC-stipulated emission limit from any single UWB radio is rather small, the aggregate effect from multiple radios can be large, both on victim narrowband systems, as well as on wideband systems, including other UWB systems. We have derived the conditional BER of a victim narrowband radio, given interference from multiple UWB sources in addition to AWGN. This quantifies the impact in terms of an impairment factor that depends upon the the signal-to-interference bandwidth ratio as well as the traditional signal-to-interference power ratio (SINR). In order to analyze the impact of aggregate interference, we have derived the interference asymptotic pdf. We also developed expressions for the BER of narrowband signals on a UWB radio, and found Chernoff bounds on the BER of a UWB system, in the presence of interference from other UWB radios. Our analysis explicitly takes into account the effect of duty-cycling the radios. Tradeoffs in the robustness to interference of time-hopped and DS-type radios were considered.

# Bibliography

[1] IEEE 802.15 WPAN High Rate Alternative PHY Task Group 3a (TG3a), http://www.ieee802.org/15/pub/TG3a.html.

[2] IEEE 802.15 WPAN Low Rate Alternative PHY Task Group 4a (TG4a), http://ieee802.org/15/pub/TG4a.html.

[3] FCC, First Report and Order, FCC-02-48, "Revision of Part 15 of the Commission's rules regarding ultra-wideband transmission systems," Federal Communications Commission, adopted February 14, 2002.

[4] http://www.uwb.org/regulatory/regulatory.html for European regulatory information.

[5] M. Luo, D. Akos, S. Pullen and P. Enge, *Potential interference to GPS from UWB transmitters. Test Plan–Ver. 4.5*, Stanford University Report, 1 May 2000; http://www.ostpxweb.dot.gov/gps/docs/tp_4_5_final.doc.

[6] M. Luo, M. Koenig, D. Akos, S. Pullen, P. Enge, *Potential interference to GPS from UWB transmitters, phase II test results*, Stanford University Report, (version 3.0) 16 March 2001.

[7] RTCA, *Ultra-wideband technology radio frequency interference effects to GPS and interference scenario development*, First interim report to Dept of Transportation, RTCA Paper No. 289-00/PMC-108, September 2000. Available at http://www.rtca.org/comm/sc159.asp.

[8] D.S. Anderson, *et al*, *Assessment of compatibility between ultrawideband devices and GPS Receivers*, NTIA Special Publication 01-45, U.S. Dept. of Commerce, Feb 2001. Available at http://www.ntia.doc.gov/osmhome/reports/uwb.

[9] R.J. Hoffman *et al*, "Measurements to determine potential interference to GPS receivers from UWB transmission systems," NTIA Report No 01-384, U.S. Dept. of Commerce, Feb. 2001.

[10] NTIA, *GPS testing report*, available at http://www.ntia.doc.gov/osmhome/uwbtestplan/gpstestfr.htm.

[11] FCC, *Measured emissions data for use in evaluating the ultra-wideband (UWB) emissions limits in the frequency bands used by the global positioning system (GPS)*, Project TRB 02-02 Report, Office of Engineering and Technology, FCC, October 22, 2002.

[12] JHU/APL, *Final Report to FCC: UWB-GPS Compatibility Analysis Project*, Johns Hopkins University/Applied Physics Lab, 8 March 2001.

[13] M. Cardoza et al., "Final report on data collection campaign for measuring UWB/GPS compatibility effects," Center for UWB Research and Engineering, University of Texas, Austin, TL-SG-01-01, 2 Feb. 2001.

[14] L.K. Brunson et al., *Assessment of compatibility between ultrawideband devices and selected federal systems*, NTIA Publication 01-43, U.S. Dept. of Commerce, Jan 2001. Available at http://www.ntia.doc.gov/osmhome/reports/uwb.

[15] W.A. Kissick, Ed., *The temporal and spectral characteristics of ultrawideband signals*, NTIA Report 01-383, U.S. Dept. of Commerce, January 2001. Available at www.its.bldrdoc.gov, and http://www.ntia.doc.gov/osmhome/uwbreports/.

[16] R.D. Wilson, R.D. Weaver, M.-H. Chung, and R.A. Scholtz, "Ultra wideband interference effects on an amateur radio receiver," *Proc. IEEE UWBST'02*, 315-319, 2002.

[17] Jay J. Ely, Gearald L. Fuller, and Timothy W. Shaver, "Ultrawideband EM interference to aircraft radios," *Proc. 21st IEEE Digital Avionics Systems Conf.*, 2002.

[18] International Ultra Wideband conference, Washington DC, Sept. 1999.

[19] Milcom: Session 38, Ultra-Wideband Communications, *Proc. IEEE Milcom*, Atlantic City, Nov. 1999.

[20] ONR Ultra wideband Workshop, Berkeley, May 2000. ftp://ftp.onr.navy.mil/PUB/freebej/UWB.

[21] Special Issue on Ultra-Wideband Systems, *IEEE Journal on Sel. Areas in Commun.*, Dec. 2002.

[22] *Proc. IEEE Conference on Ultra Wideband Systems and Technologies*, (UWBST'03), Baltimore, MD, USA, May 2002.

[23] *Proc. IEEE Conference on Ultra Wideband Systems and Technologies*, (UWBST'03), Reston, VA, USA, Nov. 2003.

[24] *Proc. 2004 Int'l Workshop on Ultra Wideband Systems*, Kyoto, Japan, May 2004.

[25] A. Swami, B. Sadler, and J. Turner, "On the coexistence of ultra-wideband and narrowband radio systems," *Proc. IEEE-Milcom'01*, McLean, VA, Oct. 2001.

[26] Jeffrey R. Foerster, "Interference modeling of pulse-based UWB waveforms on narrowband systems," *Proc. IEEE VTC*, 1931-1935, 2002.

[27] D.K. Borah, R. Jana, A. Stamoulis, "Performance evaluation of IEEE 802.11a wireless LANs in the presence of ultra-wideband interference," *Proc. IEEE WCNC*, 83-87, March 2003.

[28] J.D. Choi and W.E. Stark, "Performance analysis of ultra-wideband spread-spectrum communications in narrowband interference," *Proc. IEEE MILCOM*, 1075-80, 2002.

[29] Li Zhao, Alexander M. Haimovich and Haim Grebel, "Performance of UWB communications in the presence of interference," *IEEE Journal on Selected Areas in Communications*, 20(9), 1684-1691, Dec. 2002. (See also, Proc. IEEE ICC, 2948-52, 2001).

[30] Maria Stella Iacobucci, Maria-Gabriella di Benedetto, and Luca De Nardis, "Radio frequency interference issues in impulse radio: multiple access communication systems," *Proc. IEEE UWBST'02*, 293-97, 2002.

[31] A. Taha and K.M. Chugg, "A theoretical study on the effects of interference on UWB multiple access impulse radio," *Proc. Asilomar Conf.*, 728-32, 2002.

[32] W. Tao, W. Yong, and C. Kangsheng, "Analyzing the interference power of narrowband jamming signal on UWB systems," *Proc. IEEE PIMRC*, 612–15, 2003.

[33] Liuqing Yang and Georgios B. Giannakis, "Unification of ultra-wideband multiple access schemes and comparison in the presence of interference," *Proc. IEEE Asilomar*, 2003.

[34] Jeffrey R. Foerster, "The performance of a DS spread UWB system in the presence of multipath, NBI, and MUI," *Proc. IEEE UWBST'02*, 87–92, 2002.

[35] I. Bergel, E. Fishler, and H. Messer, "Narrow-band interference suppression in time-hopping impulse-radio systems," *Proc. IEEE UWBST'02*, 303-06, 2002.

[36] Qinghua Li and Leslie A. Rusch, "Multisuer detection for DS-CDMA UWB in the home environment," *IEEE Journal on Sel. Areas in Commun.*, 20(9), 1701–11, Dec. 2002.

[37] N. Boubaker and K.B. Letaief, "A low complexity MMSE-RAKE receiver in a realistic UWB channel and in the presence of NBI," *Proc. IEEE WCNC*, 233-7, March 2003.

[38] A.R. Forouzan, M. Nasiri-Kenari and J.A. Salehi, "Performance analysis of time-hopping spread-spectrum multiple access systems: uncoded and coded schemes," *IEEE Trans. Wireless Commun.*, 1(4), 671-681, Oct. 2002.

[39] B.M. Sadler and A. Swami, "On the Performance of Episodic UWB and Direct-Sequence Communication Systems," to appear in *IEEE Trans. Wireless Comms*, 2004.

[40] Multiple Access Communications Ltd., "An investigation into the impact of ultra-wideband transmission systems," RA0699/TDOC/99/02, Report to the RadioCommunications Agency, UK, Feb. 2000.

[41] Jay E. Padgett, John C. Koshy, Anthony A. Triolo, "Physical-layer modeling of UWB interference effects," Telcordia Report to DARPA NETEX Program, Jan. 2003. http://www.darpa.mil/ato/solicit/netex/documents.htm.

[42] Matti Hämäläinen, Veikko Hovinen, Raffaello Tesi, Jari H. Iinatti, and Matti Latva-aho, "On the UWB system coexistence with GSM900, UMTS/WMCDA and GPS," *IEEE Journal on Sel. Areas in Commun.*, 20(9), 1712-21, Dec. 2002.

[43] B. Parr, B. Cho, K. Wallace and Z. Ding, "A novel ultra-wideband pulse design algorithm," *IEEE Commun. Letters*, 219-221, 2003.

[44] X. Luo, L. Yang, and G.B. Giannakis, "Designing optimal pulse-shapers for ultra-wideband radios," *Proc. IEEE Conf. UWB Systems & Technologies*, 349-353, 2003.

[45] R.A. Scholtz, "Multiple access with time-hopping impulse modulation," *Proc. Milcom*, Oct. 1993.

[46] M. Win and R.A. Scholtz, "Ultra-wide bandwidth time-hopping spread-spectrum impulse radio for wireless multiple-access communications," *IEEE Trans. Commun.*, 48(4), 679-689, April 2000.

[47] M.Z. Win, "Power spectral density of binary digital pulse streams in the presence of independent uniform timing jitter," *Proc. Milcom*, 1997.

[48] J.G. Proakis, *Digital Communications*, McGraw Hill, 4th Edition, 2001.

[49] S.M. Yano, "Investigating the ultra-wideband indoor wireless channel," *Proc. IEEE VTC*, Spring 2002, 1200-1204, May 2002.

[50] S.S. Ghassemzadeh, R. Jana, C.W. Rice, W. Turin and V. Tarokh, "A statistical path loss model for in-home UWB channels," *Proc. IEEE UWBST'02*, 59-64, May 2002.

[51] D. Cassioli, M.Z. Win, and A.F. Molisch, "The ultra-wide bandwidth indoor channel: from statistical model to simulations," *IEEE Journal on Sel. Areas in Commun.*, 1247-1257, Dec. 2002.

[52] J. Ilow and D. Hatzinakos, "Analytic alpha-stable noise modeling in a Poisson field of interferers or scatterers," *IEEE Trans. Sig. Proc.*, 46(6), 1601-11, June 1998.

[53] G. Samorodnitsky and M.S. Taqqu, *Stable non-Gaussian random processes: Stochastic models with infinite variance*, Chapman and Hall, 1994.

[54] W. Feller, *An introduction to probability theory and its applications*, vol 2, Wiley, 2nd Ed, 1971.

[55] A. Swami and B. Sadler, "Parameter estimation for linear alpha-stable processes," *IEEE Sig. Proc. Lett.*, 5(2), 48-50, Feb. 1998.

[56] A. Swami, "Non-Gaussian Mixture Models for Detection and Estimation in Heavy-Tailed Noise," *Proc. ICASSP'00*, vol 6, 3802–05, June 2000, Istanbul, Turkey.

[57] A. Swami and B. Sadler, "On some detection and estimation problems in heavy-tailed noise," *Signal Processing* (Eurasip), 82(2002), 1829-1846, Dec. 2002.

[58] A. Swami and B. Sadler, "Channel and intensity estimation for a class of point processes," *Proc. VIII IEEE-SP Workshop on SSAP*, 440–443, Corfu, Greece, June 1996.

[59] D.L. Snyder and M.I. Miller, *Random point processes in time and space*, Springer-Verlag, 1991.

[60] R. J. Fontana, "An insight into UWB interference from a shot noise perspective," *Proc. IEEE UWBST'02*, 309-12, 2002.

[61] A. Papoulis, "Narrow-band systems and Gaussianity," *IEEE Trans. Info. Theory*, 18(1), 20-27, Jan. 1972.

[62] R.A. Altes, "Wide-band systems and Gaussianity," *IEEE Trans. Info. Theory*, 679–82, 21(6), Nov. 1975.

[63] J.D. Choi and W.E. Stark, "Impact of bandwidth upon the performance of ultra-wideband communication systems," *Proc. IEEE MILCOM*, 2003.

[64] J.M. Cramer, R.A. Scholtz, and M.A. Win, "On the analysis of UWB communication channels," *Proc. Milcom'99*, 1191-95, 1999.

[65] J. D. Taylor, E. C. Kisenwether, "Ultra-wideband radar receivers," in *Introduction to Ultra-Wideband Radar Systems* J. D. Taylor, ed, CRC Press, 1995.

# Chapter 8

# SIMULATION

William H. Tranter,
James O. Neel,
and Christopher R. Anderson

Simulation, which is the subject of this chapter, is a both a design and an analysis tool. Through the use of repeated simulations, design alternatives can be explored, parametric studies can easily be conducted, and the impact of varying critical parameters can be quickly determined. Due to the growing complexity of communications systems and the power of simulation tools and computer platforms, simulation plays an increasingly important role in understanding the performance characteristics of complex communications systems.

Simulation is a difficult subject for a variety of reasons. First, simulation development is part art and part science; second, simulation requires a detailed knowledge of a large number of fields. To see where art enters into the process, consider the steps to developing a simulation:

1. Map the system configuration under study to a simulation model. As many solutions exist and there are innumerable dependencies to consider, it is impossible to formulate definitive guidelines for choosing the most efficient model. This makes this step more art than science.

2. Partition the system to be simulated into small segments or subsystems. Again there are an innumerable number of ways in which this partitioning can be done, and the partitioning is highly dependent on the system under study and on the platform on which the simulation is being implemented. This step of partitioning is again more art than science.

3. Choose the set of modeling, simulation, and estimation tools most appropriate to the various subsystems that comprise the system under study. Fortunately, this step is more science than art, as the set of tools is limited, and well-defined criteria exists for determining the appropriateness of each tool.

4. Finally, combine the results of the subsystem simulations in a way that allows a solution to the system-level problem to be obtained. Again, an innumerable number of combinations exist, and the determination of the best combination is an art.

It can be seen that three of the four preceding steps are more art than science. Performing these steps in the best manner requires mastering the art. Mastering the art of simulation cannot be accomplished by reading a paper or a textbook, but can only be accomplished through experience. However, the literature on the subject of simulation as applied to the design and analysis of communication systems is vast, and the reader is referred to two recently published books for a basic understanding of the subject [1, 2].

It should also be mentioned that one basic decision made in simulation development is whether to use a general-purpose language and build the simulation "from scratch" or to use a simulation package specifically targeted to the type of system being considered. A number of software packages targeted to the simulation of communication systems exists, and many of these have model libraries that can prove useful for UWB systems. Especially useful are packages that are "block diagram" orientated. These range from general-purpose packages, such as SIMULINK, to packages specifically targeted to communications systems, such as SPW. The choice of simulation language is again a function of many different considerations, such as language familiarity, the uniqueness of the system being simulated, and simulation runtime requirements, so this decision can also be deemed more art than science.

The second reason simulation is a difficult subject is the fact that simulation requires a broad and detailed knowledge of numerous areas. Certainly the device to be simulated must be well understood and appropriate models must be available. Additionally, the assumptions used in model development must be known, and the impact of these assumptions on the accuracy of the resulting simulation must be evaluated. Further, simulation development requires knowledge of computer languages and software management principles as well as a thorough grounding in digital signal processing.

As the material included here focuses on UWB systems, it is assumed that the reader has a basic understanding of the subject. Models for propagation, transmitters, receivers, and signals are all addressed elsewhere in this book and are not covered in this chapter except through examples. Rather, this chapter focuses on the simulation techniques and approaches that are well-suited for simulating UWB systems at the level presented in [1].

Specifically, this chapter covers the following subjects:

1. Particular challenges introduced by the simulation of UWB communication systems.

2. An overview of fundamental simulation methodologies and terminology.

3. Architectural approaches to simulating UWB communication systems.

4. Simulation models for UWB communication systems components.

# 8.1   What's Different about UWB System Simulations?

For one familiar with Monte Carlo simulation as applied to communication systems, an important first step is to address why it is important to revisit this subject when targeting UWB systems. In other words, what's different about UWB systems that requires us to take a second look at simulation methodology? In the following, we will see that direct/quadrature signal decomposition, which is a fundamental technique used to shorten the required simulation runtime for narrowband systems, no longer provides the dramatic savings in runtime when applied to UWB signals. In addition, data collection for channel model development is a nontrivial task because the wide bandwidth required for UWB channel sounding significantly complicates the process. Also, component modeling and simulation development introduce challenges unique to the UWB signal environment.

## 8.1.1   Direct/Quadrature Signal Decomposition

The simulation methodology for a UWB communications system is quite a bit different from that used in a traditional narrowband system. When simulating narrowband systems, the carrier is usually translated to zero frequency using direct/quadrature decomposition of the modulated carrier so that the simulation model can be based on signals having relatively small bandwidth. In other words, if a bandpass signal having center frequency $f_c$ and bandwidth $B$ is represented as

$$x\left(t\right) = x_d\left(t\right)\cos 2\pi f_c t - x_q\left(t\right)\sin 2\pi f_c t \qquad (8.1)$$

where the direct channel signal, $x_d\left(t\right)$, and the quadrature channel signal, $x_q\left(t\right)$, have bandwidth $B/2$. This allows the simulation sampling frequency to be significantly reduced compared to what would otherwise be required. This of course reduces the required execution time for the simulation [1].

UWB systems, however, are typically simulated as baseband systems, and direct/quadrature decomposition is not used. Consider Figure 8.1, which illustrates the spectrum of a bandpass signal having a carrier frequency $f_0$ Hz and a bandwidth of $B$ Hz. The highest frequency contained in the bandpass signal is

$$f_h = f_0 + \frac{B}{2} \qquad (8.2)$$

Treating this signal as a baseband signal, as done with UWB signals, gives a minimum sampling frequency of

$$f_{s,BB} = 2f_h = 2f_0 + B \qquad (8.3)$$

On the other hand, if the signal is narrowband and direct/quadrature decomposition is used, two lowpass signals are created, each of which has bandwidth $B/2$. Thus the minimum sampling frequency for both the direct and the quadrature channel signals is $B$ Hz. Thus, the minimum sampling frequency for the narrowband signal is

$$f_{s,NB} = 2B \qquad (8.4)$$

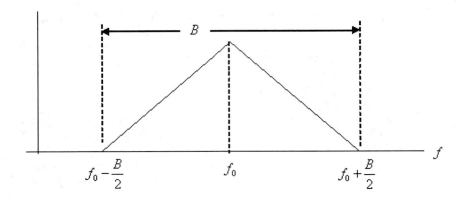

Figure 8.1: Spectrum of Wideband Signal.

because the minimum sampling frequency for both the direct and quadrature channel signals is $B$. We have interest in the ratio $R$ defined as

$$R = \frac{f_{s,BB}}{f_{s,NB}} = \frac{f_0}{B} + \frac{1}{2} \tag{8.5}$$

Large values of $R$ imply that a significant savings in simulation runtime results by using direct/quadrature decomposition, while small values of $R$ imply that a relatively small savings in simulation runtime results.

   In order to illustrate the role of the ratio $R$, we compute $R$ for two signals. Assume that the first signal is a narrowband signal having a carrier frequency of 1 GHz (1,000 MHz) and a bandwidth of 1 MHz. For this case

$$f_{s,NB} = 2B = 2MHz \tag{8.6}$$

and the value of $R$ is

$$R = \frac{1000}{1} + \frac{1}{2} = 1000.5 \tag{8.7}$$

indicating that considerable savings results from using narrowband (direct/quadrature) signal processing techniques.

   Now assume that the second signal is a UWB signal that extends from 30 MHz to 3 GHz. The center frequency (effective carrier frequency) is $(3,000 - 30)/2 =$ 1,485 MHz and the bandwidth is $(3,000 - 30) = 2,970$ MHz. For this case the value of $R$ is

$$R = \frac{1485}{2970} + \frac{1}{2} = 1 \tag{8.8}$$

indicating that there is no advantage to using direct/quadrature signal decomposition. This is clearly an extreme case but it makes the point that the reduction in sampling frequency resulting from direct/quadrature signal representations becomes less significant with increasing signal bandwidth.

Accordingly, direct/quadrature decomposition is not typically used in the simulation of UWB systems, and very high simulation sampling frequencies are often necessary. As a result, long simulation runtimes are common. It is therefore very important that the simulation be developed so that only the essential features are included in the simulation model and that the simulation code is as efficient as possible.

## 8.1.2 Model Development for UWB Systems

The development of models for UWB communication systems is made more difficult for two reasons: established models are not widely available, and the challenges associated with developing new models are often significant. Although UWB communications systems were first introduced in 1973 [3], the wireless community's interest in UWB is relatively recent. Accordingly, UWB has not had the benefit of time to develop and refine an extensive library of tested and validated models.

For narrowband systems, a large body of well-understood models has been developed. Thus, model development is often easily and quickly accomplished. In narrowband simulations, model development often simply involves identification of the model that best fits expected conditions or experimental data for the scenario under study. For instance, consider the simulation of a channel for a UMTS signal, a signal for which an astounding number of large-scale and small-scale channel models exist and have been formally standardized by the ITU. In addition to traditional channel models such as LOS, Rayleigh, and Ricean, the ITU defines simulation models for the following situations [4]:

1. Indoor office environments with parameters based on the number of floors and number and type of walls.

2. Outdoor to indoor and pedestrian environments with considerations for urban canyons, trees, and building penetration.

3. Vehicular environments with parameters for operating in urban, suburban, and rural environments.

However, when simulating a UWB system, we are not as fortunate. While a handful of channel models have been developed, most notably the Saleh-Valenzuela small-scale model [5], the Cassioli large scale model [6], and the 802.15.3a model [7], the actual number of appropriate models is limited. Similar model deficits are encountered when examining other aspects of a UWB communications system, such as amplification, signal recovery, and data conversion. Thus, when simulating UWB systems we frequently must develop new models from experimental data rather than simply extracting a previously developed and validated model from an existing model library.

For example, the multipath magnitudes for a UWB channel do not generally follow a Rayleigh or Ricean distribution. It has been observed from experimental measurements that multipath magnitudes tend to obey a Nakagami-$m$ distribution, which requires additional parameterization (the value of $m$) to model a channel. It should be noted that the Nakagami-$m$ distribution reduces to a Rayleigh distribution for $m = 1$ and can closely approximate a Ricean distribution for higher values of $m$. A more detailed treatment of the Nakgami, Rayleigh, and Ricean distributions was provided in Chapter 3.

Another difficulty encountered in UWB systems deals with phase response. UWB signals have, by definition, very large bandwidth because transmission takes place using pulses with very short duration. In order for these pulses to pass through the channel, as well as other signal-processing system components, without distortion, the amplitude response of the component in question must be constant and the phase response must be linear over the bandwidth of interest. If these conditions are not satisfied, pulse distortion occurs. Removal of this pulse distortion requires equalization. As the bandwidth increases, the constraint of constant amplitude response and linear phase response becomes increasingly difficult to satisfy. Equalization also becomes more difficult with increased bandwidth.

The condition of linear phase response across the bandwidth of interest is often most troublesome. Linear phase response implies a constant *group delay* because group delay is the negative derivative of phase response. (Recall that *phase delay* is the delay imposed at a single frequency.) Maintaining a constant group delay across a large bandwidth is very difficult.

Another critical narrowband modeling concept that fails under UWB conditions is the assumption of steady-state operation. Consider an I-UWB system with a small duty-cycle. Under this condition, it is relatively safe to assume the transmitter components return to a relaxed state between pulses. Thus, the transmitter is constantly operating in a transient mode, and the assumption of a steady-state would not be valid.

The assumption of steady-state operation vastly simplifies model development. Under steady-state conditions, the analog behavior of a circuit can be readily described with transfer function models and simple impedance-based equations. However, under transient conditions, models have to rely on the solution of differential equations, a significantly more complicated process.

### 8.1.3   UWB Simulation Development Challenges

Problems also arise during the simulation development step. Fundamental to UWB communications systems is the ultra-wide bandwidth of the signals. As previously discussed, a UWB signal has a bandwidth greater than 500 MHz with fractional bandwidth of at least 20%. If we assume a signal bandwidth of 2 GHz and apply an over-sampling factor of four to ten as would be used by traditional simulation techniques, a simulation sampling rate of 8-20 GHz is required. A detailed low-level simulation that models all aspects of the system would then be expected to run for days, if not weeks. Thus, any simulation solution must consider ways to reduce simulation run-time by simplifying the simulation model or using semi-analytic techniques, which will be discussed in the following section.

## 8.2   Developing a Simulation

Simulation is the software implementation of an experimental testbed used for gaining insight into system design alternatives and system performance characteristics. The first, and most important, step in developing a simulation program is to carefully define the required simulation products. As an example, a simulation may be developed to predict system performance over a range of receiver input signal-to-noise ratios. In this case, the simulation product would likely be a plot of the symbol error rate as a function of the signal-to-noise ratio at the receiver input. Another example might be a simulation developed to estimate the required time for a synchronizer to establish lock. The first simulation would have to be developed so it is computationally efficient because many symbol errors may have to be counted in order to estimate the symbol error rate accurately. The time required to execute the second simulation would likely be much shorter, so computational efficiency would be less of a concern. In either case, however, the simulation results will only be as good as the model upon which the simulation is based.

After the simulation products are established, the next step is to develop the simulation code. As depicted in Figure 8.2, this is a three-step process. The first step is to carefully define those attributes of the physical device being modeled that affect the required simulation products. Developing models involves making extensive measurements of the physical device being studied. The form of the model to be developed often depends upon the purpose of the simulation. For example, if the required simulation product is the symbol error rate, any system attribute affecting the symbol error rate must be included in the model. In general, the symbol error rate is considered a steady-state system attribute; therefore, transient effects

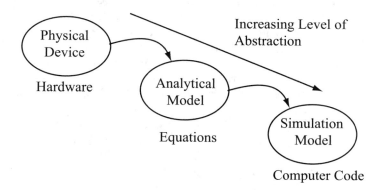

Figure 8.2: Steps in Simulation Development.

such as initial synchronization behavior are not considered. After the attributes of the physical device to be included in the model are defined, the system is then mapped to an analytical model. This analytical model takes the form of a system of equations defining the operations performed by the physical device. The last step is to determine the simulation model, which takes the form of a block of computer code (such as MATLAB or C++). Note that at each step in the process, the level of abstraction increases.

For example, suppose we wish to simulate the performance of a particular Forward Error Correction (FEC) code in the presence of Rayleigh fading. If desired, we could model the individual logic gates (or transistors) used to implement the FEC code, the parasitic inductance of the component leads used to connect the gates to a circuit board, the lossy transmission lines on the board, noise and crosstalk generated from other systems operating on the board, nonlinearities and harmonics in the mixers and amplifiers, and so forth. The result would be a very complete and accurate model. However, the simulation would take such a vast amount of time and computational power to execute that it would be essentially useless. By assuming ideal performance of logic gates, we can develop simulation code that is simple, efficient, and accurate.

It bears repeating that *the accuracy of the simulation is constrained by the accuracy of the underlying model*. Great care must be taken when deciding which parameters to include in the model and how much to trust the simulation results. A disastrous example of inaccurate modeling occurred in the design of a UWB transmitter based on a Step Recovery Diode (SRD) pulse generator (see Chapter 6, "Receiver Design Principles"). To generate a UWB pulse, a commercially available SRD was selected, modeled based on information contained in the datasheet, and simulated in a pulse generating circuit. The simulation results are shown in Figure 8.3, and a considerable amount of time and resources was expended designing the UWB transmitter based on the assumption that this SRD would produce a pulse with a peak output of 3.0 Volts. When the transmitter was constructed, the actual output pulse (shown in Figure 8.3b) was only 0.4 Volts—considerably weaker than what the simulation predicted—and resulted in an inoperative transmitter.

Another potential source of errors comes from the implementation of the simulation itself. In Chapter 1, "Introduction," a number of analytical models were presented for pulse shapes, modulation, and multiple access schemes. In order to be simulated, these continuous analytical expressions must be sampled and quantized into finite-precision sampled values. Here, the user is obviously constrained by available memory and computational power. A sampling frequency or numerical precision that is too high results in a simulation that takes an inordinate amount of time to execute. On the other hand, sampling at too low a rate or with insufficient numerical precision results in a distortion of the signal and leads to simulation errors.

The last step in the process of developing a simulation involves *validating*, or *sanity checking*, the results. There are two levels of validation. The individual component models used in the simulation must be validated, and the overall simulation must also be validated. Model validation is typically carried out by comparing the

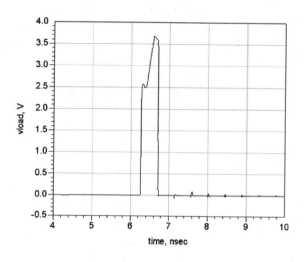

(a) Simulated

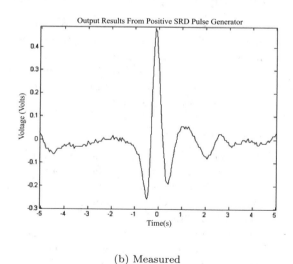

(b) Measured

Figure 8.3: Outputs from a Step Recovery Diode UWB Pulse Generator. (Note that the simulation predicts a significantly larger peak output voltage.)

SOURCE: A. Orndorff, "Transceiver Design for Ultra-Wideband Communications," M. S. Thesis [24]. © A. Orndorff, 2004. Used by permission.

operating characteristics of the model used in the simulation with measured data taken of the component being modeled or by comparing simulation results from the component with data sheets supplied by the manufacturer. Validating the overall simulation is a more complicated process and typically involves comparing the simulated system performance against field data or against an approximate theoretical analysis.

Therefore, before beginning any simulation, the engineer must be able to answer several questions. Among these are the following:

1. How much detail should be included in the analytical and simulation models, and how will these details affect any design decision based on the data received from simulation?

2. How might the differences between the physical device and the simulation model affect the design decisions, what errors might occur, and will any other factors (either in the simulation construction or in the physical implementation) overwhelm them?

3. How is the simulation to be validated (sanity checked)?

## 8.3   Simulation Methodologies–A Brief Review

Prior to beginning our study of the simulation of UWB systems, we pause to review several important concepts. The three basic simulation methodologies considered here are the basic Monte Carlo technique, a modification of the Monte Carlo technique known as semi-analytic simulation, and discrete event simulation. As we use I-UWB examples throughout the discussion of these three techniques, we conclude this section with a brief discussion of the simulation of MC-UWB.

### 8.3.1   Monte Carlo Simulation Techniques

The most basic simulation methodology is known as Monte Carlo simulation. Monte Carlo simulation is very flexible and can be applied to any system for which the signal processing operations defined by each and every functional block in the system block diagram are known. The Monte Carlo method is a game of chance (hence the name) in which a random experiment is replicated a number of times. Monte Carlo simulation has a rich history spanning several centuries. The first application of Monte Carlo simulation of which we are aware was suggested by Laplace in 1777, who used the Monte Carlo method for estimating the value of $\pi$ by using a technique known as Buffon's needle [8]. A much later application was developed in 1946 by Polish mathematician Stanislaw Ulam while pondering the probabilities of winning a game of Solitaire. Rather than mathematically computing the probability that a given 52 cards would produce a winning result, Ulam surmised that the probability could be estimated by playing the game 100 times and simply counting the number of wins [9].

All Monte Carlo simulations, including those by Laplace and Ulam, are implementations of a stochastic (random) experiment designed to estimate the probability of a particular event occurring. This is facilitated through the use of two counters in the simulation program. The first counter, known as the replication counter, measures the number of replications of the random experiment and is incremented by one each time the random experiment is repeated. The second counter, known as the event counter, is incremented by one each time the event of interest is observed. The relative frequency, or probability estimate, of the event of interest occurring is $\widehat{p} = R/N$, where $R$ and $N$ represent the values in the event and replication counters, respectively, at the end of the simulation run.

To place this in a communications context, assume that a simulation is being performed to estimate the symbol error probability of a communications system. In this situation the underlying random experiment is the passing of a digital symbol through a communications system. This is a random experiment because channel noise, interference, or other random disturbances may, or may not, cause a transmitted symbol to be received in error. Because we are executing the simulation to estimate the symbol error probability, the event of interest is a symbol being received in error. Thus, each time a symbol is passed through the system in the simulation, the replication counter is incremented by one. Each time a transmitted symbol is received in error, the event counter is incremented by one. The simulation is assumed to process $N$ symbols. After $N$ symbols have been passed through the system, we calculate the *relative frequency* of symbol error. This relative frequency is an *estimate* of the symbol error probability and is given by

$$\widehat{P}_E = \frac{N_E}{N} \tag{8.9}$$

where $N_E$ is the number of symbol errors observed in passing $N$ total symbols through the simulation. The probability of symbol error is given by

$$P_E = \lim_{N \to \infty} \left( \frac{N_E}{N} \right) \tag{8.10}$$

Because a simulation must execute in finite time in order to be useful, the underlying random experiment can only be replicated a finite number of times. As a result, the simulation is unable to determine the probability of symbol error. The best we can do is to estimate the probability through the relative frequency. In general terms, we want the estimate to be unbiased and consistent [1]. When the estimator is unbiased and consistent, accuracy improves as the number of repetitions of the experiment increases. This also holds when simulating a system that can assume a range of parameter values, as would be the case when simulating a mobile channel. Numerous different realizations should be simulated so that a given channel realization is not treated as indicative of typical system performance.

We have seen that the disadvantage of the Monte Carlo method is that the simulation run time can be quite lengthy. This is especailly true for I-UWB systems. For example, to accurately model pulse transients, pulse distortion, and timing jitter, an I-UWB signal must often be sampled at rates much higher than the Nyquist rate. To capture $\pm 10$ ps of timing jitter, an I-UWB signal would have to be sampled at a rate sufficient to model the jitter, i.e., a minimum of one sample every 10 ps. If the inter-pulse time (time between successive pulses) is 1 $\mu$s (corresponding to a pulse repetition frequency of 1 MHz), simulating 1,000 bits means processing 100 million samples. If the UWB pulse duration is on the order of 1 ns, it is obvious that the vast majority of the simulated samples has little or no impact on the simulation results. These excess samples consume and effectively waste a significant amount of processing power and system resources. Thus, a basic trade-off exists between simulation accuracy and the time required to execute the simulation. Monte Carlo simulation is, however, ideal for simulations that involve only a small number of bits or when it is necessary to investigate the transient performance of the system.

### 8.3.2  Semi-Analytic Simulation Techniques

Semi-analytic simulation is a technique that mitigates the lengthy execution times frequently encountered with Monte Carlo simulation. The semi-analytic technique is applicable to any system for which the probability density function of the decision metric, such as the output of a matched filter, is known. The semi-analytic technique is most commonly applied to systems operating in a Gaussian noise environment in which the portion of the system between the point at which noise is injected and the point where the decision statistic is made is linear. The semi-analytic technique can provide results at a fraction of the time required for a Monte Carlo simulation [1].

To illustrate the semi-analytic approach, consider a UWB system operating with a Binary Pulse Amplitude Modulation waveform where we are interested in estimating the Bit Error Rate (BER). We assume that the multipath can be modeled using a tapped delay line filter, all system components are operating in a linear region, and the only noise source is thermal noise.

To set up this problem, we'll first review a few basics. Recall from digital communications theory that the BER is a function of signal space separation and noise power. Now consider the signal constellation diagram shown in Figure 8.4. In this diagram, the transmitted signal points are denoted as $S_1$ and $S_2$, the decision regions

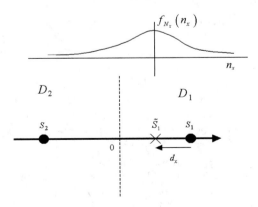

Figure 8.4: Diagram of a Semi-Analytic BER Estimation for BPSK.

are denoted as $D_1$ and $D_2$, and $f_{Nx}(x)$ is the Probability Distribution Function (PDF) of the noise in the system, modeled as a Gaussian distribution with a standard deviation of $\sigma_n$. In Figure 8.4, $S_1$ has been transmitted, and $\widetilde{S}_1$ has been received. As a result of intersymbol interference (ISI), $\widetilde{S}_1 \neq S_1$. The difference (error) between the transmitted and received constellation points is given by $d_x$. For this receiver, a correct decision is made whenever the received signal falls in the decision region corresponding to the transmitted symbol; otherwise, an error occurs.

For this system, the BER is given by

$$P_E = P_{S_1} P_{E|S_1} + P_{S_2} P_{E|S_2} \tag{8.11}$$

where $P_E$ is the probability of a bit error, $P_{S_1}$ and $P_{S_2}$ are the probabilities that $S_1$ and $S_2$ are transmitted, and $P_{E|S_1}$ and $P_{E|S_2}$ are the probabilities of error given that $S_1$ and $S_2$ are transmitted, respectively. Assuming that the system noise is Gaussian gives $P_{E|S_1}$ as

$$P_{E/S_1} = \int_{\widetilde{S}_1 \notin D_1} \frac{1}{\sqrt{2\pi}\sigma_n} e^{\frac{(n_x - \tilde{s}_1)^2}{2\sigma_n^2}} \, dn_x \tag{8.12}$$

A similar expression defines $P_{E|S_2}$. This can be evaluated by noting that this is the same as evaluating the Gaussian $Q$ function at $\frac{\widetilde{S}_1}{\sigma_n}$. Thus, to complete the semi-analytic simulation, we must simulate to find estimates for $\widetilde{S}_1$, $\widetilde{S}_2$ and $\sigma_n$ [1].

Furthermore, assume that the simulation sampling rate is 16 samples per symbol and the system uses transmitter filtering. The transmit filter is a fifth-order Butterworth filter and has a bandwidth equal to the symbol rate bandwidth (1/16). The simulation results for this example are shown in Figure 8.5. Notice in the left-hand pane the existence of two signal point clusters for both $\widetilde{S}_1$ and $\widetilde{S}_2$ due to ISI introduced in the transmitter filter (for rectangular pulse shaping a bandwidth of $2R_b$ (twice the bit rate) is required, whereas the filter only has a bandwidth $R_b$). Also notice that in this pane the two signal constellation clusters are approximately equiprobable (assuming no noise), one with a bit energy of approximately 1 and another with a bit energy of approximately $(0.61)^2$, or about 4.3 dB less than ideal. Thus, it is expected that the system would experience degradation in performance of approximately 2.15 dB, which is reflected in the right-side pane.

## 8.3.3    Discrete Event Simulation Techniques

As defined in [10], a discrete event simulation is "the modeling of systems in which the [system] changes only at a discrete set of points in time." However, many simulation architectures satisfy this definition, including the Monte Carlo simulation architecture considered in Section 8.4.1. For the purposes of this discussion, we will limit discrete event simulations to those in which the system changes only at dynamically scheduled and discrete points in time. In this approach, components of our modeled system, which we will term *entities*, schedule future *events* that affect the operation of entities in the system or the operation of the simulation.

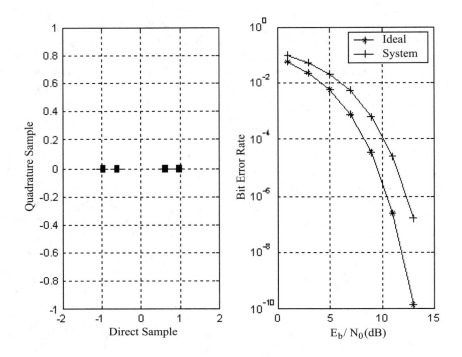

Figure 8.5: Binary PAM Semi-Analytic Simulation: (a) Signal Constellation Diagram, (b) Simulation Results.

As we shall see later in this section, a discrete event simulation has several advantages over a regularly sampled simulation, including improved sampling resolution and reduced simulation runtime. Note that a discrete event simulation architecture can incorporate many different simulation techniques, including Monte Carlo and semi-analytic techniques.

### Entity

An entity models a self-contained component of the system where the outputs of an entity are solely dependent on its inputs and internal parameters. An entity can model a component as complicated as a UWB receiver or as simple as an addition routine. It is convenient to think of an entity as a conceptual black box or as an object in the object oriented programming (OOP) sense. The OOP-discrete event simulation connection is especially strong, as object oriented languages like C++ and Java are frequently used for implementing discrete event simulations [10–12].

An example UWB entity represented as a UML (Unified Modeling Language) object is shown in Figure 8.6. This object implements the functionality of a UWB receiver through the use of various internal state machines, algorithms, and external

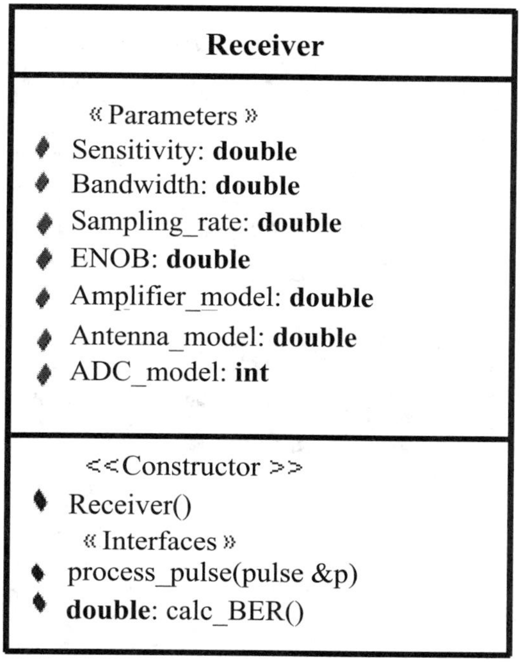

```
┌───────────────────────────────────────────────┐
│                  Receiver                        │
├───────────────────────────────────────────────┤
│            « Parameters »                        │
│   ♦ Sensitivity: double                          │
│   ♦ Bandwidth: double                            │
│   ♦ Sampling_rate: double                        │
│   ♦ ENOB: double                                 │
│   ♦ Amplifier_model: double                      │
│   ♦ Antenna_model: double                        │
│   ♦ ADC_model: int                               │
├───────────────────────────────────────────────┤
│            <<Constructor >>                      │
│   ♦ Receiver()                                   │
│            « Interfaces »                        │
│   ♦ process_pulse(pulse &p)                      │
│   ♦ double: calc_BER()                           │
└───────────────────────────────────────────────┘
```

Figure 8.6: UML Diagram of a Receiver Entity.

interfaces. Specifically, the object maintains internal models for the antenna, amplifier, and data converter, and provides mechanisms for processing pulses and calculating BER. To use this object, the internal state machines and algorithms need not be known; only the external interfaces have to be known. As implied by this example object, the internal operations of an entity can be quite complex, depending on its model and implementation, though its interfaces can remain quite simple.

However, developing complex entities has a number of drawbacks, including difficulties verifying the implementation and maintaining the code. To address these problems, entities can also be nested like objects. This allows entities that imply relatively complex operations, as in the receiver entity of Figure 8.6, to be constructed from a number of relatively simpler entities. These smaller entities can usually be independently developed, thus simplifying both the code verification process and later code maintenance.

For example, the Receiver entity diagrammed in Figure 8.6 could also be constructed from antenna, amplifier, data converter, and modem entities, as shown in Figure 8.7. Similarly, in a network simulation, the Receiver entity may be a nested component of a Transceiver entity, which is in turn part of a Cluster entity, which is in turn part of a Scatternet entity. Similarly, in a link simulation, Antenna,

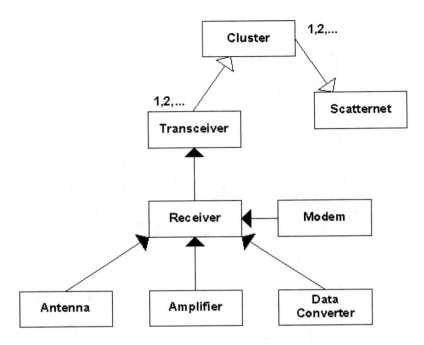

Figure 8.7: UML Diagram of a Receiver Entity with Nested Entities.

Amplifier, Data converter, and Modem entities may be comprised of a number of simpler entities.

## Event

An event models an occurrence within the system at a discrete time that affects the operation of an entity or the operation of the simulation itself. The following are examples of events that alter the operation of an entity:

1. Events that spawn an entity. For instance, a simulation may treat a pulse as an entity. The pulse could be spawned by an event generated by a Transmitter entity and be later modified by Channel and Receiver entities.

2. Events that model the arrival of an input. For instance, in the preceding example, the arrival of the pulse entity at the Channel entity would be signaled by an event.

3. Events that control the operation of an entity. Continuing this example, an event could be used to change the Channel entity's SNR during a BER sweep simulation.

Events can also affect the operation of simulation itself. Typical examples of these kinds of events include the following:

1. A simulation initiation event. This event would initialize various parameters within the entities in the simulation.

2. A simulation termination event. This could be created during the simulation or scheduled before the simulation begins. The former would occur if the simulation were designed to run for a number of bit errors. The latter would occur if the simulation were designed to run a predetermined length of time.

3. An error event. Error events usually terminate simulation. Triggers for error events include missing files and incorrectly formatted inputs.

When created, an event typically includes the following information:

1. Execution time–when the event is scheduled to occur. Most events are scheduled with an additive delay to the current simulation clock time, perhaps to model processing delay or propagation delay. However, some events are scheduled at a specific time, as would be done in a simulation that runs for a predetermined length of time.

2. Target entity or entities–the entity that should handle the event. For example, in a communications link simulation, the target of a transmitted packet entity will be a single receiver entity, but in a network simulation, the target could be every Receiver entity in the network. Except in a broadcast scenario, the packet in the network simulation would be seen as interference at a number of nodes in the network that are target entities of the event though not participating in the simulated link.

3. Payload information–data being transmitted by an event. Depending on the simulation and the event, the payload could model virtually anything; a number of samples, an IP packet, or simulation control information.

4. Associated event information–information other than the payload that is associated with an event. For example, an event may include some indication of the spawning entity or error codes, should the event fail. Both would be of particular use during the debugging phase of the simulation construction.

## Discrete Event Simulation Operation

During normal operation, each event is triggered by an entity's response to a previous event, and the simulation continues until no events are scheduled or a simulation termination event occurs. In order to initiate the simulation, the first event must be defined. This is typically accomplished by constructing the simulation so that an initiation event occurs at the beginning of the simulation. This simulation initiation event is targeted for all extant entities in the simulation. Entities then respond to this initiation event by scheduling normal events.

During a simulation, entities are idle except when responding to events. Entities can respond to events by processing data, by adjusting parameters, and by scheduling events.

Entities may schedule events targeted for other entities, targeted for themselves, or targeted for the simulation itself. Self-scheduling is frequently done when modeling a traffic source. For example, when simulating a link employing pulse position modulation, the transmitter's traffic source can self-schedule an event corresponding to the next pulse during the processing of the current pulse.

## Discrete Event Simulation Example

To illustrate the operation of a discrete event simulation, consider the canonical communications system shown in Figure 8.8 consisting of the following three components: a transmitter, a channel, and a receiver. For the purposes of this example, the system is assumed to be implementing a Binary PAM I-UWB waveform so that the transmitter generates pulses at regular intervals.

Assuming that the simulation is scheduled to run for 2.6 $\mu$s and pulses are generated at a nominal rate of 2 Mbps, Figure 8.9 illustrates the scheduling of events in the simulation. In this simulation, XMT is the name of the entity modeling the transmitter component, CHN names the entity modeling the channel, and RCV is the entity modeling the receiver.

As a result of a simulation initialization event not explicitly depicted, XMT self-schedules an event at 0.0 $\mu$s to generate a pulse. In response to the 0.0 $\mu$s event, XMT schedules an event for CHN to handle the newly generated pulse and self-schedules an event at 0.5 $\mu$s to generate the next pulse. Simulation control then passes to CHN, which modifies the pulse and schedules an event for RCV at 0.6 $\mu$s. At 0.5 $\mu$s, XMT generates the second pulse, and, as before, schedules later events for CHN and for itself. Then, at 0.6 $\mu$s, RCV responds to the event corresponding to the reception of the first pulse and schedules no new events. This process continues in this manner until a simulation termination event, which is also not explicitly depicted, occurs at 2.6 $\mu$s.

This example highlights two key differences between discrete event simulations and sampled simulations:

1. In a discrete event simulation, simulation time need not progress in regularly spaced intervals. In this example, simulation time advanced at different points by 0.2 $\mu$s, 0.3 $\mu$s, and 0.1 $\mu$s.

2. In a discrete event simulation, processing control is not linear. To handle these dynamic situations, a more complex structure is required to support the construction of discrete event simulations.

Figure 8.8: Block Diagram of a Canonical Communications System.

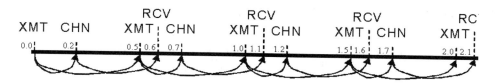

Figure 8.9: Example Discrete Event Simulation Operation.

## Discrete Event Simulation Development

In addition to events and entities, most discrete event simulations contain at least two additional components–an *event manager* and an *event list*.

An event manager is minimally responsible for the following simulation tasks:

1. Generating the initial simulation event.

2. Passing control to the proper entity at the scheduled event time.

3. Handling error events.

4. Resolving event-scheduling conflicts.

Scheduling conflicts can occur when two entities schedule events for the same time. This can be resolved in a number of different ways, but the most common technique is a first-come-first-serve approach, i.e., the event that was scheduled first is handled first.

Additionally, the event manager may provide additional services, such as supplying current simulation time and collecting statistics, provided that the current simulation time is particularly convenient for entities scheduling events, but this is not a necessity. Statistics could alternately be collected by each modeling entity or collected by special statistics entities.

To aid the event manager in these tasks, an event list is typically implemented. The event list is ordered by scheduled execution time, with the event scheduled for current execution placed at the head of the list, and the event scheduled farthest in the future placed at the tail of the list. Newly created events are inserted by timestamp into the list; completed events are deleted from the list. Although it represents poor programming practice, newly created events with timestamps before the list head are generally discarded. As a side benefit of an event list, the implementation of a simulation clock is relatively straightforward, as the current simulation time can be taken from the timestamp of the event at the head of the list.

Simulation operation with an event manager then proceeds by the repetition of the following steps until a termination event occurs or the list is empty:

1. Execute the event at the top of the event list. We call this event $A$.

2. Insert any new events spawned by $A$ into the event list.

3. Set the list head pointer to the event immediately after $A$.

**Simulation at 0.0 µs**

| Time | Event Target |
|------|-----------------|
| 0.0  | Begin Simulation |
| 2.1  | Simulation End  |

**Simulation at 0.2 µs**

| Time | Event Target |
|------|----------------|
| 0.2  | CHN           |
| 0.5  | XMT           |
| 2.1  | Simulation End |

**Simulation at 1.6 µs**

| Time | Event Target |
|------|----------------|
| 1.6  | RCV           |
| 1.7  | CHN           |
| 2.0  | XMT           |
| 2.1  | Simulation End |

Figure 8.10: Event Lists at Selected Simulation Times for a Discrete Event Simulation.

Revisiting our example from Figure 8.9, Figure 8.10 depicts how an event list for the simulation would appear at three different instances of time. At 0.0 µs the simulation initiation event and the simulation termination events are scheduled. At 0.2 µs, an event at the XMT entity, which also generated the first pulse, has scheduled events for the CHN and XMT entities at 0.2 µs and 0.5 µs respectively. (Note that these two events have been inserted into the list ahead of the previously scheduled termination event.) Then at 1.6 µs, four events have been scheduled: one for the RCV to receive the third pulse, one for the CHN entity to process the fourth pulse, one for the XMT entity to generate the fifth pulse, and the simulation termination event.

## Discrete Event Simulation Performance Considerations

A discrete event simulation architecture offers several performance advantages over a traditional sampled continuous time simulation due to the possibility of a higher sampling resolution and shorter simulation run time. To illustrate why this is so, consider the pair of pulses that might be generated from a Saleh-Valenzuela channel model as applied to a BPSK I-UWB waveform, as depicted in Figure 8.11.

In this waveform, a pulse arrives at the receiver with significant energy over the period $T_p$ that is significantly less than the frame period, $T_f$, allotted for the pulse and thus has a simulation duty cycle of $\rho = T_p/T_f$. For this particular channel realization, there are ten significant multipath components. If the multipath components arrived at regular intervals, a uniform sampling rate simulation would require a sampling rate of $T_p/10$ and have a total of $10/\rho$ samples per frame. However, a discrete event simulation[1] would require just ten samples for the entire frame. This represents, at least potentially, an enormous savings for low duty-cycle waveforms.

---

[1]The discrete event simulation could be constructed so that the channel generates 10 events for the receiver, or the channel could generate a single 10 sample event for the receiver. Ignoring simulation overhead, the nominal number of processing cycles would be the same.

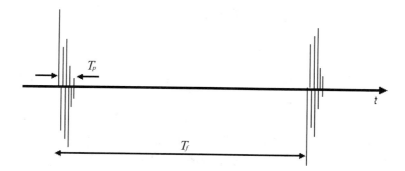

Figure 8.11: UWB Signal Generated Using Saleh-Valenzuela Channel Model.

Now consider the more likely scenario wherein the multipath components are irregularly spaced with minimum spacing of $\Delta t$. A uniform sampling architecture requires $1/\rho\Delta t$ samples to provide sufficient resolution. For a little perspective, this value will necessarily be greater than or equal to the previous total of $10/\rho$ samples per frame, and in many cases significantly greater. However, the discrete event simulation still only requires 10 samples per frame. As this example demonstrates, constructing a UWB simulation using a discrete event simulation architecture can provide significant improvements in both time resolution and simulation run-time.

However, it should be noted that this analysis overlooked some penalties that must be incurred by a discrete event simulation. In particular, processor cycles are lost handling events. In a discrete event simulation, processor cycles are lost in the following places:

1. The exchange of simulation control between the event manager and entity.

2. The handling of events within an entity.

3. The management of the event list.

Thus, the processing gain seen by implementing a discrete event simulation is lower in an actual implementation than predicted by a theoretical implementation. However, for low duty-cycle simulations, discrete event simulations retain the potential to significantly outperform a uniform sampling scheme.

Another drawback to a discrete event simulation architecture is increased development time. Many familiar communications simulation languages, such as Matlab, do not offer built-in discrete event simulation support. To use these languages, some additional work has to be performed to implement the event manager and event list. Additionally, the construction of an entity interface is slightly more complicated, as additional processing must occur to handle incoming events and to schedule outgoing events.

## 8.3.4   Multicarrier UWB (MC-UWB) Simulation

Previously, we saw that the following difficulties are encountered when simulating I-UWB systems:

1. Many traditional modeling parameters, such as phase, have an unclear meaning in a wide bandwidth signal.

2. Many traditional modeling approaches, such as complex baseband and I-Q representation, are not applicable to I-UWB.

3. Traditional channel noise modeling assumptions break down due to multipath resolvability.

4. Channel parameters, such as gain and group delay, are frequency variant.

5. The UWB necessitates an ultra-high sampling rate.

However, as an MC-UWB signal consists of numerous narrowband signals, the majority of these issues are not problems for MC-UWB simulation. Phase can be readily defined for each carrier in an MC-UWB signal. Each carrier signal in an MC-UWB can be readily broken into I-Q components and modeled using a complex baseband representation. Because of the narrow bandwidth of each carrier signal, fewer multipaths are resolvable, so traditional Gaussian, Rayleigh, and Rician noise models can often be assumed. Within each band, gain and group delay should be relatively frequency invariant. Further, as an MC-UWB signal is continuously transmitted, the familiar sampled continuous time Monte Carlo architecture can be implemented for MC-UWB simulations. Thus, a MC-UWB simulation is not significantly troubled by four of the major difficulties encountered in an I-UWB simulation.

However, as Nyquist requires, there is no avoiding the use of an ultra-high sampling rate for simulating UWB's ultra-high bandwidth. Also, because an MC-UWB signal is a continuous signal, a discrete event architecture is not appropriate and cannot be expected to produce a significant savings in processing resources.

There are also some issues that are more critical to the operation of a MC-UWB system than to an I-UWB system and serve as sources of difficulty for MC-UWB simulation. Most notable of the issues is MC-UWB's high peak–to–average power ratio (PAPR). Because of high PAPR, MC-UWB devices are repeatedly driven into nonlinear regions. This creates the following two issues for MC-UWB simulation:

1. As semi-analytic simulation techniques typically assume a linear system after the point of noise injection, there can be significant difficulty in implementing a semi-analytic simulation. Thus Monte Carlo techniques are generally preferable for MC-UWB simulations.

2. More complex component models must be used to ensure that these nonlinear effects are properly captured by the simulation. Accordingly, Section 8.4 presents an extended discussion of component models for UWB communication systems with a particular focus on modeling nonlinear effects.

## 8.4    UWB Component Simulation

This section provides examples of the previously discussed methodologies and techniques as applied to the following UWB communication systems processes and components: *pulse generation and modulation, signal amplification, antennas,* and *signal propagation.*

### 8.4.1    UWB Pulse Generation and Modulation

This section discusses the steps involved in the development of a simulation for examining how modulation and pulse shaping affects system performance as measured by the BER.

For the purposes of this experiment, the system model shown in Figure 8.12 is adopted. In this model, a randomly generated train of bits, which represent data, are modulated and pulse shaped. Then noise, labeled as $N_x$ in the diagram, is added to the signal before being demodulated. The following discusses the specific models used for each component.

#### Pulse Generation and Modulation

Several modulation techniques can be used to create UWB waveforms. Some of the most popular methods to create UWB pulse streams use mono-phase techniques, such as pulse amplitude (PAM), pulse position (PPM), or on-off keying (OOK).

UWB pulses are traditionally modeled using two different models: Gaussian pulses and Gaussian monocycles. The Gaussian pulse is a frequently used conceptual model of the pulse generated at the transmitter and can be represented as

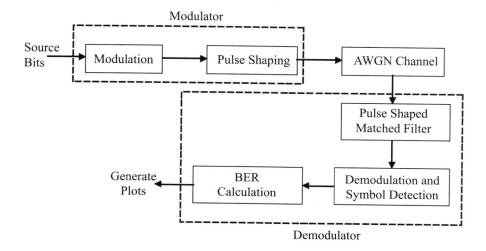

Figure 8.12: UWB Simulation System Model.

$$p(t) = Ae^{-[(t-T_c)/T_{au}]^2} \qquad (8.13)$$

where $A$ is the amplitude of the pulse, $T_c$ is the center time of the pulse, and $T_{au}$ is a pulse-shaping parameter related to the standard deviation. As the antenna-air interface imparts a derivative on any transmitted signal, the received signal is frequently modeled as a Gaussian monocycle, which takes the form

$$p(t) = 2A\sqrt{e}\,(t - T_c)\,e^{-[(t-T_c)/T_{au}]^2} \qquad (8.14)$$

Due to device limitations, it it is difficult to transmit UWB pulses that exactly match the shapes given in (8.13) or (8.14). Reasons for this limitation include frequency-dependant effects and limitations in the transmitter that impart a certain amount of pulse distortion and dispersion (discussed in Chapter 5, "Transmitter Design"), which must be accounted for in the simulation. The simplest way to account for these differences is to simply measure a generated pulse and use the measured pulse shape in the simulation. Figure 8.13 shows a measured pulse generated by an I-UWB transmitter manufactured by Multispectral Solutions, Inc.

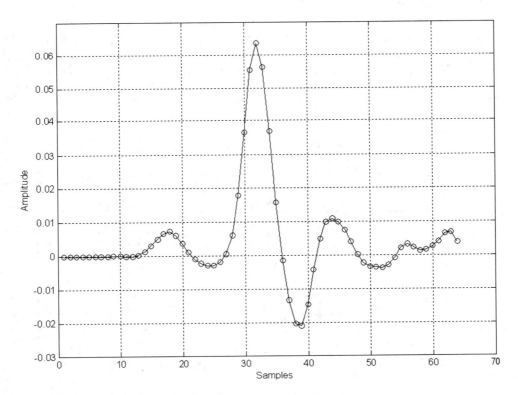

Figure 8.13: Measured Pulse Generated by an MSSI UWB Transmitter.

## Simulation Development

To simplify the construction of the simulation, the channel is assumed to be AWGN and is implemented by adding normal distributed random variables to samples of the transmitted signal. The variance of the noise is calculated from the average energy of the received symbol for a specified SNR. To perform demodulation, the received signal is passed through a filter matched to the pulse shape. The result of this operation is then mapped back to bits according to maximum likelihood criteria.

For this simulation, two different simulation techniques are considered: Monte Carlo and semi-analytic. In the Monte Carlo simulation, an independent normal variable is generated for each sample, with variance given by

$$\sigma^2 = \frac{\sum_{i=1}^{n} A_i^2}{2E_b/N_0} \tag{8.15}$$

where $n$ is the number of samples in a pulse, $A_i$ is the amplitude of a sample of the pulse, and $E_b/N_0$ is the desired SNR.

These samples are then processed, and the BER estimate is formed by taking the ratio of the number of errors in the received bits to the total number of transmitted bits.

For the semi-analytic simulation, an initial noiseless Monte Carlo simulation is performed to characterize the received signal constellation. For Binary PAM modulation, a diagram similar to the one shown in Figure 8.14 is created where $S_1$ and $S_2$ are the ideal constellation points, the decision regions are $D_1$ and $D_2$, and $\widetilde{S}_1$ is the simulated constellation point.

After calculating the noise equivalent bandwidth of the matched filter and applying side knowledge that the received signal is Gaussian distributed, bit error rates can be directly calculated for any SNR from

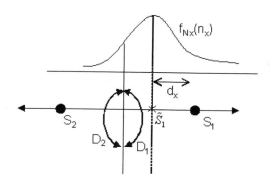

Figure 8.14: Constellation Diagram for Binary PAM UWB Signal.

$$\Pr\left\{Error/S_1\right\} = \int_{\widetilde{S}_1 \notin D_1} \frac{1}{2\pi\sigma_n} \exp\left(\frac{-\left(n_x - \widetilde{S}_1\right)^2}{2\sigma_n^2}\right) dx = Q\left(\frac{\widetilde{S}_1}{\sigma_n}\right) \qquad (8.16)$$

For OOK modulation, the signal constellation has two points, 0 and 1, in one dimension. This means that the two constellation points in PAM have moved closer by a factor of 0.5. Therefore, the same formula for calculating the BER can be used. For PPM modulation, there are two orthogonal dimensions, and the distance between the two points is $\sqrt{2E_b}$ so that the effective variance is increased by a factor of 2.

### Monte Carlo Simulation Results

The Monte Carlo simulation assumes the following parameters: binary PAM modulation, a Gaussian noise channel, Gaussian monocycles, and measured pulses. A rectangular pulse shape is also simulated to provide a means of validating the simulation. Eight samples per pulse are assumed, and pulse amplitudes are constrained to the maximum level achieved by the rectangular pulse. With respect to the rectangular pulse, energy losses of 5.94 dB, 5.54 dB, and 8.17 dB are seen for Gaussian pulses, Gaussian monocycles, and the measured response, respectively. The simulation results for the PAM modulation case are presented in Figure 8.15. Note that these energy losses are reflected in the figure. If each pulse had the same energy, under the considered LOS AWGN conditions each pulse would have achieved the same BER.

### Semi-Analytic Simulation Results

While energy-normalized pulse shaping does not have an effect on performance, the choice of modulation scheme can have a significant impact. A semi-analytic simulation was performed to study the effect of UWB modulation, particularly PAM, PPM, and OOK with Gaussian pulse shaping, and the results of this simulation are shown in Figure 8.16. Note that performance is a function of signal space separation. Recall that signal space separation of PAM is $2\sqrt{E_b}$, PPM is $\sqrt{2E_b}$, and OOK is $\sqrt{E_b}$. Note that the results match expectations as the following BER relationship is observed: PAM < PPM < OOK.

## 8.4.2    Signal Amplification

This section considers the modeling, simulation, and experiment execution and analysis of signal amplification, i.e., the processes that occur in a power amplifier and low noise amplifier (LNA) in a UWB communications system.

### Amplifier Model Development

Most traditional amplifier models, such as the $\tan^{-1}$ model, are predicated on narrowband assumptions. New models must therefore be developed for UWB applications. In these situations, the development of a behavioral model based on measured data is often a necessary approach.

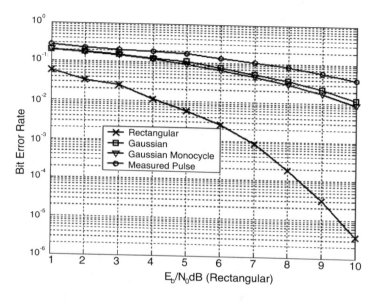

Figure 8.15: Monte Carlo Simulation Plot for Binary PAM Modulation.

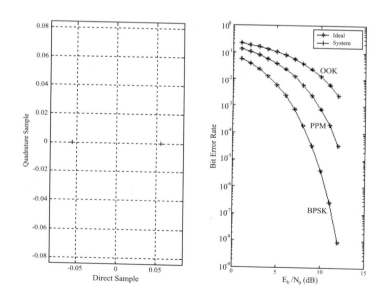

Figure 8.16: Semi-Analytic Simulation Results: Received Signal Constellation and BER Versus SNR.

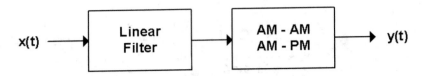

Figure 8.17: Block Diagram of a Wiener System.

When the input signal has a bandwidth comparable to the bandwidth of the amplifier, the amplifier gain should not be assumed to be flat over the entire signal bandwidth. The input signals then induce frequency-dependent effects in the amplifier, which are visible as hysteresis on the envelope transfer characteristics or input-output characteristics of the amplifier.

The modeling strategy for these amplifiers, which exhibit both nonlinearties and memory or frequency dependent effects, is a topic of active research. The most common way to introduce frequency dependencies into a system is to include a filter. A nonlinearity preceded by a linear filter is referred to as a Wiener system in control systems terminology. Figure 8.17 shows the block diagram of a Wiener system. Note that this system introduces both amplitude dependent amplitude (AM-AM) and phase (AM-PM) nonlinearities.

Another method to model a nonlinear system with memory is to use a "nonlinear filter" or a nonlinear tapped-delay line (NTDL) [13]. As the name suggests, the model consists of an FIR structure, or tapped-delay line, with input-amplitude-dependent polynomials at each tap. Thus, the dynamic system has a nonlinearity embedded in it. Figure 8.18 shows a 3-tap NTDL, where $y[n]$ is related to $x[n]$ as given by

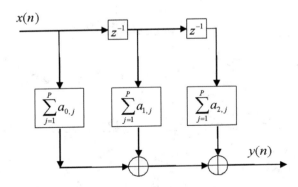

Figure 8.18: 3-Tap, $P^{th}$-Order NTDL Model of an Amplifier.

$$y[n] = \sum_{j=1}^{p} a_{0,j} x[n]^{j} + \sum_{j=1}^{p} a_{1,j} x[n-1]^{j} + \sum_{j=1}^{p} a_{2,j} x[n-2]^{j} \qquad (8.17)$$

This modeling strategy gives us two degrees of freedom for choosing the number of coefficients–filter depth and tap nonlinearity design. The advantage of this form of modeling structure is its very convenient form for least-squares coefficient estimation [14] for signal specific time-domain modeling.

For our models, the coefficients in the NTDL are determined from experiments that sounded the amplifiers to be modeled. For this study, three amplifiers from Mini-Circuits were tested and measured. Specifically, three power amplifiers (specifically the ZFL-2500 [15], ZHL-42W [16], and ZVE-8G [16] models) were measured and modeled using a NTDL approach. Upon repeated trials with different values for the filter length, tap order, and sampling frequency, the parameters listed in Table 8.1 were generated. Although increased filter lengths improve modeling accuracy as measured in terms of normalized mean-squared envelope error (NMSE), some concessions were made in trading fidelity for faster simulation execution time.

A nice feature of the NTDL amplifier model is that it is only slightly more complicated to implement than a filter. Thus, a simulation of these amplifiers can be realized by appropriately modifying the system depicted in Figure 8.18.

The models created for these three amplifiers were driven by PAM modulated data to capture the power spectral densities for both the input and output signals. The spectral densities for the ZFL-2500, ZHL-42W, ZVE-8G are plotted in Figure 8.19, Figure 8.20, and Figure 8.21.

From this simulation, it is apparent that none of the amplifiers produce the desired ideally flat frequency response and that all three amplifiers introduce some spectral regrowth, with the ZFL-2500 introducing the least spectral distortion. It should also be noted that the ZVE-8G actually attenuates the signal for low input frequencies.

## 8.4.3   Simulation of Antenna Effects

This section considers the modeling and simulation of the processes that occur in antennas in a UWB communications system. In this study, effort is focused on the simulation of the transfer function between the transmitter antenna and receiver antenna. The system can be modeled as shown in Figure 8.22, where $V_{tx}$ is the signal input to the transmit antenna and $V_{rx}$ is the signal output from the receive antenna.

Table 8.1: NTDL Amplifier Model Characteristics.

| Amplifier (Mini-Circuits) | Filter Length | Polynomial Order | NMSE (dB) |
|---|---|---|---|
| ZFL-2500 | 10 | 5 | -34 |
| ZHL-42W | 15 | 3 | -25 |
| ZVE-8G | 15 | 3 | -8 |

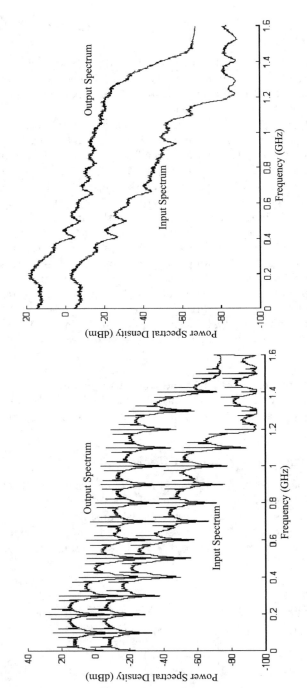

Figure 8.19: PPM (Left) and BPSK (Right) Modulated Input-Output Spectral Densities for ZFL-2500.

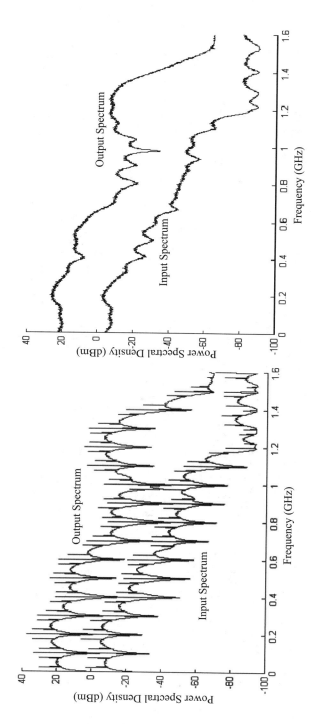

Figure 8.20: PPM (Left) and BPSK (Right) Modulated Input-Output Spectral Densities for ZHL-42W.

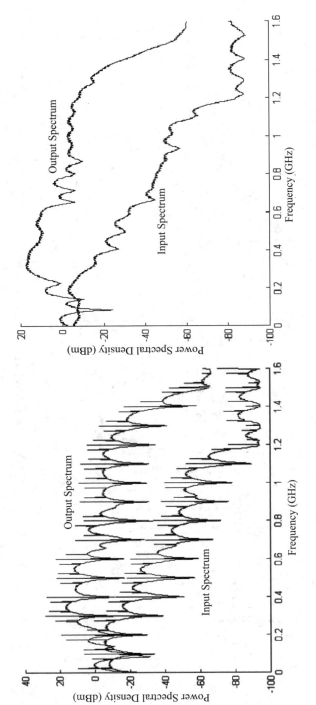

Figure 8.21: PPM (Left) and BPSK (Right) Modulated Input-Output Spectral Densities for ZVE-8G.

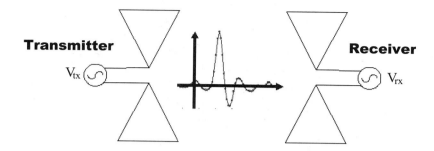

Figure 8.22: Configurations for Simulation of Antennas Transmitting a UWB Pulse.

In particular, this section considers the simulation of a TEM horn antenna. A TEM horn antenna consists of triangular conductors joined to form a V-shaped structure as shown in Figure 8.23. A TEM horn antenna is completely characterized by its side length, apex angle, $\alpha$, and elevation angle, $\beta$.

## Antenna Model Development

A block diagram for a simple antenna model is shown in Figure 8.24. Here both antennas are modeled as bandpass filters with the transmit antenna imparting a first derivative on the signal.

More complex models can be constructed by applying time-varying versions of Maxwell's equations to the antenna to characterize the propagation of the electrical and magnetic fields. An exact solution of these equations can be quite complex. However, techniques have been developed that approximate the solution by sampling both the space and time components of the solution.

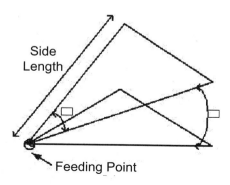

Figure 8.23: Geometric Model of a TEM Horn Antenna.

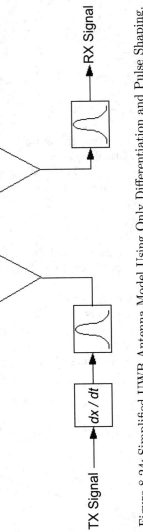

Figure 8.24: Simplified UWB Antenna Model Using Only Differentiation and Pulse Shaping.

One such technique is the FDTD (Finite Difference Time Domain) approach proposed by K.S. Yee [17], where time-dependent Maxwell's equations are sampled in both the time and space domains in unit cells to form a more tractable problem of planar geometries. The end result of the FDTD method is the creation of a time-varying spatial lattice shown in Figure 8.25 with finite difference equations for the electrical and magnetic fields associated with each plane. A more extensive treatment of the FDTD model is presented in [18–20].

An FDTD model has the following advantages:

1. The model simplifies estimation of the wideband response.

2. The model permits analysis of antennas with arbitrary structures.

3. The model permits analysis of the reciprocal effects between two materials having different physical characteristics, e.g., different values for permittivity and permeability.

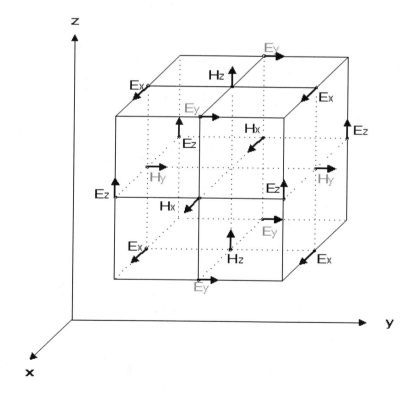

Figure 8.25: Yee Space Lattice Used in FDTD Antenna Modeling.

4. The model is applicable to other situations, including scattering effects, antenna patterns, and radar cross-sections.

5. The model yields relatively straightforward simulation code.

## Simulation Development

To construct an FDTD simulation, we first calculate all elements of the magnetic field at all points in the lattice. Time is then advanced by a specific timestep and all elements of the electric field are calculated. The process of iteratively calculating magnetic and electric fields repeats until convergence is achieved.

Because the FDTD method requires significant memory and processing time, techniques have been developed for reducing simulation runtime. One of the simplest techniques is to exploit antenna symmetry to calculate only a single symmetric component of the antenna [19]. Figure 8.26 depicts a symmetric antenna where the computational volume of the Yee lattice is reduced to an eighth of its original

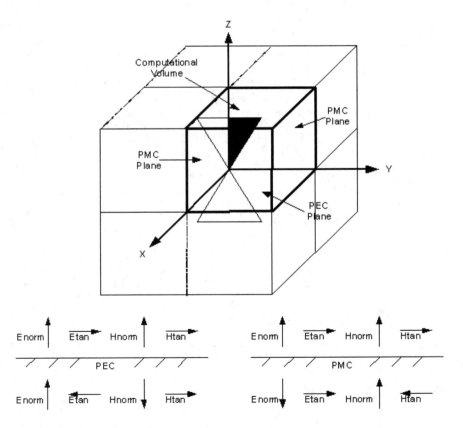

Figure 8.26: Reduced Computational Volume by Exploiting Antenna Symmetry.

volume. This then implies that memory and cycle requirements are also reduced to an eighth of the original requirements.

## Validation

To validate the FDTD simulation, the simulation results are compared with collected data and also compared with the relative performance of the FDTD simulation methodology for a Ka-band coaxial to waveguide transition structure.

Because the simulation was intended to specifically model the TEM-horn antenna suggested in [21], the data provided in [21] is compared with the simulation results. This comparison is illustrated in Figure 8.27. The errors in tails of the wave are caused by scattering and reflection in the RX antenna. Because the distance between the TX and RX antennas is relatively small compared to the far-field distance, the superposition of the propagating waves from the TX antenna and reflecting wave from the RX antenna introduces some error in the tails. These errors can be reduced by increasing the distance between the TX and RX antennas, but this also increases the memory requirements and necessitates the use of a near to far-field transformation. Therefore, under typical simulation constraints, accuracy has to be traded off for simulation runtime.

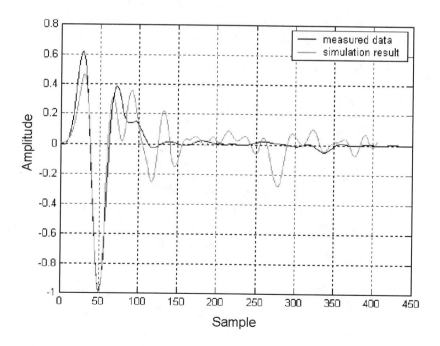

Figure 8.27: Comparison of Simulated and Measured Antenna Response to a UWB Pulse.

### 8.4.4 Simulation of UWB Channels

This section considers simulation strategies for the small-scale UWB channel with particular attention paid to the Saleh-Valenzuela model.

### Model Development

Unlike narrowband signals, in UWB the amplitude distribution of each resolved multipath in a small scale channel differs significantly from a Rayleigh distribution. This arises from the multipath components of a UWB signal being comprised of a small number of scatterers. According to [22], the best-fit distribution of the small-scale magnitude statistics is the Nakagami-$m$ distribution [23], corresponding to a Gamma distribution of energy gains. Let the gamma distribution $\Gamma\left(\Omega_k, m_k\right)$ denote the distribution that fits the energy gains of the local path delay profiles (PDPs) in the $k^{th}$ bin. The $\Omega_k$'s are given as $\Omega_k = \overline{G}_k$, and $m_k$ is related to the variance of the energy gain of the $k^{th}$ bin. The parameter $m_k$ is a random variable distributed according to the truncated Gaussian distribution given by

$$f_m\left(x\right) = \begin{cases} K_m e^{-\left((x-\mu_m)^2/2\sigma_m^2\right)}, & \text{if } x \geq 0.5 \\ 0, & \text{otherwise} \end{cases} \tag{8.18}$$

Another model proposed for the UWB channel is the Saleh-Valenzuela (S-V) model [5], which is based on indoor multipath propagation measurements using 10 ns, 1.5 GHz, radar-like pulses with a time resolution of about 5 ns in a medium-sized office building. In this model, the received signal rays arrive in clusters. Each received ray has uniformly distributed phase and Rayleigh distributed amplitude with a variance that decays exponentially with cluster and ray delays. The phase components can be treated as a uniformly distributed delay over a very narrow window of time.

The cluster arrival times, which are the arrival times of the first rays of the clusters, are modeled as a Poisson arrival process with a fixed rate, $\Lambda$. Within each cluster, subsequent rays also arrive according to a Poisson process with a different fixed rate, $\lambda$. Typically, each cluster consists of many rays, and thus $\lambda \ll \Lambda$. The arrival time of the first cluster, $T_0$, and the arrival time of the first ray within the $l^{th}$ cluster, $\tau_{0l}$, are defined as 0. Therefore, the cluster arrival time, $T_l$, and ray arrival time, $\tau_{kl}$, are described by the independent inter-arrival exponential probability density functions as given by

$$p\left(T_l | T_{l-1}\right) = \Lambda \exp\left[-\Lambda\left(T_l - T_{l-1}\right)\right], l > 0 \tag{8.19}$$

and

$$p\left(\tau_{kl} | \tau_{(k-1)l}\right) = \lambda \exp\left[-\lambda\left(\tau_{kl} - \tau_{(k-1)l}\right)\right], k > 0 \tag{8.20}$$

The channel impulse response for the S-V model is given by

$$h\left(t\right) = \sum_{l=0}^{\infty} \sum_{k=0}^{\infty} \beta_{kl} e^{j\theta_{kl}} \delta\left(t - T_l - \tau_{kl}\right) \tag{8.21}$$

where $\theta_{kl}$ is a random variable uniformly distributed over $[0, 2\pi)$, and $\beta_{kl}$ is a Rayleigh distributed random variable whose mean square value $E\left\{\beta_{kl}^2\right\}$ is defined by

$$E\left\{\beta_{kl}^2\right\} \equiv E\left\{\beta^2\left(T_l, \tau_{kl}\right)\right\} = E\left\{\beta^2\left(0, 0\right)\right\} \cdot e^{-T_l/\Gamma} e^{-\tau_{kl}/\gamma} \qquad (8.22)$$

where $E\left\{\beta^2\left(0, 0\right)\right\}$ is the average power gain of the first ray of the first cluster, and $\Gamma$ and $\gamma$ represent the power-delay time constant for the clusters and the rays, respectively.

Generally, the clusters overlap and rays and clusters extend over an infinite time as shown in (8.22). To address this issue, stopping criteria for the outer and inner sums have been developed and are $\exp\left(-T_l/\Gamma\right) \ll 1$, and $\exp\left(-\tau_{kl}/\gamma\right) \ll 1$, respectively.

### Simulation Development

To construct a simulation of a S-V channel model, the following procedure can be applied:

1. Generate the cluster arrival times based on (8.19). If the environment is LOS, the first cluster arrival time is zero; if the environment is NLOS, the first cluster arrives at a time generated from (8.19).

2. For each cluster arrival, ray arrival times are generated using (8.20) and corresponding ray magnitudes are calculated from (8.21) and (8.22).

3. The value for the ray created in step 2 is stored and the process in step 2 is repeated while $\exp\left(-\tau_k/\gamma\right) > 0.001$ is satisfied.

4. If $\exp\left(-\tau_k/\gamma\right)$ becomes smaller than 0.001, a new cluster arrival time is generated by using (8.19) and steps two and three, while the condition $\exp(-T_l/\Gamma) > 0.001$ is satisfied.

5. After these steps are completed, the channel impulse is normalized by applying large-scale fading effects across the ray magnitudes. Typically large-scale fading satisfies, at least approximately, a lognormal distribution.

The channel model is then realized with a complex filter.

### Validation

To validate the S-V model, we can use the parameter sets given in [5], which provide the median value of RMS delay spread based on 150 measurements in seven rooms. Table 8.2 shows the default parameters used for the validation and the result.

To examine the flexibility of the model, several channel realizations were simulated according to the parameter sets listed in Table 8.3.

Figures 8.28 through 8.31 depict the impulse response and the power delay profile for the Saleh-Valenzuela with these four parameter sets.

Table 8.2: Validation of Simulation Results for S-V Model.

| Parameters | cluster_lambda (1/ns) | 1/300 |
|---|---|---|
| | ray_lambda (1/ns) | 1/5 |
| | cluster_decay (ns) | 60 |
| | ray_decay (ns) | 20 |
| | LOS | 1 |
| | shadow_sigma (dB) | 0 |
| RMS Delay Spread | target value (ns) | 25 |
| | result (ns) | 31.2868 |

Table 8.3: Parameters Used in Channel Realizations.

| Description | Unit | Default set #1 | Default set #2 | Default set #3 | Default set #4 |
|---|---|---|---|---|---|
| Cluster Arrival Rate | 1/ns | 0.0233 | 0.4 | 0.0667 | 0.0667 |
| Ray Arrival Rate | 1/ns | 2.5 | 0.5 | 2.1 | 2.1 |
| Cluster Decay Factor | ns | 7.1 | 5.5 | 14.00 | 24.00 |
| Ray Decay Factor | ns | 4.3 | 6.7 | 7.9 | 12 |
| Lognormal Shadowing Term for Total Multipath | dB | 3 | 3 | 3 | 3 |

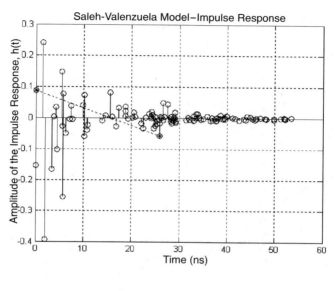

(a) Impulse Response

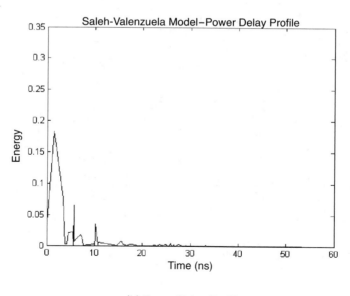

(b) Power Delay Profile

Figure 8.28: Simulated Saleh-Valenzuela Channel Model with Parameter Set #1.

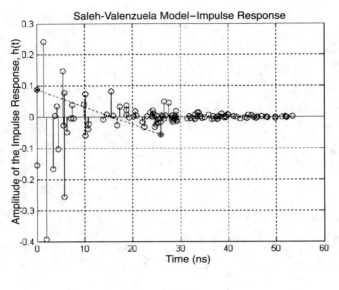

(a) Impulse Response

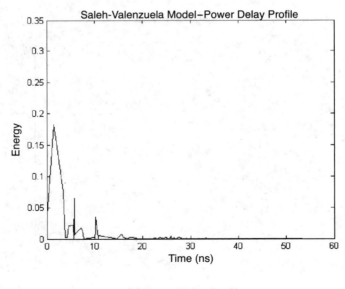

(b) Power Delay Profile

Figure 8.29: Simulated Saleh-Valenzuela Channel Model with Parameter Set #2.

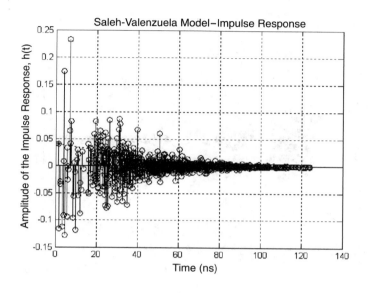

(a) Impulse Response

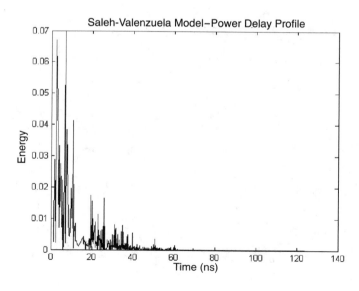

(b) Power Delay Profile

Figure 8.30: Simulated Saleh-Valenzuela Channel Model with Parameter Set #3.

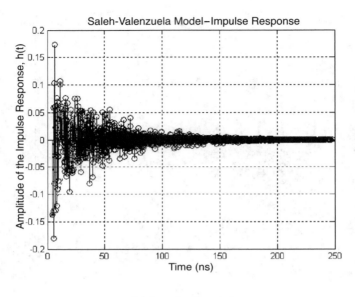

(a) Impulse Response

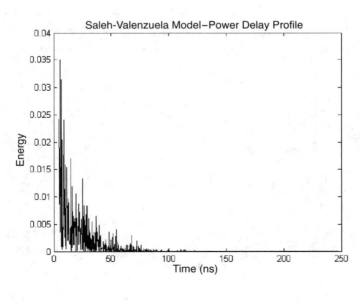

(b) Power Delay Profile

Figure 8.31: Simulated Saleh-Valenzuela Channel Model with Parameter Set #4.

## 8.5 Summary

Simulation of UWB communication systems, in contrast to the simulation of other communication systems, emphasizes the transient nature of UWB systems. Transient effects can include frequency-dependent pulse distortion imparted by RF components or the wireless channel, pulse dispersion produced by the antenna, or timing jitter generated by non-ideal oscillators. For traditional communication systems, these transient effects are only a small fraction of the symbol duration and can often be ignored. In UWB systems, they directly affect the performance of the overall communication system and must be included in the simulation. Additionally, sampling the pulse at a very high rate and then simulating every sample is an effective but highly inefficient methodology. Event-driven simulation techniques provide a much more efficient way to evaluate the performance of a UWB communication system.

This chapter has presented a discussion outlining several challenges introduced by the simulation of I-UWB systems, including architectural approaches and challenges unique to I-UWB systems, as well as the determination of models appropriate to I-UWB communication systems. Examples were provided for the simulation of pulse generation and modulation, amplification, antenna effects, and channel models. A thorough treatment of I-UWB simulation techniques would, unfortunately, require an entire book. In spite of these space limitations, we have hopefully provided sufficient coverage of the basic techniques to allow the reader to develop accurate and efficient simulations.

# Bibliography

[1] W. H. Tranter, K. S. Shanmugan, T. S. Rappaport, and K. L. Kosbar, *Principles of Communication Systems Simulation with Wireless Applications*, Upper Saddle River, New Jersey: Prentice Hall PTR, 2004.

[2] M. C. Jeruchim, P. Balaban, and K. S. Shanmugan, *Simulation of Communication Systems*, 2nd ed., New York: Kluwer Academic/Plenum Publishers, 2000.

[3] G. F. Ross, "Transmission and reception system for generating and receiving base-band duration pulse signals without distortion for short base-band pulse communication system," U.S. Patent 3,728,632, April 17, 1973.

[4] ITU ITU-R M.1225, "Guidelines for evaluations of radio transmission technologies for IMT-2000," 1997.

[5] A. A. M. Saleh, and R. A. Valenzuela, "A statistical model for Indoor multipath propagation," *IEEE Journal on Selected Areas in Communications*, vol. SAC-5, no. 2, pp. 128-137, February 1987.

[6] D. Cassioli, M. Z. Win, and A. Molisch, "The ultra-wide bandwidth indoor channel: From statistical model to simulations," *IEEE Journal on Selected Areas in Communications*, vol. 20, no. 6, pp. 1247-1257, August 2002.

[7] 802.15.3a Standard. Available: http://ieee802.org/15/pub/TG3a.html.

[8] P. Beckman, *A History of $\pi$(PI)*, New York: Barnes and Noble, 1993.

[9] R. Eckhardt, "Stan Ulam, John von Neumann, and the Monte Carlo method," *Los Alamos Science*, Special Issue (15), pp. 131-137, 1987.

[10] J. Banks, J. S. Carson, II, B. L. Nelson, and D. M. Nicol, *Discrete Event System Simulation*, 3rd ed., Prentice Hall, 2002.

[11] J. M. Garrido, *Object-Oriented Discrete-Event Simulation with Java: A Practical Introduction*, Kluwer Academic/Plenum Publishers, 2001.

[12] J. M. Garrido, *Practical Process Simulation Using Object-Oriented Techniques and C++*, Norwood, MA: Artech House, 1999.

[13] J. Kim and K. Konstantinou, "Digital predistortion of wideband signals based on power amplifier model with memory," *IEE Electronics Letters*, vol. 37, no. 23, pp. 1417-1418, November 2001.

[14] S. M. Kay, *Fundamentals of Statistical Signal Processing–Estimation Theory*, Upper Saddle River, NJ: Prentice Hall, 1993.

[15] ZFL-2500 data sheet. Available: http://www.mini-circuits.com/dg03-174.pdf.

[16] ZHL-42W and ZVE-8G data sheets. Available: http://www.minicircuits.com/dg03-176.pdf.

[17] K. S. Yee, "Numerical solution of initial boundary value problems involving Maxwell's equations in isotropic media," *IEEE Transactions on Antennas and Propagation*, vol. 14, pp. 302-307, 1966.

[18] D. M. Sullivan, "Electromagnetic Simulation Using the FDTD Method," *IEEE Press Series on RF and Microwave Technology*, 2000.

[19] A. Taflove and S. C. Hagness, *Computational Electrodynamics, the Finite-Difference Time-domain Method*, Artech House Publishers, 2000.

[20] K. S. Kunz and R. J. Luebbers, *Finite Difference Time Domain Method for Electromagnetics*, CRC Press, 1993.

[21] K. L. Shlager, G. S. Smith and J. G. Maloney, "Accurate Analysis of TEM Horn Antennas for Pulse Radiation," *IEEE Transactions on Electromagnetic Compatibility*, vol. 38, no. 3, pp. 414-423, August 1996.

[22] D. Cassioli, M. Z. Win, and A. Molisch, "The ultra-wide bandwidth indoor channel: From statistical model to simulations," *IEEE Journal on Seleced Areas in Communications*, vol. 20, no. 6, pp. 1247-1257, August 2002.

[23] M. Nakagami, "The m-distribution—A general formula of intensity distribution of rapid fading," *Statistical Method in Radio Wave Propagation*, W. C. Hoffman, ed., pp. 3-36, Oxford, U.K.: Pergamon, 1960.

[24] A. Orndorff, "Transceiver Design for Ultra-Wideband Communications," M.S. Thesis, Virginia Tech, 2004.

# Chapter 9

# NETWORKING

Michelle X. Gong
and Scott F. Midkiff

The previous chapters explored issues related to the creation of a functioning communications link. However, a link is but a small part of a communications network. The transition between establishing a functional communications link and an operational network is nontrivial and remains a subject of research even when not faced by the unique challenges offered by a UWB communications link. This chapter discusses the steps and issues involved in the creation of a network of UWB devices. As we shall see, there are aspects of UWB networking that are similar to traditional communications networks, while other aspects are quite different.

To begin our discussion of UWB networking, we review some fundamental networking concepts. A network is a collection of interconnected devices; a UWB network is collection of devices interconnected by UWB links. These links operate in the same wireless medium, so even the most rudimentary networks must define a way of sharing the wireless medium whether by coordination or by contention. More complex networks that provide indirect connections typically supplement these medium access schemes with methods for coordinating the end-to-end flow of information through the networks. Networks employ protocols to formalize the processes for medium access, end-to-end information flow, and other functions in the network.

For example, when communication occurs only between directly connected nodes, that is, over a single hop, a data link layer protocol (or a set of data link layer protocols) is needed to coordinate the transfer of information in any of several ways, such as one-to-one (unicast), one-to-many (multicast), one-to-all (broadcast), or many-to-one. Minimally, this requires the introduction of an addressing scheme to ensure that data gets to the proper recipient(s). Additionally, some scheme is needed to coordinate the use of the shared wireless channel by multiple nodes; this is typically performed by medium access control (MAC) protocols, which are discussed in more detail later in this chapter. This view of a network (direct connections between wireless devices) is analogous to the network provided by the IEEE 802.11 wireless local area network standards [1] and the WPAN specifications for a piconet [2].

However, when nodes are also connected indirectly, that is, over multiple hops, additional protocols need to be introduced. Specifically, network devices need to be able to determine how to route information between indirectly connected devices, a process that is complicated by the movement of devices, as in a mobile ad hoc network (MANET), and the lack of any network infrastructure for routing, as in a Bluetooth scatternet [3] or a MANET. This task typically falls to network layer protocols. Further, an indirect link, such as an intermediate hop, may fail, thus necessitating the use of additional protocols to ensure reliable communications between the indirectly connected nodes. Reliable end-to-end communications is typically provided by a transport layer protocol.

Ultimately, the performance of a UWB network depends on the performance of its links, its point-to-point protocols, and its end-to-end protocols and on the interaction between these protocols. Predicting the performance of a network can be quite complicated due to the mutual dependencies and interdependencies of the network protocols.

To simplify the construction of network protocols, the International Organization for Standardization (ISO) promulgated the Open Systems Interconnection (OSI) model [4]. The OSI model separates the functionality of the network protocols into seven layers—application, presentation, session, transport, network, data link, and physical. By hiding the operations of each layer behind well-defined interfaces, a process known as encapsulation, this layering approach facilitates the independent development of the protocols for each layer. Through the use of well-defined interfaces, the protocols in each layer of this model interact with the protocols only in layers immediately above and below its layer. In other words, a data link layer protocol may interact (perhaps by passing packets) with protocols in the network or physical layers, but not with protocols in the transport or application layers. Indeed, if a network is designed properly, protocols remain blissfully ignorant of the protocols that are not immediately adjacent in the stack. However, in practice, the seven-layer model is commonly simplified to a five-layer TCP/IP reference model, as shown in Figure 9.1, by aggregating the presentation and session layers into the application layer. This five-layer model is similar to the four-layer U.S. Department of Defense (DoD) network model that was used in the development of the Internet. In that model, the application layer is referred to as the process layer, the transport layer is called the host-to-host layer since it deals with end-to-end connectivity, the network layer is the internet layer, and the data link and physical layers are combined into the network access layer.

In the TCP/IP reference model of Figure 9.1 that is considered in the remainder of this chapter, the physical layer handles modulation, error coding, transmission, and reception. The purpose of the physical layer is to efficiently transmit and receive data bits with as few errors as possible. In effect, the physical layer comprises all aspects of the UWB communications covered to this point. The data link layer groups data bits into data link layer protocol data units (PDUs), called frames, that handle frame errors and control the flow of frames. The basic service provided by the data link layer is moving a frame from one node to a neighboring node via a single communication link. As previously discussed, this requires the use of a form

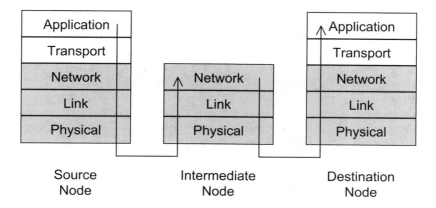

Figure 9.1: A Five-Layer Protocol Reference Model.

of addressing and, when multiple nodes must share the physical medium, a mechanism for controlling access to the medium; this is typically performed by a MAC protocol. The primary duty of the network layer is to determine appropriate paths between indirectly connected devices. The algorithm that determines these paths is appropriately called a routing algorithm, and it is realized by a routing protocol. For Internet Protocol (IP) networks, the network layer is realized by IP and associated routing protocols, such as the Border Gateway Protocol (BGP) or Open Shortest Path First (OSPF) protocol. To facilitate the network protocol's routing algorithm, another level of addressing is typically provided by the network layer. Residing between the application and the network layers, the transport layer provides an end-to-end logical connection between the source and destination nodes. As illustrated in Figure 9.1, transport layer protocols function in the end systems, the source and destination nodes, but not in the intermediate nodes that perform routing at the network layer. Transport layer protocols can also be used to ensure reliable end-to-end communications, but this is not always the case. For instance, the User Datagram Protocol (UDP) provides unreliable best-effort connectionless service to the application layer, while the Transmission Control Protocol (TCP) provides a reliable connection-oriented service. Sitting atop these layers is the application layer. The application layer runs application processes like electronic mail and web services and provides the interface presented to the user.

Broadly, the remainder of this chapter focuses on the data link layer, including medium access control, and the network layer, especially as it relates to ad hoc network formation and routing. This is not to downplay the importance of the application and transport layers, but rather is due to the relative lack of maturity of UWB networks and the fact that UWB networks are developing in a bottom-up manner. Indeed, the transport layer is important in wireless networks as the error characteristics of the wireless link can significantly influence the performance of TCP [5]. Also, novel applications may present new requirements that affect the design of the network and data link layers.

Specifically, the remainder of this chapter is organized as follows. Section 9.1 addresses the issues that make UWB networks different from more traditional networks. Section 9.2 briefly summarizes physical layer issues that affect UWB networking. Section 9.3 discusses data link layer design and several medium access schemes that are suitable for UWB systems. Section 9.4 presents two possible architectures for UWB multiple hop ad hoc networks and corresponding routing schemes. Section 9.5 provides an overview of other networking-related issues.

## 9.1   How Is UWB Networking Different?

In many ways, the networking protocols that should be used for UWB networks are similar to those appropriate for other wireless networks. However, the properties of the UWB physical channel necessitate some changes, especially in data link layer protocols. Further, to optimize performance the design of the data link and network layer protocols should account for the characteristics of UWB physical links, and likely applications and deployment scenarios for UWB networks.

For directly connected nodes, the basic requirements and objectives for the data link layer protocols do not change with the introduction of UWB technology, namely, ensuring reliable communications across a link, minimizing the likelihood of channel contention, providing flow control, and supplying basic addressing information. However, in the context of data link layer protocols, notions of a channel, methods for detecting transmission activity by other nodes, and self-interference are all affected by the choice of link parameters, including modulation schemes. The specific schemes used for these protocols in a UWB network must reflect the unique properties of UWB communications links.

The fundamental requirements of the network layer also do not change for UWB. UWB networks will likely often be ad hoc in nature. The mechanisms to best meet the requirements of the network layer for wireless ad hoc networks are a focus of current research and are certainly not well understood for UWB, which is a nascent networking technology. There are opportunities to leverage both radio link characteristics, using cross-layer design, and application requirements to optimize network layer protocols. For example, UWB devices in an ad hoc network may self-organize themselves into hierarchical clusters in ways that consider mutual interference, power conservation, and application connectivity requirements.

## 9.2   UWB Physical Layer Issues

The physical layer transmits data bits to one or more receivers and receives data bits from a transmitter using appropriate modulation and error coding schemes to maximize throughput and minimize the bit or symbol error rate. Because a UWB signal occupies a large amount of bandwidth, the data rates in a UWB system can be very high—in excess of 100 megabits per second (Mbps). In an Impulse UWB (I-UWB) [6] or a Direct Sequence UWB (DS-UWB) [7] system the pulse is so narrow, less than 10 ns and closer to 1 ns in some schemes, that synchronous transmission is essential to ensure good reception. In addition, because the bandwidth used by a UWB system is so large (several GHzs of bandwidth) the U.S. Federal Communications Commission (FCC) has imposed strict regulations on transmission power for UWB transmitters. This restriction limits the maximum effective radio range for most UWB devices to be about 10 meters for high data rates. Synchronous transmission, high data rates, and short radio range are unique characteristics of the physical layer that influence the design of higher layer protocols, especially the data link layer.

Several modulation schemes can be applied in a UWB physical layer. For a single band UWB system, there are many modulation schemes to choose from, such as impulse modulated or direct-sequence phase coded. I-UWB uses time hopping codes for signal modulation, whereas DS-UWB uses direct sequences to spread a symbol over several UWB pulses. For multichannel (MC) UWB systems, Orthogonal Frequency Division Multiplexing (OFDM) is a possible modulation scheme. The data link design depends on the particular modulation scheme used in a UWB system; thus, the design of the data link layer should differ with the modulation scheme, as discussed in Section 9.3.

## 9.3   Data Link Layer Design

The data link layer moves a frame from a transmitting node to one or more receiving nodes that are within the radio range as determined by the physical layer. The data link layer is commonly defined as having two sublayers, the logical link control (LLC) sublayer and the medium access control (MAC) sublayer. The MAC sublayer allows multiple devices to share a single medium, while the LLC is responsible for realizing a point-to-point link between endpoints and can, optionally, provide error detection and control functions. The MAC is a particularly important component of the data link layer as it is an important factor in determining the performance of communication between adjacent nodes. MAC techniques are generally well understood, but UWB introduces some new constraints and changes some underlying assumptions, as discussed in this section.

### 9.3.1   Objectives of the Data Link Layer

In general, a data link layer should meet the following four key objectives:

1. Reliable data delivery, including addressing and framing, via a single communication channel, as provided by the physical layer protocol.

2. Point-to-point flow control to prevent receiver buffer overflow at the receiver.

3. Power conservation to minimize power consumption at the sender and receiver and to reduce the likelihood of interference.

4. Fair and efficient resource sharing between participating nodes.

## Reliable Data Delivery

A wireless link, UWB or otherwise, is an unreliable medium, so packet error rates can be much higher than in wired links. If the data link layer does not provide a reliable data delivery service, errors are detected only on an end-to-end basis, and any corrupted packets have to be retransmitted on an end-to-end basis by the transport layer or, perhaps, by the application layer. This is clearly inefficient if there is a sufficiently large probability of lost packets in the path between the source and destination nodes. Error detection and retransmission at the data link layer results in corrupted packets being retransmitted more quickly and over only a single communication link. For a multiple hop network with one or more unreliable links, as in a wireless network, retransmission at the data link layer rather than at the transport layer can significantly reduce overhead and decrease latency by correcting errors at the link where they occur. Thus, the data link layer in a wireless network that has sufficiently high error rates needs to provide reliable data delivery for efficient operation.

Reliable data delivery can be achieved using acknowledgment schemes, where a receiver sends back an acknowledgment (ACK) packet to the sender to acknowledge the successful reception of one or more data packets. If no ACK is received before some time-out expires or if a negative acknowledgment (NACK) is received, the sender retransmits the lost data packet or packets. For some applications with stringent delay requirements, such as real-time audio or video, even link level retransmissions introduce too much latency and should not be used. In addition to acknowledgment schemes, forward error correction (FEC) schemes may be used at the data link layer or physical layer to reduce the effective error rate as seen by higher layer protocols. FEC can reliably correct corrupted data bits in a frame, up to a number of bits limited by the particular error correction coding scheme employed. Because FEC consumes link capacity by transmitting redundant information and increases the complexity of the transmitter and receiver, some data link layers implementations do not use FEC mechanisms.

## Point-to-Point Flow Control

If a sender transmits data frames across a link faster than the receiver can receive and process them, the receiver must store the frames in a buffer. Buffering can accommodate a short burst where the rate that frames are sent exceeds the rate at which the frames can be processed at the receiver. However, if this situation persists for too long, the buffers at the receiver fill up and buffer overflow occurs, leading to frames being lost. Lost frames result in performance degradation, possibly introducing retransmissions at the data link, transport, or application layer. To prevent buffer overflow, the link layer can introduce flow control to throttle the sender so that it can send data frames no faster than the frames can be processed at the receiver. Flow control provides a feedback mechanism to make the sender aware of how many additional packets the receiver can handle. Flow control is a relatively straightforward process for one-to-one connections, but is problematic for one-to-many and many-to-one communication patterns because coordination is required among more than two nodes.

## Power Conservation

Because wireless nodes are often battery powered, device- and network-level power conservation to extend battery life is an important design consideration. Many wireless data link layer standards define power-saving "sleep" or "snooze" modes. Such power-saving modes allow a node to temporarily turn off certain components, such as its transmitter and receiver, when it is not actively engaged in communication. Power-saving modes defined in the IEEE 802.15.3 standard are described in detail in Appendix 10.B, "UWB Standards for WPANs."

## Fair and Efficient Resource Sharing

The main responsibility of the data link layer's medium access control sublayer is to ensure fair and efficient resource sharing. Because a wireless link is, essentially, a broadcast medium, sharing radio resources among the devices in the network is an important issue. There are two major categories of medium access control schemes: contention-based protocols and collision-free channel partition protocols. In contention-based protocols, no central control node is needed for allocating channel resources to other nodes in the network. To transmit, each node must contend for radio resources. Collisions result when more than one node tries to transmit at the same time. To resolve persistent conflicts in transmission, contention-based protocols often use random backoff schemes after (or, in some cases, even before) collisions are detected. Because each node transmits at will without the benefit of global coordination, contention-based protocols are also called random access protocols. In this chapter, we use the term contention-based protocols for consistency. Examples of well-known contention-based protocols include Aloha [8], Slotted Aloha [9], Carrier Sense Multiple Access with Collision Detection (CSMA/CD) [10] as used in Ethernet, and Carrier Sense Multiple Access with Collision Avoidance (CSMA/CA) [1] as used in IEEE 802.11's MAC sublayer.

To eliminate collisions, collision-free protocols assign dedicated channel resources to each node that wishes to communicate. This works well for constant bit rate traffic, such as uncompressed voice traffic. However, for variable bit rate traffic, which is typical for data applications, channel resources will be wasted if there is no packet queued for transmission. Therefore, the utilization of channel capacity can be low for bursty data traffic. Examples of channel partition protocols include Time Division Multiple Access (TDMA), Frequency Division Multiple Access (FDMA), and Code Division Multiple Access (CDMA).

For a UWB network, interference avoidance or mitigation is another important design objective. Because a UWB signal occupies such significant bandwidth, it may interfere with other wireless devices occupying some part of the same band, thus reducing their effective bandwidth. Moreover, other radio frequency (RF) transmissions that fall in the frequency band of a UWB transmitter may corrupt communications in a UWB network. Interference mitigation may also lead to reduced transmission power and, thus, can help nodes achieve longer battery life. Several interference mitigation mechanisms have been proposed for the emerging IEEE 802.15.3 standard. For instance, when a UWB network detects either an interferer or a non-IEEE 802.15.3 network operating in the network's current channel or overlapping with the current channel, it may either change channels to an unoccupied band or reduce the maximum transmission power in the network to avoid interference. In addition to these two schemes, an IEEE 802.15.3 network can merge with another IEEE 802.15.3 network if it detects interference coming from the other network. Another proposed technique considers interference avoidance or mitigation during the network formation phase [11]. This heuristic clustering algorithm forms clusters of nodes in a manner that minimizes interference subject to constraints on radio range and multiple access capability.

A UWB network may use a variation or a combination of different medium access schemes. The following sections introduce popular contention-based and channel partition or collision-free medium access protocols and discuss medium access schemes that are appropriate for UWB networks.

## 9.3.2   Contention-Based Medium Access Control

In contention-based or random access MAC schemes, nodes independently decide when to transmit, so contention may occur, which leads to frame collisions. The simplest contention-based MAC protocol is pure Aloha, which is described next. Slotted Aloha improves on the performance of pure Aloha by synchronizing the possible transmission times for nodes and, as described next, reduces the probability of collisions. Carrier sensing, as used in Carrier Sense Multiple Access (CSMA), yields even greater improvement by requiring nodes to detect if the channel is idle or not (by sensing the transmission carrier) and to not transmit if they detect that the channel is in use [12]. CSMA is extended with schemes to further reduce the likelihood of collision in Carrier Sense Multiple Access with Collision Avoidance (CSMA/CA).

### Aloha

The earliest contention-based medium access scheme, appropriately called Aloha, was developed in the early 1970s by Abramson at the University of Hawaii [8]. The basic operation of Aloha is simple, yet elegant: stations can transmit whenever they have a packet that needs to be sent. If a collision occurs, the data packet is corrupted. The receiver can acknowledge successful receipt of the data packet. If the sender does not receive an acknowledgment within a certain time-out period, the sender assumes that there was a collision. The sender then waits a random amount of time and sends the packet again in another frame.

If one or more other frames are sent during the transmission of a frame, the frame experiences a collision, as illustrated in Figure 9.2. Since a station will transmit a frame any time that it has a packet to send, two frames will collide if packets arrive at two or more stations during an interval that is less than the time to transmit one frame. The success or failure of a frame transmission in Aloha can be thought

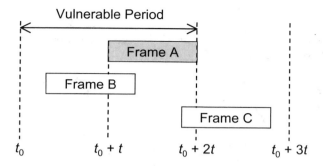

Figure 9.2: Vulnerable Period for a Frame in Aloha.

of in terms of a vulnerable period. Let $t$ be the time to transmit a frame. The vulnerable period for a given frame begins at time $t$ prior to the start of transmission of the frame and extends to the end of the time to transmit the frame. Thus, the vulnerable period is of length $2t$. Any packet that arrives or any retransmission that is scheduled within the vulnerable period will collide with the frame. As illustrated in Figure 9.2, either Frame B, which is transmitted at any time $t$ prior to the start of transmission of Frame A, or Frame C, which is transmitted during the transmission of Frame A, will cause a collision with Frame A. Assuming all frames have fixed length and packets arrive at the data link layer according to a Poisson process, we can calculate the probability that a frame is transmitted without a collision as the probability that exactly one frame is transmitted during the vulnerability period. If $G$ denotes the offered load, that is, the number of transmission attempts during each frame transmission time $t$, the throughput, $S_{Aloha}$, expressed as the fraction of transmission times is the probability that exactly one transmission occurs during two frame transmission times [8]

$$S_{Aloha} = Ge^{-2G} \qquad (9.1)$$

Note that throughput $S$ is the fraction of the channel used for successful transmission if the time to transmit one frame is normalized to $t = 1$. Figure 9.3 shows the throughput for Aloha. As seen, the maximum achievable throughput is only 18 percent of the channel capacity and occurs when the offered load is $G = 0.5$. While simple, Aloha fails to effectively use channel resources. Aloha also suffers from stability problems that can occur when a large number of stations have backlogged frames that need to be transmitted [13].

## Slotted Aloha

Slotted Aloha, as the name implies, adds the concept of time slots to Aloha [14]. Nodes are synchronized so as to implement discrete time slots, with the length of each slot being the time to transmit one frame. When a node has one or more packets to send, it must wait for the beginning of the next time slot to begin transmission.

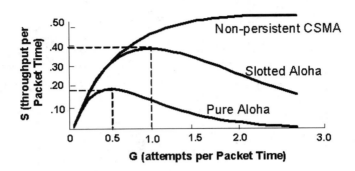

Figure 9.3: Theoretical Throughput Versus Offered Load for Aloha, Slotted Aloha, and CSMA.

By restricting the starting time of frame transmissions, collisions can occur only when two frames are transmitted in the same time slot, as illustrated in Figure 9.4. A collision will occur in slot $j$ if frames become available for transmission at two or more nodes during slot $j - 1$. Thus, the vulnerable period for slotted Aloha is one time slot or frame transmission time, versus two frame transmission times as in Aloha. Any frame that becomes ready for transmission in slot $j$ must wait for slot $j + 1$ to be transmitted. In Figure 9.4, Frames A and B both become ready for transmission during slot $j - 1$ (from time $t_0$ to $t_0 + t$), which will result in a collision. However, Frame C that becomes ready for transmission during slot $j$ (from time $t_0 + t$ to $t_0 + 2t$) will not collide with either Frame A or Frame B.

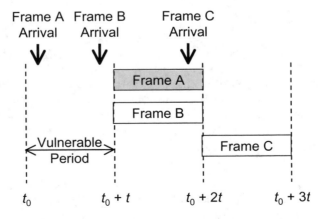

Figure 9.4: Vulnerable Period for a Frame in Slotted Aloha.

Because the vulnerable period is reduced by a factor of one-half compared to Aloha, the probability of a collision in Slotted Aloha is reduced by a factor of one-half, and the throughput, $S_{Slotted\ Aloha}$, is doubled, as indicated in the following expression and illustrated in Figure 9.3.

$$S_{Slotted\ Aloha} = Ge^{-G} \qquad (9.2)$$

However, throughput still has an exponential dependence on the offered load, $G$, so a small increase in the offered load can dramatically decrease system performance. This performance degradation occurs mainly because all nodes transmit at will without considering transmissions at other nodes.

## Carrier Sense Multiple Access Protocols

If a node can detect whether or not other nodes are currently transmitting, it can adapt its behavior accordingly. Carrier Sense Multiple Access (CSMA) is based on this idea.

A family of CSMA protocols was proposed in the 1970s by Kleinrock and Tobagi [12]. The basic idea behind a CSMA protocol is simple: a node first senses the channel to make sure it is idle before starting to transmit a frame. This behavior is sometimes called "listen before talking." If the channel is not busy, the node can transmit. If the channel is busy, the node will defer transmission. The exact behavior of a node that senses a busy channel leads to different versions of CSMA.

In nonpersistent or 0-persistent CSMA, a node with a frame ready for transmission first senses the channel. If the channel is idle, the node immediately transmits the packet. If the channel is busy, the node waits for a random amount of time, the backoff interval, and senses the channel again. The throughput, $S_{0-persistent\ CSMA}$, for 0-persistent CSMA is as follows [12].

$$S_{0-persistent\ CSMA} = \frac{Ge^{-aG}}{G\left(1 + 2a\right) + e^{-aG}} \qquad (9.3)$$

Here, $a$ is the ratio of propagation delay to packet transmission time. Long propagation delays can negatively impact the performance of CSMA protocols because long propagation delays increase the chance that a node which is ready to transmit cannot hear a transmission in progress at a distant node due to the long propagation delay. CSMA schemes may vary based on the behavior of a node when a packet arrives and the channel is busy [12]. As noted above, in 0-persistent CSMA, the node will always defer transmission of the packet for some random back-off interval that shares the same distribution as the backoff interval following a packet collision. In $p$-persistent CSMA, the distribution of the random backoff interval for new packets and the distribution of the random backoff interval for backlogged packets differ. Parameter $p$ is associated with the backoff interval distribution for new packets. In 1-persistent CSMA, a new packet arriving when the channel is busy will be transmitted at the first available idle time.

Figure 9.3 shows that CSMA usually yields better channel utilization than Aloha or Slotted Aloha. CSMA performs better than Aloha or Slotted Aloha because of the carrier sensing scheme that avoids collisions with transmitting stations.

Some more recent random access MAC schemes also use carrier sensing to increase throughput. Examples include Carrier Sense Multiple Access with Collision Avoidance (CSMA/CA) [1] and Carrier Sense Multiple Access with Collision Detection (CSMA/CD) [10].

## CSMA/CA

CSMA/CA is a commonly used protocol in wireless local area networks, including the IEEE 802.11 MAC standard [1]. CSMA/CA leverages the performance benefits of CSMA, but extends CSMA to reduce the likelihood of a collision. CSMA/CA avoids use of collision detection, as in CSMA/CD, which is used in Ethernet-wired local area networks. Collision detection is not practical in a wireless environment because a node's own transmission will typically obscure any transmissions at other nodes that may cause a collision at a receiver. Additionally, it is impossible to ensure that all transmitters detect a collision if one occurs at an intended receiver, as required in CSMA/CD.

CSMA/CA can use a request to send (RTS) and clear to send (CTS) protocol to largely avoid the "hidden terminal" problem. Because radio signals attenuate over distance, simultaneous transmissions may lead to collisions at the receiver even though both senders have sensed an idle channel.

Figure 9.5 illustrates the hidden terminal problem. As illustrated, nodes A and C are out of each other's radio range. Thus, neither node A nor node C can hear whether the other node is currently transmitting. If node A is transmitting to node B, node C may still sense that the channel is idle. Therefore, node C starts transmitting and packets from nodes A and C collide at node B. Node C is a hidden terminal with respect to node A and vice versa. However, because of the technical difficulties involved in carrier sensing in UWB systems, CSMA is not an effective choice for a random access MAC protocol. Instead, slotted Aloha may be used for exchanging control frames and for node association with a central control node. However, the inefficiencies of slotted Aloha make it unsuitable for normal, high

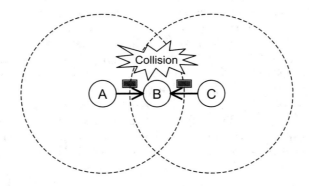

Figure 9.5: Illustration of the Hidden Terminal Problem.

data rate data transfers. A channel partition or collision-free MAC scheme is better suited for data transfer in most UWB networks.

### 9.3.3   Channel Partition Medium Access Control

TDMA, FDMA, and CDMA are commonly used and widely investigated collision-free medium access control protocols. They differ in how they partition physical layer resources among nodes. TDMA partitions physical layer channels into a set of predetermined time slots (also often called channels) and assigns different time slots to different nodes in the network. While data transmissions from different nodes are sent at different times, they share the same frequencies in a TDMA system. FDMA partitions the allocated bandwidth (frequency) into channels and assigns these channels to nodes in the network. In an FDMA system, data transmissions occur at different frequencies, but can occur at the same time. While TDMA and FDMA assign time slots and frequency channels, respectively, to nodes, CDMA assigns different spreading codes to different nodes. Therefore, CDMA allows simultaneous transmissions within the same frequency band, provided that the transmitters use different spreading codes. Note that the allocation of time, frequencies, or lengths of spreading codes can be controlled to allocate different quality of service (QoS) levels in TDMA, FDMA, and CDMA, respectively.

Transmitters and receivers must be synchronized in I-UWB and DS-UWB for efficient communication. Therefore, TDMA becomes a natural choice for the medium access scheme in I-UWB and DS-UWB systems. To use TDMA, UWB nodes need to be synchronized by a central control node. Finer synchronization is achieved through the preamble sequence transmitted along with each data packet. The IEEE 802.15.3 standard defines a TDMA scheme for data communications [15]. Appendix 10.B describes the medium access schemes used in IEEE 802.15.3 in more detail.

### 9.3.4   Multiple Access Protocols for UWB Networks

This section describes a multiple access protocol that is appropriate for UWB networks. As stated previously, a collision-free medium access control scheme is well-suited for the high data rates that can be achieved by the UWB physical layer. A random access scheme is better suited for control signaling and initial association of a node with a control node because collision-free schemes require prior coordination between a node and the control node.

#### Collision-Free Medium Access Schemes for Data Communications

In IEEE 802.15.3, a pure TDMA scheme is defined for multiple access among nodes within a piconet, regardless of the modulation scheme that is used. However, for different UWB physical layers, different multiple access schemes have been proposed to assign separate channels to simultaneously operating piconets that are in close proximity. Interpiconet channels are separated by frequency hopping codes for MC-UWB and by spreading codes for DS-UWB. Thus, current proposals for the IEEE 802.15.3 standard use TDMA within a piconet, and CDMA among different piconets.

As an alternative for consideration, a hybrid TDMA/CDMA medium access control scheme within a piconet can also be a good choice. For example, the current 802.15.3a DS-UWB proposal defines six spreading code sets that allow six independent piconets to be collocated within each other's interference range. In cases where less than six piconets are collocated, unused code sets can be used to provide higher data rates within one or more of the piconets.

The major advantages of a hybrid CDMA/TDMA scheme are greater flexibility and increased adaptability. Flexibility allows networks to be configured in different ways, while adaptability implies that the network can dynamically modify its configuration to accommodate different channel, network, and application environments. A UWB network is a likely choice for multimedia or other applications that have high data rate requirements or that require differentiated QoS. To meet different QoS requirements and to efficiently use channel resources, the medium access control scheme needs to be as flexible as possible. A pure CDMA scheme assigns one or more spreading codes to a single user for the duration of its connection, while a pure TDMA scheme only allows one user to transmit during a particular time slot. Pure CDMA or pure TDMA can achieve only one "degree of freedom," meaning channel partitioning can be based only on the assignment of spreading codes or time slots.

A hybrid TDMA/CDMA scheme is more flexible since it can achieve two degrees of freedom. This flexibility can be used to adapt to different conditions. A hybrid TDMA/CDMA scheme may assign spreading codes to a user only at certain times, such as when the user has queued packets, thus allowing multiple users to transmit during the same time slot. A central controller may assign the same spreading code to different users in different time slots or it may dynamically assign different time slots to users, which ensures great adaptability. The central control node can broadcast information about the assignment of time slots and codes to other nodes in the network to ensure coordination. Also, different sets of spreading or frequency hopping codes, different time slots, and different frequency bands (for MC-UWB systems) may be allocated to neighboring UWB networks to reduce or eliminate interference.

## Random Access Schemes for Control Signaling

While a collision-free multiple access control scheme is desirable for the transmission of data frames, it is typically not feasible for some control signaling. Before a node is associated with a central control node, it cannot obtain a time slot or a code allocation because the central control node does not know of the existence of the node. A random access scheme, such as slotted Aloha, can be used to enable control frames to be sent to establish an association between a node and a control node. Random access can also be used to "elect" a control node when a new network is formed or if the previously designated control node can no longer perform that function due to node failure or node movement. The central control node listens during all uplink time slots dedicated for random access, and detects control frames sent by any nodes that want to associate with the central control node. A suitable scheme for many applications is to use slotted Aloha for random access in the uplink (frames sent to the controller node) and TDMA for the downlink (frames sent from

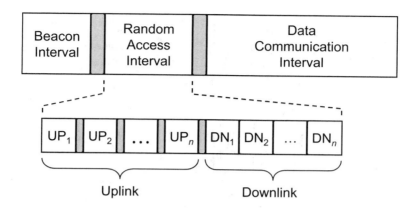

Figure 9.6: Example Communications Structure for a UWB Network.

the controller node) [16]. This scheme is illustrated in Figure 9.6. The scheme uses a beacon signal generated by the central control node to advertise the existence of the control node and to synchronize other nodes that wish to use the control node. The beacon interval is followed by the random access interval, which consists of several uplink time slots ($UP_i$) where nodes use slotted Aloha to send frames to the central control node, and the same number of downlink slots ($DN_j$) where the central control node uses TDMA to send frames to the other nodes. Normal data frames are sent during the data communication interval that could use, for example, the hybrid TDMA/CDMA scheme discussed previously. Guard times separate the three intervals to prevent overlaps in the intervals due, for example, to synchronization differences between nodes.

Using this scheme, the association procedure for a node is as follows:

1. The node first acquires downlink synchronization with the central control node using the beacon transmitted by the control node.

2. The node then randomly picks one of the $n$ uplink time slots in the random access interval and transmits its association request during the selected slot.

3. If the $k^{th}$ uplink time slot is picked, the node attempts to receive the response from the controller in the $k^{th}$ downlink time slot.

4. If a valid association response is not received in the downlink slot, the node assumes a collision has occurred, and resends its association request in the subsequent random access interval that follows the next beacon interval.

To reduce association latency and to increase channel utilization, a hybrid TDMA/CDMA scheme may be used, as shown in Figure 9.7 [16]. Slotted Aloha is used during the uplink time slots during the random access interval, as in the scheme illustrated in Figure 9.7. However, there is only one downlink time slot. The system uses CDMA to separate different channels during the downlink time slot.

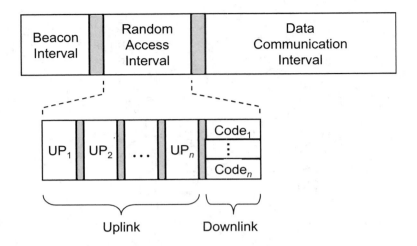

Figure 9.7: Modified Communications Structure for a UWB Network.

Note that the uplink and downlink time slots are separated using TDMA because the uplink slots must occur prior to the downlink slots.

Most steps of the association procedure for this second scheme are the same as for the first scheme, except step 3. In step 3, if the $k^{th}$ uplink time slot is picked, the node attempts to receive the response from the controller in the downlink time slot using the $k^{th}$ spreading code.

Because this modified scheme uses only one downlink time slot, it is possible that some radio resources can be freed up and channel utilization can be improved. The time for a node to associate with the central control node is also reduced because the time to receive a response is reduced. However, the central control node must be able to send multiple signals at once, which increases the complexity of the PHY layer.

### 9.3.5   Forward Error Correction and Automatic Repeat Request

Link reliability is an important consideration in the design of a wireless network. Error recovery is a critical aspect of the data link layer. Two types of error recovery schemes can be used: forward error correction (FEC), where coding is used to allow the receiver to extract the correct packet information from the frame even in the event of certain bit errors; and automatic repeat request (ARQ), where the sender and receiver coordinate to retransmit lost frames or frames received in error.

A wireless system can use error correction coding at both the physical layer and the data link layer to enhance error correction capability. However, because FEC necessarily introduces redundancy in data packets, it may result in wasted bandwidth, especially when the wireless channel conditions are good and the data error rate is low.

Compared to FEC, ARQ is easier to implement, and retransmission overhead is incurred only when there is an error. The basic idea of ARQ is quite simple. For every correctly received packet, the data link layer at the receiver side sends back a positive acknowledgment (ACK). If the received packet is corrupt, a negative acknowledgment (NACK) is sent back, or the sender times out waiting for an ACK, and the sender retransmits the lost packet. However, when the error rate is high, ARQ may introduce excessive delay and waste bandwidth.

A pure ARQ scheme may be sufficient for a small UWB network or a point-to-point UWB link due to the short radio range, robustness to multipath, and resilience to RF interference of UWB. For a larger UWB network, especially for a multihop ad hoc network, FEC combined with ARQ is likely a better choice. Cross-layer design techniques can be used to make the data link layer aware of the radio range, network size, and the interference level. One way to do this is to have the physical layer measure and predict future wireless channel conditions, for example, in terms of effective bit error rate, and then report the prediction to the data link layer. Based on the physical layer's prediction, the data link layer can adaptively choose an appropriate error coding rate, and decide whether or not ARQ should be activated. Adaptive error correction coding with ARQ has the advantages of both FEC and ARQ under different channel conditions, but comes at the cost of increased complexity at all nodes.

## 9.4    UWB Multiple Hop Ad Hoc Networks

A network is required to enable data communications beyond point-to-point link communication. Some networks, called infrastructure-based networks, rely on preexisting infrastructure to enable multiple hop communication. Other networks, called ad hoc networks, are self-organizing, require no infrastructure, and can be set up "on the fly." Under many circumstances, mobile nodes in an ad hoc network do not have direct physical links to all other nodes; that is, not all nodes are within radio range of all other nodes in the network. A multiple hop ad hoc network enables nodes without direct physical links to communicate through one or more intermediate nodes. In addition, a multiple hop ad hoc network may extend radio coverage and allow efficient radio resource sharing.

We can categorize multiple hop ad hoc network topologies into one of two types, flat topologies and hierarchical topologies. A flat topology is a strictly peer-to-peer, fully distributed network where each node can serve as both an ordinary node (a source or a destination) and a router. A node acting as a router serves as an intermediate node on the path between source and destination nodes that are not directly connected at the physical layer. Figure 9.8 illustrates a flat network topology. In the figure, a path is shown from source node A to destination node D using node B and node C as intermediate routers.

A hierarchical network topology consists of one or more localized networks or clusters of nodes called tier 1 networks and a single backbone network called the tier 2 network, as illustrated in Figure 9.9. All nodes in a cluster can be directly connected or, as shown in the figure, multiple hop routing may be needed to reach

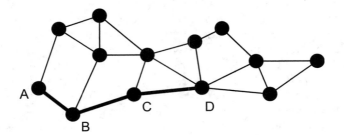

Figure 9.8: A Flat Multiple Hop Ad Hoc Network.

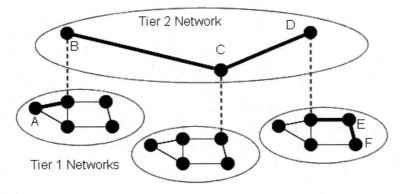

Figure 9.9: A Hierarchical Multiple Hop Ad Hoc Network.

all nodes in a cluster. Within a cluster, that is, within a tier 1 network, ordinary nodes are controlled by a cluster head that participates in a multiple hop tier 2 network. As an example, a path from node A to node F is shown in the figure. The example path goes from source node A to node B within a single cluster. The path goes from node B to node C and on to node D in the backbone network. From node D, the path goes to node E and then to destination node F within a single cluster. A fully distributed flat topology offers great flexibility and fairness among all nodes, whereas a hierarchical topology with a central control node in each cluster is often more efficient and more scalable. In a flat topology, each node acts on its own, but all the nodes in the network must participate in distributed management functions such as routing. In a hierarchical topology, a node in each cluster can be "elected" as a central control node by other nearby nodes. A network with a hierarchical topology can use a central control node to manage each cluster, and to communicate with one or more other nodes in the tier 1 network. Bluetooth, which has been standardized as IEEE 802.15.1, uses a hierarchical topology, with a star topology used in tier 1 clusters [2]. The IEEE 802.15.3 high rate wireless personal area network standard uses a hybrid topology in clusters [15]. The control is centrally managed, but data communications are distributed. UWB is being considered as a PHY layer option for IEEE 802.15.3 and, thus, the same hybrid topology will be used with a UWB physical layer.

The following sections discuss hierarchical topologies and flat topologies. Routing algorithms for each of these two types of topologies are briefly introduced.

## 9.4.1    Hierarchical Network Topologies

For a UWB multiple hop network, a hierarchical topology may be an appropriate choice because a large distributed topology introduces significant synchronization and communication overhead. The hierarchical topology allows a hybrid architecture. Nodes are self-organized into clusters and a cluster head is "elected" for each cluster. The cluster head can coordinate spreading code assignments, power control, and other localized functions within its cluster. Cluster heads also participate in a longer range, multiple hop, distributed backbone network. A scatternet in Bluetooth, as shown in Figure 9.10, is an example of a hierarchical topology [17]. In a

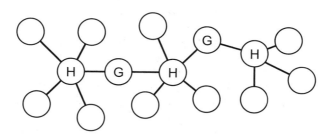

Figure 9.10: Example Scatternet with Cluster Heads (H) and Gateway Nodes (G).

Bluetooth scatternet, nodes are organized into clusters, called piconets in Bluetooth terminology. Adopting a hierarchical topology, some nodes become cluster heads or gateway nodes, while other nodes remain as ordinary nodes that rely on cluster heads and gateway nodes for communication. Cluster heads are sometimes called "masters" and ordinary nodes are sometimes called "slaves." Throughout this section, we will use the terms interchangeably. A cluster head, labeled as "H" in Figure 9.10, controls other nodes in the same cluster and relays traffic within the cluster. A gateway node, labeled as "G" in Figure 9.10, relays control information and directs data traffic from one cluster to another. Thus, multiple hop communication is achieved by connecting master nodes and gateway nodes. Even though cluster heads and gateway nodes carry a higher computational and communication burden than other nodes and may become bottlenecks, routing information propagated in the network can be reduced, especially for large-scale ad hoc networks. This is because the routing function must consider only routing to clusters and not to each individual node, as would be the case with a flat topology.

Initially, all nodes in the network are isolated, that is, a node is not aware of the existence of any of the other nodes. A scatternet formation algorithm can be used to connect all or most of the isolated nodes together into a multiple hop network. Many scatternet formation algorithms have been proposed. Existing algorithms can be classified into three categories: those that use a "super-master" node, those that construct a scatternet to fit a basic topology model, and those that use basic node clustering techniques.

The first category includes algorithms that rely on a single elected "super-master" node to interconnect the rest of the nodes. The super-master node must acquire information from all other nodes through a distributed procedure.

A two-stage distributed clustering algorithm is proposed by Ramachandran, et al. [18]. The first stage uses a randomized distributed algorithm to select master-designates and slave-designates. During this stage, all $N$ nodes in the network conduct $T$ rounds of Bernoulli trials with a probability of $p$, where $p \leq 1$, for each trial. Nodes with exactly one successful trial become master-designates and other nodes become slave-designates.

The second stage is a deterministic algorithm to select masters and a super-master node. First, all master-designates start to solicit information about slave nodes. If a slave-designate responds to the inquiry, it becomes a slave in the cluster as long as the maximum cluster size, $S$, is not exceeded. If a master-designate collects $S$ slaves in its cluster or experiences a time-out after it has collected at least one node in its cluster, it declares itself to be a master. If a master-designate fails to collect any responses before it times out, it becomes a slave-designate and starts to look for inquiries from other master-designates. If a master or a master-designate collects up to a threshold of $k$ responses from other clusters, it becomes the super-master. After a node declares itself as a super-master, it solicits information from all other clusters. After obtaining the required information, the super-master partitions the network into exactly $k$ clusters, with $k-1$ clusters having $S$ slaves and one cluster of size $N$ mod $(S+1)$. Then, the super-master selects appropriate nodes in the clusters to serve as gateway nodes. As simple as this algorithm is, it requires that each node has prior information about the total number of nodes, $N$. In addition, this algorithm needs some predefined parameters, namely $S$ and $k$, whose optimal values may not be easy to determine prior to the deployment of the network.

The Bluetooth Topology Construction Protocol (BTCP) is an asynchronous, fully distributed scatternet formation algorithm developed by Salonidis, et al. [19]. BTCP provides a good example of an approach that may be useful for a UWB network. BTCP has three phases: the leader election phase, the role determination phase, and the connection establishment phase. In the leader election phase, an asynchronous distributed election of a coordinator node is performed. At the end of the first phase, the leader node has a global view of the network. Thus, at the second phase, the leader node can assign roles to the remaining nodes. Some of the nodes are picked as master nodes and some are assigned to be gateway nodes. The rest of the nodes remain as ordinary nodes. During the last phase, all designated master nodes are required to connect to gateway nodes and ordinary nodes. This scheme does not require any *a priori* information about the network. However, it only works for a number of nodes less than or equal to 36, due to the desired properties of the resulting network [19].

Because the two-stage distributed clustering algorithm [18] and BTCP [19] depend on a single node to determine the scatternet topology and notify other nodes, this special node needs to acquire information from the other $N - 1$ nodes in the network. Thus, both algorithms have complexity $O(N)$ at the leader election phase.

Algorithms in the second category construct a scatternet based on a specific topology model, such as a spanning tree topology. Because a spanning tree topology is simple to realize and relatively efficient for packet routing, many scatternet formation algorithms are based on tree structures. Figure 9.11 shows a spanning tree topology with dark nodes as parent nodes.

A spanning tree topology tends to select the least possible number of links to form a connected scatternet. Note that the minimum number of connections is a property of a spanning tree [20]. Moreover, a spanning tree is easy to maintain because of its hierarchical structure. For example, each node maintains routing tables only to its children and grandchildren nodes. When the destination of a packet

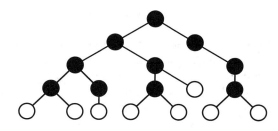

Figure 9.11: Spanning Tree Topology.

is in a node's routing table, the node will transmit the packet to the corresponding children. Otherwise, the packet is forwarded upward to the node's parent.

However, a spanning tree topology has some deficiencies [21]. First, it lacks robustness and may result in a single point of failure. After a parent node fails, all the children and grandchildren nodes are separated from the rest of the network. Second, routing in a tree topology is less efficient compared to routing in several other topologies, such as a mesh topology, because data can only move through a tree in the upward and downward directions. Therefore, the throughput or delay for a spanning tree routing algorithm can sometimes be much worse than for routing algorithms based on other topologies. Third, because all the routing paths go through parent nodes, these nodes, especially the ones higher in the hierarchy, may become heavily loaded, thus becoming performance bottlenecks in the network.

"Bluetree" is a simple spanning tree-based algorithm developed by Zaruba, et al. [21]. The assumptions for this algorithm are that there is a predesignated "blueroot" node in the network, that each node knows the identity of its neighbors, and that each node knows whether or not it is already part of a piconet. After nodes in the network power on, the "blueroot" node starts paging its neighbors one by one. If a node is paged but has not yet joined a piconet, it becomes a slave of the paging node. When a node has been assigned as a slave in a piconet, it initiates the one-by-one paging of all of its neighbors. This procedure is recursively repeated until all nodes are assigned to a piconet. A second algorithm proposed by Zaruba, et al., in the same work relaxes these assumptions. Instead of having one "blueroot" node for the whole network, any node that has the highest identification (ID) number among its neighbors can elect itself as a root and initiate the "Bluetree" procedure. The network is then spanned by connecting together disjoint, but adjacent, trees.

Algorithms in the third category do not assume any particular model for the network topology. Instead, they construct "flatter" architectures using basic clustering schemes. "Bluenet" and "BlueStars" are examples of this category of algorithm.

The "Bluenet" algorithm, developed by Wang, et al., has three phases [3]. In the first phase, isolated piconets are formed by clustering. After this phase, some separated nodes may be left out of the network. In the second phase, these separated nodes try to associate with at least one piconet. When the second phase completes, all nodes should be associated with at least one piconet. During the third phase, piconets are connected to form a scatternet. However, "Bluenet" does not guarantee the connectivity of the resulting scatternet. One of the major contributions of Wang, et al., is that they define two metrics for evaluating the performance of scatternets, the average shortest path (ASP) and the maximum traffic flow (MTF) [3].

The "BlueStars" protocol also proceeds in three phases: the topology discovery phase, the piconet formation phase, and the scatternet formation phase [17]. During the first phase, each node discovers its neighbor nodes and makes them aware of its presence. The second phase takes care of piconet formation. Based on a precomputed weight, each node decides whether it is going to be a master or a slave. One master and one or more other nodes establish communication links to form a "BlueStar." During the final phase, each master uses information gathered from neighboring nodes during the first phase to select gateway nodes. The selection of gateway nodes is performed so that the resulting scatternet is connected. The completed scatternet is a connected mesh with multiple paths between any pair of nodes, which guarantees robustness.

## 9.4.2   Flat Network Topologies

The majority of research on routing in multiple hop ad hoc networks assumes a flat topology and a fully distributed architecture. A UWB ad hoc network is sometimes called a UWB mesh network. Even though mesh networking is a relatively new area of research, significant research and development efforts have been dedicated to it. Both IEEE 802.15 and IEEE 802.11 standard groups have formed subgroups to investigate standards for wireless mesh networks. Because each node in a mesh network can act as a router, and the routing operation is completely decentralized, mesh networks offer great flexibility and can ensure fairness among nodes. In a mesh network, nodes cooperate to forward packets for each other to reach destinations that are not directly within the source node's transmission range. Thus, ad hoc routing algorithms are needed to ensure connectivity in such networks, especially where nodes can move, as is the case in mobile ad hoc networks (MANETs).

### Routing Protocols

Figure 9.12 specifies three categories of MANET routing protocols: proactive protocols, reactive protocols, and hybrid protocols. Proactive protocols are typically link-state (table-driven) protocols that maintain up-to-date routing information regardless of whether there is any request for a particular route. The Destination Sequenced Distance Vector (DSDV) protocol is a typical proactive routing protocol [22].

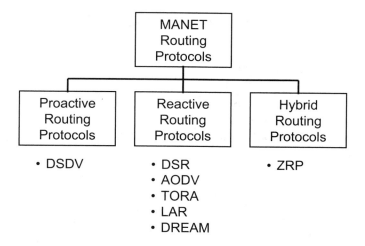

Figure 9.12: Basic Taxonomy for MANET Routing Protocols.

Reactive protocols are called source initiated on-demand routing protocols since route discovery is invoked only when a source node has a packet to send that needs a route. Examples of reactive routing protocols include Dynamic Source Routing (DSR) [23], Ad hoc On-Demand Distance Vector Routing (AODV) [24], Temporally Ordered Routing Algorithm (TORA) [25], Location Aided Routing (LAR) [26], and Distance Routing Effect Algorithm for Mobility (DREAM) [27].

Hybrid routing protocols combine features of reactive and proactive protocols. The Zone Routing Protocol (ZRP) is such a hybrid protocol [28]. Some examples of MANET routing protocols are described in the following sections.

## Destination Sequenced Distance Vector (DSDV) Protocol

DSDV is a table-driven proactive routing protocol based on the classical Bellman-Ford routing algorithm [22]. One of the distinct contributions of DSDV is that it avoids the routing loops that are often generated by the basic Bellman-Ford algorithm. Each node in the network maintains a routing table that records all reachable destinations and the number of routing hops. The method for building these routing tables is the same as the scheme used in the Bellman-Ford algorithm [13]. Bellman-Ford is a distributed algorithm, which means that each node executes the algorithm simultaneously with every other node. The Bellman-Ford algorithm does not assume that a node knows the details of the network topology. Instead, it assumes that a node initially knows only the length (or cost) of its outgoing links and the identity of every destination that is directly connected. Using information supplied by its neighbors, every node computes the shortest distance (or lowest cost) to every other node in the network. And, every node shares information about its shortest (or least cost) paths with its neighbors. Because the source node updates a sequence number every time it sends out a route request, intermediate nodes can distinguish out-of-date, or "stale," route update messages from new update messages. By associating a sequence number with the route request, DSDV avoids the introduction of any loops.

## Dynamic Source Routing (DSR)

As the name suggests, DSR is based on the source node that determines and specifies a route to the destination [23]. The DSR protocol consists of two phases: route discovery and route maintenance. Because on-demand routing protocols create routes only when desired by the source node, they all typically have these two phases. The route discovery phase is initiated only when a source node needs to send one or more packets to a destination to which the source does not have an up-to-date route. When a route has been established, it is maintained by a route maintenance procedure until the route expires.

If a node has a packet to send, it first checks its own routing cache to see whether there is an unexpired route to the destination. If no existing route is found, the source node initiates a route discovery procedure by sending a *Route Request* message that contains source and destination addresses and a unique request identifier. Each node receiving the *Route Request* message checks whether it knows an unexpired route to the destination. If no route is found, the node appends its own address to the *Route Request* packet and rebroadcasts the packet to its neighbors. If an intermediate node knows an unexpired route to the destination or the node is the destination node, a *Route Reply* message is generated. If symmetric links are assumed, the node can simply reverse the route in the route record of the *Route Request* message to send a reply back to the source. If symmetric links cannot be assumed, a node generating a *Route Reply* message may need to initiate its own route discovery procedure.

In a mobile environment, links may fail due to fading, interference, or node movements. Therefore, route maintenance is a crucial procedure in routing protocols for MANETs. In the DSR protocol, route maintenance is accomplished through the use of acknowledgments and *Route Error* messages. Acknowledgments are used to verify the correct operation of the links in a path from source to destination. A *Route Error* message is sent to the original sender of the data packet when a link fails. Upon receiving a *Route Error* message, the sender removes the failed link from its route cache. This may trigger a new route discovery procedure.

As an on-demand routing protocol, DSR is very flexible. However, DSR introduces large routing overhead and does not scale well for use in large networks, because all the routing information has to be carried in packet headers.

## Ad Hoc On-Demand Distance Vector Routing (AODV)

AODV is, in essence, a combination of DSR and DSDV [24]. However, AODV makes improvements over both protocols.

In a manner similar to DSR, AODV also employs two phases: route discovery and route reply. A node initiates a route discovery by broadcasting a Route Request (RREQ) message when it has packets for a destination node for which it has no valid route. Any node that receives the RREQ message updates its next hop table entry with respect to the preceding node in the path back to the source, thereby establishing a reverse path back to the initiator of the RREQ message. If a node does not have a route to the destination node, it rebroadcasts the RREQ message to its neighbors.

If a node knows an unexpired route to the destination or the node is the destination node, a Route Reply (RREP) message is generated and sent by unicast back to the source. Upon receiving the RREP message, each node along the route back to the source updates its next hop table entry with respect to the neighboring node along the path to the destination node. Because the RREP message is forwarded along the reverse path established by the RREQ message, AODV requires symmetric links.

Unlike DSR, AODV does not include route information in every data or control packet header, which reduces overhead. AODV also has less routing overhead than DSDV because it can minimize the number of required broadcasts by creating routes in an on-demand manner.

## Temporally Ordered Routing Algorithm (TORA)

TORA is a highly adaptive reactive routing protocol [25]. It guarantees loop-free routes and can typically provide multiple routes for any pair of source and destination nodes. It is also distributed, in the sense that nodes need only maintain information about their adjacent neighbors. Another distinctive feature of TORA is that the protocol's reaction to link failures typically involves only a localized "single pass" of the distributed algorithm to correct for a failed link.

TORA has three phases: route creation, route maintenance, and route erasure. During the route creation and route maintenance phases, a query and reply process is used to build a directed acyclic graph (DAG) rooted at the destination node. At every node, a separate copy of TORA is run for every destination. For a given destination node, each participating node is assigned a "height." No two nodes may have the same "height," i.e., the set of "heights" is totally ordered. Each link is assigned a direction, upstream or downstream, based on the relative "height" metric of the neighboring nodes. Timing is an important factor in TORA because the "height" metric is based on the logical time of a link failure. Thus, TORA assumes that all nodes in the network are synchronized (at least coarsely). Data packets can be routed from nodes with higher "height" to nodes with lower "height," just like water can only flow from a higher altitude to a lower altitude. TORA maintains a set of totally ordered "heights" at all times.

When a node needs a route to a given destination, it broadcasts a *Query* message containing the address of the destination for which it requires a route. The *Query* message travels through the network until it reaches the destination or an intermediate node that has a route to the destination node. This node then broadcasts an *Update* message listing its height with respect to the destination. As the *Update* message propagates through the network, each node updates its height to a value greater than the height of the neighbor from which it receives the Update. This results in a series of directed links from the node that originated the Query to the destination node. Thus, loop-free multipath routing to the destination can be created.

If a node discovers a destination to be unreachable, it sets a local maximum value of height for that destination. In case the node cannot find any neighbor having finite height with respect to this destination, it attempts to find a new route. If a link failure partitions the network, the node broadcasts a *Clear* message that resets all routing states and removes invalid routes from the network.

## Zone Routing Protocol (ZRP)

ZRP is a hybrid proactive and reactive routing protocol [28]. A routing zone is defined for every node in the network. The zone radius is a predetermined parameter that defines the zone in terms of the minimum number of hop counts (number of links that must be traversed) from the center node. All neighbor nodes within the radius of the zone are considered to be nodes in the routing zone. If the minimum hop count of a node from the center node is exactly the same as the zone radius, this node is called a border node. Each node in the network proactively maintains its own routing zone, using a proactive routing protocol, such as DSDV. Therefore, each node in the zone knows how to reach every other node in the same zone at any time. If a node wants to send a packet, it first checks whether the destination node is in its own zone. If the destination is in the same zone, it can send the packet directly to the destination because proactive routing protocols always maintain up-to-date routing information. If the destination is not in the same zone, the sender initiates a reactive routing protocol, such as DSR. To reduce the number of route search messages that are broadcast, a node only needs to transmit *Route Query* packets to a zone's border nodes. This operation is called "bordercasting." When the border nodes receive the query packet, they repeat the same procedure in their own zone until the destination node is reached. The zone radius is a parameter that can be tuned to fit different network requirements. ZRP becomes a purely proactive routing protocol if the zone radius is set to be equal to or larger than the network diameter (the maximum number of hops between any two nodes in the network) or if the zone radius is set to one.

## Location-Aided Routing (LAR)

Because UWB can provide precise ranging and cooperative nodes can determine location through time-of-arrival (TOA) or time-difference-of-arrival (TDOA) techniques, LAR schemes may be of particular interest for UWB networks. LAR schemes were introduced to reduce the scope of route request flooding. The basic ideas behind different versions of LAR schemes are similar. If a source node knows approximately where the destination node lies, it can restrict the propagation of route request messages only in that direction, significantly reducing the overhead for route discovery.

LAR defines two zones: the expected zone and the request zone [26]. The expected zone is defined as the region that is expected to hold the current location of the destination node. The request zone specifies a region based on the location of the source node and the expected zone. Only intermediate nodes located in the request zone can forward route request messages. First, the source node determines an expected zone based on some prior knowledge of the location and estimated speed of the destination node. Then, the source node calculates a request zone that contains the expected zone and its own location. The request zone is explicitly specified in the route request. Only nodes within the request zone forward the route request message. If a route discovery operation fails to find a route due to stale location information or incorrect estimates of node speed, a larger request zone is used for the next route discovery operation. This new request zone can include the entire network, which results in flooding a route request to the entire network.

## Distance Routing Effect Algorithm for Mobility (DREAM)

DREAM is another routing algorithm that uses location information [27]. DREAM also assumes that nodes know their physical location and can estimate their speeds. However, unlike LAR that uses control packets for route discovery, DREAM uses the flooding of data packets as a routing mechanism. Moreover, DREAM takes the distance effect into consideration when updating routes. Based on the observation that far away nodes seem to move at a lower angular speed as compared to nearby nodes, DREAM uses a time-to-live (TTL) field in location updates to control how far the information can propagate. The TTL value is calculated by estimating the distance between the source and the destination nodes in terms of the number of hops. Nodes periodically update their location information. Because of the TTL value, nearby nodes are updated more frequently than more distant nodes. This can reduce the overhead for routing updates.

Because both the DREAM and LAR algorithms use location information, they can limit the scope of route request flood and potentially reduce the overhead associated with route discovery. However, both algorithms require that nodes know their physical location and have information about their speed, so these two routing protocols may not work well for a network where nodes have limited capabilities or a heterogeneous network with some nodes not having knowledge of their location or speed.

It is worth noting that none of the routing protocols is the best under all conditions. Different routing protocols may be used for different network conditions, including different topology characteristics, different assumptions about mobility, and different traffic loads and communication patterns.

## 9.5   Other Networking Issues

In addition to link layer and network layer issues, as discussed previously, higher layer protocols and system issues must also be considered in the design and analysis of any network, including a UWB network. Here, we consider issues related to the performance of the Transmission Control Protocol (TCP), transport layer protocol [29], and QoS, which is a system issue.

### 9.5.1   TCP Performance in a Wireless Environment

Similar to other types of wireless links, the physical characteristics of the UWB physical layer may impact the performance realized by the transport layer, specifically TCP.

RF interference anywhere in the large amount of bandwidth occupied by a UWB system may lead to signal degradation and, thus, bit and frame errors. Although interference mitigation techniques can be used at the physical and data link layers, some data packets may still be lost. For example, with adaptive techniques it takes time to sense and react to changes in channel conditions. TCP responds particularly poorly to packet loss because it is designed to treat packet loss as an indication of congestion [30]. This is a reasonable assumption in typical wired networks, but is often not valid for wireless links. Multiple packet losses cause TCP to time out before retransmitting lost packets. TCP may reduce its TCP congestion window to as little as the number of bytes in one maximum segment size (MSS). This unnecessarily limits end-to-end throughput because the sender must send packets more slowly. After interference mitigation mechanisms take effect in a UWB system, the signal-to-interference ratio (SIR) may return to a desirable level. However, because the size of the TCP congestion window is reduced, a UWB sender can only send packets at a reduced data rate for a relatively long period of time, which results in wasted bandwidth.

Several improvements can be made to enhance TCP performance in UWB networks. An obvious way is to hide packet errors from TCP. If packet errors can be taken care of at the physical and/or data link layers, TCP will not be aware of the errors and, thus, will not invoke congestion control. However, retransmission at the data link layer poses another problem [31, 32]. TCP estimates the round-trip time (RTT) of a given packet based on previous samples of RTT. The value of the TCP timer, called the retransmission time-out (RTO) interval, is calculated based on the estimated value of RTT. Because errors in a wireless channel tend to be bursty, the estimated RTT may not be a good predication for a sudden increase in RTT due to retransmissions in a data link layer that uses an ARQ scheme for error recovery. In a wired network, a sudden increase in the RTT value is very likely due to congestion at intermediate routers. Therefore, it is reasonable for TCP to act accordingly, specifically by reducing congestion window size. However, in a wireless network, sudden channel degradation that is caused by fading or interference may lead to retransmissions at the data link layer, which affects estimates of the RTT value by TCP. If the errors are not frequent, the RTO value will not account for RTT variations due to data link layer retransmissions. Unaware of what is really happening at the lower layers, TCP may unnecessarily invoke congestion control after a time-out occurs. If the RTO is set to a small value, packet retransmission at the data link layer may trigger congestion control at TCP. Setting the TCP time-out to a larger value, however, may lead to a slow response, that leads to congestion in the network and, thus, slow recovery from congestion losses. Therefore, the value of RTO in a wireless environment should be calculated carefully to account for both interference losses and congestion.

The second way to improve TCP performance is to split each end-to-end TCP connection into two or more separate tandem connections [5, 33]. Because a UWB network is likely confined to a small region, a UWB radio link is likely to be the last (or first) hop in an end-to-end path from sender to receiver. Thus, for a UWB network, a single TCP connection can be split into two tandem connections, one on the wired portion of the route and one over the wireless portion.

By splitting a single TCP connection into two separate connections, each can have different flow control and congestion control mechanisms. Also, maximum packet sizes and time-out values can be different. Therefore, the implementation of TCP or some other transport protocol over the wireless connection can be tuned to perform well in a wireless environment. In addition, error recovery can be much faster due to the relatively short RTT on the wireless link. However, this scheme may require modifications to TCP, which often is not desirable or feasible.

The third scheme is to use explicit notification. Much research has focused on improving the performance of TCP using explicit notifications [34, 35]. If a node determines that the packet loss is due to a transmission error instead of congestion in the network, the node can inform the sender using an explicit notification. Thus, the sender just retransmits the packet without invoking congestion control.

All of these schemes require modifications to the conventional data link layer or to TCP. Because such modification of core network protocols requires updates of base stations and/or routers, all three schemes mentioned have yet to be implemented on a large scale.

## 9.5.2   Quality of Service Management

A UWB network can be designed to carry different types of data traffic with different QoS requirements. At the application layer, QoS is hard to quantify because it generally refers to the application quality as perceived by the user; for instance, the visual quality and/or the sound quality of streaming video content. QoS provisioning maps application-level QoS requirements onto a unique set of network-level QoS parameters or QoS metrics [36]. Conventional QoS metrics include throughput, packet loss rate, end-to-end delay, and delay jitter. QoS metrics for mobile wireless networks may include more parameters, such as power consumption and network coverage. Power saving is important because a network of battery powered devices will not be able to provide any service if the batteries are exhausted. Similarly, an ad hoc network may want to expand its coverage as much as possible to interconnect more devices. Network-level QoS parameters usually have quantifiable values or bounds that can be determined from the application's technical specifications or from experiments.

In wired networks, the types of network services can be classified into the following three major categories:

1. Best effort service provides no performance guarantees, and the system treats all traffic equally.

2. Better than best effort service does not have any deterministic guarantees, but makes a best effort to support the requested QoS requirements. For example, an application that belongs to a higher priority class will receive better service than an application that belongs to a lower priority class. Differentiated Service (DiffServ) provides such better than best effort services [37].

3. Guaranteed service delivers the highest quality of service and guarantees network performance metrics in deterministic or statistical terms. For example, the network may guarantee a certain minimum bandwidth provided to an application or guarantee a delay within a specified value. Integrated Service (IntServ) provides guaranteed services [38].

A good QoS management scheme should not only provide QoS, but should also attempt to maximize the utilization of channel resources. QoS management schemes support QoS by considering different aspects of a network, such as resource reservation, admission control, QoS routing, and packet scheduling.

Even though many QoS management strategies have been proposed for wired networks, they may not be suitable for use in a wireless network due to the challenges posed by the wireless environment. Because of the inherent unreliability of a wireless link, it is impossible to guarantee at any time the fulfillment of QoS requirements at the physical layer. Under this condition, adaptation mechanisms need to be implemented at the data link layer or, possibly, also at higher layers to reduce the impact of an unreliable physical layer on QoS as much as possible. In spite of these mechanisms, QoS requirements still may not be guaranteed deterministically in a wireless environment. Instead, QoS will most likely be provided to applications in a qualitative fashion, where applications with higher priority enjoy generally better service than applications with lower priority. Alternatively, the network performance could be guaranteed in a statistical manner. For example, in the long run and under certain channel conditions, packets of a real-time multimedia application can meet their deadlines with a certain probability.

To meet QoS requirements, all protocol layers and network components must cooperate. However in practice, the introduction of QoS in the application layer places the most demands on the data link and network layers. To meet a given QoS requirement, the MAC sublayer needs to solve the problem of medium contention and provide adaptive scheduling and resource allocation, while the LLC sublayer needs to provide reliable communication over the link that can compensate for impairments at the physical layer. Many MAC protocols based on CSMA schemes have been proposed to provide QoS in a distributed wireless network. Examples include the Group Allocation Multiple Access with Packet-Sensing (GAMA-PS) protocol [39] and the Black-Burst (BB) contention mechanism [40]. According to the current IEEE 802.15.3 standard draft, a piconet coordinator (PNC) acts as the central controller and manages QoS requirements within the piconet. Centralized control schemes usually out-perform distributed schemes, because the central control node has a global view of the network. This type of piconet architecture is similar to that of a traditional cellular network where a base station manages scheduling and resource reservation in a cell. However, there is a major difference between a UWB piconet and a cell in a cellular network. In a UWB piconet, a PNC only provides control functions and does not relay data traffic for devices, whereas a base station in a cellular system both controls scheduling and resource allocation and relays traffic for mobile terminals. Due to this major difference, the scheduling and resource allocation protocols for a UWB piconet can be quite different from those designed for cellular networks. For instance, Giancola, et al., investigated dynamic resource allocation schemes for UWB networks [41].

The network layer should be adaptive enough to accommodate different data traffic characteristics and QoS requirements. Much research has focused on QoS routing, which refers to the discovery and maintenance of routes that can satisfy QoS requirements under given resource constraints. Several QoS routing algorithms have been proposed with a variety of QoS requirements and resource constraints. Some examples include CEDAR [42], Predictive Location-Based QoS Routing [43], and Localized QoS Routing [44]. However, despite this research, QoS routing remains a challenging problem. QoS routing requires frequent updates of link state information, such as delay, available bandwidth, and loss rate, to make policy decisions. The frequent updates may result in prohibitively high control overhead for an ad hoc network with limited bandwidth. In addition, the dynamic nature of wireless ad hoc networks makes maintaining precise and consistent link state information extremely difficult [45].

General QoS support in a multihop ad hoc network is difficult, if not impossible, especially if nodes are mobile or the wireless environment is highly dynamic. Much of the recent research in this area has focused on methods where lower layers provide QoS for the application layer. The resulting protocols are often complex, and practical implementation is difficult or even infeasible. A more realistic approach is based on an adaptive QoS model, which requires applications to adapt to the time varying resources provided by the network. Kumwilaisak, et al., proposed a QoS mapping architecture where QoS requirements are adapted to wireless link variations [46]. In addition, QoS requirements at the application layer can be varied according to other indicators from the network, such as congestion levels detected at the link layer or the network layer.

## 9.6    Summary

While the fundamental principles of networking are the same regardless of the underlying physical layer, UWB has unique characteristics that influence how protocols and a UWB system are designed. A UWB network can be represented by a five-layer model, compatible with the TCP/IP suite, that includes a UWB physical layer, associated data link layer, network layer, transport layer, and application layer. Each layer provides services to the layer directly above it and uses services provided by the layer beneath it. The unique characteristics of the UWB physical layer have the greatest influence on the design of the data link layer. The characteristics of the physical layer and the design of the associated data link layer may also influence the design of the network layer, transport layer, and even application layer, especially if a design is to achieve optimal performance.

A hybrid TDMA/CDMA medium access scheme has good properties for a UWB network because of its flexibility that allows, for example, differentiated services and channel allocations to minimize interference.

A hierarchical topology, rather than a flat topology, has significant benefits for a UWB network. A hierarchical topology supports a hybrid architecture where a selected node in a cluster, called a cluster head, can coordinate the assignment of spreading codes, power control, and other local functions, while also participating in a longer range, multiple hop, backbone network. The Bluetooth scatternet model uses such a hierarchical topology and serves as a good model for UWB multiple hop networks. In a scatternet, clusters are interconnected by cluster heads and gateway nodes. A number of scatternet formation protocols have been proposed.

There are a variety of different routing protocols for multiple hop ad hoc networks. Because no single routing protocol is the best for all types of networks, different routing protocols may be used for different situations.

Beyond scatternet formation and routing, there are additional network-related issues in a UWB network. In particular, there are several mechanisms to improve TCP performance over wireless links, including hiding packet losses from TCP, splitting an end-to-end TCP connection into two separate connections, and using explicit loss notifications. All three schemes have their advantages and disadvantages, and none of them have been implemented on a large scale. Quality of service and QoS management are also potential issues in a UWB network. In general, network protocols need to be able to adapt based on data traffic characteristics, QoS constraints, and link impairments.

Because UWB networking is such a new technology, many advances will be made that improve network-level and application-level performance. Of particular interest to researchers is the information that is potentially available in a UWB network that is not available in a traditional wireless network. Nodes in a UWB system can share an extremely accurate understanding of time with their peers and can determine the distance to neighboring nodes. Nodes located in one network can also cooperatively determine the location of other nodes in the same network [47]. This information can potentially be used to fundamentally alter the way that networks allocate channel resources and route packets, thus leading to novel network architectures and protocols that fully exploit the capabilities of UWB.

# Bibliography

[1] IEEE Std. 802.11, "Wireless LAN Media Access Control (MAC) and Physical Layer (PHY) Specifications," 1997.

[2] IEEE Std 802.15.1, "Wireless MAC and PHY Specifications for Wireless Personal Area Networks (WPAN)," 2002.

[3] Z. Wang, R. Thomas, and Z. Haas, "Bluenet—a New Scatternet Formation Scheme," *Proc. 35th Hawaii International Conference on System Science*, pp. 61-69, January 2002.

[4] J. D. Day and H. Zimmermann, "The OSI Reference Model," *Proc. of the IEEE*, vol. 71, no. 12, pp. 1334-1340, December 1983.

[5] R. Yavatkar and N. Bhagwat, "Improving End-to-End Performance of TCP over Mobile Inter-networks," *Workshop on Mobile Computing Systems and Applications*, pp. 146-152, December 1994.

[6] R. A. Scholtz, "Multiple Access with Time-Hopping Impulse Modulation," *Proc. Military Communications Conference*, vol. 2, pp. 447-450, October 1993.

[7] IEEE Std. 802.15.3a, "DS-UWB Proposal Specification," March 2004.

[8] N. Abramson, "The ALOHA System—Another Alternative for Computer Communications," *Proc. AFIPS Fall Joint Computer Conference*, vol. 37, pp. 281-285, November 1970.

[9] L. Roberts, "Extensions of Packet Communication Technology to a Hand Held Personal Terminal," *Proc. AFIPS Spring Joint Computer Conference*, pp. 295-298, May 1972.

[10] R. M. Metcalf and D. R. Boggs, "Ethernet: Distributed Packet Switching for Local Computer Networks," *Communications of the ACM*, vol. 19, no. 7, pp. 395-404, July 1976.

[11] M. X. Gong, S. F. Midkiff, and R. M. Buehrer, "A Self-Organized Clustering Algorithm for UWB Ad Hoc Networks," *Proc. IEEE Wireless Communications and Networking Conference*, vol. 3, pp. 1806-1811, March 2004.

[12] L. Kleinrock and F. Tobagi, "Packet Switching in Radio Channels: Part I—Carrier Sense Multiple-Access Modes and Their Throughput-Delay Characteristics," *IEEE Transactions on Communications*, vol. COM-23, no. 12, pp. 1400-1416, December 1975.

[13] D. Bertsekas and R. Gallager, *Data Networks*, Upper Saddle River, NJ: Prentice Hall, 1992.

[14] L. G. Roberts, "ALOHA Packet System with and without Slots and Capture," *ACM SIGCOMM Computer Communication Review*, vol. 5, no. 2, pp. 28-42, April 1975.

[15] IEEE Std. 802.15.3, "Wireless Medium Access Control (MAC) and Physical Layer (PHY) Specifications for High Rate Wireless Personal Area Networks (WPAN)," 2003.

[16] M. X. Gong and S. F. Midkiff, "Medium Access Control Schemes for UWB Ad Hoc Networks," *Proc. Ultra Wide Band Summit*, December 2003.

[17] C. Petrioli, S. Basagni, and I. Chlamtac, "Configuring BlueStars: Multihop Scatternet Formation for Bluetooth Networks," *IEEE Transactions on Computers*, vol. 52, no. 6, pp. 779-790, June 2003.

[18] L. Ramachandran, M. Kapoor, A. Sarkar, and A. Aggarwal, "Clustering Algorithms for Wireless Ad Hoc Networks," *Proc. 4th International Workshop on Discrete Algorithms and Methods for Mobile Computing and Communications*, pp. 54-63, August 2000.

[19] T. Salonidis, P. Bhagwat, L. Tassiulas, and R. LaMaire, "Distributed Topology Construction of Bluetooth Personal Area Networks," *Proc. IEEE Conference on Computer Communications*, pp. 1577-1586, April 2001.

[20] D. B. West, *Introduction to Graph Theory*, 2nd ed., Upper Saddle River, NJ: Prentice Hall, 2001.

[21] G. V. Zaruba, S. Basagni, and I. Chlamtac, "Bluetrees—Scatternet Formation to Enable Bluetooth-Based Ad Hoc Networks," *Proc. IEEE International Conference on Communications*, vol. 1, pp. 273-277, June 2001.

[22] C. E. Perkins and P. Bhagwat, "Highly Dynamic Destination-Sequenced Distance-Vector Routing (DSDV) for Mobile Computers," *Computer Communication Review*, pp. 234-244, vol. 24, no. 4, October 1994.

[23] D. B. Johnson and D. A. Maltz, "Dynamic Source Routing in Ad-Hoc Wireless Networks," *Mobile Computing*, T. Imielinski and H. Korth, eds., Dordrecht, The Netherlands: Kluwer Academic Publishers, 1996.

[24] C. E. Perkins and E. M. Royer, "Ad-hoc On-Demand Distance Vector Routing," *Proc. IEEE Workshop on Mobile Computer Systems and Applications*, pp. 90-100, February 1999.

[25] V. D. Park and M. S. Corson, "A Highly Adaptive Distributed Routing Algorithm for Mobile Wireless Networks," *Proc. IEEE Conference on Computer Communications*, vol. 3, pp. 1405-1413, April 1997.

[26] Y. B. Ko and N. H. Vaidya, "Location-Aided Routing (LAR) in Mobile Ad Hoc Networks," *Proc. ACM/IEEE International Conference on Mobile Computing and Networking*, pp. 307-321, November 1998.

[27] S. Basagni, I. Chlamtac, V. R. Syrotiuk, and B. A. Woodward, "A Distance Routing Effect Algorithm for Mobility (DREAM)," *Proc. ACM/IEEE International Conference on Mobile Computing and Networking*, pp. 76-84, October 1998.

[28] Z. Haas, M. Pearlman, and P. Samar, "The Zone Routing Protocol (ZRP) for Ad Hoc Networks," Internet Draft, draft-ietf-manet-zone-zrp-04.txt, July 2002.

[29] Information Sciences Institute, "Transmission Control Protocol," Internet Engineering Task Force, Request for Comments 793, September 1981.

[30] G. Holland and N. Vaidya, "Analysis of TCP Performance over Mobile Ad Hoc Networks," *Wireless Networks*, vol. 8, no. 2/3, pp. 275-288, March-May 2002.

[31] A. DeSimone, M. Chuah, and O. Yue, "Throughput Performance of Transport-Layer Protocols over Wireless LANs," *Proc. IEEE Global Communications Conference*, vol. 1, pp. 542-549, December 1993.

[32] S. Dawkins, G. Montenegro, M. Kojo, V. Magret, and N. H. Vaidya, "Performance Implications of Link-Layer Characteristics: Links with Errors," Technical Report PILC, Internet Engineering Task Force (Internet Draft), June 1999.

[33] B. S. Bakshi, "I-TCP: Indirect TCP for Mobile Hosts," *Proc. International Conference on Distributed Computing Systems*, pp. 136-143, May 1997.

[34] H. Balakrishnan and R. Katz, "Explicit Loss Notification and Wireless Web Performance," *Proc. IEEE Global Communications Internet Mini-Conference*, November 1998.

[35] S. Biaz, "Heterogeneous Data Networks: Congestion or Corruption?," Ph.D. Thesis, Texas A&M University, College Station, TX, August 1999.

[36] A. Ganz, Z. Ganz, and K. Wongthavarawat *Multimedia Wireless Networks: Technologies, Standards, and QoS*, Upper Saddle River, NJ: Prentice Hall, 2004.

[37] S. Blake, D. Black, M. Carlson, E. Davies, Z. Wang, and W. Weiss, "An Architecture for Differentiated Services," Internet Engineering Task Force, Request for Comments 2475, December 1998.

[38] R. Braden, D. Clark, and S. Shenker, "Integrated Services in the Internet Architecture: An Overview," Internet Engineering Task Force, Request for Comments 1633, June 1994.

[39] A. Muir and J. J. Garcia-Luna-Aceves, "An Efficient Packet-Sensing MAC Protocol for Wireless Networks," *ACM/Baltzer Mobile Networks and Applications (MONET)*, vol. 3, no. 2, pp. 221-234, August 1998.

[40] J. L. Sobrinho and A. S. Krishnakumar, "Quality-of-Service in Ad Hoc Carrier Sense Multiple Access Wireless Networks," *IEEE Journal on Selected Areas in Communications*, vol. 17, no. 8, pp. 1353-1368, August 1999.

[41] G. Giancola, L. De Nardis, M.G. Di Benedetto, and E. Dubuis, "Dynamic Resource Allocation in Time Varying Ultra Wide Band Channels," *Proc. International Conference on Communications*, vol. 6, pp. 3581-3585, June 2004.

[42] R. Sivakumar, P. Sinha, and V. Bharghavan, "CEDAR: A Core-Extraction Distributed Ad Hoc Routing Algorithm," *IEEE Journal on Selected Areas in Communications*, vol. 17, no. 8, pp. 1454-1465, August 1999.

[43] S. H. Shah and K. Nahrstedt, "Predictive Location-Based QoS Routing in Mobile Ad Hoc Networks," *Proc. IEEE International Conference on Communications*, vol. 2, pp. 1022-1027, April-May 2002.

[44] X. Yuan and A. Saifee, "Path Selection Methods for Localized Quality of Service Routing," *Proc. IEEE International Conference on Computer Communications and Networks*, pp. 102-107, October 2001.

[45] I. Chlamtac, M. Conti, and J. Liu, "Mobile Ad hoc Networking: Imperatives and Challenges," *Ad Hoc Network Journal*, vol. 1, no. 1, pp. 13-64, January-February-March, 2003.

[46] W. Kumwilaisak, Y. T. Hou, Q. Zhang, W. Zhu, C.-C. J. Kuo, and Y.-Q. Zhang, "A Cross-Layer Quality-of-Service Mapping Architecture for Video Delivery in Wireless Networks," *IEEE Journal on Selected Areas in Communications*, vol. 21, no. 10, pp. 1685-1698, December 2003.

[47] R. J. Fontana and S. J. Gunderson, "UWB Precision Asset Location System," *Proc. IEEE Conference on Ultra Wideband Systems and Technologies*, pp. 147-150, May 2002.

# Chapter 10

# APPLICATIONS
# AND CASE STUDIES

Nathaniel J. August,

Christopher R. Anderson,

and Dong S. Ha

This chapter explores the wide range of applications that can exploit the unique properties of UWB systems. It also reviews some of the first commercial UWB products and applications. For both high and low data rate applications, UWB provides services, such as ranging, robust operation in harsh multipath environments, radar capability, and imaging. Further, UWB may provide these services with a low-cost and low-power implementation.

Section 10.1 discusses UWB's unique set of qualifications for specialized applications, such as Ground Penetrating Radar (GPR), fire and rescue, emergency response, and medical imaging and monitoring. Section 10.2 provides an overview of the three main categories of UWB applications: communications, radar, and location aware communications (a combination of radar and communication applications). Finally, Section 10.3 provides a detailed description and performance evaluation of several commercial UWB products.

## 10.1 Specialized Applications for UWB Signals

With a low duty cycle and wide bandwidth, UWB is naturally suitable to radar applications, which were among the first applications studied for UWB [1]. As the bandwidth of a pulse increases, it provides finer resolution of radar images for improved target identification. UWB also provides low probability of intercept and detection because the wide bandwidth-to-information ratio results in low energy across the spectrum. Thus, UWB was also initially considered for secure communications in military applications. More recently, the high-speed communications capability of UWB has been exploited for wireless personal area networks (WPANs) such as IEEE 802.15.3a (refer to Appendix B), which provides a high data rate physical layer (PHY) for personal area multimedia networking. Furthermore, UWB permits coexistence with both narrowband systems and other UWB systems. Regulatory bodies limit UWB devices to low average power in order to minimize interference with narrowband systems. Thus, UWB provides a method to reuse large amounts of existing spectrum without disturbing existing users, and it should be available worldwide in the near future.

Another important property of UWB is its precision position location capability [2–5], which is explained in more detail in Appendix A. The Cramer-Rao Lower Bound (CRLB) provides a limit on the accuracy of the delay estimate (which reduces to the ranging accuracy) that increases with the bandwidth and the signal-to-noise ratio (SNR) of the received signal [6]. See Appendix A for details on the CRLB as related to ranging. Even at low SNR, UWB can resolve distances with subcentimeter accuracy using simple signal processing. When nodes share distance estimates, they can cooperatively calculate location information, which is an important feature for many sensor network applications and network protocols [7, 8]. Because UWB RF circuitry is shared for communications and ranging purposes, UWB may provide low-cost, high-resolution ranging; and unlike GPS, UWB's location capabilities can function in indoor environments. UWB's capability to provide both location awareness and communications has made it an ideal candidate for the upcoming 802.15.4a standard for low data rate, low power, and location aware applications.

As shown in Figure 10.1, UWB provides a fundamental trade-off between data rate and link distance. The performance of five modulation schemes is presented. The bottom trace represents a long-distance communications system that uses a simple 2-PAM or PPM modulation, transmitting 100 pulses for every data bit ($N = 100$). Immediately above is a trace representing a system with the same modulation scheme, but now only transmitting 10 pulses for every data bit. At the top of the graph is a trace representing a system which uses 256-ary modulation, or 8 data bits per UWB pulse. Note that systems transmiting multiple pulses per data bit are able to maintain a specific data rate for a longer link distance.

Further, the baseband nature of UWB systems allows simple hardware implementation with no intermediate mixers or downconversion stages. UWB devices may have a nearly all-digital implementation in CMOS with minimal analog RF electronics. This simple architecture can translate to low power dissipation and low cost, which opens a variety of possible mobile applications.

Finally, UWB is relatively immune to multipath induced fading effects in both indoor and outdoor environments. For an application such as maritime asset tracking, the multipath environment may be particularly harsh, with several layers of densely packed containers stacked inside a ship's metal hold. With an appropriate receiver, a UWB system may even harvest energy from the resolvable multipath signals to improve data rate or BER.

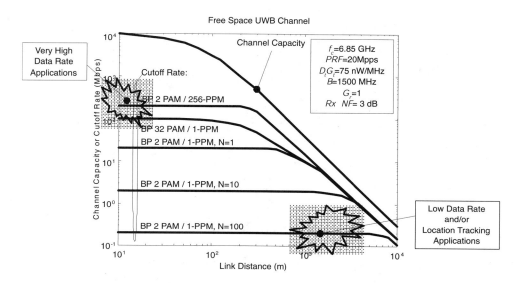

Figure 10.1: Trade-Off Between Data Rate and Distance for UWB.

SOURCE: P. Rouzet, "PULSERS Presentation to IEEE 802.15.4," *IEEE P802.15 Working Group for Wireless Personal Area Networks Publications*, [9]. © IEEE, 2003. Used by permission.

## 10.2   Applications

Before the FCC allocated spectrum for UWB devices in their First Report and Order on February 14, 2002, most UWB research was relegated to small, proprietary systems addressing military communications and radar. After the report and order, UWB has generated tremendous interest, and many new ideas for applications have come from both industry and academia. Table 10.1 presents example systems of some major UWB application spaces.

The following section reviews a wide range of applications that have been proposed for commercialization.

### 10.2.1   High Resolution Radar Applications

Radar is an important early application of UWB, and interested readers are referred to [1] for a more in-depth discussion of the subject. When radar systems use a UWB signal, the wide bandwidth and short pulse duration identify more target information, improve range accuracy, improve resilience to passive scatterers (clutter), mitigate destructive multipath effects from ground reflection, and enable a narrow antenna beam pattern [10]. The wideband pulses carry more information about the target, such as shape or material. Further, the pulse may have a low center frequency to penetrate solid structures. Finally, narrow pulses eliminate the

Table 10.1: UWB Application Spaces

| Application Space | Example Application | Proposed Standard |
|---|---|---|
| Radar | Medical imaging | N/A |
| Communications | Wireless USB; WPANs | IEEE 802.15.3a; MB-OFDM |
| Location-aware communications | Asset tracking; Sensor networks | IEEE 802.15.4a |
| Channel sounding | Wireless channel characterization | N/A |

ambiguity between polarity reversal and time delay found in a continuous wave narrowband signal, thus reducing clutter from magnetic reflectors.

In traditional radar systems, resolution is proportional to wavelength. For impulse radar, the resolution of a target approximately depends on its bandwidth (or pulse width) as

$$\Delta R = \frac{c\tau}{2} \qquad (10.1)$$

where $\Delta R$ is the target resolution (m), $c$ is the speed of light (m/s), and $\tau$ is the pulse width (sec). Thus, a Gaussian monocycle with a bandwidth of 7.5 GHz ($\sim$ 100 ps time duration) could achieve a resolution of $\Delta R = (3 \times 10^8$ m/s$\cdot 100 \times 10^{-12}$s$)/2 = 0.015$ m without extended signal processing. A single pulse is unlikely to both meet FCC regulations with regards to out-of-band emissions and have a 7.5 GHz bandwidth, so the actual resolution will be slightly larger. With added signal processing, the resolution can be improved close to the Cramer-Rao lower bound [4].

Because of its high resolution radar capability, UWB has been proposed for many novel radar applications, such as vehicular radar to enhance driver safety and comfort [11]. UWB radar can provide proximity sensing around the exterior of a car to detect any objects within range of the vehicle. Initially, applications will notify the driver of hazards. For example, a vehicular radar system would alert the driver through audio or visual output when it detects potential collisions at the side, front, or rear of the vehicle. Such a system may aid the driver by "seeing" blind spots while changing lanes, reversing, or parking. In later stages of development, multiple UWB sensors can be networked with other vehicular systems to provide autonomous control of tasks, such as parking, cruise control, transmission control, braking, airbags, safety belts, or suspension tuning. Because of its communication capability, UWB may even be applicable for complete control of autonomous vehicles on smart, networked roadways.

UWB vehicular radar has been allocated spectrum from 22 GHz to 29 GHz with the stipulation that the center frequency and the highest radiation level must be above 24.075 GHz. A further restriction is placed on the direction of radiation, which must attenuate energy above 38 degrees to the horizontal plane by 25 dB with respect to Part 15 limits. This requirement is scheduled to grow stricter year-by-year as more vehicular radar devices are included as standard equipment on commercial vehicles.

Another high resolution radar application is Ground Penetrating Radar (GPR), which detects objects buried underground [3]. The crowded infrastructure of cities places a crisscrossed web of water pipes, electrical lines, communications links, and other obstacles underground. Heavy construction equipment can damage these structures and even cause injury or loss of life. GPR provides precise location information for such obstacles, and it does not rely on records, which can be incomplete, inaccurate, inaccessible, or missing. Another valuable use of GPR is to detect abandoned land mines and unexploded ordinance [12].

GPR operates mostly in the low frequency band below 960 MHz because low frequencies penetrate substances such as soil and sand better than high frequencies. The operation of GPR is restricted to construction, law enforcement, fire and rescue, commercial mining, and scientific research. GPR is not seen as a severe source of interference to other devices because the device is positioned close to the ground and signal energy is directed into the ground.

Through-walls imaging is based on the same principle as GPR; however, signal energy is now directed horizontally. Therefore, through-walls imaging has considerable potential to interfere with existing systems, and the FCC restricts it to the public safety sector for use by firefighters and law enforcement officials. Law enforcement is particularly interested in using through-walls imaging to look through the double hulls of boats for hidden compartments and concealed contraband. The military also has significant interest in through-walls imaging for urban warfare situations. UWB can provide resolution fine enough to identify the presence of humans through walls, and it can even detect the small, involuntary motions of respiration or heartbeats.

Finally, UWB offers promise for medical imaging to enable health care professionals to look inside the body of a human or animal [13]. For example, UWB can detect movements of the heart, lungs, vocal cords, vessels, bowels, chest, bladder, or a fetus. Because UWB radar can resolve images with safe levels of radiation, it can be used even in sensitive patients, such as mothers in the final stages of pregnancy. UWB has a significant advantage over induced field devices, such as magnetic resonance imaging (MRI), which confines the patient to a small space. With the radiated field of UWB, the patient can be anywhere, and the imaging device moves while the patient remains in a comfortable position. Because UWB radar may function a few meters from the target, it can provide remote monitoring of patients through blankets and clothing. An initial medical application of UWB radar is early detection of breast cancer with space-time processing of UWB signals through an antenna array [14]. Figure 10.2 shows UWB detection of a 2 mm lesion 3.1 cm deep.

## 10.2.2 Communications Applications

Because of potential interference to and from narrowband devices, early UWB research was primarily for military communications applications. Most research was

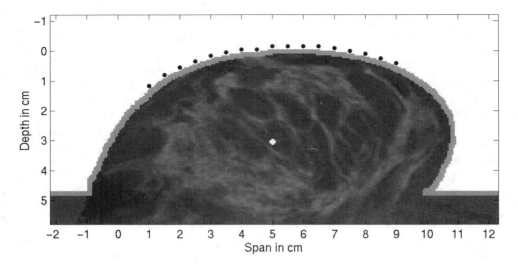

Figure 10.2: UWB Detection of 2 mm Lesion (White Dot in Center). The Black Dots on the Outside Are Antennas in an Array.

SOURCE: E. J. Bond, X. Li, S. C. Hagness, and B. D. Van Veen, "Microwave imaging via space-time beamforming for early detection of breast cancer," *IEEE Transactions on Antennas and Propagation* [14]. © IEEE, 2003. Used by permission.

performed quietly before the FCC opened the spectrum for UWB in February 2002. Thus, many military systems developed prior to 2002 provided long-range communication capability with power levels that exceed current regulations [15,16]. Examples include aircraft-to-aircraft communications, augmentation of graphic display plans (GDP) for instrument landing in inclement weather, communication and control for unmanned aerial vehicles, sensors for monitoring aerospace structures, and secure location and RF identification systems for soldiers in urban combat situations [15].

More recent applications focus on wireless personal area networks (WPANs), which connect a limited number of devices in a small coverage area (within 10 m). Current UWB radios for WPANs must meet the FCC power limits; hence, they radiate much less power than earlier systems. Taking full advantage of the high data rate of UWB, the IEEE 802.15.3a standard endeavors to define medium access and physical layers with the highest data rate possible for WPAN applications. UWB promises to deliver extremely high data rates (up to Gbps range) for multimedia applications and quick download times for large data files. To support these applications, the UWB physical layer for IEEE 802.15.3a offers data rates of 110 Mbps at 10 meters, 200 Mbps at 4 meters, and, optionally, 480 Mbps or higher at a shorter distance. Such data rates allow high-quality multimedia services and comfortable download times (seconds as opposed to minutes) for large media files [17].

Currently, the most common application for UWB WPAN communications is cable replacement for high-speed devices. Cabling is a bulky and aesthetically displeasing means of connecting devices, and it also tethers devices together to limit device mobility. XtremeSpectrum Inc.[1] proposed cable replacement for home networking [18], and Intel Corporation proposed cable replacement for office environments [19]. In fact, there are proposals for wireless Universal Serial Bus (USB) and wireless IEEE 1394 for consumer electronic (CE) devices [20]. These standards would connect peripherals, such as digital video players, projectors, MP3 players, portable disk drives, camcorders, high-resolution digital cameras, PDAs, printers, scanners, web cams, home theaters, CD players, keyboards, or mice [21]. UWB could even provide a standard interface for wireless docking of laptops [19] or wireless interchip connections.

Because cables are inconvenient in clothing, another cable replacement proposal enables wearable peripherals for health, entertainment, military, or medical purposes [22]. In Figure 10.3, UWB could clean up the awkward clutter of cables within a wearable computer system.

Time Domain Corporation has noted that environments for the proposed applications would likely be shared among multiple families or multiple businesses within a single building or confined space. Additionally, a large number of devices may be networked, so most of these applications require a standard that exploits the spatial reuse property of UWB. Operation of UWB communications in such an environment requires high aggregate data rate, low cost, low power, and small size; so performance can be measured as [18]

$$P = \frac{\text{Spatial Capacity}}{\text{Power} \cdot \text{Cost} \cdot \text{Size}} \left( \frac{bps/m^2}{W \cdot \$ \cdot m^3} \right) \qquad (10.2)$$

where $P$ is the performance metric, Spatial Capacity is in bits per second per square meter of coverage, Power is in Watts, Cost is the cost of the device, and Size is the amount of physical space the device occupies (in cubic meters). Figure 10.3 shows conceptual pictures of home, office, and wearable UWB scenarios.

## 10.2.3    Location Aware Communications Applications

UWB offers a unique blend of radar and communications applications known as Location Aware Communications. The upcoming IEEE 802.15.4a standard takes advantage of that combination in a UWB physical layer that emphasizes low data rate, low power, and location awareness instead of high throughput. The goal of the standard is to provide simple, pervasive, and seamless wireless connectivity among devices. Although this prospective standard is still in the initial stages of development, it has generated tremendous interest, with the promise to more fully exploit the unique low power and ranging characteristics of UWB. Suggested application spaces include home automation, industrial automation, and tracking of people, assets, or geographically localized phenomena. The following applications are based on the response to the IEEE 802.15.4a Call for Applications [23, 24].

---

[1]Motorola [37] bought XtremeSpectrum [38] in November 2003. XtremeSpectrum made these proposals prior to the acquisition.

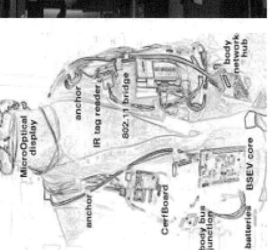

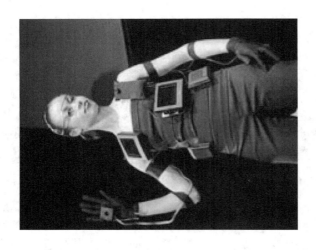

(a) Wearable

Figure 10.3: Communications Applications of UWB (continued).

SOURCE: (a). Richard DeVaul, images of MIThril 2000 System and prototype [86]. © MIT Media Lab, 2000. Used by permission.

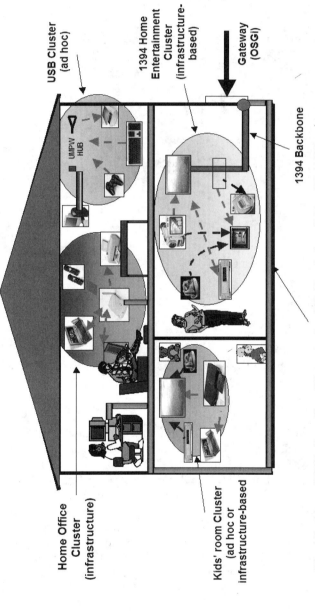

**USB Cluster (ad hoc)**

**1394 Home Entertainment Cluster (infrastructure-based)**

**Gateway (OSGi)**

UMPV HUB

**1394 Backbone**

**Home Office Cluster (infrastructure)**

**Kids' room Cluster (ad hoc or infrastructure-based)**

**Wired Backbone (HPNA, HomePlug, Ethernet, ...) or 802.11a wireless Bridge (54-108Mbps)**

(b) Home

Figure 10.3: (cont.) Communications Applications of UWB (continued).

SOURCE: (b). R. Fisher, R. Kohno, H. Ogawa, H. Zhang, M. Mc Laughlin, and M. Welborn, "DS-UWB Physical Layer Submission to 802.15 Task group 3a," doc.: IEEE 802.15-04/137r0 [18]. © IEEE, 2004. Used by permission.

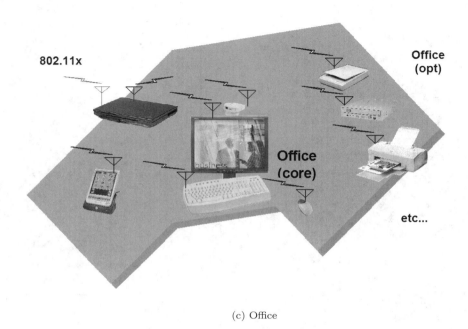

(c) Office

Figure 10.3: (cont.) Communications Applications of UWB.

SOURCE: (c). R. Kohno, H. Zhang, and H. Nagasaka, "Ultra Wideband impulse radio using free-verse pulse waveform shaping, Soft-Spectrum adaptation, and local sine template receiving," IEEE 802.15-03/097 [19]. © IEEE, 2003. Used by permission.

Many parties desire to use UWB to provide communications devices that can locate people. Aether Wire & Location, Inc. and the city of Chicago propose a location aware communications device for firefighters [24]. In a burning building, heavy smoke and darkness impair both audio and visual communications. With a UWB communications system, firefighters may monitor environmental conditions, communicate with each other, and locate injured personnel. Aether Wire proposes a similar application for soldiers to positively and quickly identify friendly soldiers to reduce accidental deaths and injuries [24]. Harris Corporation stipulates that military communications should adopt a modulation scheme with low probability of detection and intercept for stealth communications [25]. Samsung and Staccato Communications have proposed smart home applications that manage almost all aspects of the home, including doors, keys, TV, radio, and computers [24]. The system keeps track of individual users and their preferences. As individuals move throughout the smart home, a tracking system could automatically adjust room

temperatures and activate entertainment devices to a particular television channel
or website. General Atomics offers proximity sensors that provide security for cars,
computers, or homes, which automatically lock and unlock as the owner travels
in and out of range [24]. LB & J consulting suggests that UWB could be used
to track students and personnel in school buildings [24]. In schools, the tracking
system could control access to school buildings, conduct instant roll call, produce
reports to analyze incident patterns, monitor school bus transportation, and engage
parents in student supervision. Ubisense Ltd. proposes to increase productivity in
office environments by locating key personnel [24]. Privacy and security concerns,
however, would have to be addressed before such systems could be implemented.

Location-aware communications capability can also be used to manage assets.
UWB expedites the inventory control process because the network can automatically scan individual containers and report status information without physical
examination of the container's contents. Further, UWB performs well in a densely
packed environment, such as stacked pallets in trucks or stacked containers in ships.
Aether Wire has suggested precision asset location and autonomous manifesting applications for the Department of Defense, which is the largest United States trans-
porter of goods [24]. Inforange offers a similar suggested application for tracking
packages during shipment to reduce the amount of lost and misrouted packages [24].
They propose a transmit-only device to save power in the tags. General Atomics extends this idea to tracking for inventory control in warehouses and retail shops [24].
A similar technology could be used as small, discrete security tags for high-valued
items, such as leather jackets in retail stores [24]. Both Ubisense and MSSI have
suggested tracking life-saving equipment in hospitals, as it is not always possible to
locate equipment quickly in an emergency [2, 26].

Tracking mobile objects is another application that uses UWB's radar capability.
One idea is a UWB network that provides a protective security "bubble" around
a geographic area. The radar capability of UWB would detect intruders and track
their motion. In a networked environment, this could provide accurate tracking
over a large geographic area. In outer space, UWB could automatically track an
astronaut outside of a spacecraft or track the position of two spacecraft that are
docking. Time Derivative and Q-Track suggest unique applications for professional
sports to track balls, athletes, or racecars [24]. On farms and in wildlife sanctuaries,
a tracking system could automatically track the movement of cattle or wildlife.

In the proposed applications for IEEE 802.15.4a, several companies mention sensor networks, which stand to derive huge benefits from the low power and location
aware properties of UWB [26, 28]. Position location capability aids in network configuration and provides a service for the networking and application layers. Aether
Wire notes the demand for sensor networks to monitor industrial automation and
control [24]. They cite the high cost ($10 to $25 per foot) of wiring for pipe sensors,
which typically exceeds the cost of the sensors. Mobile nodes may take advantage of
UWB's location capability to organize and perform tactical maneuvers [29]. Staccato and Aether Wire suggest UWB for heating, ventilation, and air conditioning
(HVAC) control for home and office environments [24]. Extending this idea, Ech-
elon Corporation proposes to add UWB radios to water and gas meters to report

readings directly to the consumer and also to the utility company through power lines [30]. Another possibility is monitoring a human for early warning of seizures or heart beat monitoring [31]. ST Microelectronics notes that the location aware capability allows such networks to configure themselves (so that they become ad hoc networks), circumventing installation by a technician [24]. Samsung and CUNY want to augment the Global Positioning System (GPS) that is not usually receivable indoors, to enable location-assisted routing of network data. UWB is an excellent low-power physical layer for sensor networks [24]. The extremely low duty cycle of UWB allows sensors to conserve energy and operate for many years without maintenance [2]. Because of its robust operation in harsh multipath environments and penetration capability, UWB also supports sensor networks in warehouse and cargo hull environments, and can even be embedded in solid structures to provide a real-time non-invasive report on structural integrity.

### 10.2.4   Channel Sounding Applications

Another potential application of UWB technology is channel sounding or measuring the impulse response of a wireless channel [32–34]. I-UWB, with its extremely short duration pulses, provides subfoot resolution of multipath signals in a wireless channel. Such high resolution is not only useful for investigating UWB propagation mechanisms; it also provides useful information about the number, power, and relative delay of multipath signals for narrowband channels as well. More detailed information on the Channel Sounding applications of UWB technology is found in Chapter 2.

## 10.3   Case Studies

Even before the FCC authorized UWB communications, a number of commercial products had already been developed under a special waiver from the FCC. The following products provide an overview of some implementation issues not covered in the standards. Investigation of these commercial products also demonstrates the requirements of UWB systems that are critical for proper operation. For each product, we attempt to describe the system type, modulation scheme, band plan, coding scheme, spreading scheme, MAC layer, networking capability, hardware architecture, and performance. Table 10.2 summarizes the characteristics of these existing products.

### 10.3.1   XtremeSpectrum Incorporated (XSI)/Motorola Trinity Chipset

XSI demonstrated the Trinity chipset in July 2002 [35]. Trinity is the first commercial, unlicensed UWB chipset targeted to operate under the recent FCC rules regarding UWB. XSI has been active in IEEE 802.15.3a standardization efforts and champions the DS-UWB approach, which shares similar communications methods with the Trinity chipset. A description of DS-UWB can be found in Appendix B.

The Trinity chipset is named for three target features of UWB: low price, low power, and high data rate. It is marketed for WPAN applications, such as a wire replacement for streaming MPEG2 video, and for wireless access to fast Ethernet, USB2, and IEEE 1394 ports. The chipset is an I-UWB system with biphase modulation, which is also known as Antipodal Pulse Amplitude Modulation (A-PAM). Both A-PAM and biphase modulation result in the same signal constellation; however, biphase modulation has specifically been patented by XSI for UWB. Biphase modulation is the most energy-efficient binary modulation scheme. The Trinity chipset uses a high Pulse Repetition Frequency (PRF) and long spreading codes to compensate for the energy reduction in average power, and the energy reduction due to ISI. The pulse train fills the spectrum in the range of 3.1–5.0 GHz.

Trinity implements the IEEE 802.15.3 MAC protocol, which is a centralized Time Division Multiple Access (TDMA) protocol (see Appendix B) [36]. The IEEE 802.15.3 MAC supports QoS, which is necessary for video networking; multiple video streams are guaranteed the bandwidth they need for acceptable video quality. The chipset supports up to 8 nodes in a piconet. Full networking support is outside the scope of the IEEE 802.15.3 protocol, which only specifies a MAC protocol and a physical layer. Higher layers may determine the method of routing according to the demands of a particular network application. Because most applications of the Trinity chipset are for WPANs, the chips will likely not require networking support outside of a piconet.

The chipset achieves a total power consumption of 200 mW, which is suitable for battery powered components. Depending on channel conditions, the FEC rate can be traded for the data rate. The FEC options are rate $\frac{1}{2}$, rate $\frac{3}{4}$, and no coding, which support multiple data rates of 50, 75, and 100 Mbps. The maximum range is 10 meters, which is suitable for WPAN applications. Motorola markets revisions of the Trinity chipset under the name XtremeSpectrum chipset.

## 10.3.2   Time Domain Corp—PulsON Chipset

Time Domain Corporation (TDC) markets the PulsON family of UWB chips for a wide range of applications, including indoor wireless networking, video transmission, voice transmission, personnel location, asset tracking, precision measurement, through-walls sensing, industrial sensing, robotic controls, and security systems [15, 39, 40].

TDC was founded in 1987 to commercialize UWB products using the PulsON technology. The first products from TDC were mainly in response to SBIR grants and military and government contracts. Early on, TDC produced prototypes for radar and communications applications for public safety devices under a special license from the FCC. Recently, TDC also has contributed to standardization efforts for both IEEE 802.15.3a and 802.15.4a.

TDC premiered the PulsON 100 UWB chipset in 1999 and introduced the next generation PulsON 200 in 2002. These chipsets were originally designed for use in early cellular telephone systems costing under $200—a far cry from the sub-$5 WPAN applications. The PulsON 100 chipset was manufactured prior to FCC

Table 10.2: Characteristics of Various UWB Chipsets (continued).

| | Xtreme Spectrum *Trinity* | Time Domain Corp. *PulseOn 100 and 200* | MSSI *AWICS* | MSSI *PAL650* | MSSI *MAVCAS* | Aether Wire & Location *Localizer* | IEEE 802.15.4a *Standard* |
|---|---|---|---|---|---|---|---|
| System Type | I-UWB | I-UWB | I-UWB | I-UWB | I-UWB Radar | I-UWB | TBD |
| Pulse Shape | Gaussian | Gaussian Monocycle | Gaussian | Gaussian | Bandpass | Gaussian Doublet | TBD |
| Modulation Scheme | Antipodal Binary PAM | PPM / PAM | PAM | PAM | None | PAM, PPM, M-ary codes | TBD |
| Band Plan | 3.1–10.6 GHz | 100: 0.1–1.3 GHz, 200: 3.1–10.6 GHz | 1–2 GHz L-Band, 400 MHz 3 dB BW | 1–2 GHz L-Band, 400 MHz 3 dB BW; PAL650: 6–6.4 GHz | 6.10–6.60 GHz, 6.35 GHz $f_c$ | 0.1–1 GHz | TBD |
| Error Correction Coding Scheme | Convolutional Rate 1/2, 3/4, 1 | N/A | Reed-Solomon Rate 0.87 | N/A | None | N/A | TBD |
| Spreading Scheme | DS-UWB | Time Hopping & Integration | None | None | None | DS-UWB | TBD |
| MAC Layer | IEEE 802.15.3 | Time Hopping, PN Codes | TDMA | No-ACK ALOHA | None | TDMA Locally; CDMA for Clusters | CSMA/CA, TDMA |

Table 10.2: (cont.) Characteristics of Various UWB Chipsets.

| | Xtreme Spectrum *Trinity* | Time Domain Corp. *PulseOn 100* and *200* | MSSI *AWICS* | MSSI *PAL650* | MSSI *MAVCAS* | Aether Wire & Location *Localizer* | IEEE 802.15.4a *Standard* |
|---|---|---|---|---|---|---|---|
| Network Configuration | IEEE 802.15.3; Peer-to-Peer Ad Hoc | Application Dependent | Star Topology | Broadcast | None | Multi-Hop | Multi-Hop; Star; Mesh |
| Implementation | 0.18 $\mu$m CMOS & SiGe | 100: Si/SiGe, 200: SiGe | N/A | N/A | Off-the-shelf Components | 0.35 $\mu$m CMOS | < $1 Per Chip |
| Data Rate | 25, 50, 75, 100 Mbps @ 10 m | 100: 5 Mbps, 200: 9.6 Mbps @ 20 m | 2.408 Mbps within Range | 1 Mbps in Short Bursts to Meet FCC Limits | 10 kpps | 100 Mpps | Link 1: 1 Kbps; Link 2: 10 Kbps; Collector 10 Mbps |
| Range | 10 m | 100: Up to 16 km, 200: Up to 140 m | Within Aircraft; 60 m Outside Aircraft w/ Hatch Open | 600 ft. Outdoor, 200 ft. Indoor, < 1 ft. Accuracy | 20 m (low power), 300 m (high power), 1 m Accuracy | 30–60 m, 1 cm Accuracy | Max: Hundreds of Meters; Typical: 0–30 m |
| Total Power Dissipation | 200 mW | T2: 500 mW | N/A | N/A | N/A | N/A | Average of a Few Milliwatts |
| Bit Error Rate | $10^{-9}$ | $10^{-9}$ | N/A | N/A | N/A | N/A | 8% Maximum |

regulations, but the PulsON 200 chipset meets the FCC regulations for commercial UWB products.

The PulsON chipset is an I-UWB system that implements Time Modulated UWB (TM-UWB), which is also known as pulse position modulation (PPM). The transmitter is a single transistor that produces a step waveform that is filtered to produce a monocycle. The receiver uses a matched filtering technique to demodulate the received PPM signal.

To overcome the speed limitations of CMOS digital clocks, TDC uses SiGe technology for their PulsON chips. SiGe clocks have picosecond resolutions and low jitter to enable synchronization through fine-grained sliding correlation. Digital timing control achieves coarse resolution, and analog control refines the resolution to a few picoseconds or less. Because correlation is performed in the analog domain, the receiver only has to operate and lock at the pulse rate, which is much less than the time resolution for synchronization. The architecture of the PulsON T100 transmitter and receiver is shown in Figure 10.4.

The PulsON 100 chipset is divided into three blocks: a precision timing system, a correlator system, and a digital baseband signal processing and control unit. The T1 timing chip has a resolution of better than 3 ps. The T2 is a second-generation timing chip with a resolution of 1.5 ps. Both chips are a mixed signal design with analog, RF, and digital components. The T1 chip was fabricated using 0.5 $\mu$m 45 GHz Si/SiGe heterojunction bipolar transistor (HBT) technology with 0.4 $\mu$m PMOS devices and polysilicon resistors. The T2 is implemented in 0.35 $\mu$m SiGe. The correlator chip C1 is also implemented in SiGe. The baseband unit operates at the pulse rate and performs signal processing and control functions. TDC also patented one of the first wideband antennas, which has a 1.9 GHz bandwidth and dimensions of 5 cm $\times$ 7 cm $\times$ 0.7 cm. For multiple access, TDC advocates time hopping with any random pseudo noise codes that produce good autocorrelation and cross-correlation properties.

The PulsON 100 provides a moderate data rate of 5 Mbps or a long-range data rate of 16 Kbps at 300 m using directional antennas. The data rate is limited both by the pulse rate and the length of the code used for multiple access. The PulseON 200 provides data rates of up to 9.6 Mbps, but at distances of less than 20 meters. PulsON chipsets also have the ability to ascertain distance to within 1 cm for the purpose of augmenting GPS systems.

## 10.3.3   Multispectral Solutions Incorporated (MSSI)

MSSI's products include a wide range of radar and communications systems for military and civilian use [41], and this section reviews some exemplary systems in these categories. MSSI is active in standardization efforts for both IEEE 802.15.3a and 802.15.4a.

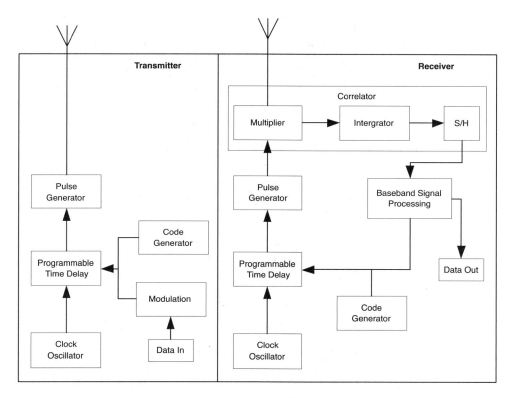

Figure 10.4: Transmitter and Receiver Architecture for PulsON 100 Chipset.

SOURCE: D. Kelley, S. Reinhardt, R. Stanley, M. Einhorn, "PulsON second generation timing chip: enabling UWB through precise timing," *2002 IEEE Conference on Ultra Wideband Systems and Technologies Digest of Papers* [39]. © IEEE, 2002. Used by permission.

## Early Communications Systems—Draco, HFUWB, and Orion

MSSI's earliest systems were implemented under a special FCC waiver and included long-range communications devices for the military [42, 43]. Three such systems are briefly reviewed here: Draco, HFUWB, and Orion.

Draco implements a fully distributed peer-to-peer, ad hoc, multihop network for long-distance communication over a few kilometers. Orthogonal frequency bands provide multiple channels to separate control and data traffic. A separate control channel is dedicated to the periodic transfer of routing and link state information. TDMA provides multiple access within a cell. Draco provides secure communications and obtains a relatively high aggregate data rate by using one bit per pulse (no spreading gain). It protects against channel errors with a Reed-Solomon code.

Another communications system, HFUWB, provides long-range communications in non-line-of-sight (NLOS) conditions. I-UWB systems become more susceptible to multipath effects as the transmitter-receiver distance increases relative to the antenna height. This is because the differential path length between multipath components becomes shorter as the distance between them increases. It becomes picoseconds for distances of a kilometer of more, so that even multipath components from short pulses may overlap and can potentially cancel each other out. Therefore, for long-distance communication, the HFUWB system operates in the 30 MHz–50 MHz military band to take advantage of low-loss surface propagation and ducting along the earth/atmosphere boundary. By operating in this band, the HFUWB system avoids potential multipath distortion that hampers long-distance communication due to ground reflections [43]. The HFUWB system operates at a data rate of 850 kbps, the average power is 6 W, and the peak power is 120 W. The antenna is quite large due to the low frequency band, but it achieves a range of 60 nautical miles (1 nautical mile = 1.1508 miles = 1.852 km) over water.

Finally, the Orion system offers both short-range and long-range communications. Short-range operation is in the 1.3–1.7 GHz band with a peak power of 0.8 W, a range of 1 km for LOS conditions, and a data rate of 128 kbps for Continuously Variable Slope Detect (CVSD) encoded voice data, and 115.2 kbps for RS-232 data. Orion also uses a modified HFUWB transceiver for long-range operation.

## Aircraft Wireless Intercommunications System (AWICS)

AWICS was created to eliminate cords in aircraft, which can impede mobility and cause dangerous entanglements in crashes [44, 45]. AWICS avoids the multipath induced fading over short distances that is inherent in narrowband communications systems. The AWICS provides reliable transmission in the presence of existing radios on large CH-53E or CH-46E helicopters. Low-power UWB will not interfere with existing aircraft systems and has low probability of detection and intercept, which is necessary to prevent hostile forces from eavesdropping or determining the aircraft position based on electromagnetic radiation. A mobile AWICS transceiver is pictured in Figure 10.5, and Figure 10.6 shows the system architecture of an AWICS transceiver.

AWICS operates by transmitting 2.5 ns wide pulses at a pulse repetition frequency (PRF) of 2.048 MHz, and because there is no spreading, the raw data rate is also 2.048 Mbps. The 2.048 MHz PRF results in an approximately 500 ns time period between pulses to combat destructive ISI from multipath components. The 500 ns window allows multipaths to "ring down" until they contribute negligible interference to the next symbol. 500 ns translates to a round-trip distance of about 500 feet and ensures that multipath reflections inside the aircraft dissipate before the next pulse.

Voice data is sent with a continuously variable slope detection (CVSD) codec for high-quality reproduction. The system uses TDMA with 500 $\mu$s long slots, and the superframe format accommodates up to 31 users and the base station. Individual packets have an 8 byte synchronization preamble followed by a 64 byte CVSD packet, a 1 byte ID, and 10 bytes of Reed-Solomon FEC. Synchronization occurs

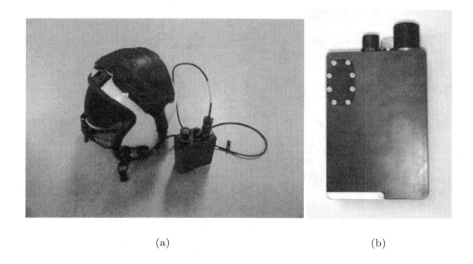

(a)                                                    (b)

Figure 10.5: MSSI AWICS System.

SOURCE: R. Fontana, E. Richley, and J. Barney, "Ultra wideband technology for aircraft wireless intercommunications systems (AWICS) design," *Proc. IEEE Conference on Ultra Wideband Systems and Technologies* [44]. © IEEE, 2003. Used by permission.

for each packet, and there is a training sequence to provide receiver gain control. The network is a star topology with a central base unit. The AWICS antenna is a dielectrically loaded wideband helix structure. With an outer shell that is waterproof, shock resistant, and G-force tolerant, the transceivers measure $5.25 \times 3.5 \times 1.5$ inches. The mobile units are small enough to be comfortably worn on a vest, and they operate for up to eight hours on battery power.

## Precision Asset Location (PAL) System

MSSI has also produced a Precision Asset Location (PAL) system that uses the precise location ability of UWB [46–48]. The *PAL650* is an FCC-compliant asset location system that locates tags in three dimensions in harsh multipath conditions. The technology was originally developed for DARPA for soldier tracking in urban terrain and for the U.S. Navy for tracking containers in naval cargo holds. The demand grew out of 1991's Operation Desert Storm, which required the shipment of over 40,000 containers; 25,000 of these containers had to be opened and inspected by hand due to inaccurate and lost manifests. The General Accounting Office (GAO) has estimated a three billion dollar loss from the lost and misplaced materials in Operation Desert Storm. Recently, the system has been commercialized for use in

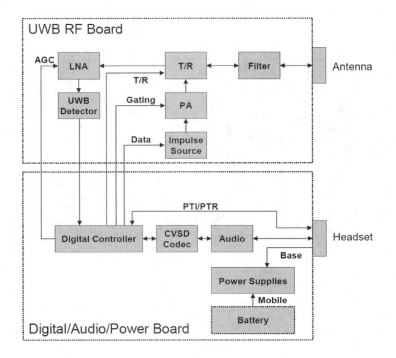

Figure 10.6: System Architecture of AWICS.

SOURCE: R. Fontana, E. Richley, and J. Barney, "Ultra wideband technology for aircraft wireless intercommunications systems (AWICS) design," *Proc. IEEE Conference on Ultra Wideband Systems and Technologies* [44]. © IEEE, 2003. Used by permission.

warehouses, shipping, supermarkets, retail establishments, robotics, manufacturing, security, and hospitals.

In a metal cargo hold, the large space combined with metal walls creates long multipath effects that may last up to 5 $\mu$s. Further, stacked containers create severe NLOS conditions. Narrowband systems are ineffective in such environments, but the $PAL650$ can resolve locations within one foot, even in severe multipath environments. The $PAL650$ system consists of four passive UWB receivers, a reference tag, a collection of active tags, and a central hub that connects to a computer. Figure 10.7 displays the following system components: a tag, a receiver, and the central processing hub. Figure 10.8 shows a typical arrangement of these components in a deployed system.

In the $PAL650$ system, the tags (which contain an I-UWB transmitter but not a receiver) are affixed to containers. The transmitter periodically broadcasts data packets, which are 40 bits in length and consist of a synchronization preamble and

(a) Asset Tag

(b) Reference Receiver

Figure 10.7: MSSI Precision Asset Location System (continued).

SOURCE: A. Ameti, R. Fontana, E. Knight, and E. Richley, "Commercialization of an ultra wideband precision asset location system," *Proc. 2003 IEEE Conference on Ultra Wideband Systems and Technologies* [47]. © IEEE, 2003. Used by permission.

(c) Central Processing Hub

Figure 10.7: (cont.) MSSI Precision Asset Location System.

SOURCE: A. Ameti, R. Fontana, E. Knight, and E. Richley, "Commercialization of an ultra wideband precision asset location system," *Proc. 2003 IEEE Conference on Ultra Wideband Systems and Technologies* [47]. © IEEE, 2003. Used by permission.

tag ID. Packets may be broadcast up to 5,200 times per second without exceeding the FCC limits. Prototype systems use an interupdate time of either 1 second or 5 seconds. With a 5 second interupdate period, the peak power can be as high as 0.25 W and still result in an average power of 5 nW (or −79 dBm/MHz). Thus, the system is able to trade the update rate for a higher peak power and longer range.

The *PAL650* system has no MAC protocol because the data rate is high, the packets are short, and the transmissions are infrequent. When collisions occur, the data is simply discarded, with the idea that devices will not be inadvertently synchronized and cause repeated collisions. A small, random phase shift programmed into each tag can help eliminate the problem of inadvertently synchronized tags.

The receivers contain two separate boards: one for RF circuitry and pulse detection and the other for control and interfacing. Note that four reference points are required to unambiguously locate a tag in three-dimensional space. Each of four receivers records the arrival time of the leading edge of a pulse via a tunnel diode detector. The four receivers are connected with Category-5 Ethernet cable to provide an absolute time reference and synchronization. The arrival times are relayed to the hub and the computer, which uses a nonlinear optimization algorithm to determine the 3-D components. This type of ranging algorithm is known as a

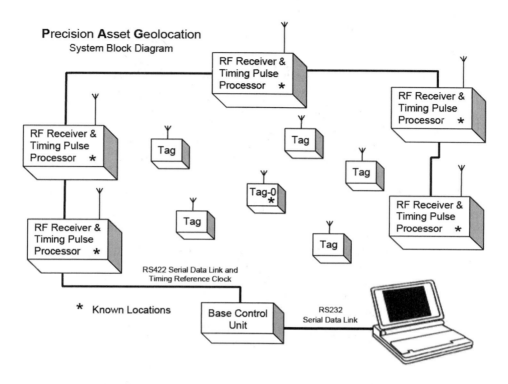

Figure 10.8: System Layout of *PAL*650.

SOURCE: R. J. Fontana and S. J. Gunderson, "Ultra-wideband precision asset location system," *IEEE Conference on Ultra Wideband Systems and Technologies
Digest of Papers* [46]. © IEEE, 2002. Used by permission.

time difference of arrival (TDOA) approach. In NLOS conditions, more than four
receivers may be used to improve the reliability of the location estimate. However,
too many receivers results in an overconstrained set of equations. The likely position for each tag is then determined with a method-of-steepest descent search to
minimize an error function. At system initialization, the reference tag is placed at
a known location and it calibrates the system.

A four-receiver prototype was tested aboard the SS Curtis in Port Hueneme,
CA, as a representative environment with severe multipath effects. The system
performed with 3–5 ft RMS errors in open space, 3 ft RMS errors with single
stacked containers, and 11–12 ft RMS accuracy with double stacked containers.

In the commercial system, a single Lithium cell battery provides an operational
lifetime of over 3.8 years. The tag measures 1 7/8 inches diameter by 7/8 inches

high. The receivers can operate with a minimum SNR of 8 dB. The precision asset location system has a maximum 2 km range outdoors and 300 feet indoors.

### Micro Air Vehicle Collision Avoidance System (MAVCAS)

One unique application of UWB radar is for small, unmanned Micro Air Vehicles (MAVs), which are defined as aircraft less than 6 inches in any dimension, possibly even as small as insects [49]. MAVs may operate cooperatively or alone to provide such applications as battlefield reconnaissance, urban surveillance, communications relays, search and rescue, remote sensing, traffic monitoring, or pollution monitoring. Until recently, MAVs have been impractical due to the strict size, weight, and power constraints of sensor devices. Because these vehicles are unmanned, they require collision avoidance and altimetry sensors to navigate complex topographies. Table 10.3 lists the strict requirements for implementation of MAVCAS and Figure 10.9 shows a prototype MAVCAS system.

The MAVCAS transmitter operates by taking a C-Band oscillator, filtering it, and time gating it to produce the result shown in Figure 10.10. At the receiver, the signal is split into two paths, one of which is delayed by 2 nanoseconds. Both signals are then sampled in parallel every 4 ns with a 250 MHz clock so that the receiver achieves a 2 nanosecond time resolution (for positioning accuracy of less than 1 foot). With additional delayed versions of the input signal processed in parallel, the receiver could attain finer and finer precision. The prototype system uses discrete RF components; however, the design can be integrated to meet the size and weight requirements of a MAV.

## 10.3.4   Aether Wire and Location Localizers

Aether Wire and Location, Inc. is developing pager-sized units that are capable of localization to centimeter accuracy over distances of kilometers in networks of hundreds or thousands of nodes [50–53]. The units, currently named Localizers, use UWB to provide low-cost, long-distance communication along with precise ranging capability. The design objective is to fabricate complete transceivers integrated on CMOS technology. The first-generation Localizer was fabricated in July 1994, and four generations have been fabricated since that time. The localizers define their position relative to other localizers, instead of a fixed point as in GPS. The devices are

Table 10.3: Requirements of MAVCAS System.

| Weight | Less than 50 g |
|---|---|
| Size | Much smaller than MAV |
| Radar Range | 50 feet minimum |
| Resolution | ± 1 foot, ± 10 degrees AOA |
| Span | 360 degrees in hover mode, ± 60 degrees in flight |
| Update Rate | 1000 times per second |

Figure 10.9: MSSI MAVCAS Prototype.

SOURCE: R. J. Fontana, E. A. Richley, A. J. Marzullo, L. C. Beard, and R. W. T. Mulloy, "An ultra wideband radar for micro air vehicle applications," *IEEE Conference on Ultra Wideband Systems and Technologies Digest of Papers* [49]. © IEEE, 2002. Used by permission.

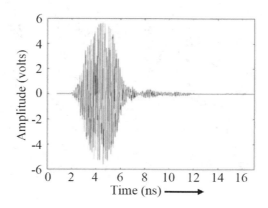

Figure 10.10: Bandpass Pulse Shape for MAVCAS in the Time Domain.

SOURCE: R. J. Fontana, E. A. Richley, A. J. Marzullo, L. C. Beard, and R. W. T. Mulloy, "An ultra wideband radar for micro air vehicle applications," *IEEE Conference on Ultra Wideband Systems and Technologies Digest of Papers* [49]. © IEEE, 2002. Used by permission.

meant to operate in urban canyons, forests, and indoors for a variety of applications, including sensor networks, soldier visualization, surveying and construction, tracking, inventory control, rescue, camera focusing, smart highways, machine control, and smart homes. Military applications include distinguishing friend from foe, logistics control, ground surveillance through sensor networks, and mine detection and clearing. Figure 10.11 displays the size of the system relative to a dime, and the layout of the transmitter and the receiver.

The localizers are I-UWB systems, but their pulse shape is unique due to the characteristics of the transmit antenna. Each "pulse" is, in fact, a series of two pulses: one with positive polarity and one with negative polarity. The pulses are sent in opposite pairs because the antenna is a Large-Current Radiator Antenna. The transmitter applies current steps to the antenna to generate pulses. Turning the current on generates a positive impulse, and turning the current off generates a negative impulse. The pairing of antipodal pulses prevents a run of identical symbols from building an impractically large current. Because the system is a DS-UWB system, each pulse pair forms a chip.

Each pulse is 1 ns wide with a 5 ns spacing between pairs. A logic 1 consists of a positive pulse followed by a negative pulse, and a logic 0 consists of a negative pulse followed by a positive pulse. One advantage of this unique pulse shape is that

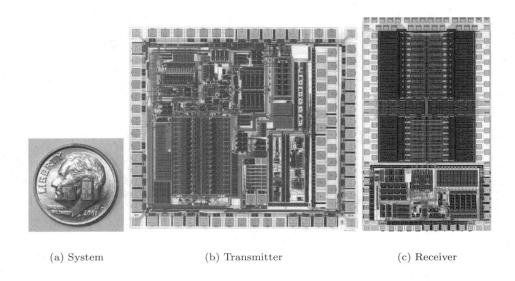

(a) System                    (b) Transmitter                    (c) Receiver

Figure 10.11: Localizer System.

Source: R. Fleming and C. Kushner, "Integrated ultra-wideband localizers," *1999 Ultra-Wideband Conference* [51]. © Aether Wire & Location, Inc., 1999. Used by permission.

thoughtful spacing between pulse pairs places nulls in strategic locations, that is, frequencies where narrowband interference may exist [54]. For 1 ns pulse width and 5 ns spacing, the energy lies between 100 MHz and 1 GHz with nulls every 200 MHz (1/spacing).

The receiver is based on a time integrating correlator (TIC), which is pictured in Figure 10.12. A TIC is the dual of a normal correlator that multiplies and then integrates. For a TIC, the received signal is stored and a reference code is moved past the analog input signal. Then, the product of the received signal and the code are summed in a bank of parallel analog integrators. Each parallel correlator uses an identical copy of the code, but shifted by a slightly different time delay. When the received signal is aligned with the proper code sequence, the integrator produces an increasing output that eventually reaches a maximum.

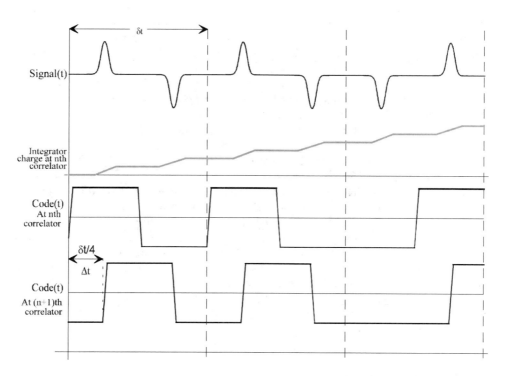

Figure 10.12: Pulse Pair, Offset Spreading Sequences, and TIC Output.

SOURCE: R. Fleming and C. Kushner, "Low-Power, Miniature, Distributed Position Location and Communication Devices Using Ultra-Wideband, Nonsinusoidal Communication Technology," *Semi-Annual Technical Report–ARPA* [50]. © Aether Wire & Location, Inc., 1995. Used by permission.

The localizers must be synchronized for accurate ranging because the ranging is based on the time of flight. During synchronization, a node searches for a particular code as well as its specific start time. Because the search process may take some time, the initiating node repeats its sequence for a sufficiently long period to ensure detection by the target node. Localizers achieve fast synchronization by using Kasami sequences, which have the special property that the side lobes of the autocorrelation are detectable with only partial overlap of the transmission and the code [55]. Therefore, if no signal is detected, the time step can advance immediately. After the target node synchronizes to the initiating node, the target then sends a synchronization code to the initiator, which ensures that both nodes are synchronized to the same reference clock and the time of flight can be accurately measured.

The distance between localizers is measured by multiplying the round-trip time by the speed of light. Note that the return signal is not an echo, but a completely new transmission, so the known circuit delays must be factored into the time of flight measurement. Then, the relative location of each localizer is determined cooperatively. Each pair notes the distance between them, and five or more localizers share this information to determine the three-dimensional positions relative to one another.

The localizers group themselves into clusters and form bridges between the clusters. Within clusters, multiple access is TDMA. Among clusters, multiple access is CDMA, where each cluster uses a different spreading code to mitigate cross-cluster interference. Multihop routing through clusters provides long-distance communication.

The fourth-generation receiver/controller chip has 32 parallel TICs, a 60 dB variable gain LNA, a real-time clock, a code generator, and a transmit antenna driver. The chip is fabricated in 0.35 $\mu$m CMOS, with single ended and differential current steering logic, and differential series pass logic families. The internally generated RF noise is reduced by placing most hardware on-chip to avoid high power connections, by using low noise logic for receiver functions, by disabling CMOS components during reception, and by using differential circuits while transmitting. The overall die size is 10 mm$^2$, including the processor, RAM, and ROM. The cost is about 50 cents, and the idling power is 30 $\mu$W. The fourth-generation chips have a positioning accuracy within 1 cm and a range of 30-60 m.

## 10.3.5    802.15.3a

At the time of this writing, the IEEE 802.15.3a standard for WPAN applications is still under consideration. It will use a MAC similar to the IEEE 802.15.3 MAC, and the physical layer will be selected from two proposals: multiband orthogonal frequency division multiplexing (MB-OFDM) or direct sequence UWB (DS-UWB). The 802.15.3 MAC and the above physical layer proposals are described in detail in Appendix B.

## 10.3.6   802.15.4a Devices

At the time of this writing, the IEEE 802.15.4a standard [24, 56] is in the early stages of development, and it should extend the capabilities of IEEE 802.15.4 with a UWB physical layer that offers ranging capability, more robust operation in harsh environments, and increased link distances. The call for proposals was released in July 2004, and should spawn innovative ideas for addressing a number of the design issues listed in Table 10.2. This section provides an overview of some possible issues that the standard may address.

IEEE 802.15.4 devices require low-power dissipation, and the additional UWB physical layer should further decrease the power consumption. For the existing physical layer, the minimum transmit power is 1 mW, whereas for UWB, the maximum transmit power allowed is 0.5 mW.

The type of application will determine many architectural considerations in the lower layers. Different applications have different traffic profiles. For low data rate applications, the traffic profiles include periodic data, such as temperature sensors; intermittent data such as perimeter sensors; or repetitive, low latency data, such as the movement of a computer mouse.

At the network level, an increased range and dense topology means that the networking layer should accommodate more nodes than the 802.15.3 WPAN, and also offer multihop routing. The existing IEEE 802.15.4 standard offers several topologies to improve multihop performance. The two topologies currently supported in IEEE 802.15.4 are star and peer-to-peer topologies, and they can be combined in a hybrid topology as shown in Figure 10.13.

In a star topology, all nodes send data through a smart, central coordinator, such as a PC or PDA. The terminal devices may be unsophisticated and designed to operate for long periods of time on battery power. In a peer-to-peer topology, nodes can communicate with any neighbor, and the resulting mesh topology is suitable for such applications as sensor networks. Large dense networks, such as sensor networks, will require a multihop routing protocol.

The hybrid topology pictured in Figure 10.13 offers the most flexibility. Local groups of nodes may be connected in either a star topology or a mesh topology, and the local groups are connected via a cluster tree topology. In this topology, there is the possibility of some nodes being full function devices (FFD) and others being reduced function devices (RFD). An FFD can coordinate an entire WPAN or a cluster, whereas an RFD is a simple device that communicates only through FFDs. An FFD could be a DVD, TV, computer, or PDA, whereas an RFD could be a light switch or a passive infrared sensor.

In the existing 802.15.4 standard, multiple access is mainly through Carrier Sense Multiple Access with Collision Avoidance (CSMA/CA). Because many devices do not have QoS constraints, they can operate without a beacon or central coordinator to reduce overhead. Power reduction is obtained through a very low duty cycle operation (for example, the device may be in an "awake" state for less than 0.1% of the time), so some devices may require MAC level synchronization; this case, a contention access period (CAP) follows a beacon. Finally, devices

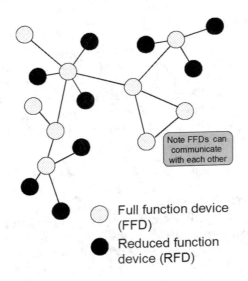

Note FFDs can
communicate
with each other

Full function device
(FFD)

Reduced function
device (RFD)

Figure 10.13: Example IEEE 802.15.4 Topology.

SOURCE: P. Gorday, J. Martinez, and P. Jamieson, "IEEE 802.15.4 overview," *IEEE P802.15 Working Group for Wireless Personal Area Networks Publications* [56]). © IEEE, 2001. Used by permission.

QoS constraints can take advantage of both the optional beacon and the optional guaranteed time slots (GTS). The frame structure is shown in Figure 10.14 [56].
The dependence on CSMA/CA presents a problem for UWB systems, namely, method of clear channel assessment (CCA). In I-UWB, CCA is difficult be- a receiver must find a narrow pulse in a relatively long time window. Existing methods for pulses, such as peak detection, matched filtering, sliding cor-, and Interleaved Periodic Correlation Processing (IPCP) [57] may take a vely long time or require expensive hardware. The 802.15.4 MAC requires ing methods of CCA [58]:

in-band energy above threshold.

802.15.4-like modulation and spreading.

2.15.4-like modulation and spreading above threshold.

been suggested to provide CCA in I-UWB [63]. Squaring circuits rrier-based UWB with BPSK signaling.

d a number of the first-generation of UWB products. UWB ties and it promises many new and exciting applications

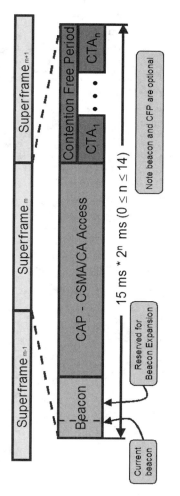

Figure 10.14: Frame Structure of IEEE 802.15.4 MAC.

with its high data rate, low power, low-cost, and precision radar capabilities. Implementations of UWB systems can be broken down into three broad categories: communications, radar, and location-aware communications. In the communications category, several companies have chosen to use UWB to implement Wireless Personal Area Networks. In other cases, the precision radar and ranging capabilities of UWB provide high-precision position location, collision avoidance, and through-walls imaging. Location-aware communications systems combine both UWB communications and radar into a system that allows a wireless network to benefit from knowing the position of nodes. Additionally, this chapter discussed the ability of UWB to provide a physical layer for the IEEE 802.15.3a high-speed Personal Area Network and the 802.15.4a low power distributed network.

Currently, UWB products provide an enhanced version of Bluetooth—offering data rates akin to USB over a 10 meter wireless link. It is expected that as UWB communications devices diminish in size and cost they will become commonplace in both consumer electronics as well as industrial communication systems. Future UWB devices will likely achieve more integration, with the ultimate goal of implementing a communications system on a single chip. Additionally, future systems are likely to employ more sophisticated customized MAC layers, which will use the location information provided by UWB.

# Appendix 10.A

# RANGE ANALYSIS OF UWB SIGNALS USING TIME OF ARRIVAL

## Nathaniel J. August

UWB can determine location by measuring the Time of Arrival (TOA) of signals [4]. Note that there are many other methods for determining range, such as time difference of arrival (TDOA), angle of arrival (AOA), and their combined radiolocation methods. The Cramer-Rao Lower Bound (CRLB) indicates the low bound on a TOA estimate as [6]

$$\sigma_{\hat{\tau}}^2 \geq \frac{1}{8\pi^2 \beta_f^2 SNR} \qquad (10.A.1)$$

where $\sigma_{\hat{\tau}}^2$ is the variance (or error) of the TOA estimates, $\beta_f$ is the bandwidth of the received signal, and $SNR$ is the energy per bit divided by the noise power ($E_b/N_0$). The CRLB for the ranging distance can be obtained as the product $c \cdot \sigma_{\hat{\tau}}$, where $c$ is the speed of light ($= 3 \times 10^8$ m/sec). The equation indicates that the impact of the SNR to CRLB is linear, while the impact of the bandwidth is quadratic, thus making UWB a good candidate for accurate ranging.

Figure 10.A.1 shows CRLBs on the ranging error in terms of SNR for four different UWB bandwidths: 0.5 GHz, 0.75 GHz, 1 GHz, and 3.3 GHz. The figure indicates that precise location information can be obtained even for UWB signals near a minimum bandwidth of 500 MHz and at moderate-to-low SNR values.

The basic process for ranging is as follows. First, the nodes synchronize to a reference clock. The reference clock may be a fixed universal reference like MSSI's $PAL650$ system [46], or the nodes may negotiate a local reference like Aether Wire's Localizers [50]. After synchronization, a TOA estimate for a received signal is obtained by detecting the peak of either (a) the original received signal or (b) the signal correlated with a template. A simple method of distance estimation is to multiply the time difference between two devices by the speed of light. However,

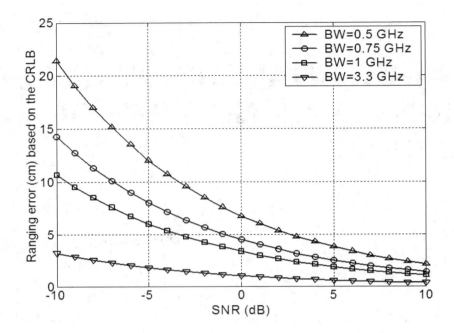

Figure 10.A.1: Lower Bound of Ranging Errors.

SOURCE: W. C. Chung and D. S. Ha, "An accurate ultra wideband (UWB) ranging for precision asset location," *Proc. 2003 IEEE Conference on Ultra Wideband Systems and Technologies* [4]). © IEEE, 2003. Used by permission.

due to noise and timing jitter, a single pulse may not provide an accurate estimation of distance. To overcome this error, one may estimate the average TOA of a train of pulses instead of a single pulse [4]. The time average of the received pulses reduces AWGN to enhance the accuracy. The use of multiple pulses increases the processing time, but the overall time required may still be a fraction of second.

# Appendix 10.B

# UWB STANDARDS FOR WPANS

Nathaniel J. August
and Michelle X. Gong

This section reviews UWB standards for wireless personal area networks (WPANs). At the time of publication, the standardization process is still underway, but two proposals are emerging as possible standards. Both proposals hope to define the physical layer (PHY) for the IEEE 802.15.3a standard, which has an existing medium access control (MAC) layer defined in the IEEE 802.15.3 standard. Because the 802.15.3a standardization process has been deadlocked for some time, special interest groups and market forces may instead decide which of the two becomes the dominant standard. This section reviews the existing IEEE 802.15.3 MAC standard and two proposed standards for a UWB PHY: multiband orthogonal frequency division multiplexing (MB-OFDM) and direct sequence ultra wideband (DS-UWB).

## 10.B.1   IEEE 802.15.3/3a MAC

This section is based on the IEEE 802.15.3 Draft Standard [60], which defines a MAC protocol and a frequency hopping PHY that operates in the unlicensed 2.4 GHz band. For a detailed guide to this standard, see [61]. The 2.4 GHz PHY is based on Bluetooth, and it specifies raw data rates of 11, 22, 33, 44, and 55 Mbps. Because many multimedia applications require a higher data rate than 55 Mbps, the 802.15.3a task group is defining an alternate, high data rate PHY based on UWB for the existing 802.15.3 MAC. Table 10.B.1 displays a short history of 802.15 WPAN standards. The IEEE 802.15.3 MAC supports the following objectives:

1. Self-organized ad hoc networks: An 802.15.3 network is an ad hoc network formed without preplanning or infrastructure.

Table 10.B.1: IEEE 802.15 History.

| Date | Event |
|------|-------|
| 1999 | 802.15.1 standard started. First MAC/PHY for WPANs. |
| 1999 | 802.15.3 standard started. Higher data rate 2.4 GHz MAC/PHY for WPANs. |
| 2000 | XtremeSpectrum attempts UWB PHY proposal, but no regulatory support. |
| 2001 | 802.15.3a standard started. Highest data rate UWB PHY with the existing 802.15.3 MAC, which was designed with the possibility of a UWB PHY. |
| 2002 | FCC Report and Order on UWB sets spectral power limits on UWB and allows for standardization to begin. |
| 2003 | 802.15.3 standard approved. |
| 2004 | 802.15.3a standardization process is ongoing. MB-OFDM and DS-UWB are the last two remaining proposals for the 802.15.3a PHY. Both are attempting to define a *de facto* industry standard in parallel with 802.15.3a. |

2. Fast connection time: The association procedure of an 802.15.3 device (DEV) is less than one second.

3. Dynamic membership: Up to 245 devices can join or leave a piconet anytime.

4. Power conservation: Three different power-saving modes accommodate different power conservation needs.

5. Asynchronous data transfer: Asynchronous data transfer efficiently uses resources for data that have no Quality of Service (QoS) requirement.

6. Multimedia QoS: The time division multiple access (TDMA) architecture supports QoS for isochronous data streams, which can range from PC mouse data to high-definition video streams.

7. Low header overhead: The source and destination IDs are only 8 bits long.

8. Security: The standard supports authentication, encryption, and integrity.

## 10.B.1.1    Topology and Components of a Piconet

An 802.15.3 network consists of one or more piconets. Each piconet consists of one or more logically associated devices sharing the same channel (see Figure 10.B.1).

One device in a piconet takes the role of the piconet coordinator (PNC), which provides centralized synchronization and manages QoS requirements, power saving modes, and access to the piconet. 802.15.3 supports peer-to-peer communication within a piconet, so data communications between devices is not required to pass through the PNC.

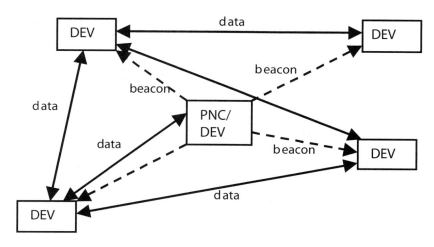

Figure 10.B.1: 802.15.3 Piconet Components [60].

SOURCE: The Institute of Electrical and Electronics Engineers, Inc., 802.15.3 Draft Standard for Telecommunications and Information Exchange Between Systems— LAN/MAN Specific Requirements—Part 15: Wireless Medium Access Control (MAC) and Physical Layer (PHY) Specifications for High Rate Wireless Personal Area Networks [60]. © IEEE, 2003. Used by permission.

## 10.B.1.2  802.15.3 MAC Superframe and Channel Time Management

Figure 10.B.2 shows an 802.15.3 superframe, which consists of three parts: the beacon, the contention access period (CAP) and the channel time allocation period (CTAP).

### Beacon

The PNC transmits a beacon at the beginning of each superframe. The beacon provides a timing reference and communicates management information to devices. The beacon also defines the start of a superframe and allocates the time slots for the remaining part of the superframe.

### Contention-Based Channel Access

The CAP can be used for asynchronous data transfers and/or command messages from the PNC to devices. Authentication, device association/disassociation, and parameter negotiation for isochronous streams can all occur in the CAP.

| Superframe #m-1 | Superframe #m | Superframe #m+1 |
|---|---|---|

| Beacon #m | Contention Access Period | Channel Time Allocation Period | | | | | |
|---|---|---|---|---|---|---|---|
| | | MCTA 2 | CTA1 | CTA2 | ▪▪▪▪ | CTA n-1 | CTA n |

Figure 10.B.2: 802.15.3 Superframe [60].

SOURCE: The Institute of Electrical and Electronics Engineers, Inc., 802.15.3 Draft Standard for Telecommunications and Information Exchange Between Systems— LAN/MAN Specific Requirements—Part 15: Wireless Medium Access Control (MAC) and Physical Layer (PHY) Specifications for High Rate Wireless Personal Area Networks [60]. © IEEE, 2003. Used by permission.

Because the wireless channel is a shared media, collisions may occur if multiple stations transmit simultaneously. To manage collisions in the contention access period, devices use CSMA/CA in 802.15.3, but UWB devices may have to use Slotted Aloha in 802.15.3a.

In CSMA/CA, a device first senses the medium before it starts transmitting. If the medium is idle after a random period of time, the device sends a short Request To Send (RTS) packet to the destination. Upon receiving the RTS, the destination responds with a Clear To Send (CTS) message. On receipt of CTS, the source sends its data packet(s). The three-step process for collision avoidance is shown in Figure 10.B.3, and it is called a dialogue. For a detailed description of CSMA/CA, interested readers can refer to the IEEE 802.11b standard [62].

Due to the difficulties involved in carrier sense for a UWB system, CSMA/CA may not be a suitable medium access protocol to use in the CAP [63]. The Slotted Aloha protocol allows a node to randomly pick a time slot for transmission without the intervention of a central control node. The source is informed of a successful

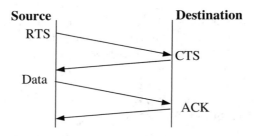

Figure 10.B.3: RTS/CTS/Data/ACK Exchange Process.

transmission through an acknowledgement packet; otherwise, the source node assumes there was a collision and retransmits. The beacon synchronizes the nodes to the start of each time slot.

### Channel Time Allocation Period Channel Access

The medium access mechanism in CTAP is based on TDMA. Because TDMA is not a contention-based protocol, it is suitable for a UWB network. The PNC divides a superframe into small time slots and schedules transmissions in these time slots. Therefore, power sensitive devices know when to listen and when to sleep. This is useful because a UWB device uses more power to listen than to transmit (unlike IEEE 802.11 devices) [64]. Therefore, limiting listening time saves power.

The CTAP period is divided into time slots called channel time allocations (CTAs), which have a guaranteed start time and duration. The PNC allocates CTAs to individual devices. When allocated, the device has sole use of the CTA and no other devices will compete for the channel during the duration of the CTA. Thus, there are no collisions during a CTA.

The standard defines two types of CTAs: dynamic CTAs and pseudo static CTAs. Dynamic CTAs can move within successive superframes, and they may be used for both asynchronous and isochronous streams. In contrast, pseudo static CTAs are used only for isochronous streams. As its name suggests, a pseudo static CTA has relatively constant duration and position, and a PNC cannot reassign a pseudo static CTA for four superframes. Thus, a device can still transmit even when four beacons are missed, which exponentially reduces the probability of losing its slot.

A private CTA is a special type of CTA, which has the same device as both the source and the destination. A private CTA reserves channel time for a dependent piconet.

Management CTAs (MCTA) are identical to CTAs except that MCTAs contain command messages to or from the PNC. MCTAs are not required because a PNC also sends commands in the beacon. There are two special types of MCTAs: open MCTAs and association MCTAs. Open MCTAs are reserved for devices that are currently associated with the piconet, whereas association MCTAs provide an opportunity for devices to join the piconet. Open MCTAs and association MCTAs use Slotted Aloha as the random access mechanism.

## 10.B.1.3   Piconet Operations
### Child and Neighbor Piconets

There are two types of dependent piconets: child piconets and neighbor piconets. Both are formed under an established piconet. The established piconet then becomes the parent piconet.

Figure 10.B.4 shows an example topology of an 802.15.3 WPAN. A child/neighbor piconet operates on the same channel as the parent piconet, whereas an independent piconet within range operates on a different channel. A child piconet is mainly used to extend the coverage area of the piconet or to shift some computational

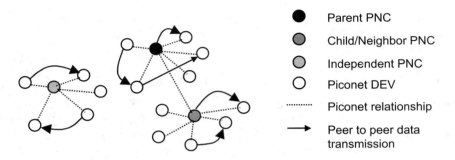

Figure 10.B.4: 802.15.3 WPAN Topology [65].

SOURCE: J. Barr, "IEEE 802.15 TG3 and SG3a," Presentation to FCC Technological Advisory Council [65]). © Motorola, 2002. Used by permission.

or memory requirements to another PNC-capable device. Direct transfer between members of different piconets is not provided. However, because the child PNC is a member of both the parent piconet and the child piconet, it is able to exchange data with any device in the parent piconet and the child piconet. The child PNC acts as a gateway node to relay packets between the two piconets.

In order to form a child piconet, a PNC-capable device in the piconet requests a pseudo static, private CTA from the current PNC. Upon receiving the private CTA, the device becomes a child PNC and starts sending its beacon in the private CTA. A parent piconet can have more than one child piconet, and it is also possible for a child piconet to have its own child piconet.

When there are no more vacant channels, a newly formed piconet may share the channel as a neighbor piconet. A PNC-capable device in the new piconet requests an established PNC to start a neighbor piconet. The neighbor PNC is not a member of the parent piconet, so it may not exchange information with any device in the parent piconet.

Because child and neighbor piconets share channel resources with the parent piconet, there is a practical limit on the number of child and neighbor piconets. Therefore, these are not the best approach for extending network coverage or size. A study group in the 802.15.3 standard body is currently examining mesh networking within the existing 802.15.3 structure.

## Starting, Maintaining, and Stopping a Piconet

To start a piconet, a PNC-capable device scans for an available channel. Upon finding a channel, it starts a piconet by sending a beacon. After a PNC establishes a piconet, devices may attempt to join the piconet. After these devices become members of the piconet, the PNC allocates resources to them. The PNC continues its role until it determines that another device in the piconet is more capable. Then the PNC may initiate a PNC handover procedure.

## PNC Handover

A PNC may attempt to choose a PNC-capable device as its successor if it finds a more capable PNC. If the successor device is currently the PNC of a dependent piconet, it may decline the request. The current PNC transfers the necessary information to the new PNC. After the new PNC has signaled that it is ready to take over, the old PNC stops generating beacons. The new PNC then broadcasts its first beacon at the time the beacon would have been sent by the old PNC.

## DEV Association and Disassociation

A device joins a piconet using the association process by sending requests to the PNC in either the CAP or in an association MCTA. The PNC may deny the association under one of the following conditions:

1. The device's address is not included in the list of device addresses maintained by the PNC.

2. There are not enough resources available to support the new device.

If the association process with the piconet is successful, the PNC provides the device with a unique identifier and broadcasts information about all the devices in the piconet, including the new device. This gives other devices in the piconet information about the new device, as well as giving the new device information about the other piconet members.

When a PNC wants to remove a device from the piconet or when a device wants to leave a piconet, the PNC or device initiates the disassociation process. In some cases, a device may be disassociated from a piconet involuntarily. For instance, the PNC will disassociate a device if the PNC does not receive any frames from the device within a predefined period. Or, a device will consider itself disassociated from the piconet if it does not receive beacons from the PNC for a predefined period. A disassociated device may reassociate with the piconet.

## 10.B.1.4  Data Communications Between DEVs

### Fragmentation and Defragmentation

In order to handle large frames, the MAC layer supports fragmentation and defragmentation. Devices may also adjust the size of MAC frames to control the frame error rate (FER) of a link.

### ACK and Retransmission

The 802.15.3 MAC provides three different acknowledgement policies for different applications: no-ACK, delayed-ACK, and immediate-ACK.

The no-ACK policy applies to frames that do not require guaranteed delivery, either because the upper layer will handle retransmission or because the retransmitted frame will be useless due to a missed deadline. For example, a receiver would simply drop a corrupted voice packet without requesting retransmission. All broadcasting and multicasting traffic uses the no-ACK policy.

The delayed-ACK policy decreases the transmission overhead by allowing the destination to acknowledge multiple frames in a single response to the source. The delayed-ACK policy is used only for isochronous connections.

With the immediate-ACK policy, the destination individually ACK's each frame. The ACKs are more timely, but the communication overhead is greater than that of delayed-ACK or no-ACK.

## 10.B.1.5 Interference Mitigation

Because an 802.15.3 network operates in an unlicensed band, it both interferes and is subject to interference from 802.15.3 piconets and other unlicensed devices. The PNC mitigates interference between piconets with the following procedures:

1. To reduce interference from a non-802.15.3 piconet, the PNC performs dynamic channel selection; it finds the channel with the least interference.

2. To reduce interference from an 802.15.3 piconet, the PNC forms a dependent piconet.

3. To reduce interference to other devices, the PNC may set a maximum transmit power. Devices may also request other devices to adjust their transmission power.

## 10.B.1.6 Power Management

The IEEE 802.15.3 draft standard defines four power management modes: ACTIVE mode, asynchronous power save (APS) mode, piconet synchronized power save (PSPS) mode, and device synchronized power save (DSPS) mode.

Power management modes are closely associated with the power states AWAKE and SLEEP. In the SLEEP state, a device is neither transmitting nor receiving. In the AWAKE state, a device is either transmitting or receiving. A device may enter the SLEEP state during any CTA for which it is neither the source nor the destination, regardless of its PM mode.

### Asynchronous Power Save (APS) Mode

A device in APS mode achieves maximum power savings because it conserves power by remaining in the SLEEP state for extended periods of time. A device in APS mode need not listen for beacons or traffic. Instead, it negotiates with the PNC to avoid being disassociated from the piconet. The PNC terminates all transactions that involve an APS device.

### Piconet Synchronized Power Save (PSPS) Mode

Whereas an APS device does not need to listen for any traffic, a device in PSPS mode listens to all system wake beacons (a special type of beacon) as decided and announced by the PNC.

### Device Synchronized Power Save (DSPS) Mode

By grouping devices that have similar power save requirements into DSPS sets, DSPS mode allows a device to synchronize its AWAKE state with other devices. The PNC manages DSPS sets, but the devices determine the parameters of the sets. In order to enter DSPS mode, a device must first join a DSPS set or form a new DSPS set. A device may register in more than one DSPS set because it may support multiple applications with different requirements.

Although devices in PSPS mode and DSPS mode listen to wake-up beacons, there is a major difference between these two modes. The PNC determines the wake beacon interval in PSPS mode, whereas the initiating device determines the wake beacon interval in DSPS mode.

## 10.B.1.7   Security

Compared to wired networks, wireless networks face unique security challenges and require stronger security protection. The wireless medium is a shared medium, so eavesdropping and "spoofing" are common problems for wireless networks.

The 802.15.3 standard provides security mechanisms for authentication, encryption, and integrity. Authentication controls the admission of a device into a secured relationship to prevent spoofing and unwanted access to critical data and functions. Encryption makes the data and control messages unintelligible to prevent eavesdropping. Integrity can prevent data or command messages from being modified by other parties. The standard supports two security modes: no security (mode 0) and the use of strong cryptography (mode 1). Mode 0 devices will not perform any cryptographic operations on MAC frames and will not accept any MAC frames that have been encrypted. Mode 1 devices use symmetric-key cryptography to protect frames by using encryption and integrity.

## 10.B.2   IEEE 802.15.3a PHY

The 802.15.3a standard defines a high-speed (up to Gbps range) UWB PHY to enhance the existing 802.15.3 standard for imaging and multimedia applications. Although the leading proposals for 802.15.3a, MB-OFDM and DS-UWB are based on the 802.15.3a task group recommendations, they may sidestep the standardization process and become industry standards. This section reviews the 802.15.3a selection criteria, channel model, link budget, and solution space.

## 10.B.2.1   Selection Criteria

The selection criteria for the IEEE 802.15.3a high-speed PHY were derived from the responses to the call for applications (CFA). The selection committee considers three categories: general solution criteria, PHY layer criteria, and MAC protocol supplements criteria. The selection committee also considers the ability of proposals to adapt to various international regulatory requirements.

## General Solution Criteria

The general solution criteria consist of the following:

1. Unit Manufacturing Cost—The cost includes estimates for the analog and digital die size, the semiconductor process, and the major external components. The cost and bill of materials should be similar to Bluetooth.

2. Signal Robustness—The minimum required receiver sensitivity should produce a packet error rate (PER) of less than 8% for 1024 byte packets, while meeting regulatory emission levels at the minimum data rate of 110 Mbps at 10 m.

3. Interference—Other radiators should not significantly affect the PHY, nor should the PHY interfere with other devices. The primary interference is from WLANs (2.4 GHz and 5 GHz), other WPANs (such as IEEE 802.15.1, 802.15.3/3a, and 802.15.4/4a), cordless phones (2.4 GHz and 5 GHz), and microwave ovens. The interference susceptibility is determined by the minimum separation distance between an interferer and the proposed system that maintains a PER of less than 8% when operating at 6 dB above receiver sensitivity. The desired separation distance is less than 1 meter.

4. Technical Feasibility—The device should be practical to manufacture, have reasonable time to market, and have minimum regulatory impact.

5. Scalability—Devices should have the ability to adjust important parameters without rewriting the standard. Scalable parameters include power, data throughput, range, frequencies of operation, and occupied bandwidth.

6. Location Awareness—This is an optional ability to determine the relative location of another IEEE 802.15.3a device.

## PHY Layer Criteria

The PHY layer criteria specify the following:

1. Size and Form Factor—Suitable form factors include PC card, compact flash, memory stick, and memory card. The form factor should be consistent with a camera, a PDA, a NIC card, or other size suitable for embedding in small devices.

2. Throughput—The throughput measures both the net data rate between PHYs and the payload data rate. The payload data rate is the raw bit rate of transmission. The net data rate considers additional packet overhead from the preambles, headers, error check sequences, and interframe spacings. Figure 10.B.5 shows the structure and overhead of an IEEE 802.15.3a frame. A net bit rate of at least 110 Mbps at 10 meters and 200 Mbps at 4 meters (PER of 8% for 1024 octet frame body) is required. Scalability to rates in excess of 480 Mbps is desirable at closer range.

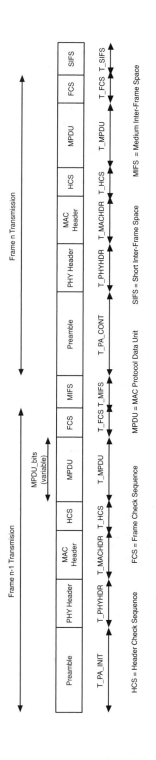

Figure 10.B.5: Packet Overhead Parameters for Data Throughput Comparison [66].

3. Simultaneously Operating Piconets—The PHY should operate in the presence of other uncoordinated piconets. Each piconet should maintain its original bit rate and transmission range, with at least three other piconets operating in close proximity.

4. Signal Acquisition—The probability for a false alarm and for a missed detect should be specified for a given preamble, channel model, and transmitter-receiver LOS distance. The target times (from the beginning of the preamble) are 6 $\mu$s for clear channel assessment (CCA) and 20 $\mu$s for acquisition.

5. System Performance—The system should average a maximum 8% PER at 110 Mbps, disregarding the worst 10% of 100 channel realizations at a distance of 10 meters.

6. Link Budget—The system should provide some implementation margin. The link budget is discussed in detail in Section 10.B.2.4.

7. Power Consumption—Portable systems should consume minimal power to extend battery life. The target power dissipation is 100 mW at 110 Mbps and 250 mW at 200 Mbps in either the transmit state or the receive state.

8. Antenna Practicality—The antenna should be consistent with the four form factors of the PC card, CompactFlash, memory disk, and SD memory.

## MAC Protocol Supplements Criteria

Some supplements and modifications to the IEEE 802.15.3 MAC may be required to support the high-speed UWB PHY.

Because the existing IEEE 802.15.3 MAC is meant to support a UWB PHY, there should be no fundamental changes. The centralized TDMA architecture is suitable for UWB. Time slots allow devices to request guaranteed data rates. Further, the beacon saves receive power, which may be equal to or greater than transmit power for UWB devices, by notifying devices of when they are not involved in a transaction. Finally, the guaranteed time slots prevent collisions between data packets, because devices are guaranteed sole ownership of a time slot.

Three major changes to the MAC may include: (a) additional synchronization time, (b) CAP access through Slotted Aloha instead of CSMA/CA, and (c) the method to allow coexistence of simultaneously operating piconets, which is PHY dependent.

## 10.B.2.2 Channel Model

The channel model for IEEE 802.15.3a devices is based on [67] and is derived from the Saleh-Valenzuela model [68]. Each multipath belongs to a cluster, which is a group of multipaths that arrive close together in time. The arrival times of each cluster and each multipath within a cluster are exponentially distributed random variables conditioned upon the previous arrival time. Each cluster, as well as each multipath within a cluster, undergoes independent fading. However, a lognormal distribution, rather than the Rayleigh distribution of the Saleh-Valenzuela model, describes the magnitudes of the multipath gains.

The channel model divides the possible channel conditions according to four scenarios: $CM1$: LOS 0-4 m, $CM2$: NLOS 0-4 m, $CM3$: NLOS 6-10 m, and $CM4$: extreme NLOS. Table 10.B.2 shows the corresponding parameters for each of these channel conditions.

For evaluation purposes, there are 100 precomputed channel realizations for each of the four classes. In each class, the power levels are normalized for an average gain of 0 dB with a standard deviation of 3 dB. The precomputed channel realizations statistically model typical impulse responses of the UWB channel. Each impulse in

Table 10.B.2: Channel Model Parameters.

SOURCE: J. Foerster, "Channel Modeling Sub-committee Report Final," IEEE doc.: P802.15-02/490r1-SG3a [67]. © IEEE, 2003. Used by permission.

| Target Channel Characteristics[a] | CM 1[b] | CM 2[c] | CM 3[d] | CM 4[e] |
|---|---|---|---|---|
| Mean Excess Delay (nsec) ($\tau_m$) | 5.05 | 10.38 | 14.18 | |
| RMS Delay (nsec) ($\tau_{rms}$) | 5.28 | 8.03 | 14.28 | 25 |
| Number of Paths (10dB) | | | 35 | |
| Number of Paths (85%) | 24 | 36.1 | 61.54 | |
| **Model Parameters** | | | | |
| $\Lambda$ - cluster arrival rate (1/nsec) | 0.0233 | 0.4 | 0.0667 | 0.0667 |
| $\lambda$ - ray arrival rate (1/nsec) | 2.5 | 0.5 | 2.1 | 2.1 |
| $\Gamma$ - cluster decay factor | 7.1 | 5.5 | 14.00 | 24.00 |
| $\gamma$ - ray decay factor | 4.3 | 6.7 | 7.9 | 12 |
| $\sigma_1$ cluster fading s.d. (dB) | 3.3941 | 3.3941 | 3.3941 | 3.3941 |
| $\sigma_2$ ray fading s.d. (dB) | 3.3941 | 3.3941 | 3.3941 | 3.3941 |
| $\sigma_x$ total shadowing s.d. (dB) | 3 | 3 | 3 | 3 |
| **Model Characteristics**[f] | | | | |
| Mean Excess Delay (nsec) ($\tau_m$) | 5.0 | 9.9 | 15.9 | 30.1 |
| RMS Delay (nsec) ($\tau_{rms}$) | 5 | 8 | 15 | 25 |
| Number of Paths (10dB) | 12.5 | 15.3 | 24.9 | 41.2 |
| Number of Paths (85%) | 20.8 | 33.9 | 64.7 | 123.3 |
| Channel Energy Mean (dB) | -0.4 | -0.5 | 0.0 | 0.3 |
| Channel Energy Std (dB) | 2.9 | 3.1 | 3.1 | 2.7 |

[a]These characteristics are based upon a 167 psec sampling time.

[b]LOS (0–4m). Channel measurements reported in [69].

[c]NLOS (0–4m). Channel measurements reported in [69].

[d]NLOS (4–10m). Channel measurements reported in [69, 70].

[e]Extreme NLOS. 25 nsec RMS delay spread.

[f]These characteristics are based upon a 167 psec sampling time.

a channel realization corresponds to a multipath component and is described with a magnitude, an arrival time, and a polarity.

### 10.B.2.3   Link Budget

Link budget is a generic term for the series of mathematical calculations that computes the performance of a communications link for specified bit rates, ranges, and bit error rates. The final output of the link budget is a link margin that describes the tolerance of a system to additional sources of loss, such as unpredicted path loss, implementation loss, waveform distortion, imperfect multipath energy capture, or amplitude fading. Table 10.B.3 identifies the necessary parameters and equations to compute the link margin.

### 10.B.2.4   Overview of Solution Space

The proposals for 802.15.3a vary mostly in the method used to fill the UWB spectrum because the FCC does not require any particular method. At one extreme, a sharp impulse fills the band as in I-UWB; and at the other extreme, many simultaneous narrowband tones fill the band as in MC-UWB [64]. Several solutions exist in between these extremes–a single band may be divided or notched into a few narrower bands, or, alternatively, several narrow bands may combine to fill increasingly larger spectrums. The solutions may or may not utilize a carrier frequency. Figure 10.B.6 displays the range of proposals (from I-UWB to MC-UWB) submitted to the 802.15.3a Task Group. The two leading proposals, DS-UWB and MB-OFDM, are shown in Figure 10.B.6 in their respective areas of the design space. DS-UWB cannot accurately be called I-UWB because it may use more than one band and it uses a carrier frequency. Likewise, MB-OFDM is a type of MC-UWB that generates multiple carriers through signal processing techniques. Some contributors to the 802.15.3a task group are listed in Table 10.B.4, along with document numbers that can provide more detail on a specific proposal. Various revisions of the documents are in a repository at http://grouper.ieee.org/groups/802/15/pub/.

The proposals specify a variety of modulation schemes, spreading schemes, and multiple access schemes (for simultaneously operating piconets); the more common schemes have been described in this book.

## 10.B.3   MB-OFDM PHY

Based on orthogonal frequency division multiplexing (OFDM), Multiband OFDM (MB-OFDM) provides a spectrally efficient means of high data rate communication with frequency diversity.[1] It is currently employed in asymmetric digital subscriber line (ADSL), IEEE 802.11a/g, IEEE 802.16a, digital audio broadcast (DAB), digital terrestrial television broadcast, Hiperlan 2, and wireless IEEE 1394.

---

[1] As of June 2004, the MB-OFDM alliance (MBOA) contains over 100 member companies, and up-to-date information can be found at http://www.multibandofdm.org.

Table 10.B.3: Link Budget Calculation.

SOURCE: J. Ellis, K. Siwiak, and R. Roberts, "P802.15. 3a Alt PHY Selection Criteria," IEEE P802-15_03031r11_TG3a-PHY-Selection-Criteria [66]. © IEEE, 2003. Used by permission.

| Parameter | Value | Value | Value |
|---|---|---|---|
| Throughput $(R_b)$ | > 110 Mb/s | > 200 Mb/s | > 480 Mb/s |
| Average Tx Power $(P_T)$ | dBm | dBm | dBm |
| Tx Antenna Gain $(G_T)$ | 0 dBi | 0 dBi | 0 dBi |
| $f'_c = \sqrt{f_{\min} f_{\max}}$: Geometric Center Frequency of Waveform ($f_{\min}$ and $f_{\max}$ are the -10 dB edges of the waveform spectrum) | Hz | Hz | Hz |
| Path Loss at 1 Meter ($L_1 = 20 \log_{10}(4\pi f'_c/c)$), $c = 3 \times 10^8$ m/s | dB | dB | dB |
| Path Loss at $d$ m ($L_2 = 20 \log_{10}(d)$) | 20 dB at $d =$ 10 meters | 12 dB at $d =$ 4 meters | Presenter Specified |
| Rx Antenna Gain $(G_R)$ | 0 dBi | 0 dBi | 0 dBi |
| Rx Power ($P_R = P_T + G_T + G_R - L_1 - L_2$ (dB)) | dBm | dBm | dBm |
| Average Noise Power per Bit ($N = -174 + 10 * log_{10}(R_b)$) | dBm | dBm | dBm |
| Rx Noise Figure Referred to the Antenna Terminal $(N_F)$[a] | dB | dB | dB |
| Average noise power per bit ($P_N = N + N_F$) | dBm | dBm | dBm |
| Minimum $E_b/N_0$ $(S)$ | dB | dB | dB |
| Implementation Loss[b] $(I)$ dB | dB | dB | dB |
| Link Margin ($M = P_R - P_N - S - I$) | dB | dB | dB |
| Proposed Min. Rx Sensitivity Level[c] | dBm | dBm | dBm |

[a]The NF is the ratio of the SNR at the antenna output with respect to the SNR at the demodulator input. The NF should include not only the LNA but also cascaded stages as per Friis's equation. The default value is 11 dB [66].

[b]Implementation loss is defined for the AWGN channel only and could include such impairments as filter distortion, phase noise, frequency errors, etc. [66].

[c]The minimum Rx sensitivity level is defined as the minimum required average Rx power for a received symbol in AWGN and should include effects of code rate and modulation [66].

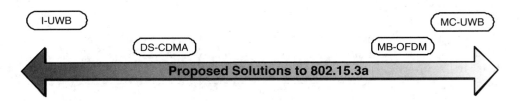

Figure 10.B.6: Overview of Design Space for 802.15.3a Proposals.

Table 10.B.4: Contributors to the IEEE 802.15.3a Alternate PHY.

| Company | Document Number | Type of Proposal [64] |
|---|---|---|
| Broadcom | 03095 | N/A |
| Discrete Time | 03099 | Multi-Band |
| Femto Devices Inc. | 03101 | Multi-Band |
| FOCUS | 03103 | Multi-Band |
| General Atomics | 03105 | Multi-Band |
| Institute for Infocomm Research | 03107 | Multi-Band |
| Intel | 03109 | Multi-Band |
| Mitsubishi | 03111 | Single Band |
| Oki Electric Industry Co., Ltd | 03119 | N/A |
| Panasonic | 03121 | N/A |
| ParthusCeva Inc. | 03123 | Single Band |
| Phillips | 03125 | Multi-Band |
| PulseLink | 03127 | Single Band |
| Rypinski | 03131 | N/A |
| Samsung | 03133, 03135 | Multi-Band |
| Sony | 03137 | Single Band |
| ST Microelectronics | 03139 | Single Band |
| Taiyo Yuden | 03145 | N/A |
| Texas Instruments | 03141 | Multi-Band |
| Time Domain | 03143 | Multi-Band |
| University of Minnesota | 03147 | Multi-Band |
| Wisair | 03151 | Multi-Band |
| XtremeSpectrum, Inc. | 03153 | Single Band |

A multiband system transmits data over several subcarriers, and OFDM is a special case of multiband transmission where the frequency bands overlap, yet they are orthogonal to each other and can be perfectly separated in theory. Figure 10.B.7 shows a conceptual model for digital OFDM transmission. The Inverse Fast Fourier Transform (IFFT) is a convenient method to generate an OFDM symbol through digital signal processing. The input data to the IFFT is assigned to subcarriers (which are represented digitally), and the IFFT converts the subcarriers into a digital time domain representation. The time domain output is filtered by a square window, which produces sinc-shaped spectrums in the frequency domain. Dense spacing of sinc functions locates successive frequency bands at the zero of the previous frequency band, thus eliminating cross-channel interference. The output signal can then be modulated onto a single high-frequency carrier.

When applied to UWB, OFDM is inherently robust to narrowband interference because it can provide a large number of subcarriers or tones. Within the UWB spectrum, a narrowband interferer can only corrupt a few tones. In OFDM, corrupt tones are usually recovered by applying a forward error correction (FEC) code or redundancy across the frequency components. By disabling selected tones, MB-OFDM can avoid bands with interference. In addition, with a cyclic prefix, MB-OFDM is resilient to the long multipath delay spread of UWB.

## 10.B.3.1 Overview of MB-OFDM Proposal

MB-OFDM divides the spectrum into several 528 MHz bands that occupy 528 MHz each. Within a 528 MHz band, a 128-point IFFT operation converts 128 subcarrier tones that are spaced 4.125 MHz apart to a time domain signal. Figure 10.B.8 shows the mapping of the tones. Data is mapped onto 100 of the subcarrier tones (shown in solid lines) using a QPSK constellation. Twelve pilot tones (shown in dashed lines), spaced every ten tones, provide carrier and phase tracking. Ten guard tones (shown in gray) shape the band edges with copies of the ten tones at the edges of the data tones. Finally, six null tones, including the DC tone, result in a total of 128 tones for a convenient radix-2 IFFT and FFT implementation.

In Figure 10.B.9, there are fourteen 528 MHz bands that belong to four distinct logical channels. The frequency axis displays the center frequency of each band. Band Group 1 spans from 3.168 GHz to 4.752 GHz and is intended for first-generation devices. Devices in this band are termed "Mode 1" devices. The remaining groups are reserved for future use. Band Group 2 spans from 4.752 GHz to 6.336 GHz; Band Group 3 spans from 6.336 GHz to 7.920 GHz; Band Group 4 spans from 7.920 GHz to 9.504 GHz; and Band Group 5 spans from 9.504 GHz to 10.560 GHz.

To further combat interference and to support simultaneously operating piconets, MB-OFDM employs frequency hopping. Band Groups 1–4 in Figure 10.B.9 can support up to 4 piconets each with 4 frequency hopping codes, and Band Group 5 can support 2 piconets for a total of 18 piconets. Figure 10.B.10 shows the time and frequency mapping of a Mode 1 device whose length-six frequency hopping code follows the sequence 3, 1, 2, 3, 1, 2. On every hop, a cyclic prefix provides immunity to ISI for channels with a multipath delay spread of less than 60.6 ns. The cyclic

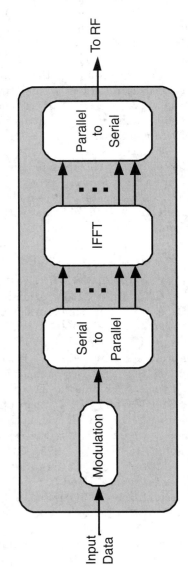

Figure 10.B.7: IFFT Model for Multiband OFDM Transmission.

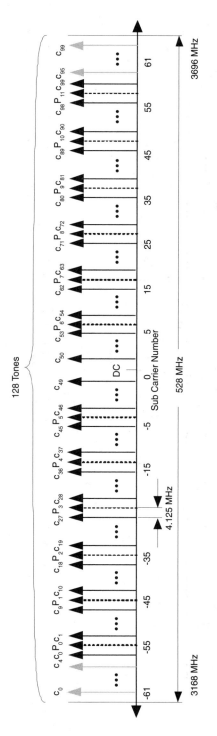

Figure 10.B.8: Subcarrier Mapping for Multiband OFDM.

SOURCE: Anuj Batra, et al, "Multi-band OFDM Physical Layer Proposal Update," IEEE 802.15-04/0220r3 [71]. © IEEE, 2004. Used by permission.

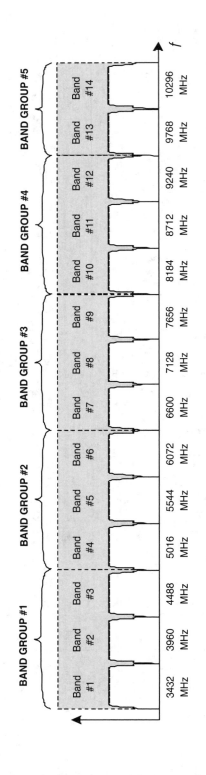

Figure 10.B.9: Band Plan for Multiband OFDM.

SOURCE: Anuj Batra, et al, "Multi-band OFDM Physical Layer Proposal Update," IEEE 802.15-04/0220r3 [71]. © IEEE, 2004. Used by permission.

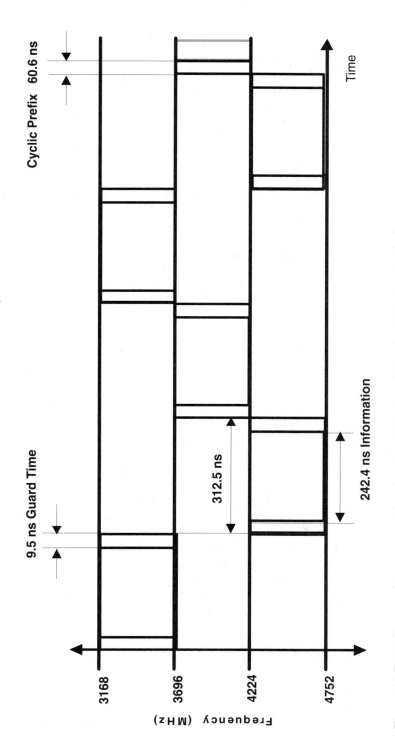

Figure 10.B.10: Time and Frequency Mapping for Mode 1 Multiband OFDM Device [64].

SOURCE: M. Welborn and B. Shvodian, "Ultra-Wideband Technology for Wireless Personal Area Networks—The IEEE 802.15.3/3a Standards UWBST Tutorial," *UWBST 2003 Conference* [64]. © IEEE, 2003. Used by permission.

prefix ensures that all samples experience the same multipath channel and maintain orthogonality. Next, there is a 242.4 ns period for transmitting an MB-OFDM symbol, and finally, a 9.5 ns guard interval provides time for the transmitter and the receiver to switch between bands. Therefore, a system spends a total of 312.5 ns in a single band for a hopping rate of 3.2 MHz.

MB-OFDM is unique among UWB modulation schemes in that it is suitable for redundancy (spreading) in both the time and frequency domains. It achieves frequency domain redundancy by applying symmetric, conjugate inputs to the IFFT. Thus a MB-OFDM system can be made more robust to frequency selective fading and interband interference. In the time domain, a MB-OFDM system provides redundancy by repeating data on OFDM symbols that occur at different times in different bands.

A Mode 1 MB-OFDM system achieves the data rates mentioned in the selection criteria through the configurations shown in Table 10.B.5. The code is a rate 1/3 convolutional code with generator polynomials $g_0 = 133_8$, $g_1 = 145_8$, and $g_2 = 175_8$. The different coding rates are achieved by puncturing the code.

More generally, the raw data rate $R$ of an MB-OFDM system can be calculated as follows:

$$R = \left( \frac{N_{SC}}{\frac{1}{\Delta_f} + T_{CP} + T_{GI}} \right) \frac{m \cdot r}{SF} \tag{10.B.1}$$

where $N_{SC}$ (number of data subcarriers) = 100, $\Delta_f$ (subcarrier frequency spacing) = 4.125 MHz, $T_{CP}$ (cyclic prefix duration) = 60.61 ns, $T_{GI}$ (guard interval duration) = 9.47 ns, $m$ (bits per symbol in base modulation) = 2 (for QPSK), $r$ is the coding rate, and $SF$ is the overall spreading gain including time and frequency spreading.

## 10.B.3.2 Simultaneously Operating Piconets

An MB-OFDM system supports simultaneously operating piconets by assigning time frequency codes (TFC) across the frequency bands. Within a band, each piconet has a unique TFC, and Table 10.B.6 displays the TFCs for each band. Band Groups 1–4 have the same number of bands, so they share identical TFCs among the band groups.

As the number of piconets increases, performance degrades gracefully. Table 10.B.7 shows the required separation distance between piconets in order for each piconet to operate without significant performance degradation. The reference distance $d_{ref}$ occurs at 0.707 of the 90% link success probability distance. The system is Mode 1 operating at 110 Mbps. Multiple access within a piconet is TDMA, as in IEEE 802.15.3.

## 10.B.3.3 Acquisition

On a larger scale, the MB-OFDM proposal specifies the frame structure in Figure 10.B.11, which includes the physical layer convergence protocol (PLCP) preamble and header. The preamble is composed of three sections. The packet synchronization sequence provides packet detection, acquisition, coarse carrier frequency estimation,

Table 10.B.5: MB-OFDM Configurations to Achieve the Data Rates in the Selection Criteria.

| Data Rate | FEC Coding Rate | Time Spreading Rate | Frequency Spreading Rate | Overall Spreading Gain | Number of Coded Bits per OFDM Symbol |
|---|---|---|---|---|---|
| 55 Mbps | 11/32 | 2 | Yes | 4 | 100 |
| 110 Mbps | 11/32 | 2 | No | 2 | 200 |
| 200 Mbps | 5/8 | 2 | No | 2 | 200 |
| 480 Mbps | 3/4 | 1 | No | 1 | 200 |

Table 10.B.6: Time Frequency Codes.

| Band Groups | Piconet Number | TFC Length | TFC | | | | | |
|---|---|---|---|---|---|---|---|---|
| 1,2,3,4 | 1 | 6 | 1 | 2 | 3 | 1 | 2 | 3 |
| | 2 | 6 | 1 | 3 | 2 | 1 | 3 | 2 |
| | 3 | 6 | 1 | 1 | 2 | 2 | 3 | 3 |
| | 4 | 6 | 1 | 1 | 3 | 3 | 2 | 2 |
| 5 | 1 | 4 | 1 | 2 | 1 | 2 | - | - |
| | 2 | 4 | 1 | 1 | 2 | 2 | - | - |

Table 10.B.7: Required Separation Distance for Simultaneously Operating Piconets.

| Channel Model | $d_{ref}$ | 1 Piconet $(d_{int}/d_{ref})$ | 2 Piconet $(d_{int}/d_{ref})$ | 3 Piconet $(d_{int}/d_{ref})$ |
|---|---|---|---|---|
| CM1 | 11.4 m | 0.40 | 1.18 | 1.45 |
| CM2 | 10.7 m | 0.40 | 1.24 | 1.47 |
| CM3 | 11.5 m | 0.40 | 1.21 | 1.46 |
| CM4 | 10.9 m | 0.40 | 1.53 | 1.85 |

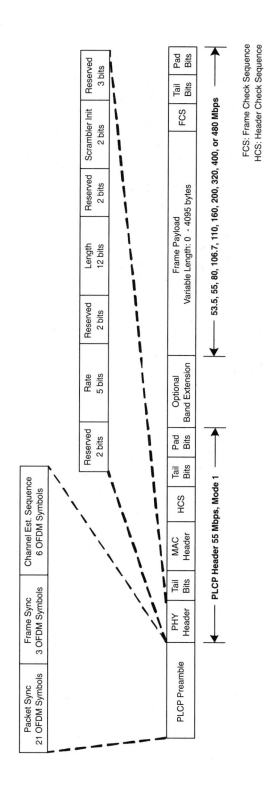

Figure 10.B.11: Frame Structure with PHY Preamble [72].

SOURCE: Anuj Batra, et al, "Multi-band OFDM Physical Layer Proposal for IEEE 802.15 Task group 3a," IEEE 802.15-03/268r3 [72]. © IEEE, 2004. Used by permission.

and coarse symbol timing. The frame synchronization sequence synchronizes the receiver algorithm within the preamble. The channel estimation sequence provides channel estimation, fine carrier frequency estimation, and fine symbol tuning. The packet frame synchronization sequence consists of 21 repetitions of a length-128 time domain sequence that varies for each piconet. In streaming mode, the packet frame synchronization sequence reduces to 9 repetitions. The packet synchronization sequence consists of 3 repetitions of the same sequence rotated by 180 degrees. The channel estimation sequence consists of 6 repetitions of the OFDM training symbol. In all cases, each symbol is pre-appended with 32 "zero samples" and appended with 5 "zero samples."

The optional band extension field indicates the mode of transmission for future Modes. Within the PHY Header, the rate field indicates the data rate of the payload, which also suggests a coding rate, puncturing pattern, and redundancy technique, as in Table 10.B.5. The length field specifies the number of bytes in the payload, excluding the FEC, and the scrambler field specifies the current state of the transmitter's scrambler.

The header is always sent at 55 Mbps, but the payload may be sent at rates of 53.5, 55, 80, 106.7, 110, 160, 200, 320, 400, and 480 Mbps. In Mode 1, the PLCP preamble lasts for 9.375 $\mu$s, and in streaming mode, the preamble is shortened to 5.625 $\mu$s because the packet sync sequence only has to provide boundary detection. The maximum payload size is 4095 octets.

## 10.B.3.4   Architecture

The proposed architecture for MB-OFDM is based on well-known architectures for OFDM systems. Figure 10.B.12 shows a typical transmitter design for an OFDM system. The input data bits are first scrambled with a length-15 pseudo random binary sequence to randomize the data. The generator polynomial for the scrambler is $g(D) = 1 + D^{14} + D^{15}$, and the transmitter sequences through four initial values. Next, a convolutional encoder and puncturing block provides the code rates as described in Table 10.B.5.

The bit interleaver maps the input bits to different IFFT inputs on different OFDM symbols to provide diversity across both tones and bands. Figure 10.B.13 shows an example of the three-stage bit interleaving process for a Mode 1 device transmitting at 55 Mbps. First, six OFDM symbols are grouped together (or $6 \cdot N_{CBPS}$ bits, where $N_{CBPS} = 100$ for systems that apply frequency spreading and $N_{CBPS} = 200$ otherwise). Second, the coded bits are interleaved using a symbol block interleaver with $N_{CBPS}$ rows and 6 columns. The output of this stage $S(j)$ is related to the input bits $U(i)$ as

$$S\left(j\right) = u\left(\left\lfloor \frac{i}{N_{CBPS}} \right\rfloor + 6 \cdot \mathrm{mod}\left(i, N_{CBPS}\right)\right) \qquad (10.B.2)$$

Third, the output bits from the symbol block interleaver are grouped into blocks of size $N_{CBPS}$ and interleaved using a block interleaver of size $N_{CBPS}/10$ rows

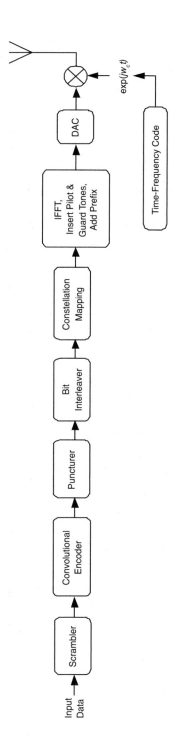

Figure 10.B.12: Transmitter Architecture for MB-OFDM [64].

SOURCE: M. Welborn and B. Shvodian, "Ultra-Wideband Technology for Wireless Personal Area Networks—The IEEE 802.15.3/3a Standards UWBST Tutorial," *UWBST 2003 Conference* [64]. © IEEE, 2003. Used by permission.

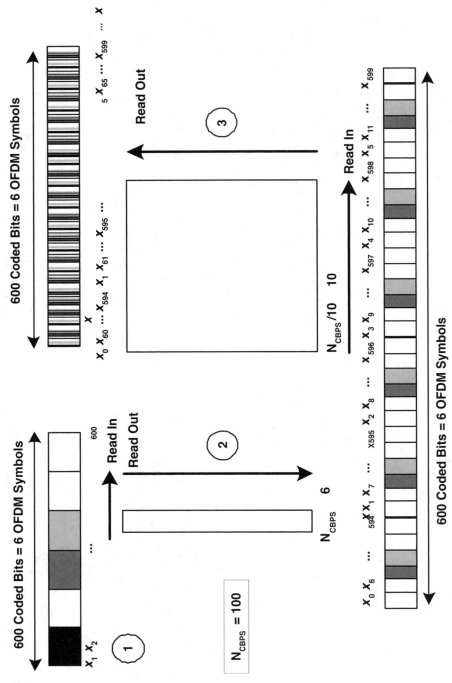

Figure 10.B.13: Bit Interleaving.

and 10 columns. The output of this interleaver $T(j)$ is related to the output of the second stage $S(j)$ as

$$T(j) = S\left(\left\lfloor \frac{10 \cdot i}{N_{CBPS}} \right\rfloor + 10 \cdot \text{mod}\left(i, \frac{N_{CBPS}}{10}\right)\right) \qquad (10.\text{B}.3)$$

Next, constellation mapping converts each group of two bits into a complex, Gray-coded QPSK constellation point. Finally, the IFFT converts the tones to a time domain output and passes them through a DAC. The output is then mixed up to one of the bands according to the current time frequency code and Band Group.

The receiver architecture in Figure 10.B.14 is also very similar to a conventional OFDM system. The front-end demodulates the received signal and mixes it to baseband. After separating the I and Q channels, the lowpass filter passes the baseband signal to an ADC. The gain control system (VGA and AGC) normalizes the receiver gain of each channel to prevent clipping. Next, the synchronization block provides frame detection and synchronization. After removing the cyclic prefix, the receiver performs the FFT to transform the time domain signals back to the original frequency domain representations. The pilot tones provide carrier frequency offset correction when the receiver and transmitter oscillators drift from each other. The deinterleaver reverses the interleaving process. Finally, a Viterbi decoder detects and corrects bit errors, and the descrambler recovers the original bits input to the transmitter.

## 10.B.3.5   Performance

This section briefly reviews the performance of the MB-OFDM proposal in terms of some important selection criteria from the IEEE 802.15.3 task group.

Firstly, MB-OFDM provides flexibility in several dimensions. The data rate scales from 55 Mbps to 480 Mbps. The occupied spectrum scales easily by band and tone selection, and power may be traded for range and data rate.

In terms of regulatory specifications, MB-OFDM offers worldwide compliance by dynamically disabling individual bands and tones that interfere with sensitive spectra, such as, the radio astronomy bands in Japan. In terms of spectral "flatness," the OFDM proposal exhibits excellent performance from the multiple, narrow bands with a zero-padded prefix (in place of a cyclic prefix). Tones lost to narrowband interference must be recovered through FEC or systems may perform handshaking to detect interference and disable the affected bands. A notched filter may be required to prevent overload from strong in-band interference.

MB-OFDM collects most of the channel energy and eliminates ISI due to the narrow bands and smaller RMS delay spread of the channel. However, the narrow bands also exhibit a Rayleigh distribution of fading magnitudes, which results in the possibility of deep fades. Further, due to the non-Gaussian probability distribution of amplitudes in OFDM modulation, MB-OFDM has a higher probability of producing signals with large peak-to-average amplitude ratio, although the peaks should not violate FCC regulations.

As computed in Table 10.B.3, the link budget specifies a link margin that describes the tolerance of a system to additional sources of loss. The receiver sensitivity

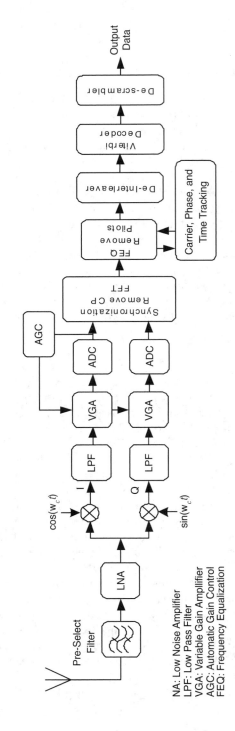

NA: Low Noise Amplifier
LPF: Low Pass Filter
VGA: Variable Gain Amplifier
AGC: Automatic Gain Control
FEQ: Frequency Equalization

Figure 10.B.14: Receiver Architecture for MB-OFDM [64].

SOURCE: M. Welborn and B. Shvodian, "Ultra-Wideband Technology for Wireless Personal Area Networks—The IEEE 802.15.3/3a Standards UWBST Tutorial," *UWBST 2003 Conference* [64]. © IEEE, 2003. Used by permission.

level specifies a corresponding minimum signal level that a receiver could detect and successfully demodulate a signal. Table 10.B.8 shows the link margin for MB-OFDM.

The MB-OFDM proposal meets the three required raw data rates of 110, 200, and 480 Mbps. However, it is also useful to calculate the data rate considering the overhead of the frame structure. Recall that in Mode 1 a device requires a 9.375 $\mu$s header. Table 10.B.9 displays the maximum achievable data rates with frame overhead for payload lengths of 1,024 and 4,024 bits. From Table 10.B.9, the maximum achievable throughput increases as the payload length increases. The data rate also increases as the number of frames increases due to the shorter header in streaming mode. Finally, higher raw data rates suffer more speed degradation from the PHY header because it is always sent at the slowest data rate of 55 Mbps, and hence lasts for a larger proportion of time as compared to the data.

The final measure of performance is the maximum distance at which the MB-OFDM system can achieve the raw data rates specified by the selection criteria. The maximum distance is defined as the distance the system achieves a maximum PER of 8% for the best 90% of channels. Table 10.B.10 includes most forms of loss, including front-end filtering, DAC clipping, ADC noise, multipath degradation, channel estimation, carrier tracking, and packet acquisition.

## 10.B.4  DS-UWB PHY

Direct sequence UWB (DS-UWB) is the name given to the UWB Forum's proposal for IEEE 802.15.3a, but DS-UWB also describes a more general UWB signal as follows.[2] In UWB, direct sequence spreading is similar to well-known narrowband direct sequence spread spectrum (DS/SS) techniques, but UWB communications systems achieve spreading gain through temporal redundancy, i.e., by transmitting multiple pulses per bit [73]. DS-UWB communication has been proposed by several researchers [73–75]. It works by transmitting a continuous train of data (one chip immediately followed by another chip) with the amplitudes of the chips modulated by a PN sequence. Multiple access is provided by assigning different users different spreading codes. The baseband DS-UWB waveform $s(t)$ in Figure 10.B.15 can be represented as [73].

$$s^{(m)}(t) = \sum_{k=-\infty}^{\infty} \sum_{j=1}^{N} w\left(t - kT_d - jT_p\right) \left(c_w\right)_j^{(m)} d_k^{(m)} \qquad (10.\text{B}.4)$$

where $w(t)$ is the baseband pulse waveform; $N$ is the number of chips per data bit (spreading gain); $T_d$ is the bit duration, equal to $NT_f$ (seconds); and $\left(c_w\right)_j^{(m)}$ is the pseudo random (PN) spreading code for the $j^{th}$ pulse. It is the $m^{th}$ PN code from a family of codes used for multiple access as discussed previously. $T_p$ is the pulse width of an individual impulse (seconds), and $d_k^{(m)}$ is the $k^{th}$ data bit in the $m^{th}$ code ($\{+1, -1\}$).

---

[2] As of June 2004, the UWB Forum, which supports DS-UWB, contains over thirty member companies, and up-to-date information can be found at http://www.uwbforum.org.

Table 10.B.8: Link Margin for MB-OFDM.

| | 110 Mbps | 200 Mbps | 480 Mbps |
|---|---|---|---|
| Received Power $P_r$ (dBm) | -74.5 @ 10 m | -66.5 @ 4 m | -60.5 @ 2 m |
| Total Noise Power $P_n$ (dBm) | -87.0 | -84.4 | -80.6 |
| Required $E_b/N_0$ $S$ (dB) | 4.0 | 4.7 | 4.9 |
| Implementation Loss $I$ (dB) | 2.5 | 2.5 | 3.0 |
| Link Margin (dBm) $M = P_r - P_n - S - I$ | 6.0 | 10.7 | 12.2 |
| Sensitivity (dBm) $= P_n + S + I$ | -80.5 | -77.2 | -72.7 |

574

Table 10.B.9: MB-OFDM Throughput Including PHY Header for Mode 1 Device.

| Number of Frames | Payload | Throughput @ 110 Mbps | Throughput @ 200 Mbps | Throughput @ 480 Mbps |
|---|---|---|---|---|
| 1 | 1024 bytes | 83.5 Mbps | 127.3 Mbps | 195.6 Mbps |
| 5 | 1024 bytes | 93.2 Mbps | 151.4 Mbps | 259.0 Mbps |
| 1 | 4024 bytes | 102.0 Mbps | 174.1 Mbps | 354.3 Mbps |
| 5 | 4024 bytes | 105.4 Mbps | 184.2 Mbps | 398.4 Mbps |

Table 10.B.10: Maximum Distance to Achieve a Given Data Rate in Different Channels.

| Data Rate | AWGN Distance | CM1 Distance | CM2 Distance | CM3 Distance | CM4 Distance |
|---|---|---|---|---|---|
| 110 Mbps | 20.5 m | 14 m | 13.2 m | 13.8 m | 13.8 m |
| 200 Mbps | 14.1 m | 8.9 m | 8.3 m | 8.5 m | 7.3 m |
| 480 Mbps | 7.8 m | 3.8 m | 3.6 m | N/A | N/A |

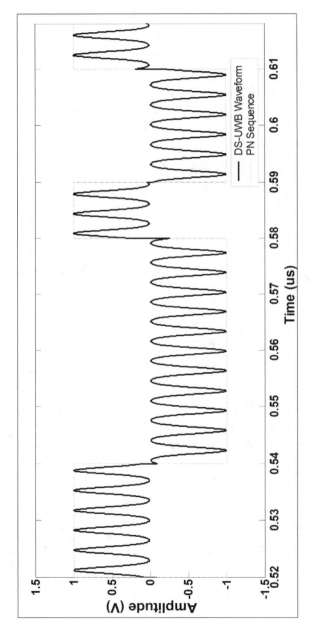

Figure 10.B.15: Direct Sequence Spreading for UWB.

The data rate $R_d$ for DS-UWB is represented as

$$R_d = \frac{1}{NT_f} \tag{10.B.5}$$

where $T_f$ is the pulse repetition time (inverse of the PRF) in seconds.

Equation 10.B.5 shows that DS-UWB systems generally have informational values of less than 1 bit per pulse. Transmitting multiple data bits per pulse is possible; however, the trade-off is a reduction in range and/or energy per bit (higher BER). Due to the high pulse rate, the continuous transmission of pulses means there is less energy available for a single pulse.

Engineers developing DS-UWB communications systems may trade several design parameters depending on system requirements, as shown in Table 10.B.11. FCC limits constrain the average transmit power, and the bandwidth depends only on the pulse width, not on the spreading gain.

## 10.B.4.1 Overview of DS-UWB

In the following sections, the term "the DS-UWB proposal" will refer to the UWB Forum's proposal for IEEE 802.15.3a, and the term "DS-UWB" will apply to any general DS-UWB system as described previously. The DS-UWB proposal achieves a wide bandwidth by modulating wideband root raised cosine (RRC) pulses on to a

Table 10.B.11: System Trade-Offs for a DS-UWB System.

| An Increase in | Means a Decrease in | While These May Remain Constant |
|---|---|---|
| Spreading Gain | Data Rate, Range | Bandwidth, Pulse Width, Average Transmit Power, PRF |
| Pulse Width | Bandwidth, Average Transmit Power | Data Rate, PRF, Spreading Gain |
| Pulse Amplitude | PRF, Data Rate | Bandwidth, Pulse Width, Average Transmit Power, Spreading Gain |
| Data Rate | Pulse Amplitude, Range | Bandwidth, Pulse Width, Average Transmit Power, Spreading Gain, Pulse Width |
| PRF | Pulse Amplitude, Range | Bandwidth, Pulse Width, Average Transmit Power, Spreading Gain, Pulse Width |
| Bandwidth | Pulse Width | Data Rate, PRF, Spreading Gain |

carrier at a high pulse rate. The single wideband signal maintains the advantages of I-UWB systems, such as multipath resolution and ranging. The modulation scheme is either BPSK or 4-ary biorthogonal keying (4-BOK), which is more energy efficient than BPSK, and hence is suitable for UWB because the maximum transmit power of UWB is constrained by regulatory limits. A family of spreading codes provides multiple access for up to 12 simultaneously operating piconets. Each piconet also has a slightly different frequency offset from the nominal center frequency. Depending on the spreading rate, FEC coding rate, and modulation scheme, the resulting data rates range from 28 Mbps for long-distance communication to 1320 Mbps for short-distance communication.

Figure 10.B.16 shows the band plan for the DS-UWB proposal. The low band extends from 3.1 GHz to 4.9 GHz, and the high band extends from 6.2 GHz to 9.7 GHz. In the figure, an RRC filter with excess bandwidth of 0.30 shapes the baseband pulse.

The base modulation scheme is BPSK, and there is also a 4-BOK modulation mode that transmitters must support and receivers may optionally support. In BPSK, one data bit determines the polarity of the spreading code. In 4-BOK, two data bits determine the choice of spreading code (from two possible choices) and the polarity of that spreading code. Figure 10.B.17 shows the baseband time domain and signal constellation representations of a length 4, 4-ary BOK code set [76]. Note that there are two orthogonal codes, $code_1$ and $code_2$, and each code also has an inverse. A device with BPSK modulation would simply use $code_1$. As described in the next section, the DS-UWB proposal uses spreading codes of lengths 24, 12, 6, 4, 3, 2, and 1, depending on the desired data rate.

A lowband system in the DS-UWB proposal achieves the data rates mentioned in the selection criteria through the configurations in Table 10.B.12. Data rates up to 1320 Mbps are possible.

## 10.B.4.2   Simultaneously Operating Piconets/Spreading

The spreading codes support multiple access for up to six piconets within each band for a total of twelve channels. Devices must support the first four channels in the low band. The chip rate $F_{chip}$ and center frequency $f_c$ of each piconet are harmonically related as $f_c = 3F_{chip}$. The offset chip rates reduce the correlation between data streams and also help to identify piconets. Table 10.B.13 shows the chip rate, center frequency, and spreading code set for each piconet channel.

Table 10.B.14 lists the different codes for each piconet channel. The piconets use the length-24 codes during acquisition (which always uses BPSK) and for BPSK data transmission with a spreading rate of 24. The length-24 codes are ternary codes, which means that they may take the values $\{-1, 0, 1\}$. 4-BOK and BPSK with spreading rates less than 12 use identical codes for each piconet during data transmission. 4-BOK uses two codes and their inverses, whereas BPSK uses only one code ($code_1$ in Table 10.B.14) and its inverse. The length-1 and length-3 codes do not support 4-BOK because there is only one code in the set. The code length depends on the desired data rate.

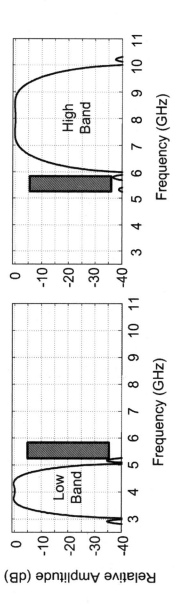

Figure 10.B.16: Band Plan for DS-UWB Proposal [76].

SOURCE: R. Fisher, R. Kohno, H. Ogawa, H. Zhang, M. McLaughlin, and M. Welborn, "DS-UWB Proposal Update," doc.: IEEE 802.15-04/140r3 [76]. © IEEE, 2004. Used by permission.

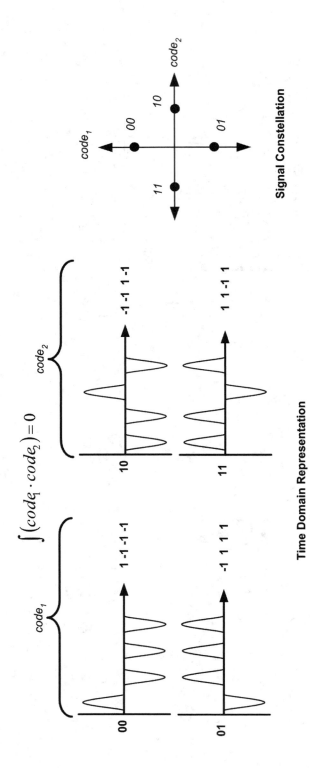

Figure 10.B.17: 4-BOK Example.

Table 10.B.12: Configurations for the DS-UWB Proposal to Achieve the Data Rates in the Selection Criteria.

| Data Rate (Mbps) | Modulation Type | FEC Rate | Spreading Code Length | Bits per Symbol | Symbol Rate (Msymb/s) |
|---|---|---|---|---|---|
| 110 | BPSK | 1/2 | 6 | 1 | $F_{chip}/6$ |
| 220 | BPSK | 1/2 | 3 | 1 | $F_{chip}/3$ |
| 500 | BPSK | 3/4 | 2 | 1 | $F_{chip}/2$ |
| 110 | 4-BOK | 1/2 | 12 | 2 | $F_{chip}/12$ |
| 220 | 4-BOK | 1/2 | 6 | 2 | $F_{chip}/6$ |
| 500 | 4-BOK | 3/4 | 4 | 2 | $F_{chip}/4$ |

Table 10.B.13: Chip Rate, Center Frequency and Spreading Code Set for Each Piconet Channel in the DS-UWB Proposal.

| Piconet Number | Chip Rate | Center Frequency | Spreading Code Set |
|---|---|---|---|
| 1 | 1313 MHz | 3939 MHz | 1 |
| 2 | 1326 MHz | 3978 MHz | 2 |
| 3 | 1339 MHz | 4017 MHz | 3 |
| 4 | 1352 MHz | 4056 MHz | 4 |
| 5 | 1300 MHz | 3900 MHz | 5 |
| 6 | 1365 MHz | 4094 MHz | 6 |
| 7 | 2626 MHz | 7878 MHz | 1 |
| 8 | 2652 MHz | 7956 MHz | 2 |
| 9 | 2678 MHz | 8034 MHz | 3 |
| 10 | 2704 MHz | 8112 MHz | 4 |
| 11 | 2600 MHz | 7800 MHz | 5 |
| 12 | 2730 MHz | 8190 MHz | 6 |

Table 10.B.14: Code Sets for the DS-UWB Proposal.

| Code Length | Piconet Number | Supports | Code | |
|---|---|---|---|---|
| 24 | 1 | BPSK | 1 0 1 -1 -1 -1 1 1 0 1 1 1 1 -1 1 -1 1 1 1 1 -1 1 -1 1 1 | |
| 24 | 2 | BPSK | -1 -1 -1 -1 1 1 -1 1 -1 -1 -1 -1 1 1 -1 1 1 1 0 -1 0 1 1 | |
| 24 | 3 | BPSK | -1 1 -1 1 1 -1 1 1 -1 1 -1 0 -1 1 1 1 1 -1 1 -1 1 1 -1 -1 -1 1 | |
| 24 | 4 | BPSK | 0 -1 -1 -1 -1 1 1 1 0 -1 1 1 -1 1 -1 -1 -1 1 1 -1 -1 -1 -1 1 | |
| 24 | 5 | BPSK | -1 1 -1 1 1 1 -1 1 0 1 1 1 -1 -1 -1 1 1 1 -1 -1 -1 -1 1 0 -1 | |
| 24 | 6 | BPSK | 0 -1 -1 0 1 -1 1 -1 -1 -1 1 1 1 1 1 -1 1 -1 1 1 -1 1 1 1 1 1 1 | |
| 12 | 1-6 | BPSK | 1 0 0 0 0 0 0 0 0 0 0 0 | *code 1* |
| | 1-6 | 4-BOK | 0 0 0 0 0 0 1 0 0 0 0 0 | *code 2* |
| 6 | 1-6 | BPSK | 1 0 0 0 0 0 | *code 1* |
| | 1-6 | 4-BOK | 0 0 0 1 0 0 | *code 2* |
| 4 | 1-6 | BPSK | 1 0 0 0 | *code 1* |
| | 1-6 | 4-BOK | 0 0 1 0 | *code 2* |
| 3 | 1-6 | BPSK | 1 0 0 | |
| 2 | 1-6 | BPSK | 1 0 | *code 1* |
| | 1-6 | 4-BOK | 0 1 | *code 2* |
| 1 | 1-6 | BPSK | 1 | |

Table 10.B.15: Required Separation Distance for Simultaneously Operating Piconets.

| Channel Model | $d_{ref}$ | 1 Piconet $(d_{int}/d_{ref})$ | 2 Piconets $(d_{int}/d_{ref})$ | 3 Piconets $(d_{int}/d_{ref})$ |
|---|---|---|---|---|
| CM1 | 15.7 m | 0.66 | 0.86 | 1.09 |
| CM2 | 15.7 m | 0.64 | 0.91 | 1.14 |
| CM3 | 15.7 m | 0.72 | 0.97 | 1.24 |

As the number of piconets increases, performance degrades gracefully. Table 10.B.15 shows the required separation distance between piconets in order for each piconet to operate without significant performance degradation. The system is operating in the lower band at 110 Mbps. Multiple access within a piconet is TDMA as in the IEEE 802.15.3 MAC.

## 10.B.4.3    Acquisition

The DS-UWB proposal also specifies the frame structure for a data packet, which includes the PLCP preamble, the 2-byte start of frame delimiter (SFD), and the 2-byte PHY header as shown in Figure 10.B.18 [77]. In the PHY header, the seed identifier provides the receiver with the seed for the data scrambler. Next, the data rate is determined by the FEC type, spreading code length, and modulation type. The FEC can have a rate of $\frac{1}{2}$, $\frac{3}{4}$, or 1 with a $K = 6$ or a $K = 4$ convolutional encoder. The modulation type is either BPSK or 4-BOK, and the spreading rate is 24, 12 6, 4, 3, 2, or 1. The interleaver type currently has only one choice of a convolutional bit interleaver. Finally, the last 14 bits specify the length in octets of the frame body, but only 11 bits are used to limit the length to 2,047 octets. The MAC header is 2 bytes and is identical to the IEEE 802.15.3 header.

The sequence of bits up to the MAC header is transmitted without FEC or interleaving at a nominal data rate of 55 Mbps in the low band and at 110 Mbps in the high band. The PHY preamble and header are not scrambled, as is the rest of the packet. The spreading rate is 24, so the chips are transmitted at 1,320 Mcps. The payload may be sent at higher bit rates depending on the spreading rate, coding rate, and modulation type. The DS-UWB proposal uses three preamble lengths, which are link quality dependant. Each preamble consists of an acquisition sequence followed by a training sequence, which estimates the channel for rake tuning and channel equalization. The transmitter modulates the length-24 spreading code with random data during the acquisition sequence, and Table 10.B.16 shows the training sequences for the medium preamble length. The short preamble length is a total of 10 $\mu$s long and is used for short-range (less than 4 meters) operation with a high data rate. The medium preamble sequence is 15 $\mu$s long and is used for medium-range

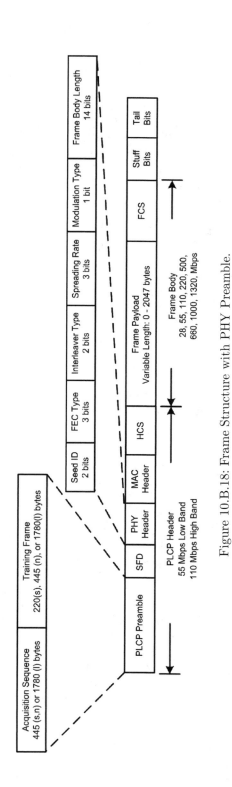

Figure 10.B.18: Frame Structure with PHY Preamble.

Table 10.B.16: Medium Length Preambles for UWB Forums DS-UWB, Base-32 Notation.

| High Band Preamble | Low Band Preamble |
|---|---|
| JNJNB5ANB6APAPCPANASASCN | JNJNB5ANB6APAPCPANASASCN |
| SK9B5K6B5K5D5D5B9ANASJPJN | JNASK9B5K6B5K5D5D5B9ANASJ |
| K5MNCPATB5CSJPMTK9MSJTC | PJNK5MNCPATB5CSJPMTK9MS |
| TASD9ASJNACTATASCSANCSAS | JTCTASD9ASCTATASCSANCSAS |
| JSJSB5ANB6JPAPD6B5ATASCPM | JSJSB5ANB6JPN5DAASB9K5MSC |
| NCSN5D5K6K5B9CND5JTJPBAM | NDE6AT3469RKWAVXM9JFEZ8C |
| NK6KAMTCNJTB5N9N6N9JNMN | DS0D6BAV8CCS05E9ASRWR914A |
| MTJSANMSD5K9K6K9JNMNMPJ | 1BR |
| SANCSN5JSK6JTJPMPJNJSASCN | |
| N5DAASB9K5MSD5B7291AT2W67 | |
| PGC9Q1FNKPHH9R64FGJZRK9T | |
| YMS2KEWFCMRY31Q8NQZ8J5Y | |
| NYTTS00Y87NKWHKV8J4YNPJR | |
| S2GEWQMJRSJGARPMKGHRRA | |
| 84GKT1Z3J50 | |

communication (approximately 10 meters). Finally, the long preamble is 30 $\mu$s long and is used for long-range communication.[3]

## 10.B.4.4 Architecture

As shown in Figure 10.B.19, a DS-UWB transmitter is similar to a direct sequence spread spectrum (DSSS) system with BPSK or 4-BOK modulation. The process begins with bit randomization through a 15-bit scrambler with $g(D) = 1+D^{14}+D^{15}$ and the transmitter sequences through four initial values. Then, the FEC encoding is either through the mandatory $K = 6$ convolutional coder or through the optional, simpler $K = 4$ convolutional encoder. Both are rate $\frac{1}{2}$ coders that also support a coding rate of $\frac{3}{4}$ through puncturing. The $K = 4$ encoder has generator polynomials $(15_8, 17_8)$, and the $K = 6$ encoder has generator polynomials $(65_8, 57_8)$. Next, the coded bits are interleaved with a convolutional interleaver to protect against burst errors. Figure 10.B.20 shows the architecture of a convolutional interleaver, which is simpler and incurs less latency than a block interleaver. In Figure 10.B.20, each chip of a coded bit is shifted into a different register from top to bottom. The top register has no delay elements, and each successively lower register provides $J = 7$ more delays than the one above. After reaching the tenth register, the process repeats again from the top. Following the interleaving process, the bits are mapped to a spreading code that depends on the spreading rate and the modulation scheme. A 30% excess bandwidth, RRC filter shapes the chips to protect against ISI, and

[3]Base 32 Notation: 0, 1, 2, 3, 4, 5, 6, 7, 8, 9, A, B, C, D, E, F, G, H, J, K, M, N, P, Q, R, S, T, V, W, X, Y, Z

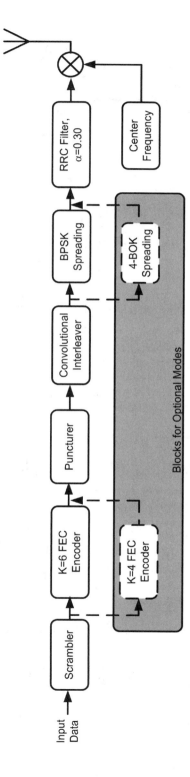

Figure 10.B.19: Transmitter Architecture for the DS-UWB Proposal [76].

SOURCE: R. Fisher, R. Kohno, H. Ogawa, H. Zhang, M. McLaughlin, and M. Welborn, "DS-UWB Proposal Update," doc.: IEEE 802.15-04/140r3 [76]. © IEEE, 2004. Used by permission.

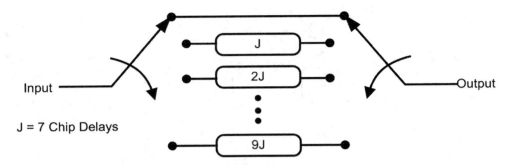

Figure 10.B.20: Convolutional Interleaver [77].

SOURCE: R. Fisher, R. Kohno, H. Ogawa, H. Zhang, M. McLaughlin, and M. Welborn, "DS-UWB Physical Layer Submission to 802.15 Task group 3a," doc.: IEEE 802.15-04/137r0 [77]. © IEEE, 2004. Used by permission.

the shaped pulses are modulated to an appropriate center frequency that depends on the piconet number.

Following demodulation, the receiver performs the reverse operations of the transmitter, as shown in Figure 10.B.21. The receiver uses both I and Q channels for the complex RAKE receiver, and it is suggested to use 16 fingers for operation at 110 Mbps and fewer fingers at higher data rates.

## 10.B.4.5   Performance

The DS-UWB proposal provides flexibility in a number of areas. The optional hardware coding and modulation modes allow a trade-off between reduced complexity and performance. Further, it accommodates flexible spectral allocation with either notched filtering techniques or soft spectrum adaptation, which controls the pulse shape to flatten the spectrum and to insert "notches" that protect sensitive frequencies [78, 79].

The spreading codes achieve up to approximately 14 dB spreading gain and result in a Gaussian amplitude distribution of signals. Further attenuation of narrowband signals can be achieved with a notched filter at the receiver input.

Most importantly, the DS-UWB proposal maintains the advantages of a single signal with ultra wide bandwidth. Because each pulse has a wide bandwidth, the DS-UWB proposal is less likely to experience deep fades. Finally, the wide instantaneous bandwidth of each pulse is more suitable for ranging purposes because range accuracy improves with bandwidth.

A unique capability of the DS-UWB proposal is its ability to provide the clear channel assessment (CCA) that distinguishes piconets with the offset carrier frequencies.

As computed in Table 10.B.3, the link budget specifies a link margin that describes the tolerance of a system to additional sources of loss. The receiver sensitivity

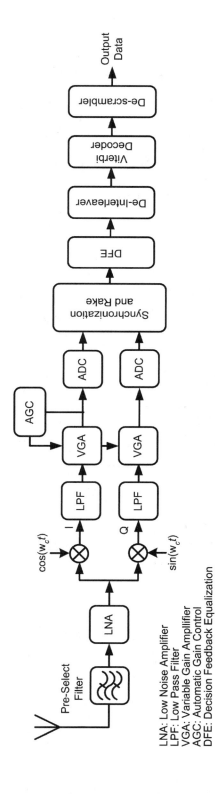

LNA: Low Noise Amplifier
LPF: Low Pass Filter
VGA: Variable Gain Amplifier
AGC: Automatic Gain Control
DFE: Decision Feedback Equalization

Figure 10.B.21: Receiver Architecture for the DS-UWB Proposal [76].

SOURCE: R. Fisher, R. Kohno, H. Ogawa, H. Zhang, M. McLaughlin, and M. Welborn, "DS-UWB Proposal Update," doc.: IEEE 802.15-04/140r3 [76]. © IEEE, 2004. Used by permission.

level specifies a corresponding minimum signal level that a receiver could detect and successfully demodulate a signal. Table 10.B.17 shows the link margin for the DS-UWB proposal.

The frame structure of the DS-UWB proposal differs from MB-OFDM in the length of time for the preamble. For the DS-UWB proposal, the preamble length depends on channel conditions so it may be slightly more or less than the MB-OFDM preamble length. Table 10.B.18 displays the maximum achievable data rates with frame overhead for a payload length of 1,024 bytes and a medium length preamble. From Table 10.B.18, the data rate increases as the number of frames increases due to the shorter burst mode header. Higher raw data rates suffer more speed degradation from the PHY header because it is sent at the slowest data rate of 55 Mbps, and hence lasts for a larger proportion of time as compared to the data, which is sent at a higher data rate.

The final measure of performance is the distance at which the system achieves each of the raw data rates specified by the selection criteria. Table 10.B.19 tabulates the distance at which the DS-UWB proposal achieves a PER of 8% for the best 90% of the channels. The figures account for a fully impaired simulation with loss from ADC degradation, multipath degradation, and channel estimation.

## 10.B.5   Summary

This appendix has introduced the 802.15.3 MAC and the requirements for a UWB PHY for the 802.15.3a standard. It has reviewed two proposals based on these requirements, but either proposal may become an industry standard for WPANs outside of the IEEE standardization process. Both proposals demonstrate good performance as compared to the selection criteria and have strong support from their respective special interest groups.

Current trends indicate that future UWB PHYs will be tailored to more applications than WPANs. Further, UWB networking standards will grow to include a MAC that supports mesh networking and better multihop routing for ad hoc networks. The upcoming IEEE 802.15.4a standard promises a PHY for more pervasive networking that also has mandatory location awareness. This appendix was based on the proposals, presentations, and articles on the Web page of the IEEE 802.15.3a Task Group: http://grouper.ieee.org/groups/802/15/pub/.

Table 10.B.17: Link Margin for the DS-UWB Proposal.

| | 110 Mbps | 220 Mbps | 500 Mbps |
|---|---|---|---|
| Received Power $P_r$ (dBm) | −74.2 @ 10 m | −66.3 @ 4 m | −60.2 @ 2 m |
| Total Noise Power $P_n$ (dBm) | −87.0 | −84.0 | −80.4 |
| Required $E_b/N_0$ $S$ (dB) | 5.0 | 5.0 | 6.0 |
| Implementation Loss $I$ (dB) | 2.5 | 2.5 | 3.0 |
| Link Margin (dBm) $M = P_r - P_n - S - I$ | 5.3 | 10.2 | 11.2 |
| Sensitivity (dBm) $= P_n + S + I$ | 79.5 | −76.5 | −71.4 |

Table 10.B.18: Throughput for Low Band DS-UWB Proposal with PLCP Header.

| Number of Frames | Payload Size | Throughput @ 110 Mbps | Throughput @ 220 Mbps | Throughput @ 495 Mbps |
|---|---|---|---|---|
| 1 | 1024 bytes | 78.30 Mbps | 124.9 Mbps | 185.1 Mbps |
| 5 | 1024 bytes | 100.4 Mbps | 188.5 Mbps | 366.2 Mbps |

Table 10.B.19: Maximum Distance to Achieve a Given Data Rate in Different Channels for the DS-UWB Proposal.

| Data Rate | AWGN Distance | CM1 Distance | CM2 Distance | CM3 Distance | CM4 Distance |
|---|---|---|---|---|---|
| 110 Mbps | 22.2 m | 16.9 m | 14.6 m | 13.4 m | 13.0 m |
| 220 Mbps | 16.2 m | 10.2 m | 8.2 m | 6.2 m | N/A |
| 500 Mbps | 8.7 m | 4.8 m | 3.2 m | N/A | N/A |

# Appendix 10.C

# UWB REGULATIONS

## Nathaniel J. August

## 10.C.1   FCC

In the United States, the FCC released its first report and order for power emissions from UWB devices on February 14, 2002 [80]. This appendix covers the latest revisions as of December 8, 2003 to the United States Code of Federal Regulations, Title 47, Section 15, which allows intentional, low-power radiation from UWB devices [81]. UWB devices provide efficient use of scarce spectrum because they may occupy and coexist with existing narrowband spectra.

The FCC limits the operating bands for a UWB device according to its application. The revision allows for both unlicensed communications applications and licensed applications, such as health monitoring, ground penetrating radar (GPR), and through-walls sensing. The FCC limits on UWB consider the effect of a UWB intruding into the sensitive communications bands located below 2 GHz. Such existing bands include TV, radio, PCS, public safety, and GPS bands. The FCC prohibits UWB communications in toys, aircraft, ships, or satellites.

### 10.C.1.1   Bandwidth Limits

The FCC defines the bandwidth limitations of a UWB device but does not specify a signal type, such as I-UWB or MC-UWB. The FCC identifies two types of UWB bandwidths: absolute bandwidth or fractional bandwidth, which are defined by the 10 dB cutoff bandwidths in Figure 10.C.1. The absolute bandwidth is the difference between the 10 dB high cutoff frequency and the 10 dB low cutoff frequency $(f_h - f_l)$. The fractional bandwidth is defined as $2*(f_h - f_l)/(f_h + f_l)$, and the center frequency is defined as $f_c = (f_h - f_l)/2$. In Figure 10.C.2, the frequency of maximum radiation $f_m$ is the same as $f_c$, but it need not be. Given these definitions, the FCC classifies a device as UWB if it either has a fractional bandwidth greater than $0.20 * f_c$ or it has an absolute bandwidth greater than 500 MHz [81].

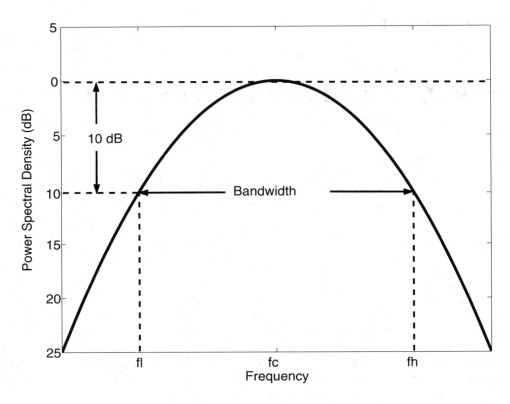

Figure 10.C.1: FCC UWB Definition.

## 10.C.1.2   EIRP Limits

There are three classes of devices covered by the FCC regulations: imaging systems, vehicular radar systems, and communications and measurement systems. This section briefly reviews the characteristics of each and the EIRP (equivalent isotropically radiated power) limits. The EIRP is equivalent to the signal power level given to the antenna multiplied by the antenna gain. Note that the EIRP can be converted to the field strength at 3 m in dB$\mu$V/m by adding 95.2 to the EIRP in dBm. The radiation limits for the three classes of devices are based on interference studies of devices likely to be victims of UWB interference.

### Imaging Systems

Some imaging systems are allowed to emit higher power than others because the object they view absorbs most of the radiated power. Note that the band that contains the center frequency and the frequency of maximum EIRP defines the operating band. Regulations for imaging systems are as follows:

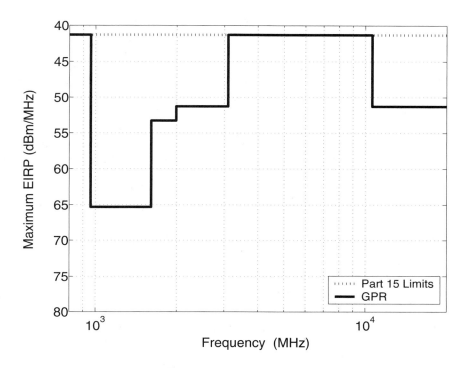

Figure 10.C.2: Average EIRP Limits for GPR Imaging Devices.

1. GPR—A GPR operates mostly in the low frequency band below 960 MHz for sufficient penetration depth, but it can also operate in the 3,100–10,600 MHz band. Operation is restricted to construction, law enforcement, fire and rescue, commercial mining, and scientific research. The device is positioned close to the ground and radiation is directed toward the ground. Figure 10.C.2 and Table 10.C.1 show the average EIRP limits for GPR devices. In addition

Table 10.C.1: Average EIRP Limits for GPR Imaging Devices.

| Frequency (MHz) | Maximum EIRP (dBm) |
|-----------------|--------------------|
| 0–960           | −41.3              |
| 960–1610        | −65.3              |
| 1610–1990       | −53.3              |
| 1990–3100       | −51.3              |
| 3100–10600      | −41.3              |
| Above 10600     | −51.3              |

to the average power limits in Figure 10.C.2, a GPR system must also limit average EIRP to –75.3 dBm/kHz in the frequency bands 1,164–1,240 MHz and 1,559–1,610 MHz. GPR systems require government coordination and individual licensing to ensure acceptable use.

2. Through-Walls Imaging—A through-walls imaging device may operate in either the low band below 960 MHz or in the mid-frequency band from 1990 MHz to 10600 MHz. Operation is limited to law enforcement and fire and rescue because of the considerable potential for interference. Figure 10.C.3 and Table 10.C.2 show the average EIRP limits for through-walls devices. In addition to the average power limits in Figure 10.C.3, a through-walls imaging system must also limit average EIRP to −75.3 dBm/kHz under low band operation and −56.6 dBm/kHz under mid band operation in the frequency bands 1,164–1,240 MHz and 1,559–1,610 MHz. Through-walls systems require government coordination and individual licensing to ensure acceptable use.

3. Surveillance Systems—A surveillance system provides a security fence around a secure area by detecting and tracking intruders. It operates in the mid frequency band from 1,990 MHz to 10,600 MHz and is for use by law enforcement, fire and rescue, public utilities, and industrial entities. Figure 10.C.4 and Table 10.C.3 show the average EIRP limits for surveillance systems. In addition to the average power limits in Figure 10.C.4, a surveillance system must also limit average EIRP to −63.3 dBm/kHz in the frequency bands 1,164–1,240 MHz and 1,559–1,610 MHz. Surveillance systems require government coordination and individual licensing to ensure acceptable use.

4. Medical Systems—Medical imaging devices operate in the frequency band from 3,100 MHz to 10,600 MHz. Use is restricted to licensed health care practitioners for the purpose of seeing inside the body of a person or animal. Figure 10.C.5 and Table 10.C.4 show the average EIRP limits for medical imaging systems. In addition to the average power limits in Figure 10.C.5, a medical imaging system must also limit average EIRP to −75.3 dBm/kHz in the frequency bands 1,164–1,240 MHz and 1,559–1,610 MHz. Medical imaging systems require government coordination, but not individual licensing, to ensure acceptable use.

## Vehicular Radar Systems

Vehicular radar systems are limited to field disturbance sensors in ground vehicles. They occupy the 22 GHz to 29 GHz band with the caveat that the center frequency and the highest radiation level must be above 24.075 GHz. Figure 10.C.6 and Table 10.C.5 show the average EIRP limits for vehicular radar systems. In addition to the average power limits in Figure 10.C.6, a vehicular radar system must also limit average EIRP to −85.3 dBm/kHz in the frequency bands 1,164–1,240 MHz and 1,559–1,610 MHz. Additionally, vehicular radar systems must attenuate energy above 38 degrees to the horizontal plane by 25 dB with respect to the requirements

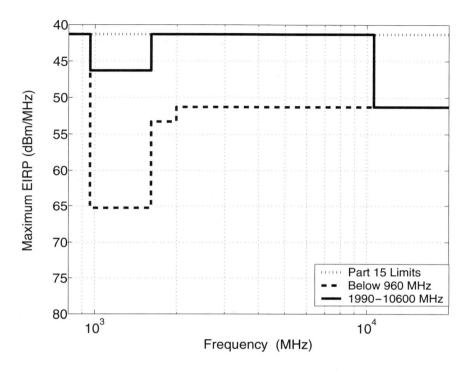

Figure 10.C.3: Average EIRP Limits for Through-Walls Imaging Devices.

Table 10.C.2: Average EIRP Limits for Through-Walls Imaging Devices.

| Frequency (MHz) | Maximum EIRP (dBm) Under 960 MHz Band | Maximum EIRP (dBm) 1990–10600 MHz Band |
|---|---|---|
| 0–960 | −41.3 | −41.3 |
| 960–1610 | −65.3 | −46.3 |
| 1610–1990 | −53.3 | −41.3 |
| 1990–10600 | −51.3 | −41.3 |
| Above 10600 | −51.3 | −51.3 |

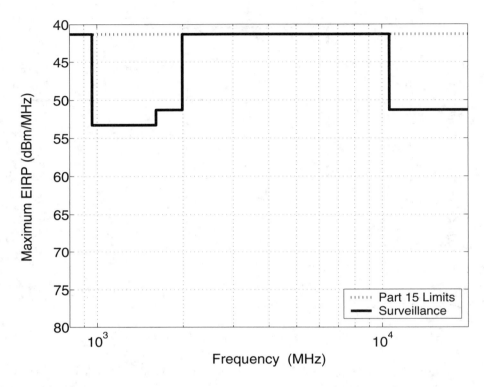

Figure 10.C.4: Average EIRP Limits for Surveillance System Imaging Devices.

Table 10.C.3: Average EIRP Limits for Surveillance System Imaging Devices.

| Frequency (MHz) | Maximum EIRP (dBm) |
|-----------------|--------------------|
| 0–960           | −41.3              |
| 960–1610        | −53.3              |
| 1610–1990       | −51.3              |
| 1990–10600      | −41.3              |
| Above 10600     | −51.3              |

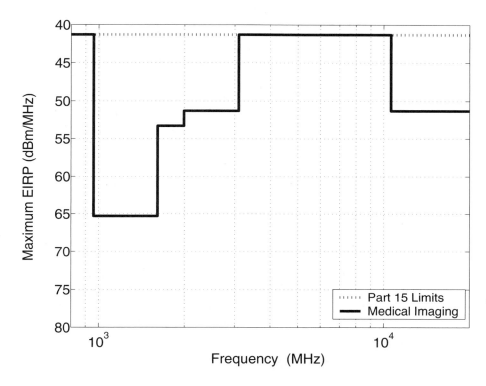

Figure 10.C.5: Average EIRP Limits for Medical Imaging Devices.

Table 10.C.4: Average EIRP Limits for Medical Imaging Devices.

| Frequency (MHz) | Maximum EIRP (dBm) |
|-----------------|--------------------|
| 0–960           | −41.3              |
| 960–1610        | −65.3              |
| 1610–1990       | −53.3              |
| 1990–3100       | −51.3              |
| 3100–10600      | −41.3              |
| Above 10600     | −51.3              |

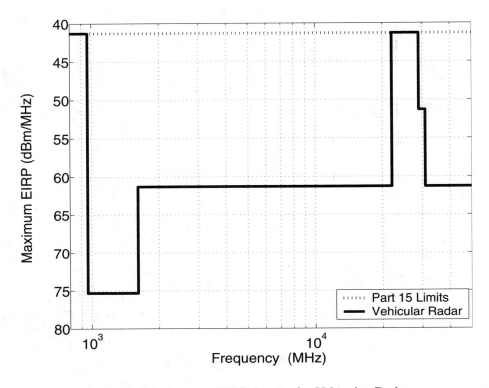

Figure 10.C.6: Average EIRP Limits for Vehicular Radar.

Table 10.C.5: Average EIRP Limits for Vehicular Radar.

| Frequency (MHz) | Maximum EIRP (dBm) |
|---|---|
| 0–960 | −41.3 |
| 960–1610 | −75.3 |
| 1610–22000 | −61.3 |
| 22000–29000 | −41.3 |
| 29000–31000 | −51.3 |
| Above 31000 | −61.3 |

in Figure 10.C.6. In 2005, the angle reduces to 30 degrees, and in 2014, the attenuation increases to 35 dB.

## Communications and Measurement Systems

Communications and measurement systems may ship in considerable volume due to their license-free spectrum allocation. Applications include wireless personal area networks (WPANs), sensor networks, precision asset location systems, among the many others listed in Chapter 10. The FCC classifies these devices as either indoor or outdoor devices. Indoor devices should be inoperable when not indoors, e.g. a device could operate on AC power only. Outdoor devices must be handheld devices and must not be supported by an outdoor UWB infrastructure. Further, outdoor devices must stop transmitting when no response is received from a receiver in a 10 second period. These systems operate in the band from 3,100 MHz to 10,600 MHz. Figure 10.C.7 and Table 10.C.6 show the average EIRP limits for both indoor and outdoor communications and measurement systems. In addition to the average power limits in Figure 10.C.7, a communications and measurement system must also limit its average EIRP to –85.3 dBm/kHz in the frequency bands 1,164–1,240 MHz and 1,559–1,610 MHz.

For emissions under 960 MHz, the average EIRP is measured using the quasi peak detector of the International Special Committee on Radio Interference (CISPR) of the International Electrotechnical Commission, which is explained in CISPR Publication 16. For radiation above 960 MHz, the RMS power is the average EIRP.

## 10.C.1.3   Peak Power Limits

Note that the previous limits are average power limits and not peak limits. For UWB devices, the peak output does not affect interference levels as much as the average output. However, the FCC does impose a peak limit, as extremely high power pulses can overload the front end of a victim receiver, and spectral lines may appear in sensitive bands. The peak limit restricts emissions at low PRFs (MHz or lower), whereas the average limit restricts emissions at high PRFs (MHz or higher).

For peak power measurements, the FCC considers the peak power across a 50 MHz resolution bandwidth (RBW) to simulate a victim receiver with large bandwidth. The maximum allowable peak in that 50 MHz bandwidth is a 0 dBm EIRP. Because it is difficult to measure EIRP with a 50 MHz RBW, the peak may also be defined by $20 * \log (RBW/50 \text{ MHz})$ dBm for any bandwidth from 1 MHz to 50 MHz. The preferred RBW is 3 MHz centered on the frequency of highest emission. The peak power limits apply to all the previously mentioned devices.

## 10.C.1.4   Unintentional Radiation

Finally, note that digital circuitry in UWB devices also emits unintentional radiation, but this radiation is governed in accordance with Part 15 and not by the previously defined limits.

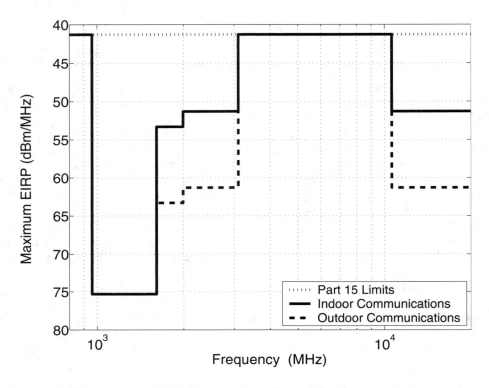

Figure 10.C.7: Average EIRP Limits for Communications and Measurement Systems.

Table 10.C.6: Average EIRP Limits for Communications and Measurement Systems.

| Frequency (MHz) | Maximum EIRP (dBm) Indoor Devices | Maximum EIRP (dBm) Hand-Held Devices |
|---|---|---|
| 0–960 | −41.3 | −41.3 |
| 960–1610 | −75.3 | −75.3 |
| 1610–1990 | −53.3 | −63.3 |
| 1990–3100 | −51.3 | −61.3 |
| 3100–10600 | −41.3 | −41.3 |
| Above 10600 | −51.3 | −61.3 |

## 10.C.2   World

Other parts of the world are currently in the process of defining their own regulatory requirements for UWB systems.

In Europe, the European Telecommunications Standardization Institute (ETSI) has commissioned Task Group 31a to develop standards and conformance-testing requirements. ETSI will provide a spectral mask much like the FCC mask to restrict out-of-band emissions. The European Conference of Postal and Telecommunications Administrations (CEPT) SE24 is dealing with regulatory issues of sharing spectrum under 6 GHz. Also, the International Telecommunication Union Radiocommunication Sector (ITU-R) is preparing recommendations in the ITU-R Task Group 1/8 for Spectrum Management [82]. The ITU recommendations will cover characteristics of UWB, compatibility between UWB and radiocommunication services, spectrum management, and UWB measurement techniques [83].

In Japan, the Ministry of Public Management, Home Affairs, Post, and Telecommunications, (MPHPT) organized a UWB regulatory committee in September 2002, and is in the process of approving a trial license and commercial regulations. The MHPHT is cooperating with the industrial sector represented by the Association of Radio Industries and Businesses (ARIB) [84].

Many other countries, such as Australia, South Korea, China, and Singapore, are beginning standardization efforts of their own [85].

## Bibliography

[1] J. D. Taylor, *Introduction to Ultra-Wideband Radar Systems*, CRC Press, Inc., 1995.

[2] R. J. Fontana, E. Richley, and J. Barney, "Commercialization of an ultra wideband precision asset location system," *Proc. IEEE Conference on Ultra Wideband Systems and Technologies*, pp. 369-373, Reston, VA, November 2003.

[3] Y.-J. Park, K.-H. Kim, S.-B. Cheo, D.-W. Yoo, D.-G. Youn, and Y.-K. Jeong, "Development of a UWB GPR system for detecting small objects buried under ground," *Proc. IEEE Conference on Ultra Wideband Systems and Technologies*, pp. 384-388, November 2003.

[4] W. C. Chung and D. S. Ha, "An accurate ultra wideband (UWB) ranging for precision asset location," *Proc. 2003 IEEE Conference on Ultra Wideband Systems and Technologies*, pp. 389-393, November 2003.

[5] G. D. Morley and W. D. Grover, "Improved location estimation with pulse-ranging in presence of shadowing and multipath excess-delay effects," *IEE Electronics Letters*, vol. 31, no. 18, pp. 1609-1610, August 1995.

[6] H. Urkowitz, *Signal Theory and Random Processes*, Artech House, 1983.

[7] S. Basagni, I. Chlamatac, V. R. Syrotiuk, and B. A. Woodward, "A Distance routing effect algorithm for mobility (DREAM)," *Proc. of Mobicom '98*, pp. 76-84, Dallas, TX, October 1998.

[8] J. Liu, P. Cheung, L. Guibas, and F. Zhao, "A Dual-Space Approach to Tracking and Sensor Management in Wireless Sensor Networks," *Proc. 1st ACM international workshop on Wireless Sensor Networks and Applications (WSNA)*, pp. 31-139, September 2002.

ouzet, "PULSERS Presentation to IEEE 802.15.4," *IEEE P802.15 Working ɔ for Wireless Personal Area Networks Publications*, doc.: IEEE 802.15-03/ January 2003. Available: http://grouper.ieee.org/groups/802/15/pub/ ʰ03/03044r0P802-15_TG4-PULSERS_IEEE_802.15.4_PULSERS- ɔn-to-802-15-4.ppt.

ʰv and D.V. Fedotov, "Ultra Wideband Radar System: Advantages ʲages," *Digest of Papers 2002 IEEE Conference on Ultra Wide- ʲnd Technologies*, pp. 201-206, May 2002.

ɲkins, R. Egri, C. Eswarappa, F. Kolak, R. Wohlert, J. Ben- ʲteri, "Ultra wide band 24GHz automotive radar front-end," ʲational Microwave Symposium Digest*, vol. 1, pp. 369-372,

[12] B. Scheers, M. Piette, and A. Vander Vorst, "The detection of AP mines using UWB GPR," *Proc. Second International Conference on the Detection of Abandoned Land Mines*, no. 458, pp. 50-54, October 1998.

[13] E. M. Staderini, "UWB radars in medicine," *IEEE Aerospace and Electronic Systems Magazine*, vol. 17, no. 1, pp. 13-18, January 2002.

[14] E. J. Bond, X. Li, S. C. Hagness, and B. D. Van Veen, "Microwave imaging via space-time beamforming for early detection of breast cancer," *IEEE Transactions on Antennas and Propagation*, vol. 51 , no. 8, pp. 1690-1705, August 2003.

[15] Available: http://www.timedomain.com.

[16] S. S. Kolenchery, J. K. Townsend, and J. A. Freebersyser, "A novel impulse radio network for tactical military wireless communications," *Proc. IEEE Military Communications Conference*, vol. 1, pp. 59-65, October 1998.

[17] M. Akahane, B. Huang, S. Sugaya, K. Takamura, "CE requirements for alternative PHY CFA," *IEEE P802.15 Working Group for Wireless Personal Area Networks Publications*, doc.: IEEE 802.15-02/043r0, January 2002. Available: http://grouper.ieee.org/groups/802/15/pub/2002/Jan02/02043r0P802-15_SG3a-CE-Requirements-for-Alternative-PHY-CFA.ppt.

[18] P. Gandolfo, "XtremeSpectrum—SG3a CFA Response," *IEEE P802.15 Working Group for Wireless Personal Area Networks Publications*, doc.: IEEE 802.15-02/031r0, January 2002. Available: http://grouper.ieee.org/groups/802/15/pub/2002/Jan02/02031r0P802-15_SGAP3-CFAReaponseAltPHY.ppt.

[19] C. Brabenac, "Intel SG3a CFA Response—Wireless Peripherals," *IEEE P802.15 Working Group for Wireless Personal Area Networks Publications*, doc.: IEEE 802.15-02/139r0, March 2002. Available: http://grouper.ieee.org/groups/802/15/pub/2002/Mar02/02139r0P802-15_SG3a-Intel-CFA-Response-Wireless-Peripherals.ppt.

[20] Intel Corporation, "Wireless USB The First High-Speed Personal Wireless Interconnect," White Paper, 2004. Available: http://www.intel.com/technology/ultrawideband/downloads/wirelessUSB.pdf.

[21] K. Siwiac, "SG3a Application Summary," *IEEE P802.15 Working Group for Wireless Personal Area Networks Publications*, doc.: IEEE 802.15-02/149r0, March 2002. Available: http://grouper.ieee.org/groups/802/15/pub/2002/Mar02/02149r0P802-15_SG3a-Application-Summary.doc.

[22] R. Aeillo, J. Ellis, and L. Taylor, "Application Opportunities for High Rate WPANs," *IEEE P802.15 Working Group for Wireless Personal Area Networks Publications*, doc.: IEEE 802.15-02/143r0, March 2002. Available: http://grouper.ieee.org/groups/802/15/pub/2002/Mar02/02143r0P802-15-SG3a-Application-Opportunities-GA.ppt.

[23] J. Ellis and L. Taylor, "IEEE P802.15.4SG4a call for applications," *IEEE P802.15 Working Group for Wireless Personal Area Networks Publications*, doc.: IEEE 802.15-03/330r1, August 2003. Available: ftp://ftp. 802wirelessworld.com/15/03/15-03-0330-01-004a-sg4a-call-applications.doc.

[24] J. Ellis and L. Taylor, "802.15.4IGa informal call for application response," *IEEE P802.15 Working Group for Wireless Personal Area Networks Publications*, doc.: IEEE 1802.15-03/537r0, July 2003. Available: ftp://ftp. 802wirelessworld.com/15/03/15-03-0537-00-004a-formal-submission-802-15-4iga-informal-cfa-response-ppt.ppt.

[25] R. Roberts, "Low probability of detection applications," *IEEE P802.15 Working Group for Wireless Personal Area Networks Publications*, doc.: IEEE 802.15.SG4a-03/430r0, November 2003. Available: ftp://ftp.802wirelessworld. com/15/03/15-03-0430-00-004a-802-15-sg4a-cfa-response-lpd-low-probability-detection.ppt.

[26] Available: http://www.ubisense.net/healthcare/index.html.

[27] W. Hirt, "Low data rate and/or positioning applications," *IEEE P802.15 Working Group for Wireless Personal Area Networks Publications*, doc.: IEEE 802.15-03/443r0, Oct. 2003. Available: ftp://ftp.802wirelessworld.com/15/03/ 15-03-0443-00-004a-sg4a-low-rate-data-and-positioning-applications.ppt.

[28] I.F. Akyildiz, S. Weilian, Y. Sankarasubramaniam, and E. Cayirci, "A survey on sensor networks," *IEEE Communications Magazine*, vol. 40, issue 8, pp. 102-114, August 2002.

[29] S. Davis, "Communication and navigation for robotic applications," *IEEE P802.15 Working Group for Wireless Personal Area Networks Publications*, doc.: IEEE 802.15-03/486r0, November 2003. Available: ftp://ftp. 802wirelessworld.com/15/03/15-03-0486-00-004a-robotic-communication-location.ppt.

[30] V. Dhingra, "Water and gas meter reading application," *IEEE P802.15 Working Group for Wireless Personal Area Networks Publications*, doc.: IEEE 802.15-03/436r0, November 2003. Available: ftp://ftp.802wirelessworld.com/15/03/15-03-0436-00-004a-echelon-sg4a-call-application-response.ppt.

[31] J. Ryckaert, "Wireless body area networks," *IEEE P802.15 Working Group for Wireless Personal Area Networks Publications*, doc.: IEEE 802.15-03/484r0, November 2003. Available: ftp://ftp.802wirelessworld.com/15/03/15-03-0484-00-004a-sg4a-cfa-response-wireless-body-area-networks.ppt

[32] T. S. Rappaport, *Wireless Communications: Principles and Practice*, 2nd ed., New Jersey: Prentice-Hall, 2002.

[33] C. R. Anderson, "Design and implementation of an ultrabroadband millimeter-wavelength vector sliding correlator channel sounder and in-building measurements at 2.5 & 60 GHz," Masters Thesis, Virginia Polytechnic Institute and State University, May 2002. Available: http://scholar.lib.vt.edu/theses/index.html.

[34] J. D. Parsons, *The Mobile Radio Propagation Channel*, New York: John Wiley and Sons, Inc., 1992.

[35] XtremeSpectrum, Inc., "TRINITY for Media-rich Wireless Applications," Product Brief of XtremeSpectrum, Inc., June 2002. Available: http://www.xtremespectrum.com/PDF/xsi_trinity_brief.pdf

[36] The Institute of Electrical and Electronics Engineers, Inc., 802.15.3 Draft Standard for Telecommunications and Information Exchange Between Systems—LAN/MAN Specific Requirements—Part 15: Wireless Medium Access Control (MAC) and Physical Layer (PHY) Specifications for High Rate Wireless Personal Area Networks, draft P802.15.3/D17, February 2003.

[37] Available: http://www.motorola.com.

[38] Available: http://www.xtremespectrum.com.

[39] D. Kelley, S. Reinhardt, R. Stanley, M. Einhorn, "PulsON second generation timing chip: enabling UWB through precise timing," *2002 IEEE Conference on Ultra Wideband Systems and Technologies Digest of Papers*, pp. 117-121, May 2002.

[40] L. W. Fullerton, "Reopening the electromagnetic spectrum with ultrawideband radio for aerospace," *Proc. IEEE Aerospace Conference*, vol. 11, pp. 201-210, March 2000.

[41] Available: http://www.multispectral.com.

[42] R. J. Fontana, "Ultra Wideband Technology—The Wave of the Future?" Keynote Address, *International Telemetry Conference*, San Diego, CA, October 2000.

[43] R. J. Fontana, A. Ameti, E. Richley, L. Beard, and D. Guy "Recent advances in ultra wideband communications systems," *IEEE Conference on Ultra Wideband Systems and Technologies Digest of Papers*, pp. 129-133, May 2002.

[44] A. Ameti, R. J. Fontana, E. J. Knight, and E. Richley, "Ultra wideband technology for aircraft wireless intercommunications systems (AWICS) design," *Proc. IEEE Conference on Ultra Wideband Systems and Technologies*, pp. 61-65, Reston, VA, November 2003.

[45] Multispectral Solutions, Inc., "UWB Wireless Intercommunications System," Product Brochure from MSSI. Available: http://www.multispectral.com/pdf/WICSp.pdf.

[46] R. J. Fontana and S. J. Gunderson, "Ultra-wideband precision asset location system," *IEEE Conference on Ultra Wideband Systems and Technologies Digest of Papers*, pp. 147-150, May 2002.

[47] A. Ameti, R. Fontana, E. Knight, and E. Richley, "Commercialization of an ultra wideband precision asset location system," *Proc. 2003 IEEE Conference on Ultra Wideband Systems and Technologies*, Reston, VA, November 2003.

[48] Multispectral Solutions, Inc., "UWB Precision Asset Localiztion (PAL650)," Product Brochure from MSSI. Available: http://www.multispectral.com/pdf/PAL.pdf.

[49] R. J. Fontana, E. A. Richley, A. J. Marzullo, L. C. Beard, and R. W. T. Mulloy, "An ultra wideband radar for micro air vehicle applications," *IEEE Conference on Ultra Wideband Systems and Technologies Digest of Papers*, pp. 187-191, May 2002.

[50] R. Fleming and C. Kushner, Principal Investigators, "Low-Power, Miniature, Distributed Position Location and Communication Devices Using Ultra-Wideband, Nonsinusoidal Communication Technology," *Semi-Annual Technical Report–ARPA*, contract J-FBI-94-058, July 1995. Available:

[51] R. Fleming and C. Kushner, "Integrated ultra-wideband localizers," *1999 Ultra-Wideband Conference*, September 1999. Available: http://www.aetherwire.com/Aether_Wire/Integrated_Ultra-Wideband_Localizers.pdf.

[52] R. Fleming, C. Kushner, G. Roberts, and U. Nandiwada, "Rapid acquisition for ultra-wideband localizers," *IEEE Conference on Ultra Wideband Systems and Technologies Digest of Papers*, pp. 245-249, May 2002.

[53] Available: http://www.aetherwire.com.

[54] M. Hamalainen, V. Hovinen, R. Tesi, J. H. J. Iinatti, and M. Latva-aho, "On the UWB system coexistence with GSM900, UMTS/WCDMA, and GPS," *IEEE Journal on Selected Areas in Communications*, vol. 20, no. 9, pp. 1712-1721, December 2002.

[55] D. V. Sarwate and M. B. Pursely, "Crosscorrelation properties of pseudorandom and related sequences," *Proc. Of the IEEE*, vol. 68, no. 5, pp. 593-619, May 1980.

[56] P. Gorday, J. Martinez, and P. Jamieson, "IEEE 802.15.4 overview," *IEEE P802.15 Working Group for Wireless Personal Area Networks Publications*, doc.: IEEE 802.15-01-0509r0, November 2001. Available: http://grouper.ieee.org/groups/802/15/pub/2001/Nov01/01509r0P802-15_TG4-Overview.ppt.

[57] I. I. Immoreev and D. V. Fedotov, "Detection of UWB signals reflected from complex targets," *IEEE Conference on Ultra Wideband Systems and Technologies*, pp. 193-196, May 2002.

[58] The Institute of Electrical and Electronics Engineers, Inc., 802.15.4 Standard for Telecommunications and Information Exchange Between Systems—LAN/MAN Specific Requirements—Part 15.4: Wireless Medium Access Control (MAC) and Physical Layer (PHY) Specifications for Low-rate Wireless Personal Area Networks (LR-WPANs), IEEE Std 802.15.4-2003, 2003.

[59] N. J. August, H. J. Lee, and D. S. Ha, "Pulse Sense: A method to detect a busy medium in ultra wideband (UWB) networks," *2004 Joint IEEE Conference on Ultra Wideband Systems and Technologies and International Workshop on UWB Systems Digest of Papers*, to be published November 2003.

[60] The Institute of Electrical and Electronics Engineers, Inc., 802.15.3 Draft Standard for Telecommunications and Information Exchange Between Systems—LAN/MAN Specific Requirements—Part 15: Wireless Medium Access Control (MAC) and Physical Layer (PHY) Specifications for High Rate Wireless Personal Area Networks, draft P802.15.3/D17, February 2003.

[61] J. P. K. Gilb, *Wireless Multimedia: A Guide to the IEEE 802.15.3 Standard*, The Institute of Electrical and Electronics Engineers, Inc., ISBN 0738136689, April 2004.

[62] The Institute of Electrical and Electronics Engineers, Inc., IEEE Std 802.11— Wireless LAN Medium Access Control (MAC) and Physical Layer (PHY) Specifications, 1999 ed.

[63] N. August, H. J. Lee, and D. S. Ha, "Pulse Sense: A method to Detect a Busy Medium in UWB Networks," *Proc. 2004 Joint UWBST Conference and IWUWB Symposium*, pp. 366-370, May 2004.

[64] M. Welborn and B. Shvodian, "Ultra-Wideband Technology for Wireless Personal Area Networks—The IEEE 802.15.3/3a Standards UWBST Tutorial," *UWBST 2003 Conference*, November 2003.

[65] J. Barr, "IEEE 802.15 TG3 and SG3a," Presentation to FCC Technological Advisory Council, April 26, 2002. Available: http://www.fcc.gov/oet/tac/april26-02-docs/FCC-TAC-802.15.3-overviewNOPICT.ppt.

[66] J. Ellis, K. Siwiak, and R. Roberts, "P802.15. 3a Alt PHY Selection Criteria," IEEE P802-15_03031r11_TG3a-PHY-Selection-Criteria, May 18, 2003. Available: http://grouper.ieee.org/groups/802/15/pub/2003/May03/03031r11P802-15_TG3a-PHY-Selection-Criteria.doc.

[67] J. Foerster, "Channel Modeling Sub-committee Report Final," IEEE doc.: P802.15-02/490r1-SG3a, March 2003. Available: http://grouper.ieee.org/groups/802/15/pub/2003/Mar03/02490r1P802-15_SG3a-Channel-Modeling-Subcommittee-Report-Final.zip.

[68] A. Saleh and R. Valenzuela, "A Statistical Model for Indoor Multipath Propagation," *IEEE Journal on Selected Areas in Communications*, vol. SAC-5, no. 2, pp. 128-137, February 1987.

[69] M. Pendergrass, "Empirically Based Statistical Ultra-Wideband Channel Model," IEEE doc.: P802.15-02/240-SG3a. Available: http://grouper.ieee.org/groups/802/15/pub/2002/Jul02/02240r1p802-15_SG3a-Empirically_Based_UWB_Channel_Model.ppt.

[70] J. Foerster and Q. Li, "UWB Channel Modeling Contribution from Intel," doc.: IEEE P802.15-02/279-SG3a. Available: http://grouper.ieee.org/groups/802/15/pub/2002/Jul02/02279r0P802-15_SG3a-Channel-Model-Cont-Intel.doc.

[71] Anuj Batra, et al, "Multi-band OFDM Physical Layer Proposal Update," IEEE 802.15-04/0220r3, May 2004. Available: ftp://ftp.802wirelessworld.com/15/04/15-04-0220-03-003a-multi-band-ofdm-physical-layer-proposal-update.ppt.

[72] Anuj Batra, et al, "Multi-band OFDM Physical Layer Proposal for IEEE 802.15 Task group 3a," IEEE 802.15-03/268r3, March 2004. Available: http://www.ieee802.org/15/pub/2003/Jul03/03268r2P802-15_TG3a-Multi-band-CFP-Document.pdf.

[73] M. Hämäläien, V. Hovinen, R. Tesi, J. H. J. Iinatti, and M. Latva-aho, "On the UWB system coexistence with GSM900, UMTS/WCDMA, and GPS," *IEEE Journal on Selected Areas in Communications*, vol. 20, no. 9, pp. 1712-1721, December 2002.

[74] J. R. Foerster, "The performance of a direct-sequence spread ultrawideband system in the presence of multipath, narrowband interference, and multiuser interference," *IEEE Conference on Ultra Wideband Systems and Technologies*, pp. 87-91, 2002.

[75] K. Siwiak, P. Withington, and S. Phelan, "Ultra-wide band radio: the emergence of an important new technology," *IEEE 53rd Vehicular Technology Conference*, vol. 2, pp. 1169-1172, Rhodes, Greece, May 2001.

[76] R. Fisher, R. Kohno, H. Ogawa, H. Zhang, M. McLaughlin, and M. Welborn,, "DS-UWB Proposal Update," doc.: IEEE 802.15-04/140r3, May, 2004. Available: ftp://ftp.802wirelessworld.com/15/04/15-04-0140-00-003a-merger2-proposal-ds-uwb-presentation.pdf.

[77] R. Fisher, R. Kohno, H. Ogawa, H. Zhang, M. McLaughlin, and M. Welborn, "DS-UWB Physical Layer Submission to 802.15 Task group 3a," doc.: IEEE 802.15-04/137r0, March 2004. Available: ftp://ftp.802wirelessworld.com/15/04/15-04-0137-00-003a-merger2-proposal-ds-uwb-update.pdf.

[78] R. Kohno, H. Zhang, and H. Nagasaka, "Ultra Wideband impulse radio using free-verse pulse waveform shaping, Soft-Spectrum adaptation, and local sine template receiving," IEEE 802.15-03/097, rev. 1, March 2003.

[79] C. Mitchell and R. Kohno, "High data rate transmissions using orthogonal modified Hermite pulses in UWB communications," *10th International Conference on Telecommunications*, vol. 2, pp. 1278-1283, March 2003.

[80] U.S. Federal Communications Commission, FCC Revision of part 15 of the commission's rules regarding ultra-wideband transmission systems: First report and order, Technical report, February 2002.

[81] U.S. Federal Communications Commission, Part 15–Radio Frequency Devices, December 8, 2003. Available: http://www.fcc.gov/oet/info/rules/part15/part15_12_8_03.pdf.

[82] B. Huang, "European UWB Regulations," Document Number 03215r0P802.15, May 14, 2003. Available: http://grouper.ieee.org/groups/802/15/pub/2003/May03/03215r0P802.15_TG3a-European-UWB-Regulations.ppt.

[83] ITU-R UWB Decision–adopted in Geneva, Switzerland, July 2002.

[84] R. Kohno, "UWB for Future Wireless-Japanese approach," *IWUWBS* Plenary Session, June 2003. Available: http://www.cwc.oulu.fi/home/iwuwbs/slides/kohno.pdf.

[85] P. Meade, "UWB in the Rest of the World," *UWB Insider*, vol. 1, no. 3, October 29, 2003.

[86] MIT Media Lab, MIThril System website, http://www.media.mit.edu/wearables/mithril/

Bradley Department of Electrical and Computer Engineering at Virginia Tech. Dr. Attiya received his B.Sc from Zagazig University in 1990, and his M.Sc. and Ph.D. degrees in 1996 and 2001, from Cairo University, all in Electronics and Communication Engineering. His research interests include UWB propagation measurements and simulations, localized waves, wideband antennas, antenna arrays, printed antennas, and antenna measurements. Dr. Attiya is a member of the IEEE. While at Virginia Tech, he contributed to UWB indoor propagation measurements carried out in both time and frequency domains and channel characterization. He also performed UWB ray tracing and simulation of indoor channels. This research contributed significantly in clarifying the fundamental aspects of UWB propagation measurements and channel modeling, as well as highlighting the potentials and the limitations of UWB communication systems.

**Nathaniel J. August** received the B.S. degree in Computer Engineering in 1998 and the M.S. degree in Electrical Engineering in 2001 from Virginia Tech. His M.S. research involved low-power VLSI design of wireless video codecs, and it has been documented in both conference and journal papers. He is currently working toward the Ph.D. degree at Virginia Tech as a Bradley Fellow and a Cunningham Fellow. In between degrees, he worked as a Validation Engineer for Intel Corporation in Portland, OR, and Sacramento, CA, on various projects including pre-silicon validation of gigabit Ethernet adapters and post-silicon validation of PCI chipsets. His current research interests include low-power VLSI design and UWB networks, systems, and hardware. He and Dr. Dong S. Ha's Virginia Tech VLSI for Telecommunications group are investigating energy efficient CMOS UWB systems that are based on their UWB radio architecture. Nathaniel's research focuses on the unique capabilities of UWB to provide efficient MAC layer services, such as clear channel assessment and full duplex communications. The UWB systems will be used in applications such as wireless personal area networks, radio frequency identification, and wireless ad hoc and sensor networks.

**Dr. R. Michael Buehrer** is an Assistant Professor with the Bradley Department of Electrical and Computer Engineering at Virginia Tech where he works with the Mobile and Portable Radio Research Group. His current projects include propagation measurement and channel modeling of UWB signals as part of the DAPRA NETEX program. Other projects include UWB system design, UWB receiver design, MIMO system design, multiuser detection, spread spectrum, software radio research, adaptive antennas, spatial channel modeling, adaptive modulation and coding and OFDM-based modulation techniques. From 1996-2001, Dr. Buehrer worked at Bell Laboratories—Lucent Technologies in the areas of CDMA receiver design, multiuser detection, adaptive antennas, and spatial channel modeling. Dr. Buehrer has co-authored more than a dozen journal and 25 conference papers and holds seven patents in the area of wireless communications. In 2001 Dr. Buehrer joined Virginia Tech, and in 2003 he was named Outstanding New Assistant Professor by the Virginia Tech College of Engineering. Currently Dr. Buehrer is a principal investigator in the DARPA NETEX program, which is designing propagation measurement equipment and models, and the NSF ITR program for which he is examining communication systems and hardware design issues for UWB. Dr. Buehrer served as co-technical program chair for the *IEEE UWB Systems and Technology Conference* in 2003.

**Dr. William A. Davis** is Director of the Virginia Tech Antenna Group with an emphasis in array technology, low-profile, broadband antennas, and UWB antenna technology, as well as measurement techniques and applications. The UWB work has focused on both measurements and characterization of antennas and propagation. He provides support to the Center for Wireless Technology in both antennas and radio/microwave frequency circuit design. He is strongly involved in numerical methods for electromagnetics as well as antennas, microwave measurements, and material characterization. Beyond Virginia Tech's ECE department, he has been actively involved with faculty in Mechanical Engineering to develop improved material processing using microwaves and microwave heating of tumors. The numerical methods work has recently centered on new stable techniques for the finite-element method in electrodynamics as well as applications in large antenna arrays. His recent research interests in radio engineering and microwaves have been oriented toward nonlinear device modeling and microwave processing of materials.

**Michelle X. Gong** received the B.S. degree from Wuhan University, China, in 1996, and the M.S. degree from University of Hawaii at Manoa, Honolulu, HI, in 2000, both in electrical engineering. She is currently working toward the Ph.D. degree at Virginia Tech as an IREAN Fellow. Prior to joining Virginia Tech, she worked as a system engineer at LinkAir, a start-up company in Silicon Valley, where her research focused on designing handoff and measurement signaling procedures and defining interfaces between physical/MAC layers and MAC/RLC layers in a 3G cellular network. Her current research interests include UWB, mobile networking, and performance evaluation. She and Dr. Scott F. Midkiff are creating MAC designs for rapid synchronization of UWB to support personal area networks and are investigating methods to create ad-hoc network formulation based on keeping limited radiated power permitted by the FCC.

**Stephen Griggs** is currently a Program Manager in DARPA's Advanced Technology Office, is involved in a wide range of communications and networking programs in DARPA, and leads the Networking in Extreme Environments, Optical Tags, Dynamic Optical Tags, and Self-Healing Minefield Programs. These programs focus on his main interests of tactical communications and ad-hoc networking in difficult and denied environments. Additionally, he is working to develop more programs in these areas and other related areas, such as handheld trusted computing and policy based network management. Mr. Griggs received his B.S. in Engineering Physics from the United States Military Academy. From 1987–1991 he served as an Armor Officer in the US Army. He has held positions as Program Manager at System Planning Corporation, Office of Special Technology, and Multispectral Solutions. He worked as an Engineer with the Naval Air Warfare Centers Aircraft Division. Before joining DARPA, Mr. Griggs was Vice President of Puritan Research Corporation.

**Dr. Dong S. Ha** received a B.S. degree in Electrical Engineering in 1974 from Seoul National University in Korea. He received M.S. and Ph.D. degrees in Electrical and Computer Engineering from the University of Iowa, Iowa City, in 1984 and 1986, respectively. Upon graduation, he joined the Department of Electrical and Computer Engineering at Virginia Tech in 1986 as an Assistant Professor and currently is a Professor and Director of Virginia Tech Information Systems Center (VISC). His main research interest lies in low-power VLSI design for wireless communications. During his recent research leave from January to June, 2003, he worked for XtremeSpectrum, Inc., which is the leading company in UWB chip development worldwide. He was involved in two tasks: UWB system design and low-power design of the baseband signal processor. In addition to the knowledge gained through the involvement in the two tasks, he also had an opportunity to learn about real world problems for UWB chip design. Dong Ha and his students have developed various building blocks for wireless communications systems including rake receivers, Viterbi decoders, Turbo decoders, and low-power ADCs. Currently, he supervises 14 students, of whom five Ph.D. students and three M.S. students are involved directly in VLSI development of UWB systems.

**Jihad Ibrahim** received his B.S. degree in Computer and Communications Engineering from the American University of Beirut in 2002. He received his M.S. degree from Virginia Tech in May 2004. He is currently pursuing his Ph.D. at Virginia Tech, where he works with the Mobile and Portable Radio Research Group as a Graduate Research Assistant. His research interests include UWB receiver design, acquisition, tracking, and interference cancellation schemes for communications and ranging applications.

**Stanislav Licul** received B.S.E.E. and M.S.E.E. degrees from Virginia Tech in 1999 and 2001, respectively. During the course of his studies, he worked at Motorola, Inc., and XM Satellite Radio, Inc. His duties included optimization of the paging receivers and design of antennas for satellite communications. He was a scholarship member of the Virginia Tech Varsity Soccer Team, recipient of the Out of State Scholar Scholarship, and founder of the Virginia Tech Croatian Student Association. Currently, he is pursuing his Ph.D. degree at Virginia Tech under the guidance of Dr. William A. Davis. In the course of his research, Stanislav Licul investigated localized waves (e.g., X-waves) using a vector valued approach and possible microwave excitation schemes. Localized waves are UWB electromagnetic waves with non-separable spatial and time characteristics. His current research is in characterization of UWB channel, antennas, and more efficient methods of measuring UWB antennas in the frequency and time-domain. He has published eight conference papers and two trade journal papers and holds one patent. He is a student member of IEEE and Eta Kappa Nu.

**Dr. Scott F. Midkiff** is a Professor of Electrical and Computer Engineering at Virginia Tech. His current research focuses on system issues related to wireless and mobile networks, including multiple access control schemes for UWB systems. Dr. Midkiff teaches courses in computer networks, telecommunications, network application design, and introductory computer engineering. He is a past recipient of an IBM University Partnership Program grant and a Digital Equipment Corporation Faculty Program/Incentives for Excellence grant. Dr. Midkiff serves on the editorial board and as the "Education and Training" Department Editor for *IEEE Pervasive Computing* magazine. He received the B.S.E. degree, summa cum laude, in Electrical Engineering and Computer Science from Duke University in 1979, the M.S. degree in Electrical Engineering from Stanford University in 1980, and the Ph.D. degree in Electrical Engineering from Duke in 1985.

**Sridharan Muthuswamy** received the B.Tech. (Hons.) degree from the Indian Institute of Technology, Kharagpur, in Electronics and Electrical Communications Engineering in 2001, and the M.S. degree from Rutgers State University, NJ, in Electrical and Computer Engineering in 2003. Currently he is a Research Assistant at the Mobile and Portable Radio Research Group at Virginia Tech working toward his Ph.D. degree. He served as a Teaching Assistant at Rutgers State University from August 2001–June 2002 and at Virginia Tech from August 2003–January 2004 in their departments of Electrical and Computer Engineering. He was a Research Assistant at the Wireless Information and Networking Laboratory at Rutgers State University from July 2002–January 2003 building a software simulator for simulation studies on UWB communication system design. From February 2003–July 2003, he was a Research Assistant at the Center for Advanced Information Processing at Rutgers State University, developing algorithms for wireless link bandwidth estimation and prediction in 802.11b wireless local area networks. Since joining Virginia Tech in August 2003, his research focuses on receiver architectures for UWB communication systems, optimized transmit diversity schemes over fading channels, synthetic MIMO and cooperative diversity systems, and cross layer optimization themes.

**James O. Neel** received his B.S.E.E. and M.S.E.E from Virginia Tech in 1999 and 2002, respectively. He is currently working toward the Ph.D. degree at the Mobile and Portable Radio Research Group at Virginia Tech as an IREAN Fellow and recipient of a Motorola University Partnership in Research grant. Since 1999, he has been a Research Assistant at MPRG, where his research has focused on software radio, communications system implementation, and analysis of distributed radio resource management. In 2003, he led a team at Virginia Tech that explored simulation techniques for UWB. He has previously coauthored two chapters of *Software Radio: A Modern Approach to Radio Engineering*.

**Dr. Sedki M. Riad** is a Professor in the Bradley Department of Electrical and Computer Engineering of Virginia Tech. He also serves as Director of the Time Domain and RF Measurements Laboratory and Director of International Programs for the College of Engineering, both of Virginia Tech. He received his B.Sc. and M. Sc. degrees, both in Electrical Engineering, from Cairo University in 1966 and 1972, respectively, and completed his Ph.D. studies at the University of Toledo, Ohio, and the National Institute for Standards and Technology in 1976. Dr. Riad has worked at the National Institute for Standards and Technology and the University of Colorado, both in Boulder, Colorado; the University of Central Florida in Orlando, Florida; King Saud University in Riyadh, Saudi Arabia; and finally at Virginia Tech, Blacksburg, Virginia (since 1979). He is a past Chairman of the U.S. National Commission A of the International Union of Radio Science. He is a registered Professional Engineer in the State of Florida. Dr. Riad is an active reviewer for several IEEE transaction societies. Dr. Riad has over 35 years of experience in academic research and development work in the areas of communication and electronic engineering, with particular emphasis in the areas of RF and microwave measurements, instrumentation, signal processing, and physical modeling. He is a Fellow of the IEEE cited for "contributions to time-domain measurements through physical modeling of sampling devices, simulation, and deconvolution." Dr. Riad has published extensively including more than 60 refereed journal articles, 15 invited presentations, and over 170 conference papers and presentations. He has supervised 18 Ph.D. and 20 M.S. students. He holds two patents and has been a director/principal investigator of about 100 research projects.

**Brian M. Sadler** is a Senior Research Scientist at the Army Research Laboratory (ARL) in Adelphi, MD, and lectures at Johns Hopkins University in signal processing and communications. He was an Associate Editor for the *IEEE Transactions on Signal Processing*, is on the editorial board for the *EURASIP Journal on Wireless Communications and Networking*, is a guest editor for the *IEEE Journal on Selected Areas in Communications* (issue on Advances in Wireless Military Communications), and is a member of the IEEE Technical Committee on Signal Processing for Communications. He organized and co-chaired the *Second IEEE Workshop on Signal Processing Advances in Wireless Communications* (*SPAWC-99*). His research interests include signal processing for mobile wireless and UWB systems, and sensor signal processing and networking, and he supports several DARPA programs in these areas.

**Dr. Ahmad Safaai-Jazi** received the Ph.D. degree (with distinction) from McGill University in 1978, the M.A.Sc. degree from the University of British Columbia in 1974, and the B.Sc. degree from Sharif University of Technology in 1971, all in electrical engineering. In 1986, he joined the electrical engineering faculty at Virginia Tech where he is now a Professor. At Virginia Tech, he teaches courses on electromagnetic fields and waves, antennas, communication systems, and fiber optics. His research interests include UWB propagation measurements and simulations, wideband antennas, and specialty optical fibers for communication and sensor applications. He is the author or co-author of more than a hundred publications, has contributed one book chapter, and holds three patents. Dr. Safaai-Jazi is a senior member of IEEE and a member of the Optical Society of America. Ahmad Safaai-Jazi has recently directed a comprehensive indoor UWB propagation measurement campaign at Virginia Tech. The effort included UWB characterization of common building materials over a frequency range of 1 to 15 GHz, time-domain and frequency-domain UWB propagation measurements in indoor environments and channel characterization, and UWB simulation of indoor channels. This research has contributed significantly to a better understanding of the potentials and the limitations of the UWB technology.

**Dr. Ananthram Swami** received the B.Tech. degree from IIT, Bombay, the M.S. degree from Rice University, Houston, and the Ph.D. degree from the University of Southern California (USC), all in electrical engineering. He has held positions with Unocal Corporation, USC, CS-3, and Malgudi Systems. He is currently with the Army Research Lab, Adelphi, MD, where he is a Fellow. Dr. Swami is a member and vice-chair of the IEEE Signal Processing Society's Technical Committee on Signal Processing for Communications, an associate editor of the *IEEE Transactions on Wireless Communications* and of the *IEEE Transactions on Signal Processing*, and a member of the Editorial Board of the *IEEE Signal Processing Magazine*. He has served as an Associate Editor for *IEEE Signal Processing Letters* and *IEEE Transactions on Circuits and Systems-II*. He was co-organizer and co-chair the 1993 *IEEE Signal Processing Society Workshop on Higher-Order Statistics* 1996 *IEEE Signal Processing Society Workshop on Statistical Signal and Array cessing*, and the 1999 *American Statistical Association—Institute of Mathe Statistics Workshop on Heavy-Tailed Phenomena*. He was a Statistical Co to the California Lottery, developed a Matlab-based toolbox for non-Gauss processing, and has held visiting faculty positions at National Institute Toulouse, France. His work is in the broad area of signal processing for cations; current interests include multi-carrier systems, UWB, sensor and cross-layer approaches.

**Dr. Dennis Sweeney** received his B.S., M.S., and Ph.D. from Vir employed by the Center for Wireless Telecommunications (CWT) His interests are unlicensed wireless applications and radio frequ

He is responsible for CWT's RF Design Lab. This lab has been involved in the design of such things as experimental UWB systems, Local Multipoint Distribution System radios, and Bluetooth applications. Dr. Sweeney worked on Global Positioning System receiving equipment and applications at the Jet Propulsion Laboratory in Pasadena, CA. Dr. Sweeney recent accomplishments include the design of a UWB channel sounder and has performed extensive channel measurements.

**Dr. William H. Tranter** received his Ph.D. from the University of Alabama in 1970 and has over 35 years experience in the design, analysis, and simulation of communication systems. He came to Virginia Tech in 1997 after 28 years at the University of Missouri—Rolla. He has lead large simulation projects for NASA, GE Aerospace, and LG Electronics. Professor Tranter has co-authored a number of leading textbooks in the areas of communication theory, analog and digital signal processing, and simulation. Tranter is a Fellow of the IEEE and was recently elected Vice President—Technical Activities of the IEEE Communications Society. He was awarded an IEEE Centennial Medal and an IEEE Third Millennium Medal. He was also awarded the IEEE Exemplary Publications Award and the IEEE Donald W. McLellan Meritorious Service Award. He served for a number of years as Editor-in-Chief of the *IEEE Journal of Selected Areas in Communications*. Currently he is the Bradley Professor of Communications in the Department of Electrical and Computer Engineering and Director of the Mobile and Portable Radio Research Group at Virginia Tech. In 2003, Dr. Tranter taught an advanced class in simulation that focused on UWB. As a part of this class, a very sophisticated UWB simulator was created and documented.

# INDEX